DATE DUE FOR RETURN

PROGRESS IN BRAIN RESEARCH

VOLUME 141

GONADOTROPIN-RELEASING HORMONE: MOLECULES AND RECEPTORS

Other volumes in PROGRESS IN BRAIN RESEARCH

PROGRESS IN BRAIN RESEARCH

VOLUME 141

GONADOTROPIN-RELEASING HORMONE: MOLECULES AND RECEPTORS

EDITED BY

ISHWAR S. PARHAR

*Department of Physiology, Nippon Medical School,
Tokyo 113-8602, Japan*

ELSEVIER

AMSTERDAM – BOSTON – LONDON – NEW YORK – OXFORD – PARIS
SAN DIEGO – SAN FRANCISCO – SINGAPORE – SYDNEY – TOKYO
2002

ELSEVIER SCIENCE B.V.
Sara Burgerhartstraat 25
P.O. Box 211, 1000 AE Amsterdam, The Netherlands

First edition 2002

Library of Congress Cataloging-in-Publication Data
Gonadotropin-releasing hormone : molecules and receptors / edited by Ishwar S. Parhar.-- 1st ed.
   p. ; cm. -- (Progress in brain research ; v. 141)
   Includes bibliographical references and index.
   ISBN 0-444-50979-8 (alk. paper) -- ISBN 0-444-80104-9 (series)
   1. Luteinizing hormone releasing hormone. 2. Luteinizing hormone releasing hormone--Receptors. 3. Neurochemistry. I. Parhar, Ishwar S., 1958- II. Series.
   [DNLM: 1. Gonadotropins--physiology. 2. Receptors, Gonadotropin--physiology. WK 900 G6378 2002]
   QP572.L85 G655 2002
   573.6'374--dc21   *1003006129*

                                                                    2002034691

British Library Cataloguing in Publication Data
A catalogue record from the British Library has been applied for.

ISBN: 0-444-50979-8 (volume)
ISBN: 0-444-80104-9 (series)
ISSN: 0079-6123

⊗ The paper used in this publication meets the requirements of ANSI/NISO Z39.48-1992 (Permanence of Paper).
Printed in The Netherlands.

List of Contributors

Y. Arai, Department of Human Sciences, University of Human and Arts, Iwatsuki, Saitama 339-8539, Japan

L. Asarian, Department of Psychology, Columbia University, College of Physicians and Surgeons, New York, NY 10032, USA

M. Blomenröhr, Department of Experimental Zoology, Utrecht University, Padualaan 8, 3584 CH Utrecht, The Netherlands

J. Bogerd, Department of Experimental Zoology, Utrecht University, Padualaan 8, 3584 CH Utrecht, The Netherlands

P.-M. Bouloux, Centre for Neuroendocrinology, Royal Free and University College Medical School, London NW3 2QG, UK

J.P. Chang, Department of Biological Sciences, University of Alberta, CW 405 Biological Sciences Centre, Edmonton, AB T6G 2E9, Canada

G. Childs, Department of Anatomy, University of Arkansas for Medical Sciences, 4301 W. Markham Street, Slot No. 510, Little Rock, AK 72205, USA

A. D'Aniello, Stazione Zoologica, Villa Comunale, 80100 Napoli, Italy

W.L. Dees, Department of Veterinary Anatomy and Public Health, Texas A&M University, University Drive, College Station, TX 77843, USA

M.M. Di Fiore, Department of Life Sciences, SUN, Via Vivaldi 43, 81100 Caserta, Italy

H.L. Eisthen, Department of Zoology, Michigan State University, East Lansing, MI 48824, USA

M. Fiorentino, Department of Zoology, University of Naples Federico II, Via Mezzocannone 8, 80134 Napoli, Italy

A. Franchi, Centro de Estudios Farmacólogicos y Botánicos CONICET, Serrano 669, 1414 Buenos Aires, Argentina

T. Funabashi, Department of Physiology, Yokohama City University School of Medicine, 3-9 Fukuura, Kanazawa-ku, Yokohama 236-0004, Japan

K. Gen, Inland Station, National Research Institute of Aquaculture, Fisheries Research Agency, 224-1, Hiruta, Tamaki, Watarai, Mie 519-0423, Japan

H. Goos, Department of Experimental Zoology, Utrecht University, Padualaan 8, 3584 CH Utrecht, The Netherlands

A.C. Gore, Kastor Neurobiology of Aging Laboratories, Mount Sinai School of Medicine, Box 1639, 1425 Madison Avenue, New York, NY 10029, USA

H.R. Habibi, Department of Biological Sciences, University of Calgary, 2500 University Drive, N.W. Calgary, AB T2N 1N4, Canada

J. Han, School of Biological Sciences, Seoul National University, Seoul, 151-742, Korea

F. Hirahara, Department of Gynecology and Obstetrics, Yokohama City University School of Medicine, 3-9 Fukuura, Kanazawa-ku, Yokohama, 236-0004, Japan

Y. Hu, Centre for Neuroendocrinology, Royal Free and University College Medical School, London NW3 2QG, UK

L. Iela, Department of Zoology, University of Naples Federico II, Via Mezzocannone 8, 80134 Napoli, Italy

L. Jennes, Department of Anatomy and Neurobiology, University of Kentucky, College of Medicine, 430 Health Science Research Building, Lexington, KY 40536, USA

H. Kagawa, National Research Institute of Aquaculture, Fisheries Research Agency, 422-1, Nakatsuhamaura, Nansai, Watarai, Mie 516-0913, Japan

S.S. Kakar, Department of Medicine, University of Louisville, Louisville, KY 40202, USA

S. Karanth, Pennington Biomedical Research Center, Louisiana State University, 6400 Perkins Road, Baton Rouge, LA 70808-4124, USA

K.L. Keen, Wisconsin National Primate Research Center, 1223 Capital Court, Madison, WI 53715-1299, USA

M. Khalil, Department of Biochemistry and Molecular Biophysics, Columbia University, College of Physicians and Surgeons, New York, NY 10032, USA

H.H. Kim, Department of Obstetrics and Gynecology, University of Chicago, 5839 South Maryland Avenue, MC 5053, Chicago, IL 60637, USA

K. Kim, School of Biological Sciences, Seoul National University, Seoul, 151-742, Korea

F. Kimura, Department of Physiology, Yokohama City University School of Medicine, 3-9 Fukuura, Kanazawa-ku, Yokohama 236-0004, Japan

C. Klausen, Department of Biological Sciences, University of Calgary, 2500 University Drive, N.W. Calgary, AB T2N 1N4, Canada

N. Kumakura, Department of Aquatic Biosciences, Tokyo University of Fisheries, Tokyo, 108-8477, Japan

S. Lakhlani, Department of Anatomy and Neurobiology, University of Kentucky, College of Medicine, 430 Health Science Research Building, Lexington, KY 40536, USA

M. Lasaga, Centro de Investigaciones en Reproducción, Fac. Medicine, UBA, Buenos Aires, Argentina

R. Leurs, Department of Pharmacochemistry, Free University, De Boelelaan 1083, 1081 HV Amsterdam, The Netherlands

W. Lin, Department of Anatomy and Neurobiology, University of Kentucky, College of Medicine, 430 Health Science Research Building, Lexington, KY 40536, USA

A. Lomniczi, Centro de Estudios Farmacólogicos y Botánicos CONICET, Serrano 669, 1414 Buenos Aires, Argentina

G. MacColl, Centre for Neuroendocrinology, Royal Free and University College Medical School, London NW3 2QG, UK

M.T. Malik, Department of Medicine, University of Louisville, Louisville, KY 40202, USA

C.A. Mastronardi, Pennington Biomedical Research Center, Louisiana State University, 6400 Perkins Road, Baton Rouge, LA 70808-4124, USA

S.M. McCann, Pennington Biomedical Research Center, Louisiana State University, 6400 Perkins Road, Baton Rouge, LA 70808-4124, USA

B. Miller, Department of Anatomy and Neuroscience, University of Texas Medical Branch, 301 University Boulevard, MRB 10 104, Galveston, TX 77555-1043, USA

D. Mitsushima, Department of Physiology, Yokohama City University School of Medicine, 3-9 Fukuura, Kanazawa-ku, Yokohama 236-0004, Japan

C. Mohn, Centro de Estudios Farmacólogicos y Botánicos CONICET, Serrano 669, 1414 Buenos Aires, Argentina

A. Mori, Department of Aquatic Biosciences, Tokyo University of Fisheries, Tokyo, 108-8477, Japan

S. Murakami, Department of Anatomy, Juntendo University, School of Medicine, Hongo, Tokyo, 113-8421, Japan

T.J. Nakamura, Department of Physiology, Yokohama City University School of Medicine, 3-9 Fukuura, Kanazawa-ku, Yokohama 236-0004, Japan

Y. Oka, Misaki Marine Biological Station, Graduate School of Science, The University of Tokyo, Kanagawa, 238-0225, Japan

K. Okuzawa, Inland Station, National Research Institute of Aquaculture, Fisheries Research Agency, 224-1, Hiruta, Tamaki, Watarai, Mie 519-0423, Japan

I.S. Parhar, Department of Physiology, Nippon Medical School, Tokyo 113-8602, Japan

D.W. Pfaff, Laboratory of Neurobiology and Behavior, The Rockefeller University, 1230 York Avenue, Box 275, New York, NY 10021, USA

S. Radovick, Department of Pediatrics, University of Chicago, 5839 South Maryland Avenue, MC 5053, Chicago, IL 60637, USA

R.K. Rastogi, Department of Zoology, University of Naples Federico II, Via Mezzocannone 8, 80134 Napoli, Italy

V. Rettori, Centro de Estudios Farmacológicos y Botánicos CONICET, Serrano 669, 1414 Buenos Aires, Argentina

T.A. Richter, Wisconsin National Primate Research Center, 1223 Capital Court, Madison, WI 53715-1299, USA

E.F. Rissman, Department of Biochemistry and Molecular Genetics, University of Virginia Medical School, P.O. Box 800733, Charlottesville, VA 22908, USA

Y. Sakuma, Department of Physiology, Nippon Medical School, Sendagi 1, Bunkyo, Tokyo 113, Japan

M. Schwanzel-Fukuda, Laboratory of Neurobiology and Behavior, The Rockefeller University, 1230 York Avenue, Box 275, New York, NY 10021, USA

C. Scorticati, Centro de Estudios Farmacológicos y Botánicos CONICET, Serrano 669, 1414 Buenos Aires, Argentina

T. Seki, Department of Anatomy, Juntendo University, School of Medicine, Hongo, Tokyo, 113-8421, Japan

J.Y. Seong, Hormone Research Center, Chonnam National University, Kwangju, 500-757, Korea

K. Shinohara, Department of Physiology, Yokohama City University School of Medicine, 3-9 Fukuura, Kanazawa-ku, Yokohama 236-0004, Japan

R. Silver, Department of Psychology, Barnard College, New York, NY 10032, USA

A.-J. Silverman, Department of Anatomy and Cell Biology, Columbia University, College of Physicians and Surgeons, New York, NY 10032, USA

G.H. Son, School of Biological Sciences, Seoul National University, Seoul, 151-742, Korea

S. Sower, Department of Biochemistry and Molecular Biology, University of New Hampshire, 46 College Road, Room 310, Durham, NH 03824, USA

S. Srinivasan, Departments of Psychiatry and Behavioral Sciences, Duke University Medical Center, Box 3497, 028 CARL Building, Durham, NC 27710, USA

K. Suyama, Department of Physiology, Yokohama City University School of Medicine, 3-9 Fukuura, Kanazawa-ku, Yokohama 236-0004, Japan

J.L. Temple, Department of Biology, University of Virginia, P.O. Box 400328, 275 Gilmer Hall, Charlottesville, VA 22903, USA

E. Terasawa, Wisconsin National Primate Research Center, 1223 Capitol Court, Madison, WI 53715-1299, USA

T. Uemura, Department of Gynecology and Obstetrics, Fujisawa City Hospital, 2-6-1 Fujisawa, Fujisawa 251-0052, Japan

P. Vissio, Centro de Estudios Farmacólogicos y Botánicos CONICET, Serrano 669, 1414 Buenos Aires, Argentina

W.C. Wetsel, Departments of Psychiatry and Behavioral Sciences, Duke University Medical Center, Box 3497, 028 CARL Building, Durham, NC 27710, USA

A.F. Wiechmann, Department of Cell Biology, University of Oklahoma Health Sciences Center, 940 S.L. Young Boulevard, Oklahoma City, OK 73104, USA

S.J. Winters, Department of Medicine, University of Louisville, Louisville, KY 40202, USA

C.R. Wirsig-Wiechmann, Department of Cell Biology, University of Oklahoma Health Sciences Center, 940 S.L. Young Boulevard, Oklahoma City, OK 73104, USA

A. Wolfe, Department of Pediatrics, University of Chicago, 5839 South Maryland Avenue, MC 5053, Chicago, IL 60637, USA

S. Yamaguchi, Department of Animal and Marine Bioresources Science, Faculty of Bioresources and Bioenvironmental Sciences, Kyushu University, Fukuoka, Fukuoka 812-8581, Japan

W.H. Yu, Pennington Biomedical Research Center, Louisiana State University, 6400 Perkins Road, Baton Rouge, LA 70808-4124, USA

Preface

Gonadotropin-releasing hormone (GnRH, previously called luteinizing-hormone releasing hormone = LHRH) represents the first step in a cascade of events coordinating the complex physiology of reproduction and reproductive behavior in vertebrates. It was initially assumed that GnRH occurred as a single molecular form, in the septo-preoptic area of the vertebrate brain, and that it served a single function. However, over the last decade, 13 structurally distinct forms of GnRHs have been isolated from different vertebrate species. Recent research has also demonstrated the existence of GnRH receptor subtypes, which have evolved in conjunction with distinct GnRH ligands. The multiplicity of GnRH molecules and receptors has recently expanded with the discovery of chicken II GnRH in the brain of humans and other mammals. It has become increasingly clear that, all vertebrate species investigated to date possess two or three different GnRH genes, and that GnRH has multiple functions in addition to stimulating the release of gonadotropins. Non-mammalian vertebrate species have played a central role in the discovery of molecular diversity of GnRHs. Furthermore, comparative studies using different vertebrate models show remarkable similarity in the distribution pattern of GnRH neurons. The embryonic origin of GnRH neurons from the olfactory placode and their subsequent migration into the septo-preoptic area is unique among brain cells, which is evolutionarily conserved among vertebrate species ranging from fish to humans.

This volume summarizes the evolution and physiology of GnRH molecules and receptors, and provides insight as to how social behavior influences cellular and molecular events in the brain from a comparative perspective. The chapters in this volume are divided into three major sections: Development and Cell Migration, GnRH Receptors, and Physiology and Regulation. The review papers arose primarily from presentations made at the Second International Symposium on the Comparative Biology of GnRH, held in Penang, Malaysia, June 2–4, 2001; a satellite symposium in conjunction with the XIV International Congress of Comparative Endocrinology, Sorrento, Italy. In addition, leading neuroscientists doing cutting-edge research in the field of GnRH were invited as authors to make this volume a valuable reference.

I would like to acknowledge the cooperation and support received from the authors. I am indebted to those who helped in organizing the symposium, in particular Dr. Tengku Sifzizul Tengku Muhammad, Dr. Malik Mumtaz, and Ms. Diljit Kaur from Universiti Sains Malaysia. The assistance given by Maureen Twaig and Tom Merriweather at the Elsevier office is most gratefully acknowledged.

Ishwar S. Parhar
Tokyo, Japan

Contents

Section III. Physiology and regulation

SECTION I

Development and cell migration

I.S. Parhar (Ed.)
Progress in Brain Research, Vol. 141
© 2002 Published by Elsevier Science B.V.

CHAPTER 1

Cell migration and evolutionary significance of GnRH subtypes

Ishwar S. Parhar *

Department of Physiology, Nippon Medical School, Tokyo 113 8602, Japan

Introduction

Gonadotropin-releasing hormone (GnRH, previously called luteinizing hormone-releasing hormone = LHRH) represents the first step in a cascade of events coordinating the complex physiology of reproduction and reproductive behavior in vertebrates. It was initially assumed that GnRH occurred as a single molecular form, in the septo-preoptic area of vertebrate brain, and that it served a single function. However, over the last decade, 13 structurally distinct forms of GnRHs have been isolated from different vertebrate species and two forms from invertebrates (Fig. 1). It has become increasingly clear that, all vertebrate species investigated to date posses two or three different GnRH forms, and that GnRH has multiple functions in addition to stimulating the release of gonadotropins (Parhar, 1997, 1999a; Sherwood et al., 1997; Weber et al., 1997). Recent research has also demonstrated the existence of GnRH receptor subtypes, which have evolved in conjunction with distinct GnRH ligands (Illing et al., 1999; Troskie et al., 2000; Wang et al., 2001; Parhar et al., 2002).

Studies using non-mammalian vertebrate species have played a central role in the discovery of molec-

ular diversity of GnRHs (Jimenez-Linan et al., 1997; King and Millar, 1997; Sherwood et al., 1997; Carolsfeld et al., 2000; Okubo et al., 2000a; Yoo et al., 2000) and the sites of embryonic origins of GnRH neurons (Muske, 1997; Parhar, 1997, 1999a). The multiplicity of GnRH and receptors has recently expanded with the discovery of chicken-II GnRH in the brain of humans and other mammals (Kasten et al., 1996; Lescheid et al., 1997; White et al., 1998; Gestrin et al., 1999; Urbanski et al., 1999). The present chapter summarizes neuroanatomical and biochemical studies from a comparative perspective with particular emphasis on the fish model to gain insights into the evolution and physiology of GnRH molecules and receptors.

Multiple GnRH genes

Some 30 years ago the structure of GnRH was first determined (Matsuo et al., 1971; Burgus et al., 1972). Since then, thirteen structurally distinct forms of GnRHs have been isolated from vertebrates and two from protochordates (Jimenez-Linan et al., 1997; Sherwood et al., 1997; Sower, 1997; Carolsfeld et al., 2000; Okubo et al., 2000a; Yoo et al., 2000). All known GnRH forms are ten amino acids in length (Fig. 1). The most common structural variation among the different forms of GnRHs occurs frequently in amino acid positions 5–8. However, the essential molecular sequences at the ends of the decapeptide, the 5′-pyroglutamyl-modified amino terminus and the 3′-amidated carboxy terminus have remained unchanged during

* Correspondence to: I. Parhar, Department of Physiology, Nippon Medical School, Tokyo 113-8602, Japan. Tel.: +81 (3)-38222131 (Ext. 5328); Fax: +81 (3)-56853055; E-mail: ishwar@nms.ac.jp

	1	2	3	4	5	6	7	8	9	10	
TUNICATE I	pGLU	HIS	TRP	SER	ASP	TYR	PHE	LYS	PRO	GLY	NH2
TUNICATE II	pGLU	HIS	TRP	SER	LEU	CYS	HIS	ALA	PRO	GLY	NH2
LAMPREY I	pGLU	HIS	TYR	SER	LEU	GLU	TRP	LYS	PRO	GLY	NH2
LAMPREY III	pGLU	HIS	TRP	SER	HIS	ASP	TRP	LYS	PRO	GLY	NH2
DOGFISH	pGLU	HIS	TRP	SER	HIS	GLY	TRP	LEU	PRO	GLY	NH2
CHICKEN II	pGLU	HIS	TRP	SER	HIS	GLY	TRP	TYR	PRO	GLY	NH2
SALMON	pGLU	HIS	TRP	SER	TYR	GLY	TRP	LEU	PRO	GLY	NH2
CATFISH I	pGLU	HIS	TRP	SER	HIS	GLY	LEU	ASN	PRO	GLY	NH2
MEDAKA	pGLU	HIS	TRP	SER	PHE	GLY	LEU	SER	PRO	GLY	NH2
HERRING	pGLU	HIS	TRP	SER	HIS	GLY	LEU	SER	PRO	GLY	NH2
SEABREAM	pGLU	HIS	TRP	SER	TYR	GLY	LEU	SER	PRO	GLY	NH2
RANA	pGLU	HIS	TRP	SER	TYR	GLY	LEU	TRP	PRO	GLY	NH2
CHICKEN I	pGLU	HIS	TRP	SER	TYR	GLY	LEU	GLN	PRO	GLY	NH2
GUINEAPIG	pGLU	TYR	TRP	SER	TYR	GLY	VAL	ARG	PRO	GLY	NH2
MAMMAL	pGLU	HIS	TRP	SER	TYR	GLY	LEU	ARG	PRO	GLY	NH2

Fig. 1. Comparison of amino acid sequences of 15 GnRH forms. The amino acids printed in bold differ from the mammalian form (see Jimenez-Linan et al., 1997; King and Millar, 1997; Sherwood et al., 1997; Sower, 1997; Carolsfeld et al., 2000; Okubo et al., 2000a; Yoo et al., 2000).

evolution. Evidence suggests that in all the major vertebrate groups, chicken-II GnRH is the most highly conserved form, while the forebrain forms vary structurally in different species (King and Millar, 1997; Sherwood et al., 1997; Parhar, 1999a). The fact that more than one GnRH molecule also exists in protocodates (tunicates) and jawless fish (lampreys: Sherwood et al., 1997; Sower, 1997), strengthens the evidence that the presence of multiple forms of GnRH in a single species is an ancient pattern in evolution (Fig. 1).

The cDNAs encoding GnRH have been determined in mammals (Adelman et al., 1986), amphibians (Hayes et al., 1994), birds (Dunn et al., 1993) and bonyfish (Bond et al., 1991; Klungland et al., 1992; Suzuki et al., 1992; Bogerd et al., 1994; White et al., 1995; Gothilf et al., 1996). Since the original discovery in birds (Miyamoto et al., 1984), a cDNA encoding a second and a third GnRH was identified in teleosts (White et al., 1994, 1995; Gothilf et al., 1996). In recent years, more than one GnRH peptide and cDNA encoding GnRH have been sequenced from the brain of several fish species and placental mammals (tree shrew, humans: White et al., 1995, 1998; Gothilf et al., 1996; Kasten et al., 1996;

Gestrin et al., 1999). Genes encoding different GnRH forms have the same architecture of four exons separated by three introns and this has been highly conserved throughout evolution, despite changes in the size and sequences of exons and introns. In each gene, the first exon encodes the 5′-untranslated region; the second and third exons encode the signal peptide, GnRH decapeptide, a proteolytic cleavage site and the GnRH-associated peptide (GAP); and the fourth exon has the carboxy terminus of GAP and the 3′-untranslated region. Comparisons between the cDNA precursors encoding the distinct members of the GnRH family show complete conservation of the proteolytic cleavage site, high homology of the decapeptide, but high structural divergence in the GAP region (see King and Millar, 1997; Sherwood et al., 1997; Yu et al., 1997).

Phylogenic distribution and ontogeny of GnRH subtypes

The brain of most vertebrate species has at least two but some advanced perciform fishes have three GnRH forms (White et al., 1995; Gothilf et al., 1996; Parhar, 1997). Localization studies have demon-

TELEOSTEAN RADIATION

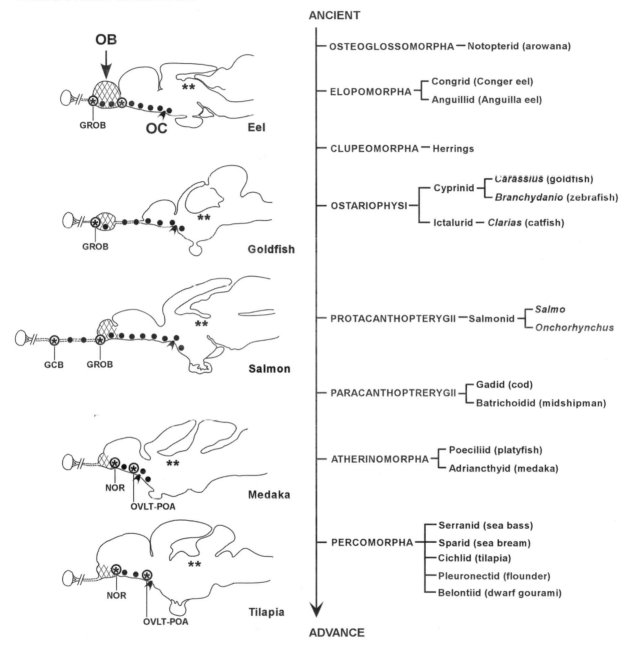

Fig. 2. Phylogenetic distribution of GnRHs in the teleostean fish radiation. Eel (Chiba et al., 1999), goldfish (Parhar et al., 2001), Salmon (Parhar and Iwata, 1993; Parhar et al., 1995, 1996), medaka (Parhar et al., 1998), Tilapia (Parhar, 1997). The evolutionary order of teleost (ancient to advanced) is modified from Meek and Nieuwenhuys (1998). OB, olfactory bulb; OC, optic chiasma; GROB, ganglia at the rostral olfactory bulbs; GCB, ganglia at the cribriform bone; NOR, nucleus olfactoretinalis; OVLT-POA cells synthesize medaka or seabream GnRH; dark dots represent scattered single GnRH cells; **, chicken-II GnRH cells in the midbrain. Note, with advancing phylogenic order: GnRH ganglia shifts intracerebral to form the NOR; medaka and seabream GnRH seen in the OVLT-POA and a decrease in GnRH cells in the basal hypothalamus (nucleus tuberis lateralis).

strated differential distribution of GnRH molecular variants in distinct brain areas in vertebrates (Fig. 2). Comparative studies using different vertebrate species show remarkable similarity in the distribution pattern of GnRH cell bodies, fiber pathways and embryonic origins, suggesting that these systems are phylogenetically conserved (Muske, 1997; Sherwood et al., 1997; Parhar, 1999a). In the following paragraphs the patterns of distribution of GnRH subtypes in vertebrate species are described in some detail.

Cells of mesencephalic origin

The mesencephalonic GnRH system is a highly conserved neuronal system present in teleosts, amphibians, reptiles, birds and mammals (see reviews by King and Millar, 1997; Muske, 1997; Sherwood et al., 1997; Parhar, 1999a). In the mesencephalon, GnRH cells are exclusively located at the caudal most part of the prosencephalon, rostral to the ocularmotor nucleus. These cells lie along the midline, close to the subependyma of the third ventricle, and close to large blood vessels (Figs. 2 and 3). They are round to triangular in shape, about 30–40 μm

in diameter and synthesize chicken-II GnRH subtype (Parhar, 1997). The embryonic origin of mesencephalonic GnRH neurons is different from the forebrain GnRH neurons. In the urodele amphibians, ablations of the olfactory placodes results in bilateral loss of GnRH neurons in the hypothalamus but not chicken-II GnRH neurons in the midbrain; demonstrating that mesencephalonic GnRH neurons do not originate from the olfactory placodes but from the ventricular ependyma of the third ventricle (Northcutt and Muske, 1994). Similarly, developmental studies in fishes have shown that chicken-II GnRH neurons differentiate from precursor cells in the ventricular ependyma (Parhar and Iwata, 1993; Parhar, 1997; Parhar et al., 1998) (Fig. 3), which might also be true for other vertebrate species.

Cells of olfactory origin

Ancient migratory route (olfactory to diencephalon)

Land vertebrates

The embryonic origin of GnRH neurons from the olfactory placode and their subsequent migration into

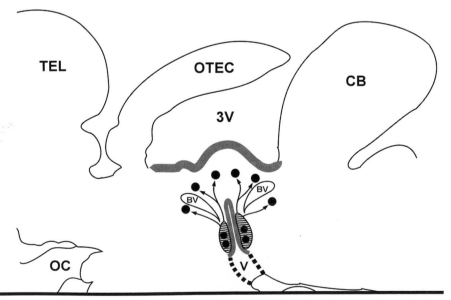

Fig. 3. Diagrammatic representation of sagittal brain section illustrating the location of midbrain chicken-II GnRH neurons in tilapia *Oreochromis niloticus*. Chicken II GnRH synthesizing neurons originate from the ventricular ependyma (shaded area). These cells migrate a short distance and are seen located around large blood vessels (Parhar and Iwata, 1993; Parhar, 1997). Dark dots represent chicken-II GnRH neurons. Tel, telencephalon; OTEC, optic tectum; CB, cerebellum; OC, optic chiasma. BV, blood vessel; V, ventricle.

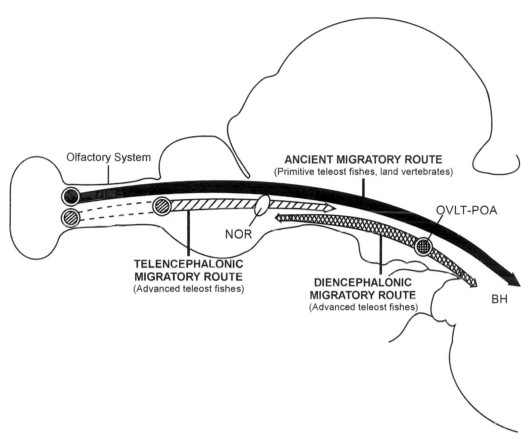

Fig. 4. Diagrammatic representation of a sagittal brain section illustrating the migratory paths taken by GnRH neurons. In primitive teleosts (salmon, catfish, mammalian GnRH) and in land vertebrates (chicken-I, mammalian GnRH) GnRH neurons migrate along the *'ancient migratory route'*, which originates from the olfactory placodes and extends to the diencephalon–basal hypothalamus (BH). Cells are seen scattered but as a continuum along this path. In advanced teleosts, salmon GnRH synthesizing neurons migrate along the *'telencephalonic migratory route'*, which originates from the olfactory placodes (or olfactory system) and extends to the nucleus olfactoretinalis (NOR), with scattered cells along the basal telencephalon. In advanced teleosts (medaka, tilapia) GnRH cells synthesizing medaka GnRH and seabream GnRH originate from the diencephalonic regions (OVLT–POA, organum vasculosum lamina terminalis–preoptic area) and probably migrate along the *'diencephalonic migratory route'*.

the forebrain–preoptic area is unique among brain cells (Fig. 4). This migratory route, which I describe here as the *'ancient migratory route'* has been recorded in mammals including humans (Schwanzel-Fukuda et al., 1989; Schwanzel-Fukuda, 1997) and in non-mammalian vertebrate species (birds: Murakami et al., 2000; amphibians: D'Aniello et al., 1994; Muske, 1997; fishes: Parhar and Iwata, 1993; Parhar et al., 1995; Parhar, 1997). In virtually all land vertebrates, GnRH neurons in the forebrain of adults appear diffusely distributed within the diencephalic region. Majority of the GnRH cells are present in the septo-preoptic area as a loose cluster of disperse cells

with few scattered cells in the basal hypothalamus (Silverman et al., 1994). In rodents and birds, GnRH neurons migrating from the olfactory placodes are strongly associated with fibers immunoreactive for cell adhesion molecules (polysialylated form of neural cell adhesion molecule, PSA-NCAM: Murakami et al., 2000; Wray, 2001). Microinjections of antibodies against NCAM into the olfactory placodes in mice (Schwanzel-Fukuda, 1997) or ablations of the olfactory placodes in amphibians and birds (Murakami et al., 1992; Northcutt and Muske, 1994) result in the loss of preoptic GnRH neurons. Therefore, these studies provide strong evidence that the

8

embryonic origin of the septo-preoptic GnRH neurons is the olfactory placodes.

Primitive teleost fishes

In teleost fishes, which belong to the evolutionary primitive order (e.g. salmonids), GnRH neurons follow the 'ancient migratory route' similar to that in land vertebrates (Fig. 4). During embryonic development, neurons that synthesize salmon GnRH are born in the basal regions of the olfactory placodes (Parhar and Iwata, 1993; Parhar et al., 1995). Small groups of cells are seen outside the central nervous system as cell clusters in the olfactory nerve close to the olfactory epithelium and at the interface of the olfactory nerve and the olfactory bulbs. These clusters of GnRH cells were described as ganglia at the cribriform bone (GCB) and ganglia at the rostral olfactory bulbs (GROB) in the salmonids for the first time (Parhar and Iwata, 1993; Parhar et al., 1994, 1995) (Fig. 2). The arrival of GnRH neurons in the basal telencephalon coincides with the completion of smoltification in salmonids (Parhar et al., 1994, 1995; Parhar and Iwata, 1996) and metamorphosis in amphibians (Hayes et al., 1994). In adults, scattered GnRH cells are seen along the basal regions of the olfactory forebrain; forming a continuum from the medial parts of the olfactory nerve (terminal nerve), basal olfactory bulbs, basal telencephalon, preoptic area and the basal hypothalamus (nucleus lateralis tuberis). No GnRH cell clusters are seen within the olfactory bulb-forebrain at any developmental stage. This pattern of GnRH cellular distribution is most prevalent in teleost fish of the order elopomorpha, clupeomorpha, ostariophysi and protacanthopterygii, which include species such as eels (Montero et al., 1994; Chiba et al., 1999), goldfish (Kim et al., 1995; Parhar et al., 2001), catfish (Zandbergen et al., 1995) and salmonids (Parhar and Iwata, 1993, 1996; Parhar et al., 1994, 1995) (Fig. 2). Interestingly, these species have long olfactory nerves (salmonids) and or displaced olfactory bulbs separated from the telencephalon by long olfactory tracts (goldfish, catfish). These primitive teleosts have been reported to have only two GnRH forms, i.e. chicken-II GnRH along with salmon GnRH (salmonids, goldfish) or catfish GnRH (catfish) or mammalian GnRH (eel) (King and Millar, 1997; Sherwood et al., 1997).

Thus, in primitive teleost fishes the distribution pattern, migratory route and the placodal origins of forebrain GnRH (salmon, catfish GnRH) is similar to mammalian and chicken-I GnRH in land vertebrates.

Telencephalonic migratory route (olfactory to telencephalon)

Advance teleost fishes

The evolutionarily advanced teleost fishes (atherinomorpha: medaka; percomorpha: tilapia) lack the 'ancient migratory route', instead they have devised an alternate GnRH migratory route which consists of a telencephalonic and a diencephalonic component (Fig. 4). The diencephalonic component gives birth to GnRH neurons seen in the organum vasculosum of the lamina terminalis–preoptic area (OVLT–POA). The 'telencephalonic migratory route', comprises GnRH neurons born in the olfactory placodes, and migrate a short distance along the olfactory nerves–basal olfactory bulbs to form the nucleus olfactoretinalis (NOR) at the caudalmost part of basal olfactory bulbs, with scattered neurons in the rostro-basal telencephalon (Fig. 4). In fact, like the 'ancient migratory route', the 'telencephalonic migratory route' has its embryonic origins in the olfactory placodes. However, in highly evolved teleosts (e.g. perciform tilapia), the olfactory nerve is short and therefore during development the presumptive olfactory bulbs and the olfactory placodes lie against each other, which gives the impression that GnRH neurons originate from or begin GnRH synthesis when they reach the rostral olfactory bulbs (Parhar, 1997). In advanced teleosts, the most distinctive feature of forebrain GnRH neuronal distribution is the clustering of GnRH neurons at the caudalmost part of the olfactory bulb–telencephalon interface to form the nucleus olfactoretinalis (NOR: Munz and Claas, 1987) (Fig. 2). Furthermore, compared to primitive teleost fishes, advanced teleosts have fewer scattered neurons along the basal olfactory bulbs, basal telencephalon and the basal hypothalamus (nucleus lateralis tuberis). In fact, cells in the basal hypothalamus appear to decrease in number with advancing evolutionary order. This pattern of GnRH cellular distribution is most prevalent in the teleostean order paracanthopterygii (midshipman: Grober et al.,

1995), atherinomorpha (platyfish: Schreibman and Margolis-Nunno, 1987; medaka: Parhar et al., 1998) and the more advanced percomorpha (tilapia: Parhar, 1997) (Fig. 2). Interestingly, these species have short olfactory nerves and their olfactory bulbs are located against the telencephalic hemispheres. These fishes have been reported to have three GnRH forms, i.e. preoptic GnRH (seabream GnRH: tilapia niloticus, haplochromis burtoni; medaka GnRH: medakafish); mesencephalonic GnRH (chicken-II GnRH) and telencephalic GnRH (salmon GnRH: tilapia, *H. burtoni*, medaka) (White et al., 1995; Gothilf et al., 1996; Parhar, 1997; Okubo et al., 2000a).

When compared along the teleostean evolutionary radiation, salmon GnRH has two distinct distribution patterns: (1) olfactory system to diencephalon, an ancient distribution pattern seen in primitive fishes; (2) olfactory system to telencephalon, an advanced distribution pattern seen in more recently evolved teleost fishes (Fig. 4).

Cells of diencephalonic origin

Unlike primitive teleosts, advanced teleosts have a distinct loose cluster of GnRH cells in the anteriobasal OVLT–POA area, with few scattered cells in the basal telencephalon and the basal hypothalamus (nucleus lateralis tuberis). These cells synthesize seabream GnRH (cichlid: White et al., 1995; tilapia: Parhar, 1997; Red seabream: Okuzawa et al., 1997; Gilthead seabream: Gothilf et al., 1996) or medaka GnRH (medakafish: Okubo et al., 2000a). They are about 20–25 µm in diameter, with unipolar or bipolar dendrites and number about 100–200 cells in cichlid fish. This OVLT–POA cell population, unlike the nucleus olfactoretinalis (NOR) of the '*telencephalonic migratory route*', does not originate from the olfactory placode; instead this cell population originates from the basal diencephalons (Parhar, 1997; Parhar et al., 1998). The OVLT–POA GnRH cells appear to migrate dorsolaterally from the basal diencephalonic regions, along what might be considered the '*diencephalonic migratory route*' (Fig. 4). As in the advanced teleost fishes, the origin of OVLT–POA GnRH neurons from the basal diencephalon has been observed in some species of birds and lampreys (Norgren and Chen, 1994; Tobet et al., 1996). However, a diencephalic origin

of forebrain GnRH neurons remains to be seen in mammals.

Two GnRH forms in the forebrain of advanced teleosts

Telencephalonic verses diencephalonic GnRH neurons

Morphological and biochemical evidence

A large body of published evidence shows that in advanced teleost fishes (paracanthoptrerygii, atherinomorpha, percomorpha), GnRH cells of the terminal nerve ganglia (nucleus olfactoretinalis, NOR) are distinct from cells of the OVLT–POA.

(1) At all developmental stages, morphologically cells of the NOR are large, round in shape (~30–40 µm) and present as clusters (ganglia) whereas the POA neurons are small and usually fusiform in shape (~20–30 µm) and scattered (Parhar, 1997; Parhar et al., 1998).

(2) The cDNA sequence of NOR specific salmon GnRH is different from the OVLT–POA specific seabream or medaka GnRH (seabream GnRH: White et al., 1995; Gothilf et al., 1996; medaka GnRH: Okubo et al., 2000a).

(3) In many species of teleost fish, the NOR but not the OVLT–POA GnRH neurons co-express molluscan cardioexcitatory tetrapeptide (FMRFamide: Stell et al., 1984). Furthermore, NOR begins GnRH expression during early embryonic life, long before gonadal development, and these cells are regulated by thyroid hormones and testosterone (Soga et al., 1998; Parhar et al., 2000). On the contrary, the expression of OVLT–POA GnRH coincides with gonadal sex differentiation (Parhar, 1997; Chiba et al., 1999), and these cells are regulated by estrogen and ketotestosterone (Soma et al., 1996; Parhar et al., 2000).

Developmental evidence

There is compelling evidence that the NOR and the OVLT–POA GnRH neurons have embryonic origins from progenitor cells situated at different locations. Developmental studies have shown that the salmon GnRH synthesizing NOR neurons originate from the olfactory placodes whereas the seabream GnRH synthesizing OVLT–POA neurons originate from the anteriobasal OVLT–POA region in the perciform tilapia and possibly medaka GnRH in the medakafish (Parhar, 1997; Parhar et al., 1998). Supportive evidence from studies in the lampreys and birds suggests that the POA GnRH neurons originate from the basal diencephalon (Norgren and Chen, 1994; Tobet et al., 1996). Furthermore, some species of lampreys lack a terminal nerve and terminal nerve GnRH neurons but not the POA GnRH neurons (Eisthen and Northcutt, 1996).

Hence, in advanced teleost fishes, morphological and biochemical studies demonstrate that the NOR and the OVLT–POA are two different GnRH cell populations in the forebrain, which synthesize distinct GnRH molecular forms and have different embryonic origins; from the olfactory placodes and within the diencephalons respectively. Likewise, is there evidence of two GnRH forms in the forebrain of other vertebrates as in advanced teleost fishes?

Evidence of a third GnRH form in land veterbrates?

Like in advanced teleost fishes there is evidence of two GnRH forms in the forebrain of primitive teleost fishes and land vertebrates. Two GnRH forms have been cloned from the forebrain of primitive fishes, such as the goldfish, catfish and salmonids. However, tetraploidy has been cited as a cause of this gene duplication (Ashihara et al., 1995; Yu et al., 1997). In mammals, there is physiological evidence of more than one population of forebrain GnRH neurons. For example, under certain physiological states only a subpopulation of POA GnRH neurons express galanin (Merchenthaler et al., 1991) or c-Fos (Lee et al., 1990), and not all GnRH neurons of the septo-preoptic area are under gonadal steroid influence (Herbison, 1998). The existence of two distinct cell populations in the forebrain, responsible for LH surge and pulse generation, has been long debated (Kimura and Funabashi, 1998). Equally debated is the issue of LH and FSH control by one or two different GnRH molecules. Recently, however, McCann and coworkers have shown that Lamprey III GnRH can function, as a FSH releasing hormone (Yu et al., 1997) and LH release is

the classical role of mammalian GnRH (= LHRH). Lamprey III GnRH has been localized in the forebrain of rats and mice (Hiney et al., 2002; Ogawa et al., 2002a), which further emphasizes the existence of more than one GnRH form in the forebrain. However, the site(s) of embryonic origin of Lamprey III GnRH is unknown, which could be the diencephalon. Furthermore, new populations of GnRH neurons have been reported in the forebrain whose origin does not appear to be placodal (Skynner et al., 1999). Thus, unlike advanced teleost fishes with two distinct GnRH molecular forms in the forebrain, land vertebrates, on the other hand, lack specific markers to allow distinction between forebrain GnRH neurons of placodal origin and subpopulations of POA GnRH neurons that might have origins in the basal diencephalon.

GnRH receptor subtypes

GnRH exerts diverse intracellular actions through binding to specific G-protein coupled receptors (Sealfon et al., 1997). The existence of several molecular forms of GnRH ligand within a single species of bonyfish and their ability to effectively stimulate gonadotropins, growth hormone, prolactin and somatolactin cells (Parhar and Iwata, 1994; Melamed et al., 1996; Weber et al., 1997) suggests that there may also be more than one form of GnRH receptor (Troskie et al., 1998). A single class of GnRH receptor has been described in some teleost species (African catfish: Blomenrohr et al., 1997; Tensen et al., 1997; striped bass: Alok et al., 2000; rainbow trout: Madigou et al., 2000) but two different class of GnRH receptors have been demonstrated in the goldfish (Illing et al., 1999), medaka (Okubo et al., 2000b), primates (Millar et al., 2001; Neill et al., 2001), which express two different mRNA. More recently, a third GnRH receptor subtype has been reported in the bullfrog (Wang et al., 2001).

In the pituitary of perciform tilapia, GnRH-Rs type IA and IB have been shown in LH cells, GnRH-R type IB in prolactin cells and GnRH-R type III in growth hormone cells. The presence of type IA, IB and type III GnRH receptors in different cells in the perciform tilapia, demonstrates that three types of GnRH-Rs exist in a single diploid species (Parhar et al., 2002). This supports the notion that the three native GnRH variants in tilapia might have their respective cognate receptors. The presence of type IA and IB GnRH-Rs in FSH and LH cells suggests that two different GnRH ligands can independently control the synthesis/secretion of FSH and LH. Alternatively, dimerization of type IA and IB GnRH-Rs might be important for their functional activation, which implies a complex interplay of GnRH ligands and receptors.

Nomenclature of GnRH subtypes

Based on DNA sequences Fernald and White (1999) classified GnRH subtypes into: hypothalamic form (Type I); mesencephalonic form (Type II) and telencephalonic form (Type III). The classification of salmon GnRH as telencephalonic form (GnRH Type III) by Fernald and White (1999) does not take into account the distribution of salmon and catfish GnRH in primitive fishes, which is not restricted to the telencephalon but is also present in the hypothalamus. On the contrary, 'salmon GnRH' in advanced teleost fishes has a different molecular structure and distribution pattern from salmon GnRH in primitive fishes (Parhar et al., 1996) (Figs. 2 and 4). Therefore, based on the distribution pattern and ontogenic origins from the olfactory system it would be better to consider: (1) salmon and catfish GnRH of primitive fishes along with mammalian GnRH as GnRH-I; (2) the highly conserved mesencephalonic GnRH as GnRH-II; (3) 'salmon GnRH' of advanced teleost as GnRH-III; (4) the newly cloned medaka and seabream GnRH, present only in recently evolved teleost fishes, as GnRH-IV (Fig. 5).

Functional significance of GnRH subtypes

Cells of olfactory origin (olfactory to diencephalon): hypophysiotropic

Primitive teleost fishes and land vertebrates (GnRH-I)

In the following discussion, based on the distribution pattern and their olfactory-related origins, salmon GnRH of primitive fishes, catfish, chicken-I and mammalian GnRH are grouped as GnRH-I. GnRH-I is the only GnRH form present in the forebrain of land vertebrates and primitive fishes. The regulation of gonadotrophs (LH/FSH) by GnRH-I is well documented in all vertebrate species (Sherwood et al., 1997; Yu et al., 1997). In primitive teleost fishes, the synthesis of GnRH-I (salmon GnRH: salmonids, Parhar and Iwata, 1994; Parhar et al., 1995; catfish GnRH: catfish, Dubois et al., 2000; mammalian GnRH: eel, Chiba et al., 1999) prior to the development of gonadotropes, the presence of GnRH-I immunoreactive fibers in the pituitary and GnRH Type IA receptors specifically in the gonadotrophs supports the role of GnRH-I in the differentiation and regulation of the pituitary gonadotropes and gonads (Parhar et al., 2002) Similarly, in land vertebrates, GnRH-I (mammalian GnRH, chicken-I GnRH) plays a vital role in the regulation of LH synthesis–secretion (Silverman et al., 1994; King and Millar, 1997). In fishes, as in land vertebrates, gonadal status and gonadal hormones are known to regulate GnRH-I activity (Gore and Roberts, 1997; Dubois et al., 1998; Herbison, 1998). In addition, thyroid hormones have been implicated in the regulation of GnRH-I during metamorphosis in salmonids (Parhar et al., 1994, 1996; Parhar and Iwata, 1996). In humans, the failure of neurons synthesizing GnRH-I to migrate from the olfactory placodes to the basal forebrain results in Kallman's Syndrome, which is hypogonadotropic hypogonadism coupled with anosmia (Schwanzel-Fukuda, 1997). That GnRH-I is involved in reproduction as well as in imprinting of odors and olfactory memory has been speculated in downstream migrating salmonids (Parhar et al., 1994, 1995; Parhar and Iwata, 1996).

Cells of mecenphalonic origin: neuromodulator

All vertebrate species (GnRH-II)

GnRH-II has remained unchanged during evolution and therefore is identical in all vertebrate species studied to date (King and Millar, 1997). Nevertheless, the function of GnRH-II is still speculative. GnRH-II may have originated as a regulator of reproduction; however, during evolution it has also acquired non-reproductive and extra-pituitary functions. For example, within the central nervous system the widespread distribution of GnRH-II fibers suggests it might act as a neuromodulator and perhaps

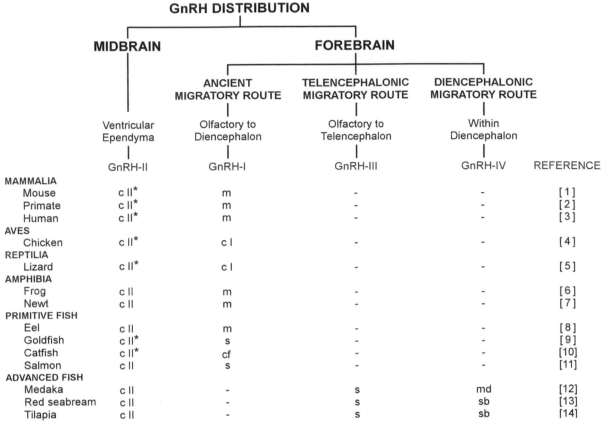

Fig. 5. The proposed nomenclature of GnRH (GnRH I–IV) is based on the sites of origin, migration and final location of GnRH cells in the midbrain and forebrain. Chicken II GnRH (cII) is the most conserved form present in the midbrain of all vertebrates. The seabream (sb) and medaka (md) GnRH forms are present in the diencephalon–preoptic area of only advanced teleost. Asterisks indicates not studied but most likely of ventricular origin. Dased lines indicate unknown/uncertain. S, salmon GnRH; cI, chicken-I GnRH; cf, catfish GnRH, m, mammalian GnRH. References: [1] Schwanzel-Fukuda, 1997; Wray, 2001; [2] Terasawa and Quanbeck, 1997; [3] Schwanzel-Fukuda et al., 1989; [4] Murakami et al., 2000; [5] D'Aniello et al., 1994; [6] Hayes et al. (1994); [7] Murakami et al., 1992; [8] Chiba et al., 1999; [9] Parhar et al., 2001; [10] Dubois et al., 1998; [11] Parhar and Iwata, 1993; Parhar et al., 1995; [12] Parhar et al., 1998; [13] Okuzawa et al., 1997; [14] Parhar, 1997.

fine-tune reproductive behaviors. In the musk shrew, a placental mammal, and the ringdove GnRH-II has been implicated in mating and courtship behaviors (Rissman, 1996; Silver et al., 1996). On the contrary, gonadal hormones have no effect on GnRH-II mRNA and peptide levels, which suggests that GnRH-II might be under different regulatory mechanism (Soga et al., 1998; Parhar et al., 2000) and might have functions other than reproduction. In fishes, GnRH-II regulates prolactin release (Weber et al., 1997) probably via Type IB GnRH receptors (Parhar et al., 2002), which suggests a possible role of GnRH-II in osmoregulation/acid-base balance.

Cells of olfactory origin (olfactory to telencephalon): hypophysiotropic, neuromodulation and reproductive behavior

Advance teleost fishes (GnRH-III)

Compared to primitive teleost fishes, 'salmon GnRH' in advanced teleost fishes has a different molecular structure and is synthesized specifically by cells of the nucleus olfactoretinalis (NOR, Parhar et al., 1996). The presence of GnRH-III immunoreactive fibers in the pituitary suggests its role as a hypophysiotropic hormone (Parhar et al., 1996; Parhar, 1997), and the widespread fibers in the brain suggests its role as a neuromodulator (Oka, 1997). Sex steroids and thyroid hormones have been implicated in the regulation of GnRH-III (Soga et al., 1998; Parhar, 1999b; Parhar et al., 2000). More recently, the role of GnRH-III, has been extended to include aggression and nest building behavior (Yamamoto et al., 1997; Soga et al., 2001; Akiyama et al., 2002; Ogawa et al., 2002b).

Cells of diencephalonic origin: hypophysiotropic hormone

Advance teleost fishes (GnRH-IV)

GnRH-IV (seabream, medaka GnRH) is present in the OVLT–POA area, only in advanced teleost fishes (cichlids: White et al., 1995; tilapia: Parhar, 1997; seabream: Gothilf et al., 1996; Okuzawa et al., 1997; medakafish: Okubo et al., 2000a). The concurrent development of GnRH-IV synthesizing neurons, GnRH fibers in the pituitary, gonadal sex differentiation and the appearance of steroid producing cells in the gonads, suggest GnRH-IV may have a role in sex differentiation (Parhar, 1997). GnRH-IV is known to regulate LH/FSH in advanced teleost fishes (Holland et al., 1998). In addition, an increase in GnRH-IV cell size and cell numbers is associated with gonadal maturation and changes in sex steroid levels (estrogen: Parhar et al., 2000; testosterone: Soga et al., 1998; Ketotestosterone: Soma et al., 1996), which further emphasizes the role of GnRH-IV in reproduction. Furthermore, reproduction-related territorial behavior has been associated with change in GnRH-IV neuronal size (Francis et al., 1993). Since advanced teleost fishes are the only vertebrate species to possess GnRH-IV, therefore, it is likely that this GnRH molecular form might have been acquired for functions specific to advanced teleost.

Summary

Hypothetically it can be assumed that in advanced teleost fishes, GnRH-III and GnRH-IV neurons migrate along the '*telencephalonic*' (anterior) and '*diencephalonic*' (posterior) migratory route, which perhaps fuses in primitive teleost fishes and land vertebrates to form the '*ancient migratory route*' (in all probability = *nervus terminalis*; *see* Von Bartheld et al., 1988) of GnRH-I neurons.

The difference in distribution pattern of GnRH forms in the vertebrate brain is due to distinct embryonic origins:

(1) *Cells of olfactory origin*, which give rise to GnRH-I (salmon, catfish, chicken I, mammalian GnRH) are distributed along the olfactory system and the basal forebrain in primitive fishes and in land vertebrates; GnRH-I might be pivotal for LH/FSH synthesis–release, olfaction and metamorphosis in lower vertebrates. In advanced teleost fishes, neurons synthesizing GnRH-III ('salmon' GnRH) originate from the olfactory system; they are distributed along the basal olfactory bulbs, with distinct ganglia (NOR) at the caudalmost part of the olfactory bulbs and few scattered cells in the basal telencephalon. The NOR might function as a neuromodulator, hypophysiotropic hormone and regulate visual associated reproductive behaviors.

(2) *Cells of mesencephalonic origin*, which give rise to GnRH-II (chicken-II GnRH) are evolutionarily conserved; might function as a neuromodulator involved in motor-associated reproductive behaviors and acid–base balance.

(3) *Cells of diencephalonic origin*, which give rise to GnRH-IV (seabream, medaka GnRH); they are localized in the anterior-basal OVLT–POA area and present only in advanced teleost fishes. GnRH-IV has been implicated in gonadal sex differentiation, gonadal maturation, LH/FSH secretion and territorial behavior. Advance teleost fishes for yet unknown functions might have acquired GnRH-IV.

Although all GnRH subtypes participate in some aspect of reproduction; the precise function of each GnRH form still remains unclear.

Acknowledgements

I would like to thank my students, T. Soga, S. Yamada, H. Tosaki and S. Ogawa for their invaluable help.

References

Adelman, J.P., Mason, A.J., Hayflick, J.S. and Seeburg, P.H. (1986) Isolation of the gene and hypothalamic cDNA for the common precursor of gonadotropin-releasing hormone and prolactin release-inhibiting factor in human and rat. *Proc. Natl. Acad. Sci. USA*, 83: 179–183.

Akiyama, G., Kato, S., Soga, T., Tamano, K., Kawai, T., Sakuma, Y. and Parhar, I.S. (2002) Social stress controls terminal nerve GnRH neurons. *Soc. Behav. Neurosci., Amherst, USA*, Abs. 179.

Alok, D., Hassin, S., Sampath Kumar, R., Trant, J.M., Yu, K. and Zohar, Y. (2000) Characterization of a pituitary GnRH-receptor from a perciform fish, *morone saxatilis*: functional expression in a fish cell line. *Mol. Cell. Endocrinol.*, 168: 65–75.

Ashihara, M., Suzuki, M., Kubokawa, K., Yoshiura, Y., Kobayashi, M., Urano, A. and Aida, K. (1995) Two differing precursor genes for the salmon-type gonadotropin-releasing hormone exist in salmonids. *J. Mol. Endocrinol.*, 15: 1–9.

Blomenrohr, M., Bogerd, J., Leurs, R., Schulz, R.W., Tensen, C.P., Zandbergen, M.A. and Goos, H.J.Th. (1997) Differences in structure–function relations between nonmammalian and mammalian gonadotropin-releasing hormone receptors. *Biochem. Biophys. Res. Commun.*, 238: 517–522.

Bogerd, J., Zandbergen, T., Andersson, E. and Goos, H. (1994) Isolation, characterization and expression of cDNAs encoding the catfish-type and chicken-II-type gonadotropin-releasing-hormone precursors in the African catfish. *Eur. J. Biochem.*, 222: 541–549.

Bond, C.T., Francis, R.C., Fernald, R.D. and Adelman, J.P. (1991) Characterization of complementary DNA encoding the precursor for gonadotropin-releasing hormone and its associated peptide from a teleosts fish. *Mol. Endocrinol.*, 5: 931–937.

Burgus, R., Butcher, M., Amoss, M., Ling, N., Monahan, M., Rivier, J., Fellows, R., Blackwell, R., Vale, W. and Guillemin, R. (1972) Primary structure of ovine hypothalamic luteinizing hormone-releasing factor (LRF). *Proc. Natl. Acad. Sci. USA*, 69: 278–282.

Carolsfeld, J., Powell, J.F.F., Park, M., Fischer, W.H., Craig, A.G., Chang, J.P., Rivier, J.E. and Sherwood, N.M. (2000) Primary structure and function of three gonadotropin-releasing hormones, including a novel form, from an ancient teleost, herring. *Endocrinology*, 141: 505–512.

Chiba, H., Nakamura, M., Iwata, M., Sakuma, Y., Yamauchi, K. and Parhar, I.S. (1999) Development and differentiation of gonadotropin hormone-releasing hormone neuronal systems and testes in the Japanese eel (*Anguilla japonica*). *Gen. Comp. Endocrinol.*, 114: 449–459.

D'Aniello, B., Pinelli, C., King, J.A. and Rastogi, R.K. (1994) Neuroanatomical organization of GnRH neuronal systems in the lizard (*Podarcis s. sicula*) brain during development. *Brain Res.*, 657: 221–226.

Dubois, E.A., Florijn, M.A., Zandbergen, M.A., Peute, J. and Goos, H.J.Th. (1998) Testosterone accelerates the development of the catfish GnRH system in the brain of immature African catfish (*Clarias gariepinus*). *Gen. Comp. Endocrinol.*, 112: 383–393.

Dubois, E.A., Zandbergen, M.A., Peute, J., Hassing, I., Dijk, W.v., Schulz, R.W. and Goos, H.J.Th. (2000) Gonadotropin-releasing hormone fibers innervate the pituitary of the male

African catfish (*Clarias gariepinus*) during puberty. *Neuroendocrinology*, 72: 252–262.

Dunn, I.C., Chen, Y., Hook, C., Sharp, P.J. and Sang, H.M. (1993) Characterization of the chicken preprogonadotropin-releasing hormone-I gene. *J. Mol. Endocrinol.*, 11: 19–29.

Eisthen, H.L. and Northcutt, R.G. (1996) Silver lampreys (*Ichthyomyzon unicuspis*) lack a gonadotropin-releasing hormone- and FMRFamide-immunoreactive terminal nerve. *J. Comp. Neurol.*, 370(2): 159–172.

Fernald, R.D. and White, R.B. (1999) Gonadotropin-releasing hormone genes: phylogeny, structure and function. *Front. Neuroendocrinol.*, 20: 224–240.

Francis, R.C., Soma, K.K. and Fernald, R.D. (1993) Social regulation of the brain–pituitary–gonadal axis. *Proc. Natl. Sci. USA.*, 90: 7794–7798.

Gestrin, E.D., White, R.B. and Fernald, R.D. (1999) Second form of gonadotropin-releasing hormone in mouse: immunocytochemistry reveals hippocampal and periventricular distribution. *FEBS Lett.*, 448: 289–291.

Gore, A.C. and Roberts, J.L. (1997) Regulation of gonadotropin-releasing hormone gene expression in vivo and in vitro. *Front. Neuroendo.*, 18: 209–245.

Gothilf, Y., Munoz-Cueto, J.A., Sagrillo, C.A., Selmanoff, M., Chen, T.T., Kah, O., Elizur, A. and Zohar, Y. (1996) Three forms of gonadotropin-releasing hormone in a perciform fish (*Sparus aurata*): complementry deoxyribonucleic acid characterization and brain localization. *Biol. Reprod.*, 55: 636–645.

Grober, M.S., Myers, T.R., Marchaterre, M.A., Bass, A.H. and Myers, D.A. (1995) Structure, localization, and molecular phylogeny of a GnRH cDNA from a paracanthopterygian fish, the plainfin midshipman (*Porichthys notatus*). *Gen. Comp. Endocrinol.*, 99(1): 85–89.

Hayes, W.P., Wray, S. and Battey, J.F. (1994) The frog gonadotropin-releasing hormone-I (GnRH-I) gene has a mammalian-like expression pattern and conserved domains in GnRH-associated peptide, but brain onset is delayed until metamorphosis. *Endocrinology*, 134:1835–1845.

Herbison, A.E. (1998) Multimodal influence of estrogen upon gonadotropin-releasing hormone neurons. *Endocr. Rev.*, 19(3): 302–330.

Hiney, J.K., Sower, S.A., Yu, W.H., McCann, M. and Dees, W.L. (2002) Gonadotropin-releasing hormone neurons in the preoptic–hypothalamic region of the rat contain lamprey gonadotropin-releasing hormone III, mammalian luteinizing hormone-releasing hormone, or both peptides. *Proc. Natl. Acad. Sci. USA.*, 99: 2386–2391.

Holland, M.C.H., Gothilf, Y., Meiri, I., King, J.A., Okuzawa, K., Elizur, A. and Zohar, Y. (1998) Levels of the native forms of GnRH in the pituitary of the gilthead seabream, Sparus aurata, at several characteristic stages of the gonadal cycle. *Gen. Comp. Endocrinol.*, 112: 394–405.

Illing, N., Troskie, B., Nahorniak, C., Hapgood, J., Peter, R. and Millar, R.P. (1999) Two gonadotropin-releasing hormone receptor subtypes with distinct ligand selectivity and differential distribution in brain and pituitary in the goldfish. *Proc. Natl. Acad. Sci. USA.*, 96: 2526–2531.

Jimenez-Linan, M., Rubin, B.S. and King, J.C. (1997) Examination of guinea pig luteinizing hormone-releasing hormone gene reveals a unique decapeptide and existence of two transcripts in the brain. *Endocrinology*, 138: 4123–4130.

Kasten, T.L., White, S.A., Norton, T.T., Bond, C.T., Adelman, J.P. and Fernald, R.D. (1996) Characterization of two new pre-proGnRH mRNAs in the tree shrew: first direct evidence for mesencephalic GnRH gene expression in a placental mammal. *Gen. Comp. Endocrinol.*, 104: 7–19.

Kim, M.-H., Oka, Y., Amano, M., Kobayashi, M., Okuzawa, K., Hasegawa, Y., Kawashima, S., Suzuki, Y. and Aida, K. (1995) Immunocytochemical localization of sGnRH and cGnRH-II in the brain of goldfish, *Carassius auratus. J. Comp. Neurol.*, 356: 72–82.

Kimura, F. and Funabashi, T. (1998) Two subgroups of gonadotropin-releasing hormone neurons control gonadotropin secretion in rats. *News Physiol. Sci.*, 13: 225–231.

King, J.A. and Millar, R.P. (1997) Coordinated evolution of GnRHs and their receptors. In: I.S. Parhar and Y. Sakuma (Eds.), *GnRH Neurons: Gene to Behavior*. Brain Shuppan Publishers, Tokyo, pp. 51–77.

Klungland, H., Lorens, J.B., Andersen, O., Kisen, G.O. and Alestrom, P. (1992) The Atlantic salmon prepro-gonadotropin releasing hormone gene and mRNA. *Mol. Cell. Endocrinol.*, 84: 167–174.

Lee, W.-S., Smith, M.S. and Hoffman, G.E. (1990) Luteinizing hormone-releasing hormone neurons express fos protein during the proestrous surge of luteinizing hormone. *Proc. Natl. Acad. Sci. USA.*, 87: 5163–5167.

Lescheid, D.W., Terasawa, E., Abler, L.A., Urbanski, H.F., Warby, C.M., Millar, R.P. and Sherwood, N.M. (1997) A second form of gonadotropin-releasing hormone (GnRH) with characteristics of chicken GnRH-II is present in the primate brain. *Endocrinology*, 138: 5618–5629.

Madigou, T., Mananos-Sanchez, E., Hulshof, S., Anglade, I., Zanuy, S. and Kah, O. (2000) Cloning, tissue distribution, and central expression of the gonadotropin-releasing hormone receptor in the rainbow trout (*Oncorhynchus mykiss*). *Biol. Reprod.*, 63: 1857–1866.

Matsuo, H., Baba, Y., Nair, R.M.G., Arimura, A. and Schally, A.V. (1971) Structure of the porcine LH- and FSH-releasing hormone. I. The proposed amino acid sequence. *Biochem. Biophys. Res. Commun.*, 43:1334–1339.

Meek, J. and Nieuwenhuys, R. (1998) Holosteans and teleosts. In: R. Nieuwenhuys, H.J. Ten Donkelaar and C. Nicholson (Eds.), *The central nervous System of Vertebrates*. Springer-Verlag, Berlin, Vol. 2, pp. 759–937.

Melamed, P., Gur, G., Elizur, A., Rosenfeld, H., Rentier-Delrue, F. and Yaron, Z. (1996) Differential effects of gonadotropin releasing hormone, dopamine and somatostatin and their second messengers on the mRNA levels of gonadotropin IIB subunit and growth hormone in the teleost fish, tilapia. *Neuroendocrinology*, 64: 320–328.

Merchenthaler, I., Lopez, F.J., Lennard, D.E. and Negro-Vilar, A. (1991) Sexual differences in the distribution of neurons coexpressing galanin and luteinizing hormone-releasing hormone in the rat brain. *Endocrinology*, 129: 1977–1988.

Millar, R., Lowe, S., Conklin, D., Pawson, A., Maudsley, S., Troskie, B., Ott, T., Millar, M., Lincoln, G., Sellar, R., Faurholm, B., Scobie, G., Kuestner, R., Terasawa, E. and Katz, A. (2001) A novel mammalian receptor for the evolutionarily conserved type II GnRH. *Proc. Natl. Acad. Sci. USA*, 98: 9636–9641.

Miyamoto, K., Hasegawa, Y., Nomura, M., Igarashi, M., Kangawa, K. and Matsuo, H. (1984) Identification of the second gonadotropin-releasing hormone in chicken hypothalamus: Evidence that gonadotropin secretion is probably controlled by two distinct gonadotropin-releasing hormones in avian species. *Proc. Natl. Acad. Sci. USA*, 81: 3874–3878.

Montero, M., Vidal, B., King, J.A., Tramu, G., Vandesande, F., Dufour, S. and Kah, O. (1994) Immunocytochemical localization of mammalian GnRH (gonadotropin-releasing hormone) and chicken GnRH-II in the brain of the European silver eel (*Anguilla anguilla* L.). *J. Chem. Neuroanat.*, 7: 227–241.

Munz, H. and Claas, B. (1987) The terminal nerve and its development in the teleost fishes. In: L.S. Demski and M. Schwanzel-Fukuda (Eds.), *The Terminal Nerve (Nervus Terminalis) Structure, Function and Evolution. Ann. N.Y. Acad. Sci.*, 519: 50–59.

Murakami, S., Kikuyama, S. and Arai, Y. (1992) The origin of the luteinizing hormone-releasing hormone (LHRH) neurons in the newts (*Cynops pyrrhogaster*): the effect of olfactory placode ablation. *Cell Tissue Res.*, 269: 21–27.

Murakami, S., Seki, T., Rutishauser, U. and Arai, Y. (2000) Enzymatic removal of polysialic acid from neural cell adhesion molecule perturbs the migration route of luteinizing hormone-releasing hormone neurons in the developing chick forebrain. *J. Comp. Neurol.*, 420: 171–181.

Muske, L.E. (1997) Ontogeny, phylogeny and neuroanatomical organization of multiple molecular forms of GnRH. In: I.S. Parhar and Y. Sakuma (Eds.), *GnRH Neurons: Gene to Behavior*. Brain Shuppan Publishers, Tokyo, pp. 145–180.

Neill, J.D., Duck, L.W., Sellers, J.C. and Musgrove, L.C. (2001) A gonadotropin–releasing hormone (GnRH) receptor specific for GnRH II in primates. *Biochem. Biophys. Res. Commun.*, 282: 1012–1018.

Norgren, R.B. Jr. and Chen, G. (1994) LHRH neuronal subtypes have multiple origins in chickens. *Dev. Biol.*, 165: 735–738.

Northcutt, R.G. and Muske, L.E. (1994) Multiple embryonic origins of gonadotropin-releasing hormone (GnRH) immunoreactive neurons. *Dev. Brain Res.*, 78: 279–290.

Ogawa, S., Mirasoll, E.G., Mesalo, R., Pfaff, D.W. and Parhar, I.S. (2002a) Existence of multiple GnRH forms and their co-localization with estrogen receptor B in the male mouse brains. *Soc. Neurosci. USA*, Abs. (in press).

Ogawa, S., Soga, T., Yamamoto, N., Sakuma, Y. and Parhar, I. (2002b) Modulation of reproductive behaviors by antisense GnRH in cichlid fish. *Soc. Behav. Neurosci.*, Amherst, USA, Abs. 148.

Oka, Y. (1997) GnRH neuronal system of fish brain as a model system for the study of peptidergic neuromodulation. In: I.S. Parhar and Y. Sakuma (Eds.), *GnRH Neurons: Gene to Behavior*. Brain Shuppan Publishers, Tokyo, pp. 245–276.

Okubo, K., Amano, M., Yoshiura, Y., Suetake, H. and Aida, K. (2000a) A novel form of gonadotropin-releasing hormone in

16

the medaka, *Oryzias latipes. Biochem. Biophys. Res. Commun.*, 276(1): 298–303.

Okubo, K., Nagata, S., Ko, R., Kataoka, H., Yoshiura, Y., Mitani, H., Kondo, M., Naruse, K., Shima, A. and Aida, K. (2000b) Identification and characterization of two distinct GnRH receptor subtypes in a teleost, the medaka *Oryzias latipes. Endocrinology*, 142: 4729–4739.

Okuzawa, K., Granneman, J., Bogerd, J., Goos, H.J.Th., Zohar, Y. and Kagawa, H. (1997) Distinct expression of GnRH genes in the red seabream brain. *Fish Physiol. Biochem.*, 17: 71–79.

Parhar, I.S. (1997) GnRH in tilapia: three genes, three origins and their roles. In: I.S. Parhar and Y. Sakuma (Eds.), *GnRH Neurons: Gene to Behavior*. Brain Shuppan Publishers, Tokyo, pp. 99–122.

Parhar, I.S. (1999a) Multiple gonadotropin-releasing hormone neuronal systems in vertebrates. *Korean J. Bio. Sci.*, 3: 1–7.

Parhar, I.S. (1999b) Hormonal regulation of salmon-GnRH gene. In: B. Norberg, O.S. Kjesbu, G.L. Taranger, E. Andersson and S.O. Stefansson (Eds.), *Proceedings of 6th International Symposium on the Reproductive Physiology of Fish*. John Grieg AS, Bergen, Norway, pp. 43–46.

Parhar, I.S. and Iwata, M. (1993) Molecular forms of gonadotropin-releasing hormone (GnRH): origin, migration and gene expression in teleosts. In: M.R.P. Varavudhi and S. Lorlowhakarn (Eds.), *Progress in Comparative Endocrinology. Proceedings of the Second Intercongress Symposium of the Asia and Oceania Society for Comparative Endocrinology, Chiangmai, Thailand*, pp. 111–114.

Parhar, I.S. and Iwata, M. (1994) Gonadotropin releasing hormone (GnRH) neurons project to growth hormone and somatolactin cells in the steelhead trout. *Histochem. J.*, 102: 195–203.

Parhar, I.S. and Iwata, M. (1996) Intracerebral expression of gonadotropin-releasing hormone and growth hormone-releasing hormone is delayed until smoltification in the salmon. *Neurosci. Res.*, 26: 299–308.

Parhar, I.S., Koibuchi, N., Sakai, M., Iwata, M. and Yamaoka, S. (1994) Gonadotropin-releasing hormone (GnRH): expression during salmon migration. *Neurosci. Lett.*, 172: 15–18.

Parhar, I.S., Iwata, M., Pfaff, D.W. and Schwanzel-Fukuda, M. (1995) Embryonic development of gonadotropin-releasing hormone neurons in the sockeye salmon. *J. Comp. Neurol.*, 362: 256–270.

Parhar, I.S., Pfaff, D.W. and Schwanzel-Fukuda, M. (1996) Gonadotropin-releasing hormone gene expression in teleosts. *Mol. Brain Res.*, 41: 216–227.

Parhar, I.S., Soga, T., Ishikawa, Y., Nagahama, Y. and Sakuma, Y. (1998) Neurons synthesizing gonadotropin-releasing hormone mRNA subtypes have multiple developmental origins in the medaka. *J. Comp. Neurol.*, 401: 217–226.

Parhar, I.S., Soga, T. and Sakuma, Y. (2000) Thyroid hormone and estrogen regulate brain region-specific messenger ribonucleic acids encoding three gonadotropin-releasing hormone genes in sexually immature male fish, *Oreochromis niloticus. Endocrinology*, 141(5): 1618–1626.

Parhar, I.S., Tosaki, H., Sakuma, Y. and Kobayashi, M. (2001). Sex differences in the brain of goldfish: Gonadotropin-

releasing hormone and vasotocinergic neurons. *Neuroscience*, 104: 1099–1110.

Parhar, I.S., Soga, T., Sakuma, Y. and Millar, R.P. (2002) Spatiotemporal expression of gonadotropin-releasing hormone receptor subtypes in gonadotropes, somatotropes and lactotropes in the cichlid fish. *J. Neuroendocrinol.*, 14: (in press).

Rissman, E.F. (1996) Behavioral regulation of gonadotropin-releasing hormone. *Biol. Reprod.*, 54: 413–419.

Schreibman, M.P. and Margolis-Nunno, H. (1987) Reproductive biology of the terminal nerve (nucleus olfactoretinalis) and other LHRH pathways in teleost fishes. In: L.S. Demski and M. Schwanzel-Fukuda (Eds.), *The Terminal Nerve (Nervus Terminalis): Structure, Function and Evolution. Ann. N.Y. Acad. Sci.*, 519: 60–68.

Schwanzel-Fukuda, M. (1997) The origin and migration of LHRH neurons in mammals: a comparison between species including humans. In: I.S. Parhar and Y. Sakuma (Eds.), *GnRH Neurons: Gene to Behavior*. Brain Shuppan Publishing, Tokyo, pp. 221–242.

Schwanzel-Fukuda, M., Bick, D. and Pfaff, D.W. (1989) Luteinizing hormone-releasing hormone (LHRH)-expressing cells do not migrate normally in an inherited hypogonadal (Kallmann) syndrome. *Mol. Brain Res.*, 6: 311–326.

Sealfon, S.C., Weinstein, H. and Millar, R.P. (1997) Molecular mechanisms of ligand interaction with the gonadotropin-releasing hormone receptor. *Endocr. Rev.*, 18: 180–205.

Sherwood, N.M., Von Schalburg, K. and Lescheid, D.W. (1997) Origin and evolution of GnRH in vertebrates and invertebrates. In: I.S. Parhar and Y. Sakuma (Eds.), *GnRH Neurons: Gene to Behavior*. Brain Shuppan Publishing, Tokyo, pp. 3–25.

Silver, R., Silverman, A.J., Vitkovic, L. and Lederhendler, I. (1996) Mast cells in the brain: evidence and function significance. *Trends Neurosci.*, 19(1): 25–31.

Silverman, A.-J., Livne, I. and Witkin, J.W. (1994) The gonadotropin-releasing hormone (GnRH) neuronal systems: immunocytochemistry and in situ hybridization. In: E. Knobil and J.D. Neill (Eds.), *The Physiology of Reproduction*. Raven Press, New York, NY, pp. 1683–1709.

Skynner, M.J., Slater, R., Sim, J.A., Allen, N.D. and Herbison, A.E. (1999) Promoter transgenics reveal multiple gonadotropin-releasing hormone-I-expressing cell populations of different embryological origin in mouse brain. *J. Neurosci.*, 19(14): 5955–5966.

Soga, T., Sakuma, Y. and Parhar, I.S. (1998) Testosterone differentially regulates expression of GnRH messenger RNAs in the terminal nerve, preoptic and midbrain of male tilapia.*Mol. Brain Res.*, 60: 13–20.

Soga, T., Tamano, K., Kawai, T., Sakuma, Y. and Parhar, I.S. (2001). Social Status controls terminal nerve GnRH isoform. *Second International Symposium on the Comparative Biology of GnRH: Molecular Forms and Receptors*. Penang, Malaysia, Abstract SC6, p. 50.

Soma, K.K., Francis, R.C., Wingfield, J.C. and Fernald, R.D. (1996) Androgen regulation of hypothalamic neurons containing gonadotropin-releasing hormone in a cichlid fish: integration with social cues. *Horm. Behav.*, 30: 216–226.

Sower, S.A. (1997) Evolution of GnRH in fish of ancient origins.

In: I.S. Parhar and Y. Sakuma (Eds.), *GnRH Neurons: Gene to Behavior*. Brain Shuppan Publishing, Tokyo, pp. 27–49.

Stell, W.K., Walker, S.E., Chohan, K.S. and Ball, A.K. (1984) The goldfish nervus terminalis: A luteinizing hormone-releasing hormone and molluscan cardioexcitatory peptide immunoreactive olfactoretinal pathway. *Proc. Natl. Acad. Sci. USA*, 81: 940–944.

Suzuki, M., Hyodo, S., Kobayashi, M., Aida, K. and Urano, A. (1992) Characterization and localization of mRNA encoding the salmon-type gonadotropin-releasing hormone precursor of the masu salmon. *J. Mol. Endocrinol.*, 9: 73–82.

Tensen, C., Okuzawa, K., Blomenrohr, M., Rebers, F., Leurs, R., Bogerd, J., Schulz, R. and Goos, H. (1997) Distinct efficacies for two endogeneous ligands on a single cognate gonadoliberin receptor. *Eur. J. Biochem.*, 243: 134–140.

Terasawa, E. and Quanbeck, C. (1997) Two types of LHRH neurons in the forebrain of the rhesus monkey during embryonic development. In: I.S. Parhar and Y. Sakuma (Eds.), *GnRH Neurons: Gene to Behavior*. Brain Shuppan Publishing, Tokyo, pp. 197–219.

Tobet, S.A., Chickering, T.W. and Sower, S.A. (1996) Relationship of gonadotropin-releasing hormone (GNRH) neurons to the olfactory system in developing lamprey (*Petromyzon marinus*). *J. Comp. Neurol.*, 376(1): 97–111.

Troskie, B., Illing, N., Rumbak, E., Sun, Y.-M., Hapgood, J., Sealfon, S., Conklin, D. and Millar, R. (1998) Identification of three putative GnRH receptor subtypes in vertebrates. *Gen. Comp. Endocrinol.*, 112: 296–302.

Troskie, B.E., Hapgood, J.P., Millar, R.P. and Illing, N. (2000) Complementary deoxyribonucleic acid cloning, gene expression, and ligand selectivity of a novel gonadotropin-releasing hormone receptor expressed in the pituitary and midbrain of *Xenopus laevis*. *Endocrinology*, 141: 1764–1771.

Urbanski, H.F., White, R.B., Fernald, R.D., Kohama, S.G. and Densmore, V.S. (1999) Regional expression of mRNA encoding a second form of gonadotropin-releasing in the macaque brain. *Endocrinology*, 140: 1945–1948.

Von Bartheld, C., Claas, B., Munz, H. and Meyer, D.L. (1988) Primary olfactory projections and the nervus terminalis in the African lungfish: implications for the phylogeny of cranial nerves. *Am. J. Anat.*, 182: 325–334.

Wang, L., Bogerd, J., Choi, H.S., Seong, J.Y., Soh, J.M., Chun, S.Y., Blomenrohr, M., Troskie, B.E., Millar, R.P., Yu, W.H., McCann, S.M. and Kwon, H.B. (2001) Three distinct types of GnRH receptor characterized in the bullfrog. *Proc. Natl. Acad. Sci. USA*, 98: 361–366.

Weber, G.M., Powell, J.F.F., Park, M., Fischer, W.H., Craig, A.G., Rivier, J.E., Nanakorn, U., Parhar, I.S., Ngamvongchon, S., Grau, E.G. and Sherwood, N.M. (1997) Evidence that gonadotropin-releasing hormone (GnRH) functions as a prolactin-releasing factor in a teleost fish (tilapia *Oreochromis Mossambicus*) and primary structures for three native GnRH molecules. *J. Endocrinol.*, 155: 121–132.

White, S.A., Bond, C.T., Francis, R.C., Kasten, T.L., Fernald, R.D. and Adelman, J.P. (1994) A second gene for gonadotropin-releasing hormone: cDNA and expression pattern in the brain. *Proc. Natl. Acad. Sci. USA*, 91: 1423–1427.

White, S.A., Kasten, T.L., Bond, C.T., Adelman, J.P. and Fernald, R.D. (1995) Three gonadotropin-releasing hormone genes in one organism suggest novel roles for an ancient peptide. *Proc. Natl. Acad. Sci. USA*, 92: 8363–8367.

White, R.B., Eisen, J.A., Kasten, T.L. and Fernald, R.D. (1998) Second gene for gonadotropin-releasing hormone in humans. *Proc. Natl. Acad. Sci. USA*, 95: 305–309.

Wray, S. (2001) Development of luteining hormone releasing hormone neurones. *J. Neuroendocrinol.*, 13: 3–11.

Yamamoto, N., Oka, Y. and Kawashima, S. (1997) Lesion of gonadotropin-releasing hormone-immunoreactive terminal nerve cells: Effect the reproductive behavior of male dwarf gouramis. *Neuroendocrinology*, 65: 403–412.

Yoo, M.S., Kang, H.M., Choi, H.S., Kim, J.W., Troskie, B.E., Millar, R.P. and Kwon, H.B. (2000). Molecular cloning, distribution and pharmacological characterization of a novel gonadotropin-releasing hormone ([Trp8]GnRH) in frog brain. *Mol. Cell. Endocrinol.*, 164: 197–204.

Yu, K.L., Lin, X.-W., Bastos, J.C. and Peter, R.E. (1997) Neural regulation of GnRH in teleost fishes. In: I.S. Parhar and Y. Sakuma (Eds.), *GnRH Neurons: Gene to Behavior*. Brain Shuppan Publishers, Tokyo, pp. 277–312.

Zandbergen, M.A., Kah, O., Bogerd, J., Peute, J. and Goos, H.J.Th. (1995) Expression and distribution of two gonadotropin-releasing hormones in the catfish brain. *Neuroendocrinology*, 62: 571–578.

I.S. Parhar (Ed.)
Progress in Brain Research, Vol. 141

CHAPTER 2

GnRH in the invertebrates: an overview

Rakesh K. Rastogi [1,*], Maria M. Di Fiore [2], Antimo D'Aniello [3], Luisa Iela [1],
Maria Fiorentino [1]

[1] *Department of Zoology, University of Naples Federico II, Via Mezzocannone 8, 80134 Napoli, Italy*
[2] *Department of Life Sciences, SUN, Via Vivaldi 43, 81100 Caserta, Italy*
[3] *Stazione Zoologica, Villa Comunale, 80100 Napoli, Italy*

Introduction and comparative historical perspective

The gonadotropin-releasing hormone [GnRH = luteinizing hormone-releasing hormone (LHRH)] represents a pivotal peptide in animal reproduction. GnRH is a decapeptide originally isolated from the porcine and ovine hypothalamus and characterized for its ability to enhance the release of pituitary gonadotropins, FSH and LH (Amoss et al., 1971; Matsuo et al., 1971). This pioneering research has led to the identification of at least 11 different forms of GnRH in vertebrates, 2 new forms in tunicates, and 1 in a mollusk (see Powell et al., 1996; Jimenez-Linan et al., 1997; Carolsfeld et al., 2000; Iwakoshi et al., 2002). These GnRH forms can be identified as follows: mammalian (mGnRH), guinea pig GnRH, chicken-I (cGnRH-I), chicken-II (cGnRH-II), salmon (sGnRH), seabream (sbGnRH), catfish (cfGnRH), herring (hrGnRH), dogfish (dfGnRH), lamprey-I (lGnRH-I), lamprey-III (lGnRH-III), tunicate-I (tGnRH-I), tunicate-II (tGnRH-II, a disulfide-linked dimer), and octopus (octoGnRH, a dodecapeptide). Based on amino acid homology, i.e., sharing part of their sequence, Gn-RHs in the animal kingdom are now represented as a family of structurally related peptides. All known

forms of GnRH peptides have in common a pyroglutamyl residue at the amino terminus and an amidated glycine at the carboxy terminus. Except for the octoGnRH, all other known forms are decapeptides and differ by one or more amino acids; amino acids 4 and 9 are conserved in all forms known to date.

Since it triggers the release of pituitary gonadotropins in all vertebrates, its current denomination as 'GnRH' is fully justified. During the late 1960s and early 1970s, the isolation of a hypothalamic hypophysiotropic peptide, its characterization as GnRH, and the determination of its amino acid sequence, marked the beginning of an era in GnRH research. Since that time many investigators have focused their attention to what seems to be an ever increasing number of its molecular variants, their distribution in the neural and nonneural tissues, and their physiological roles as neurotransmitter, neuromodulator, neurohormone, local hormone, and so forth, placing the GnRH at an important junction between endocrine and nervous systems.

Since the late 1970s to date, a vast assortment of review articles and brief updates has been published: Silverman and Zimmerman (1978), Nozaki and Kobayashi (1979), King and Millar (1980, 1987, 1991, 1992, 1995, 1997), Peter (1983), Nozaki et al. (1984), Demski (1984, 1987), Barry et al. (1985), Sherwood (1987), Silverman (1988), Clayton (1989), Chieffi et al. (1991), Andersen et al. (1992), Muske (1993, 1997), Sherwood et al. (1993, 1997), Silverman et al. (1994), Rastogi and Iela (1994), Parhar et al. (1995), Rissman (1997), Demski et

* Correspondence to: R.K. Rastogi, Department of Zoology, University of Naples Federico II, Via Mezzocannone 8, 80134 Napoli, Italy. Fax: +39-081-2535210;
E-mail: rakesh.rastogi@unina.it

20

al. (1997), Kim et al. (1997), Rastogi et al. (1998). These have dealt variously with the distribution of GnRH-like immunoreactivity in the central nervous system (CNS) and nonneural tissues, primary structure of GnRH-like molecule, phylogenetic diversity, endocrine/nonendocrine and behavioral regulation of GnRH synthesis and secretion (also see Rissman, 1996; D'Aniello et al., 2001), genes encoding GnRHs, GnRH neuronal migration from olfactory placode into the developing basal forebrain in jawed vertebrates, bioactivity, mechanisms of action of GnRH and its receptors, regulation of GnRH gene expression, and its evolution either within a specific group of animals or across the entire vertebrate and/or invertebrate lineage. In the early 1970s attention was projected to higher vertebrates (particularly mammals, and less frequently birds). Research on GnRH-like peptides in nonmammalian vertebrates, including reptiles, amphibians and fishes, gained momentum during the late 1970s, and it was in the 1980s that the term GnRH was largely accepted to replace LHRH, at least in nonmammalian vertebrates. During a time span of little more than 30 years, little over 100 species of vertebrates (17 species of mammals, 8 birds, 18 reptiles, 22 amphibians, little over 45 bony, cartilaginous and agnathan fishes) and only 17 species of invertebrates have been examined (see Demski et al., 1997; King and Millar, 1997; Parhar, 1997; Sower, 1997; Carolsfeld et al., 2000). The invertebrates were totally ignored until 1980 when Georges and Dubois, for the first time, ventured into the realm of protochordates and observed GnRH-like immunoreactivity in the nervous system of an ascidian tunicate, *Ciona intestinalis* (Georges and Dubois, 1980). This finding, however, did not provide an impetus strong enough to proceed with active research on the presence of GnRH-like peptides in the invertebrates. However, with the availability in successive years of more sophisticated instrumentation and biochemical methodology, and the understanding that GnRH is a molecule not restricted to the nervous system, a greater attention was focused on invertebrates during the last 6–7 years. These studies have provided surprising results (see Powell et al., 1996; Di Fiore et al., 2000; Zhang et al., 2000; Iwakoshi et al., 2002), in as much as in some invertebrates vertebrate-type GnRH forms have been isolated and characterized.

Immunocytochemical localization, isolation and characterization of GnRH-like peptides

Cnidarian nervous system

According to a recent study, using anti-cGnRH-II and anti-mGnRH, on the sea pansy (*Renilla koellikeri*) and starlet sea anemone (*Nematostella vectensis*) by Anctil (2000), GnRH-immunoreactive (GnRHir) neurons are distributed individually in the endodermal and myoepithelial layers throughout, as well as in mesenteric filaments bearing gametes. In the mesenteric filaments such neurons usually occur in clusters. In the sea pansy, GnRHir neurons are relatively more abundant in the tentacles, and those located in the wall of the gastrovascular cavity often possess a cilium. In the starlet anemone, on the other hand, GnRHir fibers innervate the follicular layer of the ovocytes. All GnRHir neurons in the two species represent a subset of neurons sharing similarities of morphology, distribution and innervation targets. High-performance liquid chromatography (HPLC) and radioimmunoassay (RIA) of whole sea pansy extracts yielded two elution peaks which were immunoreactive with antisera against dfGnRH or mGnRH (Anctil, 2000).

Flatworms

No reports are available.

Nematodes

Brownlee et al. (1993) made a brief mention of the presence of GnRH-like immunoreactivity in some parts of the central nervous system of the roundworm, *Ascaris suum*.

Annelids

Dhainaut-Courtois et al. (1985) failed to observe GnRH-like immunoreactivity in the polychaet worm, *Nereis diversicolor*. However, as cited in Al-Yousuf (1990), GnRHir material is present in the neural tissue of the leech, and in some nonneural tissues of earthworms.

Mollusks

Within this taxonomic group, the second largest, of protostome coelomates, only 5 species have been investigated. The first account of GnRH-like immunoreactivity in a mollusk was made available in 1993, when Goldberg et al. described the distribution of GnRHir neurons in the CNS (circumesophageal ganglia: cerebral, pleural, parietal, visceral or abdominal, and pedal) of a gastropod, the hermaphroditic pond snail *Helisoma trivolvis*. It was observed that, using two different antibodies viz. anti-mGnRH and anti-cGnRH-II, mGnRH-like immunoreactive (ir) neurons were observed in the cerebral and pedal ganglia, whereas cGnRH-II-like ir cells were observed in the parietal and visceral ganglia. No GnRHir cell bodies were described in the pleural ganglia. A differential anatomical localization of the two types of GnRH-like neurons in the mollusk has close analogy with a differential distribution of multiple forms of GnRH in neuroanatomically distinct regions in the vertebrate brain (King and Millar, 1997; Muske, 1997; Rastogi et al., 1998). In a subsequent study on *Helisoma trivolvis* and another pond snail, *Lymnaea stagnalis*, these results were confirmed with slight insignificant differences, and the distribution of GnRHir neurons in the CNS of *Lymnaea* was consistent with that in *Helisoma* (Young et al., 1999). In both studies, however, it was emphasized that the cerebral ganglia are most heavily laid with GnRHir neurons, and that the reproductive tract in both species shows GnRHir innervation. In the earlier study, furthermore, HPLC of CNS homogenate yielded multiple peaks, the major one coeluting with mGnRH (Goldberg et al., 1993). That the CNS of *Helisoma trivolvis* secretes a GnRH-like factor was further corroborated by a bioassay study in which crude CNS extract from *Helisoma* was capable of stimulating gonadotropic hormone release from dispersed goldfish pituitary cells in static incubation (Goldberg et al., 1993). Studies on GnRH-related or -like peptides in mollusks have been extended to cephalopod mollusks as well. One such study showed the localization of GnRHir perikarya in the subpedunculate lobe and the optic gland and ir fibers along the optic tract in *Octopus vulgaris* (Di Cristo et al., 1995; Di Cosmo and Di Cristo, 1998). This was later confirmed by Iwakoshi

et al. (2002). These authors cloned a precursor protein cDNA and observed that the RT-PCR transcripts were expressed in the supra- and subesophageal ganglia, optic peduncle complex, and optic gland. These observations on the presence of a GnRH-like peptide in different regions of the CNS were substantiated by enzyme-linked immunosorbent assay and TOF-mass spectrometric analysis. A dodecapeptide, with similar amino and carboxy termini as in all known vertebrate GnRH decapeptides, was characterized. Iwakoshi et al. (2002) have proposed it as octopus GnRH, and its GnRH-like activity was confirmed by a dose-dependent stimulation of LH release in primary cultures of quail anterior pituitary cells. *Loligo vulgaris*, another cephalopod mollusk, showed distinct GnRHir perikarya in the CNS (using a polyclonal anti-mGnRH, LR1, courtesy G. Benoit), including the optic gland (Di Fiore, D'Aniello A. and Rastogi, unpublished). A preliminary, three-step, purification of the optic gland homogenate (10 g sample) was performed and the extract was subjected to HPLC which revealed one major peak (retention time 36.5 min) and two minor peaks (retention times 34.5 and 45.8 min), eluting in correspondence respectively with synthetic mGnRH, cGnRH-I and sGnRH (Fig. 1). Proceeding further, a sample extract, after additional purification with Sep-Pak, was incubated overnight at 4°C together with anti-mGnRH, at a dilution of 1 : 25. Chromatographed under similar conditions this extract failed to show the 36.5 min major peak (Fig. 1) and thus strengthened the supposition that *Loligo* optic glands contain, at least, an mGnRH-like peptide. The presence of an mGnRH-like peptide can be suggested also in the cerebral and pedal ganglia of a bivalve mollusk (*Mytilus edulis*) in which anti-GnRH immunostains a great number of perikarya and fibers (Pazos and Mathieu, 1999)

In striking contrast with the above reports on mollusks, Zhang et al. (2000) failed to detect GnRHir elements in the CNS of an opisthobranch gastropod mollusk, *Aplysia californica*. Although no ir perikarya were observed anywhere within the CNS by using anti-mGnRH, anti-cGnRH-II, anti-lGnRH-III and anti-tGnRH-I, there was strong immunoreactivity in the connective sheath of all ganglia. RIAs also did not detect any GnRHir in the CNS. In this mollusk, a unique distribution of GnRHir material has been described: in the ovotestis, hemocytes,

22

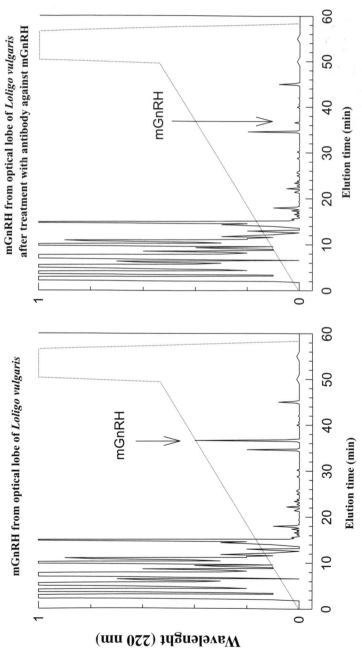

Fig. 1. High-pressure liquid chromatography elution profile of the optic lobe extract (filtered through Ultrafree Millipore-05 centrifugal filter membrane 30 kDa) from *Loligo vulgaris*, prior (left panel) and after anti-mGnRH overnight preabosrption (right panel). Arrow indicates the elution time of synthetic mGnRH, not shown in this figure.

and hemolymph (Zhang et al., 2000). Reverse-phase HPLC, coupled with specific RIAs, showed that the GnRH-like material in these sites cross-reacts with mGnRH and tGnRH-I, and that the hemolymph GnRH is biochemically and immunologically distinct from that extracted from the ovotestis and hemocytes. The timing and the width of the HPLC elution peaks indicated that *Aplysia* may possess multiple forms of GnRH-like peptides.

Arthropods

Hansen et al. (1982) described GnRHir material in the neural tissue of an insect, *Leucophaea maderae*, whereas Andriès and Tramu (1984) described it in the mesenteron of an odonate insect. Verhaert et al. (1990) also mentioned the existence of GnRHir material in the nervous system of an arthropod, the cockroach. It is puzzling that the largest animal taxon has been largely ignored about GnRH.

Echinoderms

Among echinoderms, we have detected, by immunocytochemistry, GnRHir material in the oocytes of the sea urchin, *Paracentrotus lividus*. Preabsorbed overnight with specific antigen (cGnRH-I and mGnRH), each antibody failed to immunostain oocytes any further. Gonadal extracts were purified and subjected to HPLC which allowed us to separate one major peak whose chromatographic characteristics, mass spectrometry and sequence analysis revealed that the sea urchin gonads secrete a cGnRH-I-like peptide (Rastogi, Di Fiore, Ceciliani, D'Aniello S., Branno, D'Aniello B. and D'Aniello A., unpublished). A second, minor peak elutes near mGnRH retention time, and needs further study. In vitro bioassay analysis showed that native (purified) and synthetic cGnRH-I stimulate the secretion of testosterone, estradiol and progesterone from the sea urchin gonad (unpublished).

Hemichordates

From an evolutionary point of view, hemichordates represent most probably a base group of the chordate lineage. GnRHir cells in two hemichordates, *Saccoglossus bromophenolosus* and *Ptychodera ba-*

hamensis, have been revealed by using anti-mGnRH, anti-tGnRH-I, and anti-sGnRH. These GnRHir cells are the well-known epidermal gland cells, the Mulberry cells, which are associated with neural tissue and project their basal processes into the nerve fiber layer. Thus the GnRHir Mulberry cells were found distributed widely in the ectoderm of the proboscis, collar, and anterior portion of the trunk region (Cameron et al., 1999). Ultrastructural features of the GnRHir cell types were also described and HPLC analysis, coupled with RIA, indicated the presence of a lGnRH-like molecule.

Cephalochordates

Within this chordate subgroup, there is only immunocytochemical evidence of the presence of GnRH-like peptide(s) in *Amphioxus* sp. and *Branchiostoma belcheri* (see Chang et al., 1983; Schriebman et al., 1986).

Tunicates

Of all invertebrate taxa, this group of invertebrate chordates has received the greatest attention. In fact, not only 4 different species have been investigated, but in two of them the GnRH-like peptides have been isolated, purified and sequenced. A total of four different GnRH forms are now known from this taxon: two GnRHs are vertebrate type, i.e., cGnRH-I and mGnRH, and two are unique to tunicates and were named as tGnRH-I and tGnRH-II (Powell et al., 1996; Craig et al., 1997; Di Fiore et al., 2000). To begin with, Georges and Dubois (1980) had for the first time detected the occurrence of GnRHir neurons around the cerebral ganglion and along the dorsal strand in the sea squirt, *Ciona intestinalis*. Quite some time later, a GnRH-like factor in the *Ascidiella aspersa* extract was detected by RIA, and it was pointed out that it had a molecular weight little higher than that of mGnRH molecule (Dufour et al., 1988). This was soon followed by an immunological study for the detection of GnRH-like immunoreactivity in *Chelyosoma productum* (Kelsall et al., 1990). Using anti-lGnRH, immunostained material was detected in the neural ganglion and some of its roots. HPLC–RIA analysis indicated the presence of a GnRH peak strongly cross-reactive with anti-lGnRH. A more recent immunohistochemical study, using anti-lGnRH, in this sea squirt had revealed GnRHir neurons forming a plexus around the dorsal strand and lining the wall of the blood sinus close to the gonoducts and gonads (Powell et al., 1996). GnRHir fibers also enter into the posterior nerve roots leading into the cerebral ganglion; no GnRHir cells were observed in the 'CNS'. This same study identified two novel decapeptides in this tunicate: tGnRH-I differing from mGnRH in amino acid residues 5 to 8, and tGnRH-II which differs from mGnRH and tGnRH-I not only in amino acid residues 5 to 8, but also because it is a disulfide-linked dimer (Powell et al., 1996; Craig et al., 1997). Yet in another ascidian, *Halocynthia roretzi* (Ohkuma et al., 2000), anti-mGnRH immunohistochemistry revealed ir neurons in the cerebral ganglion. In *Ciona intestinalis*, further studies using either anti-cGnRH-II or anti-lGnRH confirmed the distribution pattern of GnRH-immunoreactivity on the surface area of the cerebral ganglion, along the inner wall of the dorsal blood sinus, and on the ovarian surface (anti-cGnRH-II, Mackie, 1995; anti-lGnRH, Tsutsui et al., 1998). More recently, Di Fiore et al. (2000) have isolated and purified from the *Ciona intestinalis* gonadal extract two vertebrate-type GnRHs, whose primary structure was determined as cGnRH-I and mGnRH. Anti-mGnRH and anti-cGnRH-I, both yielded a relatively strong immunoreaction in developing follicles.

Biological activities

A highly diversified neural and nonneural distribution pattern of GnRH-like material in the invertebrate groups points to the possibility that GnRH may play a variety of roles. In fact, in order to clarify the possible function(s) of GnRH in invertebrates, different synthetic GnRH forms as well as GnRH-containing tissue extract or purified native GnRH-like peptide have been used both in vivo and in vitro bioassays. The multiple functions of different forms of GnRH in a single species, among invertebrates, can be deduced partially from studies in the ascidian, *Ciona intestinalis*, which is the only invertebrate in which four forms of GnRH have been identified by determination of primary structure. In this species tGnRH-I and -II have been localized in the neural complex and along the blood sinus

and thus these forms may plausibly function as neurotransmitter or neuromodulator, whereas cGnRH-I and mGnRH produced in the gonads may function as local hormones. In vitro treatment of gonads of *Ciona intestinalis* with synthetic and native mGnRH and cGnRH-I has been shown to enhance the synthesis and release of sex steroids (Di Fiore et al., 2000). In addition to this, native mGnRH, like the synthetic molecule, induced LH release from rat pituitary fragments in vitro. Different GnRH forms (mGnRH, sGnRH, cGnRH-II, tGnRH-I, and tGnRH-II) injected into the ascidian body also have been shown to induce gamete release (probably acting as neuromodulator of other neurons innervating the gonoducts), synthetic tGnRH-I being far more effective, in a dose-dependent manner, than other GnRH forms (Terakado, 2001); in this paper, on page 282, there is the mention of tGnRH-induced spawning in yet another ascidian, *Molgula manhattensis* and on page 283 it is reported that vertebrate-type GnRHs and tGnRHs do not induce the maturation of oocytes in vitro in *Halocynthis roretzi*. In contrast with Terakado's observations, Craig et al. (1997) had observed that synthetic tGnRH-I or tGnRH-II do not induce gamete release in mature *Corella*, another ascidian genus. Both tGnRHs, nevertheless, resulted in a doubling of the gonadal content of estradiol after 24 h in *Chelyosoma productum* (see fig. 12 in Craig et al., 1997). In *Ciona intestinalis* gonads, in vitro production of estradiol and testosterone was stimulated by native and synthetic mGnRH and cGnRH-I (Di Fiore et al., 2000). A relatively similar finding was earlier reported in the amphioxus in which injection of an mGnRH agonist into the body cavity resulted in an increase in the production of estradiol and testosterone (Chang et al., 1983). It is thus clear that GnRH-like molecules do play a role in protochordate reproduction. What remains to be done is a detailed analysis of which regulatory molecule(s) controls which aspect(s) of reproduction.

Consistent with the notion of a role of GnRH in protochordate reproduction, mGnRH and cGnRH-I peptides have been observed to stimulate the sea urchin gonadal production in vitro of estradiol and testosterone (unpublished data). Whether GnRH-like peptides also play a neuromodulatory or neurotransmitter role in these deuterostomes is a theme for future research.

Among mollusks, in the water snail *Planorbis corneus*, synthetic mGnRH had been demonstrated to exert an excitatory effect on serotoninergic and giant dopaminergic neurons, thus convalidating its role as a neurotransmitter or neuromodulator (Steiner and Felix, 1989). Consistent with this hypothesis, induction of electrophysiological response was observed in *Helisoma trivolvis* following injection of its nervous system extract (Goldberg et al., 1993). Data confirming a neuromodulatory role of GnRH in invertebrates are also coming from the sea pansy in which mGnRH and cGnRH-II had been observed to cause a reduction of the amplitude and frequency of rachidial peristaltic waves in a dose-dependent manner, mGnRH being 10 times more potent than cGnRH-II (Anctil, 2000). In this species, furthermore, chromatographically purified native GnRH-like peptides (fractions 1 and 2) showed a similar activity in the sea pansy peristalsis bioassay. As to the effects of GnRH-like peptides in processes related to reproduction in mollusks, Young et al. (1997) had observed that synthetic mGnRH can induce an increase in egg-laying in *Helisoma trivolvis*. GnRHir cells in ganglia innervating the reproductive system, and GnRHir innervation of the reproductive tracts in snails suggest that reproductive activities may be regulated, at least in part, by GnRH (see Young et al., 1999). Lateral lobes in the CNS of snails are the control centers of egg-laying and the presence of GnRHir cells in and around this area makes it likely that this peptide has a neuromodulatory influence on neurons of these lobes. In a recent study it was demonstrated that some vertebrate-type GnRHs exert a strong mitogenic action on dissociated gonadal cells of marine bivalves (Pazos and Mathieu, 1999). The bag cell neurons of *Aplysia* are well known as neuroendocrine neurons regulating reproductive functions and are functionally analogous to vertebrate gonadotropic cells of the pituitary; the bag cells release an egg-laying peptide hormone, in response to a characteristic electrical firing (called afterdischarge), which induces ovulation and egg-laying behavior (see Conn and Kaczmarek, 1989). In *Aplysia californica*, Zhang et al. (2000) have demonstrated that cGnRH-II can significantly decrease the duration of afterdischarge. It was suggested that GnRH may be a factor released by the ovotestis and hemocytes into the circulation to alter

neural function and the GnRH produced in hemocytes may serve as a novel mediator not only of neural functions but also of immune functions in this gastropod.

Very little is known on the regulation of GnRH synthesis and release in neural or nonneural sites in invertebrates. A recent random report in the ascidian, *Halocynthia roretzi*, which spawns at a fixed latency after sunrise, indicates that light may be involved (Ohkuma et al., 2000).

Concluding remarks

Immunocytochemical, chromatographic–radioimmunological and sequence techniques have unveiled that GnRH exists in different forms throughout the animal kingdom not only within the CNS but also outside it, in different tissues and cell types. Gonadal site of production of GnRH-like peptides is no more a prerogative of vertebrates. In fact, there is confirmatory evidence that the sea squirt and sea urchin gonads contain vertebrate-type GnRH(s).

The still valid view that GnRH acts in a diffuse manner, like a 'neurohormone', has made detailed anatomical studies recurrent in vertebrates. Morphological evidence strongly suggests that, for a proper understanding of GnRH's function(s) in the nervous system and outside it, we need to know both the anatomical details of GnRH pathways and the precise location pattern. In recent years, scientific community has also switched attention to studies of GnRH receptors, cloning and sequencing of cDNAs. The presence of different forms of GnRH and their influence on different tissues in invertebrates implies the existence of multiple cognate receptor types all of which are members of the G-protein coupled receptor family (see Blomenröhr et al., 1997; Troskie et al., 1998; Safarian et al., 2001). It is conceivable that GnRH acts upon target cells through receptor-mediated process. In their paper on the effects of GnRHs on gonial DNA synthesis in a bivalve mollusk, Pazos and Mathieu (1999) have cited unpublished data, on page 117, about the occurrence of receptors for GnRH-like peptides in the gonads. The specificity of a biological action of GnRH has in many cases been ascertained by use of GnRH antagonists (see Pazos and Mathieu, 1999).

Studies involving an evolutionary approach to

GnRH have adopted one particular strategy; attention has in recent years been focused on understudied invertebrate taxonomic groups. Invertebrate species that vary in their phylogenetic relationships are likely to reveal differences (as well as similarities) in GnRH form and function. Studies on invertebrate species are beginning to provide further insights on the structural and functional similarities and differences. The perspective that evolution has acted to shape GnRH molecule effectively, leads to investigations in a comparative approach. The new explorations in the invertebrates rely on the substantial foundation provided by an enormous literature in vertebrates on the form and role of GnRH-like molecules in reproduction, although, GnRH-like peptides are also considered to act as neurotransmitter and neuromodulator (see Tsutsui et al., 1998; Zhang et al., 2000; Terakado, 2001). Molecular variation in GnRH form within a single species is also leading investigators to dissociate the specific functions of each GnRH form. In other words, multiple GnRH forms in a single species may imply multiple functions of GnRH. To sum up, the emerging evolutionary perspective of GnRH peptide(s) in vertebrates is likely to be greatly influenced by recent studies on GnRH-like peptides in the invertebrates and, as the cnidarian GnRH author (Anctil, 2000) puts it, "the evolutionary history of the GnRH family is traceable to the earliest invertebrate known to possess a nervous system" (see Fig. 2).

Whereas in tetrapod vertebrates there are only three additional forms (cGnRH-I, cGnRH-II, and guinea pig GnRH), besides the mammalian form (mGnRH), in fish groups there seems to be tendency for the ancestral decapeptide to radiate into a variety of GnRH forms, such as the lGnRH-I and II, dfGnRH, sGnRH, cfGnRH, hrGnRH, sbGnRH. In addition, some fish species do possess either the mammalian form of GnRH or cGnRH-II. This latter form was considered until recently to be the most antique form of vertebrate-type GnRH. In addition, cGnRH-I was limited to reptiles and birds. While a cGnRH-II-like form has not yet been described in an invertebrate, cGnRH-I molecular form appeared long before, in the echinoderm gonad, and successively in tunicates in which mGnRH was also isolated and characterized. It thus becomes evident that mGnRH and cGnRH-I-like molecules appeared in deuteros-

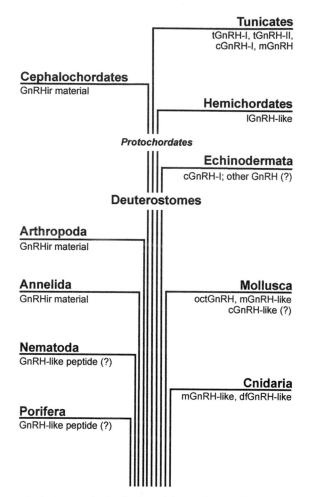

Fig. 2. Taxonomic distribution of GnRH-like peptides in invertebrates.

tomes long before the onset of vertebrate lineage (Fig. 2). As of today, the primary structure of only one new form of GnRH has been revealed in a protostome invertebrate, the octopus. Although a couple of investigations have accounted for the eventual presence of a GnRH-like peptide in two different species of insects, it is surprising that arthropods have not yet been convincingly shown to possess a GnRH-like peptide. Research in more species of invertebrates would certainly yield interesting results.

Acknowledgements

Our unpublished research data cited herein were financially supported by the Stazione Zoologica and the Università degli Studi di Napoli Federico II.

References

Al-Yousuf, S. (1990) Neuropeptides in annelids. In: A. Epple and M.H. Stetson (Eds.), *Progress in Comparative Endocrinology.* Wiley-Liss Inc., New York, NY, pp. 232–241.

Amoss, M., Burgus, R., Blackwell, P., Vale, W., Fellows, R. and Guillemin, R. (1971) Purification, amino acid composition and N-terminus of the hypothalamic luteinizing hormone releasing factor (LRF) of ovine origin. *Biochem. Biophys. Res. Commun.,* 44: 205–210.

Anctil, M. (2000) Evidence for gonadotropin-releasing hormone-like peptides in a cnidarian nervous system. *Gen. Comp. Endocrinol.,* 119: 317–328.

Andersen, A.C., Tonon, M.-C., Pelletier, G., Conlon, J.M., Fasolo, A. and Vaudry, H. (1992) Neuropeptides in the amphibian brain. *Int. Rev. Cytol.,* 138: 89–210.

Andriès, J.-C. and Tramu, G. (1984) Détection immunohistochimique des substances apparentées à des hormones peptidiques de Mammifères dans le mésentéron *d'Aeshna cyanea* (Insecte, Odonate). *C.R. Acad. Sci. Paris,* 299: 181–184.

Barry, J., Hoffman, G.E. and Wray, S. (1985) LHRH-containing systems. In: A. Björklund and T. Hökfelt (Eds.), *Handbook of Chemical Neuroanatomy, Vol. 4.* Elsevier, Amsterdam, pp. 166–215.

Blomenröhr, M., Bogerd, J., Leurs, R., Schulz, R.W., Tensen, C.P., Zandbergen, M.A. and Goos, H.J.Th. (1997) Differences in structure–function relations between nonmammalian and mammalian gonadotropin-releasing hormone receptors. *Biochem. Biophys. Res. Commun.,* 238: 517–522.

Brownlee, D.J.A., Rairweather, I., Johnston, C.F., Smart, D., Shaw, C. and Halton, D.W. (1993) Immunocytochemical demonstration of neuropeptides in the central nervous system of the roundworm, *Ascaris suum* (Nematoda: Ascaroidea). *Parasitology,* 106: 305–316.

Cameron, C.B., Mackie, G.O., Powell, J.F.F., Lescheid, D.W. and Sherwood, N.M. (1999) Gonadotropin-releasing hormone in Mulberry cells of *Saccoglossus* and *Ptychodera* (Hemichordata: Enteropneusta). *Gen. Comp. Endocrinol.,* 114: 2–10.

Carolsfeld, J., Powell, J.F.F., Park, M., Fischer, W.H., Craig, A.G., Chang, J.P., Rivier, J.E. and Sherwood, N.M. (2000). Primary structure and function of three gonadotropin-releasing hormones, including a novel form, from an ancient teleost, herring. *Endocrinology,* 141: 505–512.

Chang, C.-Y., Liu, Y. and Zhu, H. (1983) Steroid sex hormones and their functional regulation in amphioxus (*Branchiostoma belcheri* Gray). In: B. Lofts and W.N. Holmes (Eds.), *Current Trends in Comparative Endocrinology.* Hong Kong University Press, Hong Kong, pp. 205–207.

Chieffi, G., Pierantoni, R. and Fasano, S. (1991) Immunoreactive GnRH in hypothalamus and extrahypothalamic areas. *Int. Rev. Cytol.,* 127: 1–55.

Clayton, R.N. (1989) Gonadotropin-releasing hormone: its actions and receptors. *J. Endocrinol.,* 120: 11–19.

Conn, P.J. and Kaczmarek, L.K. (1989) The bag cell neurons of *Aplysia. Mol. Neurobiol.,* 3: 237–273.

Craig, A.G., Fischer, W.H., Park, M., Rivier, J.E., Musselman, B.D., Powell, J.F.F., Reska-Skinner, S.M., Prakash, M.O., Mackie, G.O. and Sherwood, N.M. (1997) Sequence of two gonadotropin-releasing hormones from tunicate suggest an important role of conformation in receptor activation. *FEBS Lett.,* 413: 215–225.

D'Aniello, A., Di Fiore, M.M. and Rastogi, R.K. (2001) NMDA can modulate mGnRH secretion from rat hypothalamus. In: H.J.Th. Goos, R.K. Rastogi, H. Vaudry and R. Pierantoni (Eds.), *Perspective in Comparative Endocrinology: Unity and Diversity.* Monduzzi Editore (International Proceedings Division), Bologna, pp. 541–546.

Demski, L.S. (1984) The evolution of neuroanatomical substrates of reproductive behavior: sex steroid and LHRH-specific pathways including the terminal nerve. *Am. Zool.,* 24: 809–830.

Demski, L.S. (1987) Phylogeny of luteinizing hormone-releasing hormone system in protochordates and vertebrates. In: L.S. Demski and M. Schwanzel-Fukuda (Eds.), *The Terminal Nerve (Nervus Terminalis) Structure, Function, and Evolution. Ann. N.Y. Acad. Sci.,* 519: 1–14.

Demski, L.S., Beaver, J.A., Sudberry, J.J. and Custis, J.R. (1997) GnRH systems in cartilaginous fishes. In: I.S. Parhar and Y. Sakuma (Eds.), *GnRH Neurons: Genes to Behavior.* Brain Shuppan, Tokyo, pp. 123–143.

Dhainaut-Courtois, N., Dubois, M.P., Traumu, G. and Masson, M. (1985) Occurrence and coexistence in *Nereis diversicolor* O.F. Muller (Annelida Polychaeta) of substances immunologically related to vertebrate neuropeptides. *Cell Tissue Res.,* 242: 97–108.

Di Cosmo, A. and Di Cristo C. (1998) Neuropeptidergic control of the optic gland of *Octopus vulgaris*: FMRFamide and GnRH immunoreactivity. *J. Comp. Neurol.,* 398: 1–12.

Di Cristo, C., Di Cosmo, A., D'Aniello, A., D'Aniello, B. and Rastogi, R.K. (1995) GnRH and FMRFamide in the brain and optic gland of *Octopus vulgaris.* In: *Proceedings of 56th National Meeting of the Italian Zoological Union.* Abstracts, pp. 241–243.

Di Fiore, M.M., Rastogi, R.K., Ceciliani, F., Messi, E., Botte, V., Botte, L., Pinelli, C., D'Aniello, B. and D'Aniello, A. (2000) Mammalian and chicken-I forms of gonadotropin-releasing hormone in the gonads of a protochordate, *Ciona intestinalis. Proc. Natl. Acad. Sci. USA,* 97: 2343–2348.

Dufour, S., Monniot, F., Monniot, C., Baloche, S., Kerdelhue, B. and Fontaine, Y.-A. (1988) Dosage radioimmunologique, chez *Ascidiella aspersa*, d'un facteur de type gonadolibérine (GnRH) de poids moléculaire apparent supérieur celui du décapeptide mammalien. *C.R. Acad. Sci. Paris,* 306: 253–256.

Georges, D. and Dubois, M.P. (1980) Mise en évidence par des techniques d'immunofluorescence d'un antigène de type LH-RH dans le système nerveux de *Ciona intestinalis* (tunicier ascidiacé). *C.R. Acad. Sci. Paris,* 290: 29–31.

28

Goldberg, J.I., Garofalo, R., Price, C.J. and Chang, J.P. (1993). Presence and biological activity of a GnRH-like factor in the nervous system of *Helisoma trivolvis*. *J. Comp. Neurol.*, 336: 571–582.

Hansen, B.L., Hansen, G.N. and Scharrer, B. (1982) Immunoreactive material resembling vertebrate neuropeptides in the *corpus cardiacum* and *corpus allatum* of the insect *Leucophaea maderae*. *Cell Tissue Res.*, 225: 319–329.

Iwakoshi, E., Takuwa-Kuroda, K., Fujisawa, Y., Hisada, M., Ukena, K., Tsutsui, K. and Minakata, H. (2002) Isolation and characterization of a GnRH-like peptide from *Octopus vulgaris. Biochem. Biophys. Res. Commun.*, 291: 1187–1193.

Jimenez-Linan, M., Rubin, B.S. and King, J.C. (1997) Examination of guinea pig luteinizing hormone-releasing hormone gene reveals a *Endocrinology* unique decapeptide and existence of two transcripts in the brain. *Endocrinology*, 138: 4123–4130.

Kelsall, R., Coe, I.R. and Sherwood, N.M. (1990). Phylogeny and ontogeny of gonadotropin-releasing hormone: comparison of guinea pig, rat, and a protochordate. *Gen. Comp. Endocrinol.*, 78: 479–494.

Kim, K., Seong, J.Y., Kang, S.S. and Sun, W. (1997) Neuroendocrine regulation of GnRH gene expression. In: I.S. Parhar and Y. Sakuma (Eds.), *GnRH Neurons: Genes to Behavior.* Brain Shuppan, Tokyo, pp. 401–427.

King, J.A. and Millar, R.P. (1980) Comparative aspects of luteinizing hormone-releasing hormone structure and function in vertebrate phylogeny. *Endocrinology*, 106: 707–717.

King, J.A. and Millar, R.P. (1987) Phylogenetic diversity of LHRH. In: B.H. Vickery and J.J. Nestor (Eds.), *LHRH and its Analogs: Contraceptive and Therapeutic Applications.* MTP, Lancaster, pp. 53–73.

King, J.A. and Millar, R.P. (1991) Gonadotropin-releasing hormone. In: P.K.T. Pang and M.P. Schreibman (Eds.), *Vertebrate Endocrinology: Fundamentals and Biomedical Applications, Vol. IV.* Academic Press, New York, NY, pp. 1–31.

King, J.A. and Millar, R.P. (1992) Evolution of gonadotropin-releasing hormones. *Trends Endocrinol. Metabol.*, 3: 339–346.

King, J.A. and Millar, R.P. (1995) Evolutionary aspects of gonadotropin-releasing hormone and its receptor. *Cell. Mol. Neurobiol.*, 15: 5–23.

King, J.A. and Millar, R.P. (1997) Coordinated evolution of GnRHs and their receptors, In: I.S. Parhar and Y. Sakuma (Eds.), *GnRH Neurons: Genes to Behavior.* Brain Shuppan, Tokyo, pp. 51–77.

Mackie, G.O. (1995) On the 'visceral nervous system' of *Ciona. J. Mar. Biol. Assoc. UK*, 57: 141–151.

Matsuo, H., Baba, Y., Nair, R.M.G., Arimura, A. and Schally, A.V. (1971) Structure of the porcine LH- and FSH-releasing hormone. I. The proposed amino acid sequence. *Biochem. Biophys. Res. Commun.*, 43: 1334–1339.

Muske, L.E. (1993) Evolution of gonadotropin-releasing hormone (GnRH) neuronal systems. *Brain Behav. Evol.*, 42: 215–230.

Muske, L.E. (1997) Ontogeny, phylogeny and neuroanatomical organization of multiple molecular forms of GnRH. In: I.S.

Parhar and Y. Sakuma (Eds.), *GnRH Neurons: Genes to Behavior.* Brain Shuppan, Tokyo, pp. 145–180.

Nozaki, M. and Kobayashi, H. (1979) Distribution of LHRH-like substance in the vertebrate brain as revealed by immunohistochemistry. *Arch. Histol. Jpn.*, 42: 201–219.

Nozaki, M., Tsukahara, T. and Kobayashi, H. (1984) Neuronal systems producing LHRH in vertebrates. In: K. Ochiai, Y. Arai, T. Shioda and M. Takahashi (Eds.), *Endocrine Correlates of Reproduction.* Springer-Verlag, Berlin, pp. 3–27.

Ohkuma, M., Katagiri, Y., Nakagawa, M. and Tsuda, M. (2000). Possibile involvement of light regulated gonadotropin-releasing hormone neurons in biological clock for reproduction in cerebral ganglion of the ascidian, *Halocynthia roretzi. Neurosci. Lett.*, 293: 5–8.

Parhar, I.S. (1997) GnRh in tilapia: three genes, three origins and their roles. In: I.S. Parhar and Y. Sakuma (Eds.), *GnRH Neurons: Genes to Behavior.* Brain Shuppan, Tokyo, pp. 99–122.

Parhar, I.S., Pfaff, D. and Schwanzel-Fukuda, M. (1995) Genes and behavior as studied through gonadotropin-releasing hormone (GnRH) neurons: comparative and functional aspects. *Cell. Mol. Neurobiol.*, 15: 107–116.

Pazos, A.J. and Mathieu, M. (1999) Effects of five natural gonadotropin-releasing hormones on cell suspensions of marine bivalve gonad: stimulation of gonial DNA synthesis. *Gen. Comp. Endocrinol.*, 113: 112–120.

Peter, R.E. (1983) Evolution of neurohormonal regulation of reproduction in lower vertebrates. *Am. Zool.*, 23: 685–695.

Powell, J.F.F., Reska-Skinner, S.M., Prakash, M.O., Fischer, W.H., Park, M., Rivier, J.E., Craig, A.G., Macie, G.O. and Sherwood, N.M. (1996) Two new forms of gonadotropin-releasing hormone in a protochordate and the evolutionary implications. *Proc. Natl. Acad. Sci. USA*, 93: 10461–10464.

Rastogi, R.K. and Iela, L. (1994) Gonadotropin-releasing hormone: present concepts, future directions. *Zool. Sci.*, 11: 363–373.

Rastogi, R.K., Meyer, D.L., Pinelli, C., Fiorentino, M. and D'Aniello, B. (1998) Comparative analysis of GnRH neuronal systems in the amphibian brain. *Gen. Comp. Endocrinol.*, 112: 330–345.

Rissman, E.F. (1996) Behavioral regulation of gonadotropin-releasing hormone. *Biol. Reprod.*, 54: 413–419.

Rissman, E.F. (1997) Behavioral regulation of the GnRH system. In: I.S. Parhar and Y. Sakuma (Eds.), *GnRH Neurons: Genes to Behavior.* Brain Shuppan, Tokyo, pp. 343–366.

Safarian, H., Gur, G., Rosenfeld, H., Yaron, Z. and Levavi-Sivan, B. (2001) Regulation of GnRH receptor mRNA in Tilapia pituitary. In: H.J.Th. Goos, R.K. Rastogi, H. Vaudry and R. Pierantoni (Eds.), *Perspective in Comparative Endocrinology: Unity and Diversity.* Monduzzi Editore (International Proceedings Division), Bologna, pp. 639–645.

Schriebman, M.P., Demski, L.S. and Margolis-Nunno, H. (1986) Immunoreactive (ir-) LHRH in the 'brain' of amphioxus. *Am. Zool.*, 26: 30A.

Sherwood, N. (1987) The GnRH family of peptides. *Trends Neurosci.*, 10: 129–132.

Sherwood, N.M., Lovejoy, D.A. and Coe, I.R. (1993) Origin of mammalian gonadotropin-releasing hormones. *Endocr. Rev.*, 14: 241–254.

Sherwood, N.M., von Schalburg, K. and Leschneid, D.W. (1997) Origin and evolution of GnRH in vertebrates and invertebrates. In: I.S. Parhar and Y. Sakuma (Eds.), *GnRH Neurons: Genes to Behavior*. Brain Shuppan, Tokyo, pp. 3–25.

Silverman, A.J. (1988) The gonadotropin-releasing hormone (GnRH) neuronal systems: immunocytochemistry. In: E. Knobil and J. Neill (Eds.), *The Physiology of Reproduction*. Raven Press, New York, NY, pp. 1283–1303.

Silverman, A.J. and Zimmerman, E.A. (1978) Pathways containing luteinizing hormone-releasing hormone (LHRH) in the mammalian brain. In: D.E. Scott, G.P. Kozlowski and A. Weindl (Eds.), *Brain–Endocrine Interaction III. Neural Hormones and Reproduction*. Karger, Basel, pp. 83–96.

Silverman, A.J., Livne, I. and Witkin, J.W. (1994) The gonadotropin-releasing hormone (GnRH) neuronal systems: immunocytochemistry and in situ hybridisation. In: E. Knobil and E. Neill (Eds.), *The Physiology of Reproduction*. Raven Press, New York, NY, pp. 1683–1710.

Sower, S.A. (1997) Evolution of GnRH in fish of ancient origin. In: I.S. Parhar and Y. Sakuma (Eds.), *GnRH Neurons: Genes to Behavior*. Brain Shuppan, Tokyo, pp. 27–49.

Steiner, F.A. and Felix, D. (1989) Effects of hypothalamic releasing hormones and biogenic amines on identified neurones in the circumoesophageal ganglia of the water snail (*Planorbis corneus*). *Comp. Biochem. Physiol.*, 92C: 301–307.

Terakado, K. (2001) Induction of gamete release by gonado-tropin-releasing hormone in a protochordate, *Ciona intestinalis*. *Gen. Comp. Endocrinol.*, 124: 277–284.

Troskie, B., Illing, N., Rumbak, E., Sun, Y.-M., Hapgood, J., Sealfon, S., Conklin, D. and Millar, R. (1998) Identification of three putative GnRH receptor subtypes in vertebrates. *Gen. Comp. Endocrinol.*, 112: 296–302.

Tsutsui, H., Yamamoto, N., Ito, H. and Oka, Y. (1998) GnRH-immunoreactive neuronal system in the presumptive ancestral chordate, *Ciona intestinalis* (Ascidian). *Gen. Comp. Endocrinol.*, 112: 426–432.

Verhaert, P., Ma, M. and De Loof, A. (1990) Immunocytochemistry and comparative insect (neuro) endocrinology. In: A. Apple, C.G. Scanes and M.H. Stetson (Eds.), *Progress in Comparative Endocrinology*. Wiley-Liss Inc., New York, NY, pp. 315–322.

Young, K., Zalitach, R., Chang, J.P. and Goldberg, J.I. (1997) Distribution and possible reproductive role of a gonadotropin-releasing hormone-like peptide in the pond snail, *Helisoma trivolvis. Soc. Neurosci. Abstr.*, 23: 696.

Young, K.G., Chang, J.P. and Goldberg, J.I. (1999) Gonadotropin-releasing hormone neuronal system of the fresh water snails *Helisoma trivolvis* and *Lymnaea stagnalis*: possible involvement in reproduction. *J. Comp. Neurol.*, 404: 427–437.

Zhang, L., Wayne, N.L., Sherwood, N.M., Postigo, H.R. and Tsai, P.S. (2000) Biological and immunological characterization of multiple GnRH in an opisthobranch mollusk, *Aplysia californica. Gen. Comp. Endocrinol.*, 118: 77–89.

I.S. Parhar (Ed.)
Progress in Brain Research, Vol. 141
© 2002 Elsevier Science B.V. All rights reserved

CHAPTER 3

Structural and chemical guidance cues for the migration of GnRH neurons in the chick embryo

S. Murakami [1,*], T. Seki [1], Y. Arai [1,2]

[1] *Department of Anatomy, Juntendo University School of Medicine, Hongo, Tokyo 113-8421, Japan*
[2] *Department of Human Sciences, University of Human Arts and Sciences, Iwatsuki, Saitama 339-8539, Japan*

Introduction

A body of evidence indicates that neurons producing gonadotropin-releasing hormone (GnRH) or luteinizing hormone-releasing hormone (LHRH) originate in the olfactory placode and migrate along the terminal and/or vomeronasal nerves or the olfactory nerve into the forebrain (Schwanzel-Fukuda and Pfaff, 1989; Wray et al., 1989a,b; Daikoku-Ishido et al., 1990; Muske and Moore, 1990; Ronnenkleiv and Resco, 1990; Murakami et al., 1991; Norgren and Lehman, 1991; Sullivan and Silverman, 1993; Chiba et al., 1994; D'Aniello et al., 1994; Parhar et al., 1995; Schwanzel-Fukuda et al., 1996; Amano et al., 1998). In the chick embryo, the migration of GnRH neurons from the olfactory placode starts at embryonic day (ED) 3.5. After that, GnRH neurons are generated continuously in the olfactory epithelium and migrate into the forebrain along the olfactory nerve by about ED 8. From ED 11 to the day of hatching, the majority of GnRH neurons tend to move into their adult position, whereas GnRH neurons in the olfactory epithelium have almost disappeared (Fig. 1, Murakami et al., 1991). GnRH immunoreactive (ir) terminals are detected in

the median eminence by ED 14 (Sullivan and Silverman, 1993), indicating the development of the hypothalamo–pituitary–gonadal axis.

Experimental studies with dye-labeling or chick-quail chimeric transplantation have provided direct evidence for the actual migration of GnRH neurons from the nose to the brain (Murakami and Arai, 1994a; Yamamoto et al., 1996). Labeling olfactory placodal cells with DiI resulted in sequential appearance of DiI-labeled cells expressing GnRH in the olfactory nerve, the rostral forebrain and the septo-preoptic area. Furthermore, olfactory placodectomy resulted in the absence of GnRH neurons in the nasal–forebrain axis of newts and chick embryos (Akutsu et al., 1992; Murakami et al., 1992). These results indicate that GnRH neurons arise from the epithelium of the olfactory placode.

In the central nervous system in general, neurons that are born in the ventricular layer migrate to their adult location. At least, two modes of migration are recognized in the developing brain. One mode of neuronal migration is that neuroblast migrate along the specialized glial cells known as a radial glia (Rakic, 1990). Another mode of migration is non-radial or tangential, in which neurons migrate parallel to the surface of the brain. In this type of cell migration, neuroblasts often move along the axons of preexisting neurons (Ono and Kawamura, 1989; Rakic, 1990; Kawano et al., 1995; Phelps and Vaughn, 1995) or disperse tangentially with cellular processes or with no neural structural supports (Gray et al., 1990; Ryder and Cepko, 1994; O'Rourke et

---

*Correspondence to: S. Murakami, Department of Anatomy, Juntendo University School of Medicine, Hongo, Tokyo 113-8421, Japan. Tel.: +81-3-5802-1025; Fax: +81-3-5800-0245;
E-mail: smuraka@med.juntendo.ac.jp

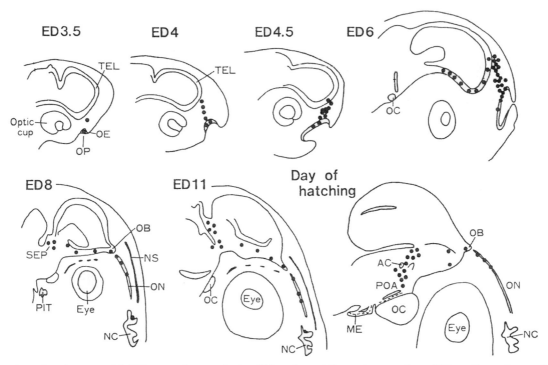

Fig. 1. Development of GnRH neuronal system in the chick embryo. AC; anterior commisure, ME; median eminence, NC; nasal cavity, NS; nasal septum, OB; olfactory bulb, OC; optic chiasm, OE; olfactory epithelium, ON; olfactory nerve, OP; olfactory placode, PIT; pituitary, POA; preoptic area, SEP; septum, TEL; telencephalon. (From Arai et al., 1997.)

al., 1995; Goleden et al., 1997; Tamamaki et al., 1997). These gliophilic and neurophilic interaction between migrating neurons and the neural structural elements are essential for the neuronal migration (Rakic, 1990).

The migration of GnRH neurons is a unique example of neuronal migration which crosses regional boundaries within the nasal–forebrain axis. There seems some differences in mode of their migration between GnRH neurons migrating before and after the entry to the forebrain. In this article, we describe and discuss, on the basis of experimental studies, the mode of migration of GnRH neurons and the possible role of highly polysialylated NCAM (PSA-NCAM) in this process.

Migration pathway of GnRH neurons in chick embryos before the entry to the forebrain: migratory guide

The migration process of GnRH neurons from the nose to the brain is associated closely with the ol-

factory system. A variety of molecules expressed on the fibers emerging from the olfactory epithelium have been used as markers for the pathway of GnRH neurons. In chick embryos, cell adhesion molecules such as NCAM, PSA-NCAM and NgCAM delineate the pathway (Murakami et al., 1991; Norgren and Brackenbury, 1993). Transient and specific immunoreactivity to somatostatin (SST) also delineates the olfactory pathway (Murakami and Arai, 1994b). Double labeling study showed that SST-ir neural substrates are a different neuronal population from GnRH neurons. Therefore, SST is a maker for the olfactory nerve fibers in chick embryos. In rodents, NCAM, PSA-NCAM, glycoconjugates identified by CC2 antibody and the intermediate filament marker, peripherin are expressed on the axons along which GnRH neurons migrate (Schwanzel-Fukuda et al., 1992; Tobet et al., 1992; Wray et al., 1994; Yoshida et al., 1995, 1999; Toba et al., 2001).

As shown in Fig. 2, from the olfactory epithelium to the base of the forebrain, GnRH neurons course along the olfactory nerve which express PSA-

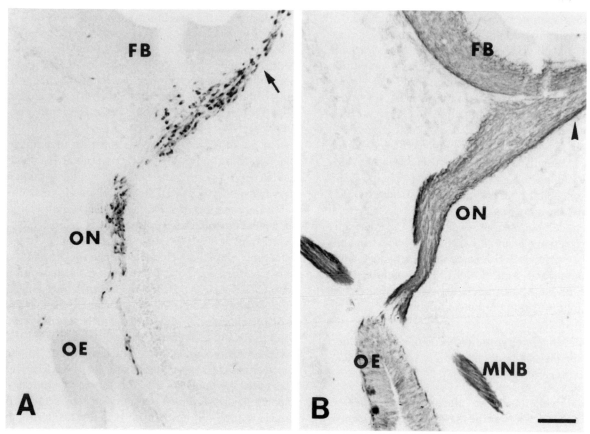

Fig. 2. Comparison of GnRH (A) and PSA-NCAM (B) immunoreactivity in the olfactory–forebrain area at ED 6.5. (A) GnRH neurons migrate along the medial portion of the olfactory nerve (ON). They enter the medial forebrain at a slightly caudal level to the junction with the olfactory nerve (arrow). (B) Section adjacent to A. A strong PSA-NCAM immunoreactivity is observed in the developing neural tissues. The olfactory nerve branches medially and its subset fibers extend to the medial forebrain (arrowhead). FB; forebrain, MNB; medial nasal branch of the ophthalmic nerve of the trigeminal nerve, OE; olfactory epithelium. Scale bar = 100 μm.

NCAM, preferentially concentrate in the medial part of the olfactory nerve. In mammals such as rats and mice including human, the path taken by GnRH neurons is the vomeronasal and/or terminal nerves (Schwanzel-Fukuda and Pfaff, 1989; Tobet et al., 1992; Wray et al., 1994; Yoshida et al., 1995; Schwanzel-Fukuda et al., 1996). The vomeronasal organ is absent in chick (Wenzel, 1987), but it is supposed that the terminal nerve component is contained in the nasal region of adult avian species from immunohistochemistry and neural fiber tracing studies (von Bartheld et al., 1987; Wirsig-Wiechmann, 1990; Norgren et al., 1992). Although the terminal nerve cannot be discriminated from the olfactory nerve bundle in chick embryo, it is possible that the

medial part of the olfactory nerve containing GnRH neurons may correspond partly to the terminal nerve component in the chick embryo.

Migratory guide: within the forebrain

Once GnRH neurons enter the forebrain, most of them are found to be associated with caudally projecting olfactory related fibers. It is reported that these fibers express NCAM, PSA-NCAM, NgCAM, and transiently express axonal surface glycoprotein (TAG1), peripherin and SST (Schwanzel-Fukuda et al., 1992; Norgren and Brackenbury, 1993; Murakami and Arai, 1994b; Wray et al., 1994; Yoshida et al., 1995). The down-regulation of these molecules coincides spatiotemporally with the completion of GnRH cell movement. Therefore, it has been suggested that this subset of olfactory fibers provide guiding substrate for GnRH neuronal migration within the brain (Norgren and Brackenbury, 1993; Murakami and Arai, 1994b; Wray et al., 1994). In rats, DiI labeling and immunohistochemical studies have revealed the existence of a projection that diverges from the main vomeronasal nerve, and reaches caudally into the rostral forebrain (Yoshida et al., 1995; Toba et al., 2001). The origin of caudally extending fibers is not known. Besides GnRH neurons, the olfactory placode gives rise to several different cell populations, one of which is transiently SST-expressing neurons found to migrate from the olfactory placode and located in the olfactory nerve (Murakami and Arai, 1994b). These neurons may contribute the SST-ir medial branch of the olfactory fibers that extend to the caudal forebrain.

Axon-dependent migration of GnRH neurons in the nasal region: effect of olfactory placodectomy and olfactory nerve transection

Complete ablation of the olfactory placode induces the complete loss of the nasal cavity. No GnRH neurons and fibers are detected in either the nasal region or in the forebrain on the operated side, suggesting that GnRH neurons arise from the olfactory placode (Akutsu et al., 1992; Murakami et al., 1992). The importance of the olfactory nerve as migratory guide of GnRH neurons into the forebrain has been directly shown by the olfactory nerve transection study (Murakami et al., 1998). When the pathway of the olfactory nerve is physically interrupted by a membrane filter, both the central projection of the olfactory nerve and GnRH neurons aggregate at the site of a membrane filter (Fig. 3). GnRH neurons cannot enter the brain because of the failure of the olfactory nerve targeting to the brain. Availability of a central projection of the olfactory nerve is prerequisite for the normal migration of GnRH neurons into the forebrain.

In this regard, it is of particular significance to note a report of Kallmann syndrome where GnRH neurons stay outside the brain and do not make contact with the forebrain, because of discontinuity of the olfactory–terminal nerve complex between the nasal and the forebrain (Schwanzel-Fukuda et al., 1989). Analysis of KAL gene and its encoded protein, anosmin, has supported the possibility that genetic defect of KAL gene disrupts the terminal navigation of the early olfactory axons or olfactory bulb differentiation (Franco et al., 1991; Legouis et al., 1991, 1993; Rugarli et al., 1993; Hardelin et al., 1999).

On the other hand, in embryos with incomplete placodectomy, migrating GnRH neurons are found to deviate from their regular course along the trigminal nerve elements (Fig. 4). This phenomenon is also observed in the case of the olfactory nerve transection in which the olfactory epithelium is almost intact (Murakami et al., 1998). Considerable number of the olfactory nerve fibers elongated along the ophthalmic nerve of the trigeminal nerve bundle in association with GnRH neurons. However, in some cases, GnRH neurons were found to migrate separately from SST-ir olfactory axons. The deviated GnRH neurons were found to migrate both the central direction and the peripheral direction of the ophthalmic nerve. In our recent findings, many GnRH neurons migrate along the spinal nerve from the olfactory placode transplanted in the forelimb region. The occurrence of GnRH neurons in anterior horn, posterior horn and spinal ganglion suggests that GnRH neurons invade the spinal cord using not only sensory axons but also motor axons as migratory guides (Murakami and Arai, 2002). These results suggest that the migration of GnRH neurons is principally axon-dependent and their identity to axonal choice for migration is not specific to the olfactory nerve in the peripheral environment.

The role of PSA-NCAM in the migration of GnRH neurons

Contact interaction between migrating neurons and structural guiding substrates plays an important role for migratory process (Rakic, 1990). It has been suggested that cell adhesion molecules (CAMs) mediate cell to cell interaction during neuronal migration (Rakic et al., 1994). The most notable molecules related to GnRH neuronal migration is neural cell adhesion molecule (NCAM). NCAM can be modified by the addition of polysialic acid (PSA). The degree of polysialation of NCAM is regulated developmen-

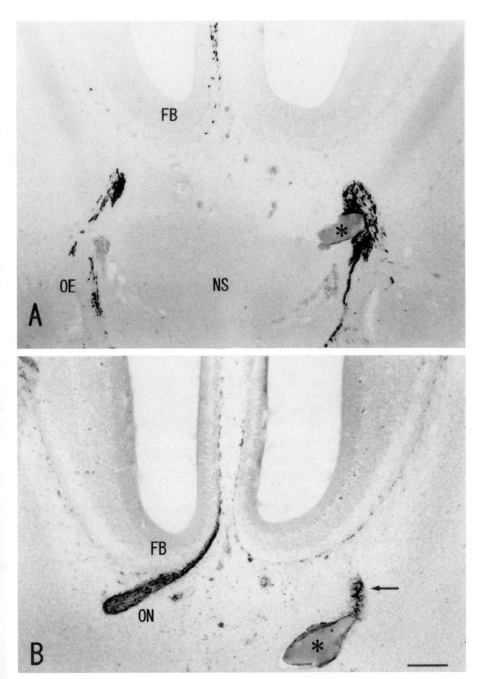

Fig. 3. The olfactory–forebrain region of an embryo treated with a membrane filter (*) sacrificed at ED 7. (A) On the operated side, GnRH-ir neurons are seen to aggregate at the site of a membrane filter (*). Note the absence of GnRH neurons in the forebrain (FB). (B) Section taken slightly dorsal level to A. SST-ir olfactory fibers cease to grow further near the membrane filter on the operated side (arrow). These remnant fibers do not contact the forebrain. NS; nasal septum, OE; olfactory epithelium. Scale bar = 250 μm. (From Murakami et al., 1998.)

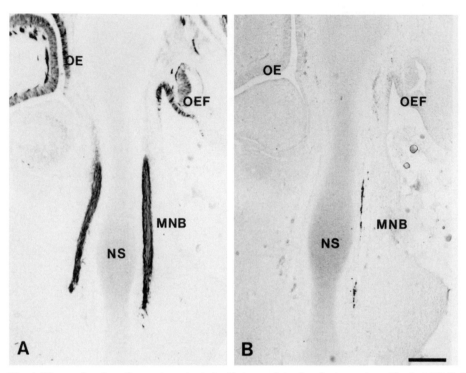

Fig. 4. The nasal region of an embryo treated with incomplete placodectomy at sacrificed at ED 7. (A) PSA-NCAM-ir fiber bundle of the ophthalmic nerve of the trigeminal nerve (MNB) are seen bilaterally. (B) GnRH-ir neurons are found migrating from the olfactory epithelium fragment (OEF) along the PSA-NCAM-ir ophthalmic nerve bundle in the peripheral direction along the nasal process. Note the absence of GnRH-ir neurons in the ophthalmic nerve on the unoperated side. NS; nasal septum, OE; olfactory epithelium. Scale bar = 250 μm. (From Murakami et al., 1995.)

tally. PSA-NCAM has long α2-8-linked sialic acid polymer, which have been shown to serve as an overall negative regulator of cell interactions and to be associated with cell migration and axon guidance in a number of systems during development of the nervous system (Rutishauser et al., 1988; Rutishauser and Landmesser, 1996). The developing neural tissue including the olfactory system expresses NCAM and PSA-NCAM (Seki and Arai, 1993). In chick embryos, both migrating GnRH neurons and the olfactory nerve have been found to express PSA-NCAM. Strong immunoreactivity for PSA-NCAM on GnRH neurons disappeared after these cells reached their adult positions. Therefore, it has been suggested that PSA-NCAM play an important role in the migration of GnRH neurons (Murakami et al., 1991). A close association with GnRH neurons and NCAM and/or PSA-NCAM positive olfactory/vomeronasal nerve also has been demonstrated in the olfactory placode culture system (Koide and Daikoku, 1995;

Murakami et al., 1995). Furthermore, PSA-NCAM is also positive to the trigeminal nerve into which GnRH neurons migrate when the developing olfactory nerve are physically interrupted.

To examine the role of PSA-NCAM in the migration of GnRH neurons, PSA-specific endoneuraminidase N (endoN) which selectively removes PSA from NCAM was applied in chick embryos (Murakami et al., 1998, 2000). Despite the absence of PSA, GnRH neurons were able to migrate along the olfactory nerve to arrive at the forebrain following endoN treatment. The proportion of migrating GnRH neurons in the nasal region was tend to be high in endoN treated embryos, compared to that in control embryos, suggesting that, in the absence of PSA, GnRH neurons may have some difficulty penetrating into the forebrain. However, PSA removal disrupted the normal migration pattern of GnRH neurons within the forebrain (Fig. 5). The proportion of migrating GnRH neurons in the medial forebrain

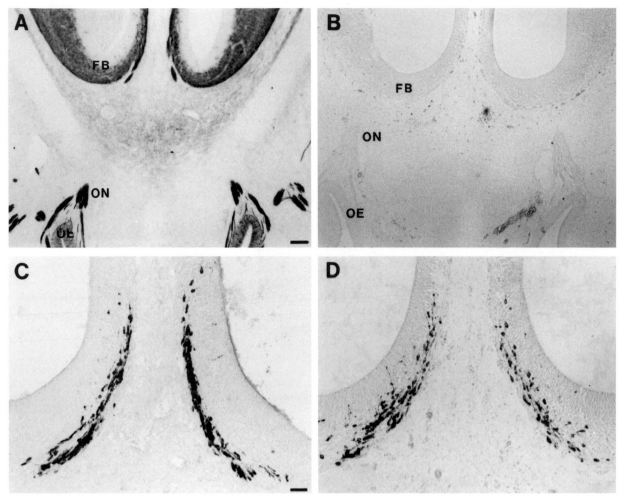

Fig. 5. Effect of PSA removal on the migration pattern of GnRH neurons in the forebrain at ED 6.5. Comparison of a control embryo (A, C) and an endoN-treated embryo (B, D). (A) Strong PSA-NCAM expression is observed in the nasal–forebrain region. (B) No PSA immunoreactivity is observed in the nasal–forebrain region after the enzymatic removal of PSA. (C) A cluster of GnRH neurons directly enter the medial forebrain and migrate along the pial surface of the medial forebrain in a control embryo. (D) After removal of PSA, the distribution of GnRH neurons in the medial forebrain surface is more dispersed compared with that of the control embryo. FB; forebrain, OE; olfactory epithelium, ON; olfactory nerve. Scale bars = 125 μm in A; 50 μm in C. (From Murakami et al., 2000.)

was significantly decreased in endoN-treated embryos compared to that in control embryos (Fig. 6). In mice, absence of PSA can affect the initial migration of GnRH neurons in the nasal region with a significant accumulation of these cells (Yoshida et al., 1999).

In the embryo with physical interruption of the olfactory nerve, however, enzymatic PSA removal did not only interfere with the migration of GnRH neurons into the ophthalmic nerve bundle of the trigeminal nerve, but also did not inhibit the migration of GnRH neurons into the forebrain on the unoperated side. In mouse nasal explants, some GnRH neurons are found to migrate along the NCAM-negative olfactory axons (Wray et al., 1994). Outside the central nervous system, other molecules appear to play a role in GnRH cell movement along the nerve fibers. For example, cell surface-associated extracellular matrix such as phosphacan is found in the GnRH neurons and the olfactory nerve in chick embryos (Nishizuka et al., 1996). Since phosphacan is known to bind NCAM and NgCAM (Grumet, 1990), heterophilic

38

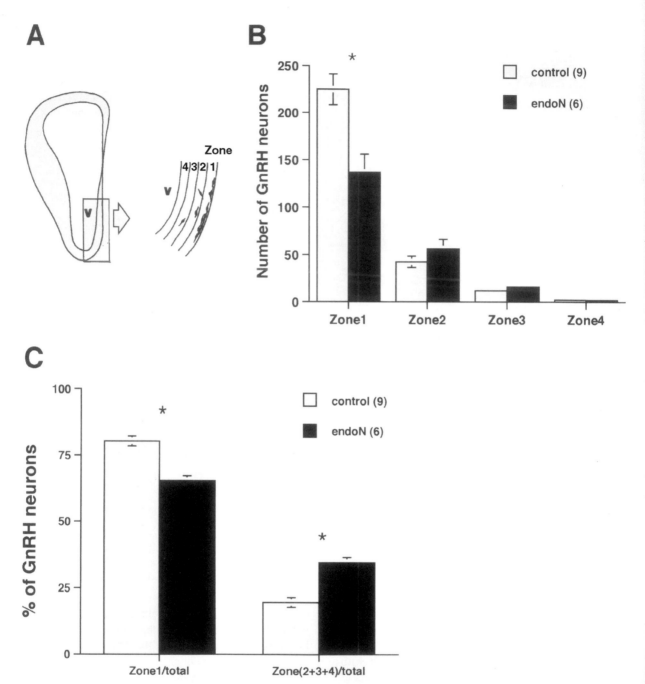

Fig. 6. Quantitation of the effect of PSA removal on the distribution of GnRH neurons in the medial forebrain at ED 6.5. (A) relative distribution of GnRH neurons was measured by subdividing the medial forebrain into four zones (1–4) parallel to the brain surface. (B) Histograms of the mean GnRH neurons number in each zone. Although most GnRH neurons are located in zone 1 for both endoN-treated embryos and control embryos, endoN treatment resulted in a significant reduction in the number of GnRH neurons in this region (asterisk, $P < 0.004$). (C) The percentage of total GnRH neurons in zone 1 and zones 2–4. Note that, after PSA removal, the relative location of GnRH neurons shifts significantly from zone 1 to zones 2–4 (asterisks; $P < 0.0001$). V, lateral ventricle. (From Murakami et al., 2000.)

binding with NCAM between GnRH neurons and the olfactory axons may be involved in the migration of GnRH neurons.

Development of migration pathway within the forebrain

The exact pathway of GnRH neurons to their final location within the brain is unclear to date. We observed the progressive distribution changes of GnRH neurons with the extension of a subset of olfactory fibers, which is specifically immunoreactive to SST in chick embryos (Murakami et al., 2000). At ED 4.5, a few pioneer GnRH neurons enter the rostral forebrain where the olfactory nerve contacts with the presumptive olfactory bulb. No detectable guiding structural substrates for their migration are found in the rostral forebrain at that stage. A small number of GnRH neurons continue to enter the brain at the junction with the olfactory nerve, and they disperse into the presumptive olfactory bulb and in the direction caudal to the medial forebrain. These GnRH neurons appear to migrate in no association with any neural structures by ED 5, since a well defined migration pathway of GnRH neurons in the brain is not yet recognized. Considerable change occurs at ED 5.5–6. One component of the olfactory nerve branches to form a small bundle of axons before entry to the forebrain. These fibers can be identified by SST and PSA-NCAM antibodies as a subset of the olfactory nerve and extend toward the ventromedial forebrain (Fig. 7B, C). These fibers then invade directly the medial forebrain, accompanying with many GnRH neurons (Fig. 7A). By ED 6.5, migration pathway of GnRH neurons in the brain becomes easily recognizable (Fig. 7D–F). Large numbers of GnRH neurons in small clusters entered the medial forebrain and then migrated with the medial branch of the olfactory nerve. Most of these GnRH neurons migrate along the brain surface beneath the pia mater. The migration direction is primarily tangential.

During ED 6.5–7.5, this pathway extends near the preseptal area. The course of medial branch of the olfactory nerve is also specifically identified by antibody to axonin-1, which is the chick homolog of TAG-1 and can promote neurite outgrowth in vitro (Stoeckli et al., 1991; Hasler et al., 1993). Axonin-1-ir fibers seem to be identical to a transient expression

of TAG-1 on the caudal branches of the vomeronasal nerve in rats (Yoshida et al., 1995). In our materials, axonin-1-ir fibers extended to the parenchyma near the roof of the third ventricle, where many GnRH neurons were apart from these fibers and turn ventrally toward the preoptic area (Fig. 8A–C; unpublished data). At ED 11, GnRH neurons almost reached their adult positions, whereas SST and axonin-1 immunoreactivity on the medial branch of the olfactory nerve disappeared.

On the basis of these observations, it is proposed that at least two major pathways of GnRH neuronal migration exist in the brain. The main pathway is characterized by a small cluster of GnRH neurons that are aligned continuously along the medial forebrain surface dorsocaudally. It has been implicated that axophilic migration of GnRH neurons occurs in this pathway that largely corresponds to the course of a subset of the olfactory fibers. The migration of GnRH neurons on this pathway appears to be regulated by PSA-NCAM. PSA removal disrupted the specific pattern of GnRH neurons within the medial forebrain. Significant declustering of GnRH neurons and deviation of migrating GnRH neurons from the main pathway occurred (Figs. 5 and 6). Because the axonal pattern of the medial branch of the olfactory nerve, migration substrate within the brain was not affected, a defect on the GnRH neurons could not be a secondary effect.

Another pathway is independent of the medial branch of the olfactory nerve. For example, before these fibers invade the forebrain, a small number of GnRH neurons enter the forebrain and disperse in the presumptive olfactory bulb and the medial forebrain. After a subset of olfactory fibers entered the medial forebrain in association with GnRH neurons, some GnRH neurons still disperse at multiple levels in the main pathway, especially, the diverse dispersion of GnRH neurons occurs at the most caudal level of the main pathway. In this case, many GnRH neurons seem to leave the main pathway and to change their course in the direction of the ventral preoptic area. However, it might be possible that the pioneer GnRH neurons that entered the forebrain in the earlier stage join a group of these deviated GnRH neurons. In either cases, it remains unclear the structural and chemical guiding substrates for the migratory routes of these GnRH neurons.

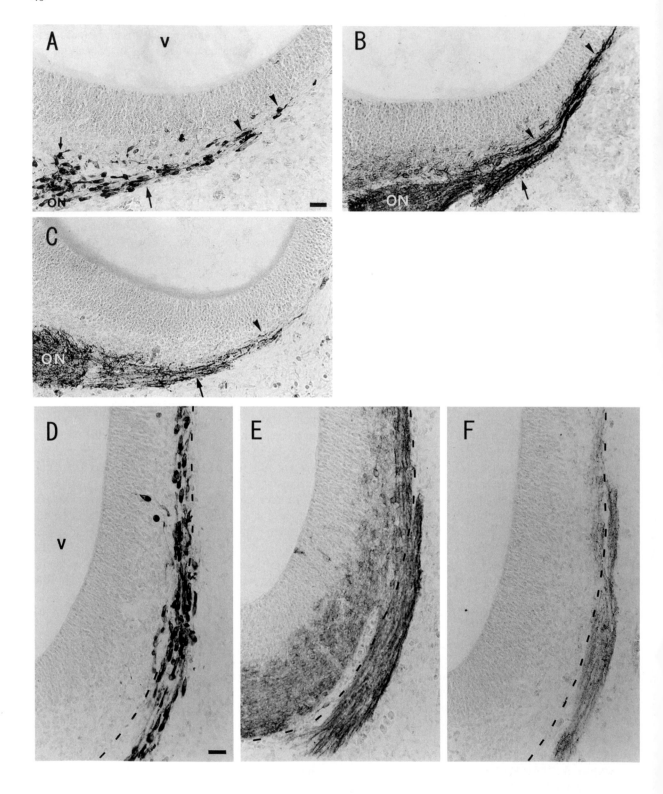

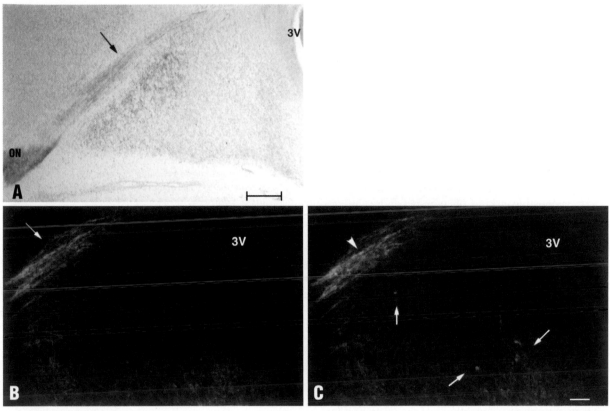

Fig. 8. Sagittal sections through the forebrain at ED 7 embryos. Axonin-1 immunostaining (A) and double fluorescence for GnRH and axonin-1 (B, C). Rostral is to the left. A: A subset of olfactory fibers that is immunoreactive to axonin-1 branches from the olfactory nerve (ON) and extend into the forebrain dorsocaudally (arrow). (B) Axonin-1-ir fibers (arrow) are seen to extend along the medial forebrain surface beneath the pia mater, and appear to terminate the parenchyma near the roof of the third ventricle (3V). (C) The merged image of the same section as in B, with axonin-1 (green) and GnRH (red). Many GnRH neurons are seen to associate with axonin-1-ir fibers (arrowhead). At the most caudal level of the course of axonin-1-ir fibers, some LHRH neurons are seen to scatter in direction to the ventral forebrain (arrows). These GnRH neurons seem to deviate from the main pathway and turn to the ventral area. Note the absence of axonin-1-ir fibers in the ventral region where GnRH neurons disperse. Scale bars = 200 μm in A; 30 μm in B.

Conclusion

In the nasal region, evidence indicates that the olfactory nerve guides the migration of GnRH neurons into the forebrain during normal development. However, a straying phenomenon of GnRH neurons into the ophthalmic nerve of the trigeminal nerve suggests that GnRH neurons could use any type of

Fig. 7. Coronal sections of the rostral forebrain at ED 5.5 (A–C) and ED 6.5 (D–F). (A) Many GnRH-ir neurons migrate in direction to the medial forebrain (large arrow) and enter the ventromedial forebrain. These cells migrate along the region close to the pial surface of the brain (arrowheads). A few GnRH-ir cells enter the brain at the junction of the olfactory nerve (ON, small arrows). (B) Section adjacent to A. PSA-ir fibers that are branching medially from the main olfactory nerve bundle extend toward the ventromedial forebrain (arrow). A portion of PSA-ir fibers enter the medial forebrain and extend along the pial surface (arrowhead). (C) Section adjacent to B. The SST-ir fibers branch medially from the olfactory nerve and extend toward the ventromedial forebrain (arrow), accompanying by migrating GnRH neurons. A subset of fibers enters the forebrain (arrowhead). (D) A small cluster of GnRH-ir cells directly enter the medial forebrain and course along the medial forebrain surface. (E) Section adjacent to D. PSA-ir fibers branch from the olfactory nerve accompanied by PSA-positive GnRH-ir cells and enter the forebrain. (F) Section adjacent E. SST-ir fibers branch from the olfactory nerve and extend along the migration pathway of GnRH neurons. The dashed line in D–F indicates the pial surface. V; lateral ventricle. Scale bars = 30 μm in A; 25 μm in D. (From Murakami et al., 2000.)

42

axons as guidance substrate for their migration. This conjecture is well reinforced by the fact that GnRH neurons were detectable in the spinal nerve and also in the spinal cord following placode graft transplantation in the forelimb region. The presence of structural cues seems to be of importance primarily in the migration of GnRH neurons in the peripheral region.

Within the forebrain, examination of the normal development of GnRH neurons indicates that the main pathway of GnRH neuronal migration largely corresponds to the course of a subset of the olfactory fibers, suggesting a kind of axon-dependent migration of these GnRH neurons. PSA-NCAM is more likely to regulate the interaction of the migrating neurons with their axonal substrates in the brain, whereas it may not be important for the migration of GnRH neurons along the peripheral neural elements. Another migration mode taken by GnRH neurons is independent of the olfactory related fibers. The substrates and chemical cues for their migration remain largely unknown. Although the final migration pathway of GnRH neurons to their adult positions is not yet clear, the dispersion of GnRH neurons from the main pathway (Fig. 8B, C) implicates that GnRH neurons may show a final phase of the migration targeting on the preoptic–hypothalamic area. A subset of the olfactory fibers as guiding substrate may act as a kind of temporary scaffolding for GnRH neurons to move to the subsequent pathways that lead to the final site. More detail analysis of the migration pathway of GnRH neurons to their permanent location is needed for the understanding the structural and molecular substrata that guide them.

Acknowledgements

We would like to thank Dr. K. Wakabayashi for generous supply of LHRH monoclonal antibody, LRH13. This study was supported by Grants-in Aid from the Ministry of Education, Science and Culture of Japan and by Japan Private School Promotion Foundation.

References

Akutsu, S., Takada, M., Ohki-Hamazaki, H., Murakami, S. and Arai, Y. (1992) Origin of luteinizing hormone-releasing hormone (LHRH) neurons in the chick embryo: effect of the olfactory placode ablation. *Neurosci. Lett.*, 142: 241–244.

Amano, M., Oka, Y., Kitamura, S., Ikuta, K. and Aida, K. (1998) Ontogenic development of salmon GnRH and chicken GnRH-II systems in the brain of masu salmon (*Oncorhynchus masou*). *Cell. Tissue. Res.*, 293: 427–434.

Arai, Y., Murakami, S., Seki, T., Miyakawa, M. and Kamiya, M. (1997) Origin and migration of GnRH neurons in the chick embryo: an experimental study. In: I.S. Parhar and Y. Sakuma (Eds.), *GnRH Neurons: Gene to Behavior*. Brain Shuppan, Tokyo, pp. 181–195.

Chiba, A., Oka, S. and Honma, Y. (1994) Ontogenetic development of gonadotropin releasing hormone-like immnoreactive neurons in the brain of the chum salmon, *Oncorhynchus keta*. *Neurosci. Lett.*, 178: 51–54.

Daikoku-Ishido, H., Okamura, Y., Yanaihara, N. and Daikoku, S. (1990) Development of the hypothalamic luteinizing hormone-releasing hormone-containing neuron system in the rat: in vivo and in transplantation studies. *Dev. Biol.*, 140: 374–387.

D'Aniello, B., Masucci, M., di Meglio, M., Iela, L. and Rastogi, R.K. (1994) Immunohistochemical localization of GnRH in the crested newt (*Triturus carnifex*) brain and terminal nerve: a developmental study. *J. Neuroendocrinol.*, 6: 167–172.

Franco, B., Guioli, S., Pragliola, A., Incerti, B., Bardoni, B., Tonlorenzi, R., Carrozzo, R., Maestrini, E., Pieretti, M., Taillon-Miller, P., Brown, C.J., Willard, HF., Lawrence, C., Persico, M.G., Camerino, G. and Ballabio, A. (1991) A gene deleted in Kallmann's syndrome shares homology with neural cell adhesion and axonal path-finding molecules. *Nature*, 353: 529–536.

Goleden, J.A., Zitz, J.C., McFadden, K. and Cepko, C.L. (1997) Cell migration in the developing chick diencephalon. *Development*, 124: 3525–3533.

Gray, G.E., Leber, S.M. and Sanes, J.R. (1990) Migratory patterns of clonally related cells in the developing central nervous system. *Experientia*, 46:929–940.

Grumet, M. (1990) The neuronal chondroitin sulfate proteoglycan neurocan binds to the neural cell adhesion molecules Ng-CAM/L1/NILE and N-CAM and inhibits neural adhesion and neurite outgrowth. *J. Biol. Chem.*, 125: 669–680.

Hardelin, J.-P., Julliard, A.K., Moniot, B., Soussi-Yanicostas, N., Verney, C., Schwanzel-Fukuda, M., Ayer-Le Lievre, C. and Petit, C. (1999) Anosmin-1 is a regionally restricted component of basement membranes and interstitial matrix during organogenesis implications for the developmental anomalies of X chromosome-linked Kallmann syndrome. *Dev. Dyn.*, 215: 26–44.

Hasler, T.H., Rader, C., Stoeckli, E.T., Zuellig, R.A. and Sonderegger, P. (1993) cDNA cloning, structural features, and eucaryotic expression of human TAG-1/axonin-1. *Eur. J. Biochem.*, 211: 329–339.

Kawano, H., Ohyama, K., Kawamura, K. and Nagatsu, I. (1995) Migration of dopaminergic neurons in the embryonic mesencephalon of mice. *Dev. Brain Res.*, 86: 101–103.

Koide, I. and Daikoku, S. (1995) In vitro analysis of the centripetal migration mechanisms of developing LHRH neurons. *Arch. Histol. Cytol.*, 58: 265–283.

Legouis, R., Hardelin, J.P., Levilliers, J., Claverie, J.-M., Compain, S., Wunderle, V., Millasseau, P., Le Paslier, D., Cohen, D., Caterina, D., Bougueleret, L., Delemarre-Van de Waal, H., Lutfalla, G., Weissenbach, J. and Petit, C. (1991) The candidate gene for the X-linked Kallmann syndrome encodes a protein related to adhesion molecules. *Cell*, 67: 423–435.

Legouis, R., Ayer-Le Lievre, C., Leibovici, M., Lapointe, F. and Petit, C. (1993) Expression of the *KAL* gene in multiple neuronal sites during chicken development. *Proc. Natl. Acad. Sci. USA*, 90: 2461–2465.

Murakami, S. and Arai, Y. (1994a) Direct evidence for the migration of LHRH neurons from the nasal region to the forebrain in the chick embryo: a carbocyanine dye analysis. *Neurosci. Res.*, 19: 331–338.

Murakami, S. and Arai, Y. (1994b) Transient expression of somatostatin immunoreactivity in the olfactory–forebrain region in the chick embryo. *Dev. Brain Res.*, 82: 277–285.

Murakami, S. and Arai, Y. (2002) Migration of LHRH neurons into the spinal cord: evidence for axon-dependent migration from the transplanted chick olfactory placode. *Eur. J. Neurosci.*, 16: 684–692.

Murakami, S., Seki, T., Wakabayashi, K. and Arai, Y. (1991) The ontogeny of luteinizing hormone-releasing hormone (LHRH) producing neurons in the chick embryo: possible evidence for migrating LHRH neurons form the olfactory epithelium expressing a highly polysialylated neural cell adhesion molecule. *Neurosci. Res.*, 12: 421–431.

Murakami, S., Kikuyama, S. and Arai, Y. (1992) The origin of the luteinizing hormone-releasing hormone (LHRH) neurons in newts (*Cynops pyrrhogaster*): the effect of olfactory placode ablation. *Cell Tissue Res.*, 269: 21–27.

Murakami, S., Kamiya, M., Akutsu, S., Seki, T., Kuwabara, Y. and Arai, Y. (1995) Straying phenomenon of migrating LHRH neurons and highly polysialylated NCAM in the chick embryo. *Neurosci. Res.*, 22: 109–115.

Murakami, S., Seki, T., Rutishauser, U. and Arai, Y. (1998) LHRH neurons migrate into the trigeminal nerve when the developing olfactory nerve fibers physically interrupted in chick embryos. *Gen. Comp. Endocrinol.*, 112: 312–321.

Murakami, S., Seki, T., Rutishauser, U. and Arai, Y. (2000) Enzymatic removal of polysialic acid from neural cell adhesion molecule perturbs the migration route of luteinizing hormone-releasing hormone neurons in the developing chick forebrain. *J. Comp. Neurol.*, 420: 171–181.

Muske, L.E. and Moore, F.L. (1990) Ontogeny of immnoreactive gonadotropin-releasing hormone neuronal systems in amphibians. *Brain Res.*, 534: 177–187.

Nishizuka, M., Ikeda, S., Arai, Y., Maeda, N. and Noda, M. (1996) Cell surface-associated extracellular distribution of a neural proteoglycan, 6B4 proteoglycan/phosphacan, in the olfactory epithelium, olfactory nerve, and cells migrating along the olfactory nerve in chick embryos. *Neurosci. Res.*, 24: 345–355.

Norgren, R.B. and Lehman, M.N. (1991) Neurons that migrate from the olfactory epithelium in the chick express luteinizing hormone-releasing hormone. *Endocrinology*, 128: 1676–1678.

Norgren, R.B. and Brackenbury, R. (1993) Cell adhesion molecules and the migration of LHRH neurons during development. *Dev. Biol.*, 160: 377–387.

Norgren, R.B., Lippert, J. and Lehman, M.N. (1992) Luteinizing hormone-releasing hormone in the pigeon terminal nerve and olfactory bulb. *Neurosci. Lett.*, 135: 201–204.

Ono, K. and Kawamura, K. (1989) Migration of immature neurons along tangentially oriented fibers in the subpial part of the fetal mouse medulla oblongata. *Exp. Brain Res.*, 78: 290–300.

O'Rourke, N.A., Sullivan, D.P. Kaznowski, C.E., Jacobs, A.A. and McConnell, S.K. (1995) Tangential migration of neurons in the developing cerebral cortex. *Development*, 121: 2165–2176.

Parhar, I.S., Iwata, M., Pfaff, D.W. and Schwanzel-Fukuda, M. (1995) Embryonic development of gonadotropin-releasing hormone neurons in the sockeye salmon *J. Comp. Neurol.*, 362: 256–270.

Phelps, P.E. and Vaughn, J.E. (1995) Commissural fibers may guide cholinergic neuronal migration in developing rat cervical spinal cord. *J. Comp. Neurol.*, 355: 38–50.

Rakic, P. (1990) Principles of neural cell migration. *Experientia*, 46: 882–891.

Rakic, P., Cameron, R.S. and Komoro, H. (1994) Recoginition, adhesion, transmembrane signaling and cell motility in guided neuronal migration. *Curr. Opinion Neurobiol.*, 4: 63–69.

Ronnekleiv, O.K. and Resco, J.A. (1990) Ontogeny of gonadotropin-releasing hormone-containing neurons in early fetal development of rhesus macaques. *Endcrinology*, 126: 498–511.

Rugarli, E.I., Lutz, B., Kuratani, S.C., Wawersik, S., Borsani, G., Ballabio, A. and Eichele, G. (1993) Expression pattern of the Kallmann syndrome gene in the olfactory system suggests a role in neuronal targeting. *Nat. Genet.*, 4: 19–26.

Rutishauser, U. and Landmesser, L. (1996) Polysialic acid in the vertebrate nervous system: a promoter of plasticity in cell–cell interaction. *Trends. Neurosci.*, 19: 422–427.

Rutishauser, U., Acheson, A., Hall, A.K., Mann, D.M. and Sunshine, J. (1988) The neural cell adhesion molecule (NCAM) as a regulator of cell–cell interactions. *Science*, 240: 53–57.

Ryder, E.F. and Cepko, C.L. (1994) Migration patterns of clonally related granule cells and their progenitors in the developing chick cerebellum. *Neuron*, 12: 1011–1028.

Schwanzel-Fukuda, M. and Pfaff, D.W. (1989) Origin of luteinizing hormone-releasing hormone neurons. *Nature*, 338: 161–164.

Schwanzel-Fukuda, M., Bick, M. and Pfaff, D.W. (1989) Luteinizing hormone-releasing hormone (LHRH)-expressing cells do not migrate normally in an inherited hypogonadal (Kallmann) syndrome. *Mol. Brain Res.*, 6: 311–326.

Schwanzel-Fukuda, M., Abraham, S., Crossin, K.L., Edelman, G.M. and Pfaff, D.W. (1992) Immunocytochemical demonstration of neural cell adhesion molecule (NCAM) along the migration of route of luteininzing hormone-releasing hormone (LHRH) neurons in mice. *J. Comp. Neurol.*, 321: 1–18.

Schwanzel-Fukuda, M., Crossin, K.L., Pfaff, D.W., Bouloux, P.M.G., Hardelin, J.-P. and Petit, C. (1996) Migration of luteinizing hormone-releasing hormone (LHRH) neurons in early human embryos. *J. Comp. Neurol.*, 366: 547–557.

Seki, T. and Arai, Y. (1993) Distribution and possible roles of the highly polysialylated neural cell adhesion molecule (NCAM-H) in the developing and adult central nervous system. *Neurosci. Res.*, 17: 265–290.

Stoeckli, E.T., Kuhn, T.B., Duc, C.O., Ruegg, M.A. and Sonderegger, P. (1991) The axonally secreted protein axonin-1 is a potent substratum for neurite growth. *J. Cell Biol.*, 112: 449–455.

Sullivan, K.A. and Silverman, A.-J. (1993) The ontogeny of gonadotropin-releasing hormone neurons in the chick. *Neuroendocrinology*, 58: 597–608.

Tamamaki, N., Fujimori, K.E. and Takauji, R. (1997) Origin and route of tangentially migrating neurons in the developing neocortical intermediate zone. *J. Neurosci.*, 17: 8313–8323.

Toba, Y., Ajiki, K., Horie, M., Sango, K. and Kawano, H. (2001) Immunohistochemical localization of calbindin D-28k in the migratory pathway from the rat olfactory placode. *J. Neuroendocrinol.*, 13: 683–694.

Tobet, S.A., Crandall, J.E. and Schwarting, G.A. (1992) Relationship of migrating luteinizing hormone-releasing hormone neurons to unique olfactory system glycoconjugates in embryonic rats. *Dev. Biol.*, 155: 471–482.

von Bartheld, C.S., Lindörfer, H.W. and Meyer, D.L. (1987) The nervus terminalis also exists in cyclostomes and birds. *Cell Tissue Res.*, 250: 431–434.

Wenzel, B.M. (1987) The olfactory and related system in birds. In: L.S. Demski and M. Schwanzel-Fukuda (Eds.), *The Terminal Nerve (Nervus Terminalis): Structure, Function and Evolution*. Annals of New York Academy of Science, New York, pp. 137–149.

Wirsig-Wiechmann, C.R. (1990) The nervus terminalis in the chick: a FMRFamide-immunoreactive and AchE-positive nerve. *Brain Res.*, 523: 175–179.

Wray, S., Grant, P. and Gainer, H. (1989a) Evidence that cells expressing luteinizing hormone-releasing hormone mRNA in the mouse are derived from progenitor cells in the olfactory placode. *Proc. Natl. Acad. Sci. USA*, 86: 8132–8136.

Wray, S., Nieburgs, A. and Elkabes, S. (1989b) Spatiotemporal cell expression of luteinizing hormone-releasing hormone in the prenatal mouse: evidence for an embryonic origin in the olfactory placode. *Dev. Brain Res.*, 46: 309–318.

Wray, S., Key, S., Qualls, R. and Fuesko, S.M. (1994) A subset of peripherin positive olfactory axons delineates the luteinizing hormone releasing hormone neuronal migratory pathway in developing mouse. *Dev. Biol.*, 166: 349–354.

Yamamoto, N., Uchiyama, H., Ohki-Hamazaki, H., Tanaka, H. and Ito, H. (1996) Migration of GnRH-immunoreactive neurons from the olfactory placode to the brain: a study using avian embryonic chimeras. *Dev. Brain Res.*, 95: 234–244.

Yoshida, K., Tobet, S.A., Crandall, J.E., Jimenez, T.P. and Schwarting, G.A. (1995) The migration of luteinizing hormone-releasing hormone neurons in the developing rat is associated with a transient, caudal projection of the vomeronasal nerve. *J. Neurosci.*, 15: 7769–7777.

Yoshida, K., Rutishauser, U., Crandall, J.E. and Schwarting, G.A. (1999) Polysialic acid facilitates migration of luteinizing hormone-releasing hormone neurons on vomeronasal axons. *J. Neurosci.*, 19: 794–801.

I.S. Parhar (Ed.)
Progress in Brain Research, Vol. 141

CHAPTER 4

What defines the nervus terminalis? Neurochemical, developmental, and anatomical criteria

Celeste R. Wirsig-Wiechmann [1,2,*], Allan F. Wiechmann [1,3], Heather L. Eisthen [4]

[1] *Department of Cell Biology, University of Oklahoma Health Sciences Center, Oklahoma City, OK 73104, USA*
[2] *Department of Zoology, University of Oklahoma, Norman, OK 73019, USA*
[3] *Department of Opthalmology, University of Oklahoma Health Sciences Center, Oklahoma City, OK 73104, USA*
[4] *Department of Zoology, Michigan State University, East Lansing, MI 48824, USA*

Introduction

The nervus terminalis — a neuromodulatory system

The *nervus terminalis*, or terminal nerve, is a diffusely organized system of neurons that lie within the nasal cavity and rostral forebrain of all jawed vertebrates, including humans (Fig. 1). Its most significant feature is that some of its neurons contain the reproductive neuropeptide, gonadotropin-releasing hormone (GnRH). External to the forebrain, the cell bodies of the nervus terminalis are typically embedded within chemosensory nerves in the nasal cavity or they may be congregated in compact ganglia, commonly in the region of the olfactory bulbs. The neural processes of the plexus project to peripheral sensory structures (e.g. olfactory and/or vomeronasal mucosa, Wirsig-Wiechmann, 1993; Wirsig-Wiechmann and Wiechmann, 2001; retina in fish, Stell et al., 1984; Oka et al., 1986) and to multiple areas of the brain (Oka, 1992). While there is some variation in the extent and complexity of the projections among species, the projections

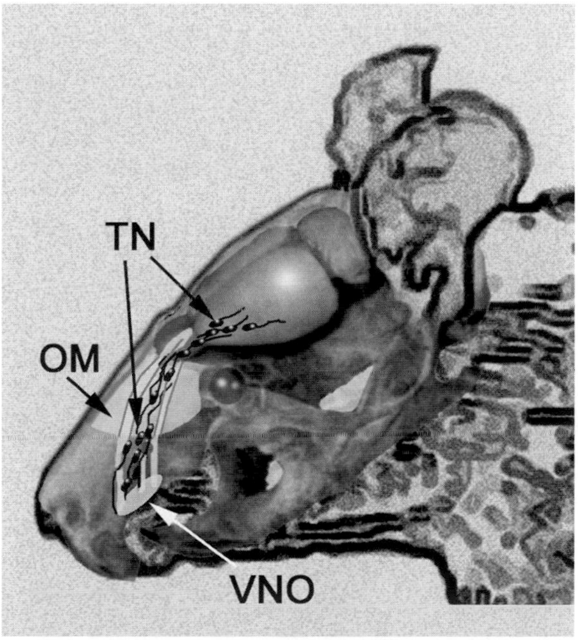

Fig. 1. Diagram of a rodent head showing the location of the *nervus terminalis* (TN, black bipolar cells) in relation to the brain and nasal cavity. In rodents, these TN neurons are embedded within vomeronasal nerves on the nasal septum and along the rostromedial surface of the olfactory bulbs. More caudally the plexus separates from the vomeronasal nerve and lies along the caudoventral surface of the olfactory bulb before penetrating the brain substance in the region of the anterior olfactory nucleus. OM, olfactoy mucosa; TN, *nervus terminalis*; VNO, vomeronasal organ.

* Correspondence to: C.R. Wirsig-Wiechmann, Department of Cell Biology, University of Oklahoma Health Science Center, 940 S.L. Young Boulevard, Oklahoma City, OK 73104, USA. Tel.: +1 405-271-2377;
E-mail: celeste-wirsig@ouhsc.edu

46

to peripheral olfactory receptive regions and fore-brain are consistently present in all species studied. Also consistent is the presence of GnRH in some or all neurons within the nervus terminalis complex (Schwanzel-Fukuda and Silverman, 1980).

History: the discovery of the nervus terminalis

The nervus terminalis was first described as an 'uberzahliger Nerv' (supernumerary nerve) distinct from the olfactory system in *Galeus canis*, the smooth dogfish, by Fritsch (1878). This discovery was made possible by the fact that the nervus terminalis system in sharks is anatomically separate from the olfactory tracts, i.e. it can be visually identified in gross dissected preparations of selachian rostral cranium as a separate ganglionated nerve bundle lying along the olfactory tract (Fig. 2). This initial discovery was followed by numerous gross and microscopic examinations of the nervus terminalis neural plexus in various vertebrate species. In most of these studies nonspecific labeling techniques (e.g. Nissl, Golgi, pyridine silver) were used that allowed a somewhat incomplete determination of

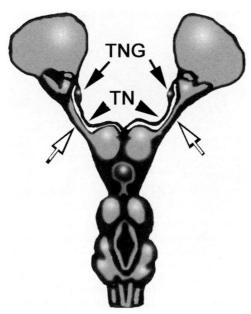

Fig. 2. Diagram of the dorsal aspect of a shark brain demonstrating the location of the nervus terminalis (TN) nerve and ganglion (TNG). In sharks, the nervus terminalis is physically separated from the olfactory tracts (open arrows).

the location of cell bodies and axonal projections within peripheral olfactory structures and forebrain. In 1980, Schwanzel-Fukuda and Silverman reported that a population of neurons within the nervus terminalis system contains GnRH. This discovery made it possible to study the distribution and projections of nervus terminalis neurons in a much more detailed manner. It also led to the speculation that the nervus terminalis is involved in some aspect of reproduction.

Recent studies have strongly suggested that the GnRH-containing neurons of the nervus terminalis are neuromodulatory to peripheral chemosensory neurons (Eisthen et al., 2000), to retina (Stell et al., 1984) and to brain (Oka and Matsushima, 1993). It is likely that this neuromodulatory system is important for reproduction. Gonadotropin-releasing hormone is known to facilitate sexual behaviors when GnRH receptors are stimulated in specific forebrain and midbrain areas (Moss and McCann, 1973; Moss and Foreman, 1976; Moore et al., 1982). The concentration of GnRH in the nervus terminalis has been shown to fluctuate during courtship and mating behavior (Propper and Moore, 1991) suggesting that the nervus terminalis is activated during sensory-induced reproductive behaviors. Likewise, lesions of the nervus terminalis alter reproductive behaviors (Wirsig and Leonard, 1987; Wirsig-Wiechmann, 1997; Yamamoto et al., 1997).

The GnRH neurons of the nervus terminalis form a system that appears to be functionally separate from the preoptic GnRH neurons in most animals, i.e. the nervus terminalis is generally not involved with gonadotropin release. This is supported by the fact that in many species there are separate forms of GnRH (Lovejoy et al., 1992; Parhar et al., 1998b), separate precursor cells (Whitlock et al., 2000) and, in some cases, differences in the timing of initial GnRH synthesis by the two populations of neurons (Parhar, 1997). However, the common migratory route from the olfactory placodal region into the brain of the nervus terminalis and preoptic/hypothalamic group of GnRH neurons (see below) suggests a close relationship between these two systems. Thus, the nervus terminalis system may represent a coordinating system for reproductive behavior and physiology that may act in concert with the preoptic/hypothalamic GnRH system.

While a substantial amount of information has been gathered over the past 20 years on the GnRH component of the nervus terminalis, there has been no formal analysis of which specific and consistent features can be used to define the nervus terminalis system as a functional unit. The purpose of this review is to propose: (1) developmental, (2) neurochemical, and (3) anatomical features that combined can be used to define the nervus terminalis in all species. Interspecies differences that may signify evolutionary trends will also be discussed.

Development of the nervus terminalis and preoptic/hypothalamic GnRH system

The GnRH neurons that come to lie along the nervus terminalis in the nasal cavity and rostral forebrain as well as those that reside in the preoptic/hypothalamic areas migrate out of the olfactory placodal region during development (Schwanzel-Fukuda and Pfaff, 1989; Wray et al., 1989). These neurons migrate along a path on the nasal septum where the mature nervus terminalis will eventually lie (Schwanzel-Fukuda et al., 1981; Schwanzel-Fukuda and Pfaff, 1990). Some GnRH neurons of the nervus terminalis migrate into the ventral forebrain but many remain in the nasal cavity and in peripheral autonomic ganglia associated with the nasal cavity (Wirsig-Wiechmann and Lepri, 1991; Wirsig-Wiechmann, 1993). There is no evidence that any nervus terminalis cells arise from the neural tube.

Numerous studies have demonstrated that removal of the olfactory placode at certain stages of development eliminates both nervus terminalis and preoptic/hypothalamic GnRH systems (Akutsu et al., 1992; Murakami et al., 1992; Northcutt and Muske, 1994). While it was originally thought that GnRH neurons originated from the same precursor group as olfactory neurons, there is now evidence that GnRH neurons arise from neural crest cells that migrate very early into the region of the olfactory placode (Whitlock et al., 2000). The groups of GnRH neurons destined for the nervus terminalis and preoptic/hypothalamic groups, respectively, arise from separate areas of the neural plate before their migration into the placode.

Other neurochemically separate components of the nervus terminalis (see below) also arise from the

olfactory placodal region, (e.g. molluscan cardio-excitatory peptide/neuropeptide Y (FMRFamide/NPY)-like immunoreactive neurons, Northcutt and Muske, 1994; Hilal et al., 1996). In addition to the nervus terminalis and preoptic GnRH neurons, it appears that the olfactory placode gives rise to groups of neurons that migrate even further into the brain, such as tyrosine hydroxylase-immunoreactive neurons (Verney et al., 1996). The fate of these latter neurons is not known. Whether these neurons are functionally connected with the nervus terminalis system is also not known.

Based on this developmental information, we propose that one of the defining characteristics of the nervus terminalis is that its neurons arise from the neural crest region of the neural plate, move into the region of the olfactory placode and migrate from there into peripheral nasal areas and ventral forebrain areas.

Neurochemical components of the nervus terminalis

Multiple forms of GnRH are present in vertebrate brains. Typically there are at least two forms of GnRH: a varying form (e.g. chicken-I GnRH, King and Millar, 1982a,b; salmon GnRH, Sherwood et al., 1983) within the forebrain and a 'chicken-II' form most commonly found within the midbrain (Miyamoto et al., 1984). In addition, there can be neurochemically dissimilar forms of GnRH between the nervus terminalis and preoptic/hypothalamic neural groups (Parhar et al., 1998b; Chiba et al., 1999) that are differentially regulated. This supports the hypothesis that the nervus terminalis and preoptic/hypothalamic neurons are functionally separate groups (Soga et al., 1998). In addition, the nervus terminalis itself may contain multiple forms of GnRH (Lovejoy et al., 1992; Forlano et al., 2000).

Besides GnRH, other neuropeptides, neurotransmitters or associated enzymes are co-localized in GnRH neurons or found in a separate population of neurons within the nervus terminalis plexus. Compounds that have been consistently identified in nervus terminalis across multiple classes of animals are (1) a neuropeptide tyrosine (NPY)-like peptide (that can also be visualized by antibodies against molluscan cardioexcitatory peptide, FMRF-

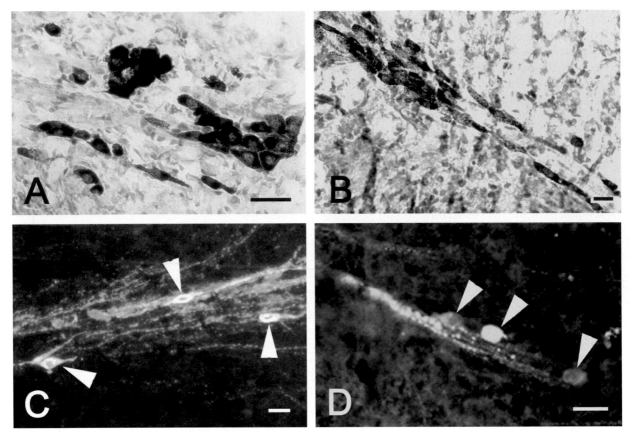

Fig. 3. Micrograph of nervus terminalis neurons on the medial surface of the olfactory bulbs of hamster labeled with (A, B) acetylcholinesterase histochemistry and (C, D) GnRH immunocytochemistry. (A) These neurons can be grouped into tight clusters, or (B) distributed along the nervus terminalis nerve fibers. (C, D) There are fewer GnRH containing neurons (arrowheads) along the nervus terminalis. Bars = 20 μm.

amide; Stell et al., 1984; Muske and Moore, 1988; Wirsig-Wiechmann, 1990; Oelschlager et al., 1998; D'Aniello et al., 2001) and (2) cholinergic related enzymes (choline acetyltransferase, Schwanzel-Fukuda et al., 1986; and acetylcholinesterase, Wirsig and Getchell, 1986; Wirsig and Leonard, 1986; Wirsig-Wiechmann, 1990; White and Meredith, 1995). Most frequently, GnRH and the NPY-like peptide are present in separate populations of nervus terminalis neurons (White and Meredith, 1995), but these peptides can be co-localized (Stell et al., 1984; Wirsig-Wiechmann and Oka, 2002). The presence of cholinergic enzymes suggests that neurons along the nervus terminalis also contain acetylcholine. White and Meredith (1993, 1995) have supplied further evidence for a cholinergic component. They

have shown that nervus terminalis ganglion neurons in sharks are responsive to acetylcholine. Heavy acetylcholinesterase labeling, which is suggestive of cholinergic function rather than cholinoceptive function (receiving cholinergic input), usually occurs in the population of nervus terminalis neurons that does not contain GnRH (Wirsig and Leonard, 1986; White and Meredith, 1995). These potential cholinergic neurons seem to be much more numerous than GnRH neurons along the nervus terminalis of rodents (Wirsig and Leonard, 1986; Fig. 3). Therefore, in many species, it is most likely that acetylcholine and the NPY-like peptide are commonly present in the same group of neurons and that the GnRH neurons represent a separate population.

49

There have also been reports of the presence of vasoactive intestinal peptide (Schwanzel-Fukuda et al., 1986) and glutamate (Yamamoto et al., 2001) in nervus terminalis. As will be discussed below, the presence of compounds in addition to GnRH in neurons of the nervus terminalis varies among species. For example, most non-mammalian species appear to possess an NPY-like neural component of the nervus terminalis, but few mammalian species have demonstrated such a component (Oelschlager et al., 1998). This may represent the unavailability of antisera to label the mammalian homologue of the NPY-like compound found in fish, reptiles, amphibians and birds.

Based on this neurochemical information, we propose that a second defining characteristic of the nervus terminalis is that at least one population of its neurons contains a form of GnRH, and another or same population of neurons contains acetylcholine and/or an NPY-like compound.

Neuroanatomical characteristics of the nervus terminalis

Nervus terminalis cell body location and characteristics

The location and distribution of nervus terminalis neural cell bodies differs between species. However, in most species, the neural cell bodies are diffusely distributed along the fibers of the nervus terminalis neural plexus both within the nasal cavity and within the forebrain (see Fig. 1; Wirsig and Getchell, 1986; Wirsig and Leonard, 1986; Wirsig-Wiechmann, 1993). In addition, the entire plexus is embedded within other nerve fascicles in the nasal cavity. In a few species (sharks and teleosts), the neurons are packed into a ganglion located external to (see Fig. 2; Fritsch, 1878; White and Meredith, 1995) or internal to the forebrain at the junction between the olfactory bulb and telencephalon (Grober et al., 1987; Oka, 1992). In mammals (Wirsig-Wiechmann and Lepri, 1991) and amphibians (Wirsig-Wiechmann, 1993; Wirsig-Wiechmann and Ebadifar, 2002), GnRH neurons have also been found in the pterygopalatine and palatine ganglia, respectively (Fig. 4). Most recently we have also observed the presence of FMRFamide/NPY-like neu-

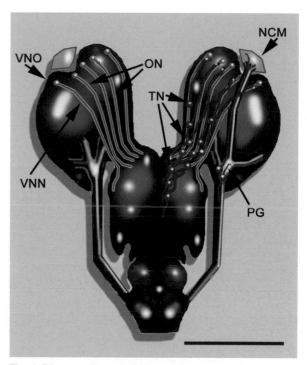

Fig. 4. Diagram of a ventral view of tiger salamander brain and nasal cavity illustrating the location of GnRH nervus terminalis neurons (TN, red neurons) in the olfactory and palatine neural systems. GnRH neurons of the palatine system project to the naris constrictor muscle (NCM) located between the vomeronasal organ (VNO) and rostral tip of the main olfactory chamber. ON, olfactory nerve fascicles; NCM, naris constrictor muscle; PG, palatine ganglion; TN, nervus terminalis; VNN, vomeronasal nerve; VNO, vomeronasal organ. Bar = 500 μm.

rons within the palatine ganglion of the salamander (Wirsig-Wiechmann and Ebadifar, 2002), which strongly suggests that these two neural populations belong to the nervus terminalis system.

In those species with distributed neurons, the neural cell bodies are frequently fusiform/bipolar cells (Fig. 5A) and are embedded within olfactory (Wirsig and Getchell, 1986), vomeronasal (Wirsig and Leonard, 1986) and autonomic nerve fascicles (Wirsig-Wiechmann and Lepri, 1991) in the nasal cavity. In the brain, the neural cell bodies are diffusely distributed along nervus terminalis fiber projections. Neurons in compact ganglia tend to be rounder and frequently larger (Oka, 1992; White and Meredith, 1995; Fig. 5B).

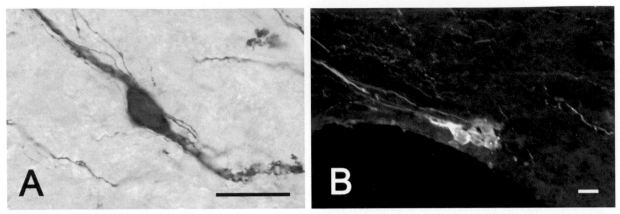

Fig. 5. Micrographs of GnRH nervus terminalis neurons from (A) tiger salamander, and (B) dwarf gourami illustrating the morphological differences of these neurons between the two species. (A) Salamander GnRH neuron from the ventral forebrain. Salamander GnRH neurons are usually medium-sized (roughly 20 μm) fusiform-shaped bipolar neurons with thick invaginated dendritic processes tapering from the cell body. (B) Gourami GnRH neurons in the nervus terminalis ganglion located at the ventrocaudal border of the olfactory bulb. The largest of the gourami nervus terminalis ganglionic neurons are round and 30 μm in diameter. Bars = 30 μm.

Nervus terminalis projections

Nervus terminalis neurons generally have two axons, one that projects peripherally toward the chemosensory mucosa, and one that projects toward the brain. The peripherally directed nervus terminalis axons project to a number of nasal structures including olfactory and vomeronasal mucosa, Bowman's glands, smooth naris muscle and autonomic parasympathetic ganglia (Wirsig and Getchell, 1986; Wirsig-Wiechmann and Lepri, 1991; Wirsig-Wiechmann, 1993; Wirsig-Wiechmann and Holliday, 2002). These processes do not appear to contact specific cells; rather they end within the lamina propria of chemosensory mucosa. A few processes appear to lie in proximity to Bowman's glands in the very rostral regions of the olfactory mucosa in salamanders (Wirsig-Wiechmann, 1993). Like the GnRH processes in the median eminence, the nervus terminalis fibers are frequently varicose, suggesting that GnRH is secreted along the length of the process. Ultrastructural evidence also suggests that GnRH may be secreted from the neural cell body of nervus terminalis neurons (Oka and Ickikawa, 1991). These data suggest that GnRH is acting as a paracrine hormone/modulator in the nasal cavity, autonomic ganglion and perhaps the brain.

Evidence that GnRH is secreted from the nervus terminalis in the nasal cavity is based on the observation that (1) olfactory and vomeronasal neurons possess GnRH receptors (Wirsig-Wiechmann and Jennes, 1993; Wirsig-Wiechmann and Wiechmann, 2001), and (2) GnRH modulates the manner in which chemosensory neurons respond to odors (Eisthen et al., 2000; Wirsig-Wiechmann et al., 2000). Gonadotropin-releasing hormone may reach the chemosensory neurons via diffusion directly from the nervus terminalis fibers locally, through intravascular transport and diffusion, or uptake and secretion by nasal glands (Fig. 6). The fact that chemosensory neurons do not show a response to GnRH if their cilia are removed (Rona Delay, personal communication) strongly suggests that GnRH must reach the surface of the mucosa. Nervus terminalis fibers do not project to the surface of the epithelium (which is sealed by tight junctions between chemosensory neuron dendrites) but are restricted to the lamina propria. We hypothesize that glandular secretion is the mechanism of GnRH transport to the cilia (Fig. 6).

Within the brain, nervus terminalis fibers generally project diffusely to widespread areas. In gourami fish, Oka and his colleagues (Oka, 1992; Oka and Matsushima, 1993) have shown via intracellular tracer injection that a single nervus terminalis neuron (located near the olfactory bulb) has a massive projection to multiple areas of brain and even into the brainstem. In some species of fish, neurons of

Fig. 6. Illustration of the chemosensory mucosa depicting two mechanisms by which GnRH could reach the chemosensory neurons (green neurons): (1) Diffusion of GnRH (pink stars) into the chemosensory epithelium from the GnRH neuron (pink neuron) located in the lamina propria (lowest black region). In this case GnRH would interact with receptors on the neural cell bodies or proximal dendrites. (2) Uptake of GnRH by nasal glands (blue gland) in the lamina propria and release from gland ducts along with glandular mucus (white mucus) onto the surface of the chemosensory epithelium (upper white region). In this case, GnRH would stimulate receptors on the dendritic knob or cilia/microvilli. On the surface of the chemosensory epithelium, which is adjacent to the lumen of the nasal cavity, the odor molecules (yellow stars) also interact with the olfactory receptor neuron cilia (depicted here) or vomeronasal neuron microvilli to stimulate odor receptors.

the nervus terminalis ganglion project through the brain, into the optic nerve and to the retina (Stell et al., 1984; Grober et al., 1987). In many studies the nervus terminalis neurons that project to the retina have been termed the *nucleus olfactoretinalis*. There is some controversy as to whether this group of neurons belongs to the nervus terminalis system (Szabo et al., 1991). In catfish, nervus terminalis neurons have been reported to project to the pituitary gland (Krishna et al., 1992). Several studies in amphibians have utilized tract tracers such as horseradish peroxidase to trace nervus terminalis projections to the brain (Hofmann and Meyer, 1989). However, most of the results remain inconclusive since extrabulbar

olfactory projections (Hofmann and Meyer, 1992) have also been labeled. Hofmann and Meyer (1992) have demonstrated labeled nervus terminalis neurons in the nasal cavity following injections of tracer into the diencephalon.

In mammals, very few definitive tracing studies have been carried out to determine how extensively an individual neuron projects. Jennes (1987) has conducted the only study that has shown the projections of an individual nervus terminalis neuron to the amygdala. Additionally, from numerous immunocytochemical studies, it is fairly clear that the nervus terminalis projects diffusely to olfactory bulb, rostral olfactory regions of the forebrain and septum (Schwanzel-Fukuda and Silverman, 1980; Witkin and Silverman, 1983; Jennes, 1986; Wirsig-Wiechmann and Wiechmann, 2001). More caudally it is not possible to differentiate nervus terminalis from preoptic GnRH neuron projections.

Neural interactions

Anatomical and electrophysiological data from fish and amphibians suggest that nervus terminalis neurons interact with one another (e.g. GnRH neurons can stimulate other GnRH neurons' activity; Abe and Oka, 2000) as well as their target organs (e.g. olfactory epithelium; Eisthen et al., 2000). In amphibians, we have observed a close relationship between GnRH neurons and NPY-like neurons (Fig. 7). This has also been observed in the nervus terminalis of a mammal, the brown bat (Oelschlager et al., 1998). Whether the NPY-like peptide influences GnRH neural function and vice versa in the nervus terminalis is not known. However, there is evidence that NPY and GnRH neurons of the preoptic/hypothalamic system interact with one another as well as on their target, the pituitary gland. Neuropeptide Y increases GnRH secretion in the preoptic/hypothalamic system (Crowley and Kalra, 1987). It also influences luteinizing hormone secretion by itself, and pre-exposure to NPY has been shown to augment pituitary cell responses to GnRH (Evans et al., 2001). Interestingly, in catfish, the NPY-like fibers project to the pituitary (Krishna et al., 1992) and may interact with the GnRH fibers in control of gonadotropin release. In relation to the innervation of sensory targets by the nervus terminalis system,

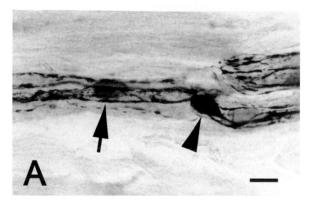

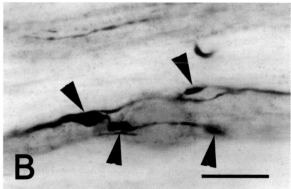

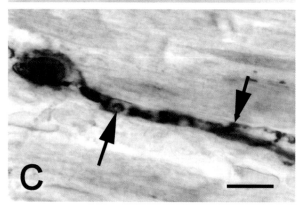

Fig. 7. Micrographs of tiger salamander olfactory nerve fascicles including nervus terminalis neurons labeled with GnRH (brown) and FMRFamide (blue) antisera. (A) Two nervus terminalis neurons within the plexus, one containing GnRH (arrow) and the other an NPY/FMRFamide-like peptide (arrowhead). (B) Apparent contacts (arrowheads) by an NPY/FMRFamide-like neuron onto a GnRH neuron. (C) Apparent contacts (arrows) by a GnRH neuron onto an NPY/FMRFamide-like neuron. Axonal contacts between neurons generally occur on the cell body (B) or dendrite (C). Bars = 20 μm.

GnRH and the NPY-like peptide individually can modify the activity of peripheral sensory systems (olfactory receptor neurons: Eisthen et al., 2000, 2001; and retinal ganglion cells: Stell et al., 1987).

In some fish (e.g. goldfish, dwarf gourami) the GnRH and NPY-like peptide are in the same nervus terminalis neurons and therefore could be released together (Wirsig-Wiechmann and Oka, 2002). There may be some intra-neuronal control over the relative amount of each peptide released by factors such as the firing frequency of the neuron. The separation of GnRH and NPY-like neural populations in other species undoubtedly enables greater neural and/or hormonal control of release of the two peptides, giving the system greater flexibility in its response to inputs from different sources (White and Meredith, 1995).

Nervus terminalis neurons receive inputs from multiple brain areas (Fernald and Finger, 1984; Ya-mamoto and Ito, 2000). The brain areas that project to the nervus terminalis ganglion mainly include sensory processing regions, especially for olfactory information, and projections from brainstem modu-latory systems. It has been shown that the nervus terminalis ganglion in sharks specifically receives noradrenergic inputs from central sources (Fernald and Finger, 1984) and that these inputs influence the activity of the nervus terminalis ganglion neurons (White and Meredith, 1995). Axonal inputs to the nervus terminalis mainly contact the proximal por-tion of the dendrite and the cell body (Figs. 7 and 8). These inputs have been shown to involve synaptic contacts with nervus terminalis neurons (White and Meredith, 1987; Oka and Ickikawa, 1991). While it is known that adjacent GnRH nervus terminalis neu-rons can influence the activity of one another (Abe and Oka, 2000; Oka and Abe, 2001), very little is know about the specific stimuli that control the activ-ity level of the nervus terminalis. Nervus terminalis GnRH neurons do not appear to be activated at the time of the luteinizing hormone surge and ovulation (Lee et al., 1990). However, the synthesis and/or release of GnRH in these neurons are influenced by steroids (Wirsig-Wiechmann and Lee, 1999).

Based on neuroanatomical data collected on the location of nervus terminalis cell bodies, axonal pro-jections and interactions with other neural systems, we propose the following minimal characteristics in

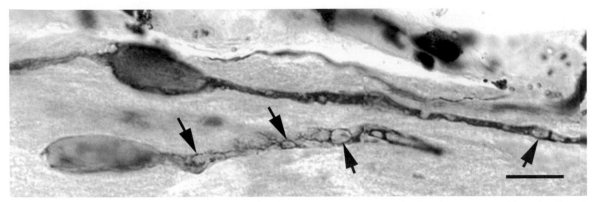

Fig. 8. Micrographs of two tiger salamander nervus terminalis neurons located within the olfactory nerve illustrating invaginations (arrows) within the dendrites. These invaginated areas appear more lightly labeled because there is little cytoplasm in this area. At these invagination sites presumed axonal contacts occur. The upper neuron contains GnRH and the lower neuron contains an NPY/FMRFamide-like peptide. Bar = 20 μm.

defining the nervus terminalis: (1) its cell bodies lie either within the nasal cavity or within the region of the olfactory bulbs, (2) its axons project peripherally to the chemosensory and/or nasal mucosa and centrally to ventral forebrain areas, especially olfactory/limbic/hypothalamic areas, and (3) its cell bodies receive neural inputs from central brain regions.

Species diversity

The functional studies of the nervus terminalis system all suggest that it is a neuromodulatory system involved in reproduction. While certain features of the nervus terminalis are consistent across all species (e.g. presence of GnRH), there are some obvious species differences in the location, density, morphology and size of nervus terminalis cell bodies, relative proportion of GnRH containing neurons compared to non-GnRH containing neurons, and axonal projections. These differences may not reflect significant functional differences of the system, but rather differences in species' body forms and specific behavioral/endocrine strategies related to reproduction.

There is variation in the proportion of nervus terminalis neurons that contain GnRH as opposed to other neurotransmitters among different species. While in certain species of fish (e.g. dwarf gourami), the GnRH to non-GnRH ratio of terminal nerve neuron number is roughly 1:2 (Oka et al., 1986), in rodents the ratio is almost five times as great

(Schwanzel-Fukuda and Silverman, 1980; Wirsig and Leonard, 1986). Our studies in hamsters suggest that approximately 10% of nervus terminalis neurons contain GnRH (see Fig. 3). However, in tiger salamanders, it appears that equal numbers of neurons contain GnRH or a FMRFamide/NPY-like peptide (unpublished observations; Fig. 7).

The size of the neurons along the nervus terminalis also varies between species. In dwarf gourami the neurons of the terminal nerve ganglion can be as large as 30 μm (Oka and Ickikawa, 1991; Fig. 5), whereas in many other species the neuronal diameter averages from 10 μm in rodents (Fig. 3) to 20 μm in amphibians (Fig. 7). The cell size may reflect the extensiveness of the projections of individual nervus terminalis neurons. In fish, a single neurons can send projections to many parts of the brain, including the brainstem (Oka, 1992). In mammals, it appears that the nervus terminalis projects to more rostral brain areas including the olfactory, limbic and hypothalamic regions rather than brain stem areas (Jennes, 1987). In many species, the non-GnRH containing neurons generally tend to be larger (1.5×) than the GnRH containing neurons (Schwanzel-Fukuda and Silverman, 1980), suggesting a difference in function or projections compared to the GnRH containing neurons.

The nasal regions to which the nervus terminalis projects differ somewhat between species. In rodents, the nervus terminalis is associated with the vomeronasal system (Fig. 1). The GnRH contain-

54

ing fibers of the nervus terminalis project mainly to the lamina propria of the vomeronasal organ. In addition, GnRH fibers from neurons located in the pterygopalatine ganglion project along the nasopalatine nerve to innervate the caudal region of the vomeronasal organ. This association of the nervus terminalis with the vomeronasal organ correlates with the fact that this chemosensory organ is very important in rodent reproduction (Wysocki, 1979). In tiger salamanders, while some GnRH fibers project to the vomeronasal organ, most nervus terminalis processes are confined to ventromedial olfactory nerve bundles that project to the rostromedial region of the olfactory chamber (Fig. 4). It is unclear whether this region of the chemosensory mucosa is involved in reproduction, such as the detection of pheromones.

It is interesting to note that, in tiger salamanders, the rostral olfactory epithelium is thicker, and demonstrates a much slower cell turnover rate and is less responsive to odors than the caudal olfactory epithelium (MacKay-Sim and Patel, 1984). A positive correlation between GnRH administration and inhibition of cell growth has been demonstrated in cancer therapies (Ben-Yehudah et al., 2001; Grundker et al., 2001). It is possible that an increase in GnRH secretion from the nervus terminalis during mating could influence cell turnover in the chemosensory epithelium in addition to its immediate effect on olfactory receptor neuron responses to odors (Eisthen et al., 2000). It could be argued that high cell turnover might not be advantageous during courtship and mating when an animal needs to conserve energy and use the sensory system extensively for pheromonal detection. Dawley and coworkers (Dawley, 1998; Dawley et al., 2000) have shown that there are seasonal changes in chemosensory epithelium thickness and vomeronasal organ size in the salamander, *Plethodon cinereus*. Cell turnover is highest at the end of the mating season. However, it is not clear whether these changes are related directly to GnRH secretion from the nervus terminalis.

In certain species of fish, the nervus terminalis sends projections through the brain to the retina. These projections include both the GnRH and FMRFamide/NPY-like components of the nervus terminalis system. While such direct projections to the retina do not appear to be very common in

other classes of animals, there have been reports of GnRH fibers in the optic nerves of mammals (Witkin, 1987, in primates; Wirsig-Wiechmann and Wiechmann, 2002, in voles). These GnRH fibers do not seem to project directly into the retina. The retina of voles does express GnRH receptors (Wirsig-Wiechmann and Wiechmann, 2002) suggesting that the GnRH reaches the retina by diffusion along the optic nerve or through the vasculature. In *Rana pipiens*, the FMRFamide/NPY-like (but not the GnRH) component of the nervus terminalis sends fibers to the retina (Wirsig-Wiechmann and Basinger, 1988).

In the majority of animals, the fibers of the nervus terminalis are unmyelinated, secretory fibers. However, in sharks and dolphins, many of the nervus terminalis fibers are myelinated (White and Meredith, 1987; Demski et al., 1987). It has been proposed that the non-GnRH population of nervus terminalis neurons may be autonomic (Buhl and Oelschlager, 1986; Meredith and White, 1987), which is supported by the fact that these neurons appear to be cholinergic (Schwanzel-Fukuda et al., 1986; Wirsig and Leonard, 1986).

It is interesting to note that while whales and dolphins completely lack a nasal chemosensory system, they still possess a nervus terminalis. With respect to these animals, the potential autonomic component of the nervus terminalis may have become adapted for diving, and/or controlling nasal gland secretion or vascular tone rather than for modulating olfaction (Buhl and Oelschlager, 1986). The intracerebral GnRH component may still be functional in cetaceans since it is thought to modulate brain areas subserving reproductive functions.

Conclusion

The nervus terminalis is a neurosecretory modulatory system that receives input from the brain and modulates peripheral sensory systems and brain nuclei probably related to reproduction. This system does exist in the human nasal cavity. It may participate in apparent alterations in olfactory sensitivity related to changes in circulating hormones (Doty et al., 1981; Hummel et al., 1991; Graham et al., 2000) as well as gender-specific activation of reproductive brain areas by pheromonal stimulation (Savic et al., 2001; Sobel and Brown, 2001). Understanding the manner in which the nervus terminalis might influence our responses to social odors may shed light on the subtle aspects of human behaviors such as the choice of a partner.

Based on the research conducted to date on this system in animals as well as in humans, we propose the following defining characteristics for the nervus terminalis neurons: The neurons: (1) arise from the neural crest and migrate from the olfactory placode region to nasal and rostral brain areas, (2) in the mature organism, lie within the nasal cavity or superficially in the substance of the brain near the olfactory bulbs and receive inputs from the brain, (3) project to nasal mucosa and rostroventral areas of the brain, especially olfactory/limbic areas, and (4) contain GnRH, acetylcholine and an NPY-like peptide. It is clear from the studies conducted over the past 20 years, that the nervus terminalis projects to and modulates a number of peripheral and central structures involved with sensory processing and reproduction. The peripheral systems that receive the greatest and most direct projections from the nervus terminalis appear to be the most important for the reproductive strategies of the species (e.g. in fish that use visual information for mating the nervus terminalis projects extensively to the retina; in rodents that use pheromonal information for mating the nervus terminalis projects extensively to the vomeronasal organ). The purpose of this initial attempt to define the characteristics of the nervus terminalis as a functional unit is to set a basic framework for designing experiments that can test hypotheses regarding the mechanisms by which the nervus terminalis influences and is influenced by other neural systems.

Abbreviations

GnRH   gonadotropin-releasing hormone
FMRFamide molluscan cardioexcitatory peptide
NPY   neuropeptide Y

Acknowledgements

The research cited here by Wirsig or Wirsig-Wiechmann was supported largely by grants from NIH (NS27586 and DC04270), NSF (IBN-9896098), Presbyterian Health Foundation and the Oklahoma Center for the Advancement of Science and Technology (HR00-078). Heather L. Eisthen is supported by NSF grant IBN 9982934.

References

Abe, H. and Oka, Y. (2000) Modulation of pacemaker activity by salmon gonadotropin-releasing hormone (sGnRH) in terminal nerve (TN)-GnRH neurons. J. Neurophysiol., 83: 3196–3200.

Akutsu, S., Takada, M., Ohki-Hamazaki, H., Murakami, S. and Arai, Y. (1992) Origin of luteinizing hormone-releasing hormone (LHRH) neurons in the chick embryo: effect of the olfactory placode ablation. Neurosci. Lett., 142: 241–244.

Ben-Yehudah, A., Prus, D. and Lorberboum-Galski, H. (2001) I.V. Administration of L-GNRH-PE66 efficiently inhibits growth of colon adenocarcinoma xenografts in nude mice. Int. J. Cancer, 92: 263–268.

Braun, C.B., Wicht, H. and Northcutt, R.G. (1995) Distribution of gonadotropin-releasing hormone immunoreactivity in the brain of the Pacific hagfish, Eptatretus stouti (Craniata: Myxinoidea). J. Comp. Neurol., 353: 464–476.

Buhl, E.H. and Oelschlager, H.A. (1986) Ontogenetic development of the nervus terminalis in toothed whales. Evidence for its non-olfactory nature. Anat. Embryol., 173: 285–294.

Chiba, H., Nakamura, M., Iwata, M., Sakuma, Y., Yamauchi, K. and Parhar, I.S. (1999) Development and differentiation of gonadotropin hormone-releasing hormone neuronal systems and testes in the Japanese eel (Anguilla japonica). Gen. Comp. Endocrinol., 114: 449–459.

Crowley, W.R. and Kalra, S.P. (1987) Neuropeptide Y stimulates the release of luteinizing hormone-releasing hormone from medial basal hypothalamus in vitro: modulation by ovarian hormones. Neuroendocrinology, 46: 97–103.

D'Aniello, B., Fiorentin, M., Pinelli, C., Guarino, F.M., Angelini, F. and Rastogi, R,K. (2001) Localization of FMRFamide-like immunoreactivity in the brain of the viviparous skink (Chalcides chalcides). Brain Behav. Evol., 57: 18–32.

Dawley, E.M. (1998) Species, sex, and seasonal differences in VNO size. Microsc. Res. Tech., 41: 506–518.

Dawley, E.M., Fingerlin, A., Hwang, D., John, S.S. and Stankiewicz, C.A. (2000) Seasonal cell proliferation in the chemosensory epithelium and brain of red-backed salamanders, Plethodon cinereus. Brain Behav. Evol., 56: 1–13.

Demski, L.S., Schwanzel-Fukuda, M. and Ridway, S.H. (1987) The terminal nerve of dolphins: general anatomy and LHRH-immunocytochemistry. Soc. Neurosci. Abstr., 13: 528.

Doty, R.L., Snyder, P.J., Huggin, G.R. and Lowry, L.D. (1981) Endocrine, cardiovascular and psychological correlates of olfactory sensitivity to changes during the human menstrual cycle. J. Comp. Physiol. Psych., 95: 45–60.

Eisthen, H.L. and Northcutt, R.G. (1996) Silver lampreys (Ichthyomyzon unicuspis) lack a gonadotropin-releasing hormone- and FMRFamide-immunoreactive terminal nerve. J. Comp. Neurol., 370: 159–172.

Eisthen, H.L., Delay, R.J., Wirsig-Wiechmann, C.R. and Dionne, V.E. (2000) Neuromodulatory effects of gonadotropin-releasing hormone on olfactory receptor neurons. J. Neurosci., 20: 3947–3955.

Eisthen, H.L., Fawley, J.A. and Zawacki, S.R. (2001) Olfactory receptor cell physiology is modulated by peptides structurally similar to those found in the terminal nerve. *Soc. Neurosci. Abstr.*, 27: Abstr. 62.5.

Evans, J.J., Pragg, F.L. and Mason, D.R. (2001) Release of luteinizing hormone from the anterior pituitary gland in vitro can be concurrently regulated by at least three peptides: gonadotropin-releasing hormone, oxytocin and neuropeptide Y. *Neuroendocrinology*, 73: 408–416.

Fernald, R.D. and Finger, T.E. (1984) Catecholaminergic neurons of locus coeruleus project to the ganglion cells of the nervus terminalis (NT) in goldfish. *Soc. Neurosci. Abstr.*, 10: 50.

Forlano, P.M., Maruska, K.P., Sower, S.A., King, J.A. and Tricas, T.C. (2000) Differential distribution of gonadotropin-releasing hormone-immunoreactive neurons in the stingray brain: functional and evolutionary considerations. *Gen. Comp. Endocrinol.*, 118: 226–248.

Fritsch, G. (1878) *Untersuchungen über den deineren Bau des Fischgehirns.* Berlin.

Graham, C.A., Janssen, E. and Sanders, S.A. (2000) Effects of fragrance on female sexual arousal and mood across the menstrual cycle. *Psychophysiology*, 37: 76–84.

Grober, M.S., Bass, A.H., Burd, G., Marchaterre, M.A., Segil, N., Scholz, K. and Hodgson, T. (1987) The nervus terminalis ganglion in *Anguilla rostrata*: an immunocytochemical and HRP histochemical analysis. *Brain Res.*, 436: 148–152.

Grundker, C., Volker, P. and Emons, G. (2001) Antiproliferative signaling of luteinizing hormone-releasing hormone in human endometrial and ovarian cancer cells through G protein alpha(I)-mediated activation of phosphotyrosine phosphatase. *Endocrinology*, 142: 2369–2380.

Hilal, E.M., Chen, J.H. and Silverman, A.J. (1996) Joint migration of gonadotropin-releasing hormone (GnRH) and neuropeptide Y (NPY) neurons from olfactory placode to central nervous system. *J. Neurobiol.*, 31: 487–502.

Hofmann, M.H. and Meyer, D.L. (1989) The nervus terminalis in larval and adult *Xenopus laevis. Brain Res.*, 498: 167–169.

Hofmann, M.H. and Meyer, D.L. (1992) Peripheral origin of olfactory nerve fibers by-passing the olfactory bulb in *Xenopus laevis. Brain Res.*, 589: 161–163.

Hummel, T., Gollesch, R., Wildt, G. and Kobal, G. (1991) Changes in olfactory perception during the menstrual cycle. *Experientia*, 47: 712–715.

Jennes, L. (1986) The olfactory gonadotropin-releasing hormone immunoreactive system in mouse. *Brain Res.*, 386: 351–363.

Jennes, L. (1987) Sites of origin of gonadotropin releasing hormone containing projections to the amygdala and the interpeduncular nucleus. *Brain Res.*, 404: 339–344.

Jennes, L. and Stumpf, W.E. (1980a) LHRH-neuronal projections to the inner and outer surface of the brain. *Neuroendocrinol. Lett.*, 2: 241–246.

Jennes, L. and Stumpf, W.E. (1980b) LHRH-systems in the brain of the golden hamster. *Cell Tissue Res.*, 209: 239–256.

Key, S. and Wray, S. (2000) Two olfactory placode derived galanin subpopulations: luteinizing hormone-releasing hormone neurones and vomeronasal cells. *J. Neuroendocrinol.*, 12: 535–545.

King, J.A. and Millar, R.P. (1982a) Structure of chicken hypothalamic luteinizing hormone-releasing hormone, II. Isolation and characterization. *J. Biol. Chem.*, 257: 10729–10732.

King, J.A. and Millar, R.P. (1982b) Structure of chicken hypothalamic luteinizing hormone-releasing hormone, I. Structural determination on partially purified material. *J. Biol. Chem.*, 1257: 10722–10728.

Krishna, N.S., Subheda, N. and Schreibman, M.P. (1992) FMRFamide-like immunoreactive nervus terminalis innervation to the pituitary in the catfish, *Clarias batrachus* (Linn.): demonstration by lesion and immunocytochemical techniques. *Gen. Comp. Endocrinol.*, 85: 111–117.

Lee, W.S., Smith, M.S. and Hoffman, G.E. (1990) Luteinizing hormone-releasing hormone neurons express Fos protein during the proestrous surge of luteinizing hormone. *Proc. Natl. Acad. Sci. USA*, 187: 8183–8187.

Lovejoy, D.A., Stell, W.K. and Sherwood, N.M. (1992) Partial characterization of four forms of immunoreactive gonadotropin-releasing hormone in the brain and terminal nerve of the spiny dogfish (Elasmobranchii; *Squalus acanthias*). *Regul. Pept.*, 37: 39–48.

MacKay-Sim, A. and Patel, U. (1984) Regional differences in cell density and cell genesis in the olfactory epithelium of the salamander, *Ambystoma tigrinum. Exp. Brain Res.*, 57: 99–106.

Meredith, M. (1983) Sensory physiology of pheromone communication. In: Vandenbergh, J.G. (Ed.), *Pheromones and Reproduction in Mammals.* Academic Press, New York, NY, pp. 199–252.

Meredith, M. and White, J. (1987) Interactions between the olfactory system and the terminal nerve. Electrophysiological evidence. *Ann. N.Y. Acad. Sci.*, 519: 349–368.

Miyamoto, K., Hasegawa, Y., Nomura, M., Igarashi, M., Kangawa, K. and Matsuo, H. (1984) Identification of the second gonadotropin-releasing hormone in chicken hypothalamus: evidence that gonadotropin secretion is probably controlled by two distinct gonadotropin-releasing hormones in avian species. *Proc. Natl. Acad. Sci. USA*, 81: 3874–3878.

Moore, F.L., Miller, L.J., Spielvogel, S.P., Kubiak, T. and Folkers, K. (1982) Luteinizing hormone-releasing hormone involvement in the reproductive behavior of a male amphibian. *Neuroendocrinology*, 35: 212–216.

Moss, R.L. and Foreman, M.M. (1976) Potentiation of lordosis behavior by intrahypothalamic infusion of synthetic luteinizing hormone-releasing hormone. *Neuroendocrinology*, 20: 176–181.

Moss, R.L. and McCann, S.M. (1973) Induction of mating behavior in rats by luteinizing hormone-releasing hormone. *Science*, 181: 177–179.

Murakami, S., Kikuyama, S. and Arai, Y. (1992) The origin of the luteinizing hormone-releasing hormone (LHRH) neurons in newts (*Cynops pyrrhogaster*): the effect of olfactory placode ablation. *Cell Tissue Res.*, 269: 21–27.

Muske, L.E. and Moore, F.L. (1988) The nervus terminalis in amphibians: anatomy, chemistry and relationship with the hypothalamic gonadotropin-releasing hormone system. *Brain Behav. Evol.*, 32: 141–150.

Northcutt, R.G. and Muske, L.E. (1994) Multiple embryonic origins of gonadotropin-releasing hormone (GnRH) immunoreactive neurons. *Brain Res. Dev. Brain Res.*, 78: 279–290.

Oelschlager, H.A., Helpert, C. and Northcutt, R.G. (1998) Coexistence of FMRFAMIDE-like and LHRH-like immunoreactivity in the terminal nerve and forebrain of the big brown bat, *Eptesicus fuscus. Brain Behav. Evol.*, 52: 139–147.

Oka, Y. (1992) Gonadotropin-releasing hormone (GnRH) cells of the terminal nerve as a model neuromodulator system *Neurosci. Lett.*, 142: 119–122.

Oka, Y. (1995) Tetrodotoxin-resistant persistent Na^+ current underlying pacemaker potentials of fish gonadotrophin-releasing hormone neurones. *J. Physiol.*, 482: 1–6.

Oka, Y. and Abe, H. (2001) Physiology of GnRH neurons and modulation of their activities by GnRH. In: R. Handa, S. Hayashi, E. Terasawa and M. Kawata (Eds.), *Neuroplasticity, Development, and Steroid Hormone Action.* CRC Press, Boca Raton, FL, in press.

Oka, Y. and Ickikawa, M. (1991) Ultrastructure of the ganglion cells of the terminal nerve in the dwarf gourami (*Colisa lalia*). *J. Comp. Neurol.*, 304: 161–171.

Oka, Y. and Matsushima, T. (1993) Gonadotropin-relcasing hormone (GnRH)-immunoreactive terminal nerve cells have intrinsic rhythmicity and project widely in the brain. *J. Neurosci.*, 13: 2161–2176.

Oka, Y., Munro, A.D. and Lam, T.J. (1986) Retinopetal projections from a subpopulation of ganglion cells of the nervus terminalis in the dwarf gourami (*Colisa lalia*). *Brain Res.*, 367: 341–345.

Parhar, I.S. (1997) GnRH in tilapia: three genes, three origins and their roles. In: I.S. Parhar and Y. Sakuma (Eds.), *GnRH Neurons: Genes to Behavior.* Brain Shuppan, Tokyo, pp. 99–122.

Parhar, I.S., Soga, T., Ishikawa, Y., Nagahama, Y. and Sakuma, Y. (1998a) Neurons synthesizing gonadotropin-releasing hormone mRNA subtypes have multiple developmental origins in the medaka. *J. Comp. Neurol.*, 401: 217–226.

Parhar, I.S., Soga, T. and Sakuma, Y. (1998b) Quantitative in situ hybridization of three gonadotropin-releasing hormone-cncoding mRNAs in castrated and progesterone-treated male tilapia. *Comp. Endocrinol.*, 112: 406–414.

Pause, B.M., Sojka, B., Krauel, K., Fehm-Wolfsdorf, G. and Ferstl, R. (1996) Olfactory information processing during the course of the menstrual cycle. *Biol. Psych.*, 44: 31–54.

Propper, C.R. and Moore, F.L. (1991) Effects of courtship on brain gonadotropin hormone-releasing hormone and plasma steroid concentrations in a female amphibian (*Taricha granulosa*). *Gen. Comp. Endocrinol.*, 81: 304–312.

Sarkar, S. and Subhedar, N. (2000) Beta-endorphin and gonadotropin-releasing hormone in the forebrain and pituitary of the female catfish, *Clarias batrachus*: double-immunolabeling study. *Gen. Comp. Endocrinol.*, 118: 39–47.

Savic, I., Berglund, H., Gulyas, B. and Roland, P. (2001) Smelling of odorous sex hormone-like compounds causes sex-differentiated hypothalamic activations in humans. *Neuron*, 31: 661–668.

Schwanzel-Fukuda, M. and Pfaff, D.W. (1989) Origin of luteinizing hormone releasing hormone neurons. *Nature*, 338: 161–164.

Schwanzel-Fukuda, M. and Pfaff, D.W. (1990) The migration of luteinizing hormone-releasing hormone (LHRH) neurons from the medial olfactory placode into the medial basal forebrain. *Experientia*, 46: 956–962.

Schwanzel-Fukuda, M. and Silverman, A.J. (1980) The nervus terminalis of the guinea pig: a new luteinizing hormone-releasing hormone (LHRH) neuronal system. *J. Comp. Neurol.*, 191: 213–225.

Schwanzel-Fukuda, M., Robinson, J.A. and Silverman, A.J. (1981) The fetal development of the luteinizing hormone-releasing hormone (LHRH) neuronal systems of the guinea pig brain. *Brain Res. Bull.*, 7: 293–315.

Schwanzel-Fukuda, M., Morrell, J.I. and Pfaff, D.W. (1986) Localization of choline acetyltransferase and vasoactive intestinal polypeptide-like immunoreactivity in the nervus terminalis of the fetal and neonatal rat. *Peptides*, 7: 899–906.

Sherwood, N., Eiden, L., Brownstein, M., Spiess, J., Rivier, J. and Vale, W. (1983) Characterization of a teleost gonadotropin-releasing hormone. *Proc. Natl. Acad. Sci. USA*, 80: 2794–2798.

Sobel, N. and Brown, W.M. (2001) The scented brain: pheromonal responses in humans. *Neuron*, 31: 512–514.

Soga, T., Sakuma, Y. and Parhar, I.S. (1998) Testosterone differentially regulates expression of GnRH messenger RNAs in the terminal nerve, preoptic and midbrain of male tilapia. *Brain Res. Mol. Brain Res.*, 60: 13–20.

Stell, W.K., Walker, S.E., Chohan, K.S. and Ball, A.K. (1984) The goldfish nervus terminalis: a luteinizing hormone-releasing hormone and molluscan cardioexcitatory peptide immunorcactive olfactoretinal pathway. *Proc. Natl. Acad. Sci. USA*, 81: 940–944.

Stell, W.K., Walker, S.E. and Ball, A.K. (1987) Functional–anatomical studies on the terminal nerve projection to the retina of bony fishes. *Ann. N.Y. Acad. Sci.*, 519: 80–96.

Szabo, T., Blahser, S., Denizot, J.P. and Ravaille-Veron, M. (1991) The olfactoretinalis system = terminal nerve? *Neuroreport*, 2: 73–76.

Verney, C., el Amraoui, A. and Zecevic, N. (1996) Comigration of tyrosine hydroxylase- and gonadotropin-releasing hormone-immunoreactive neurons in the nasal area of human embryos. *Brain Res. Dev. Brain Res.*, 97: 251–259.

von Bartheld, C.S., Rickmann, M.J. and Meyer, D.L. (1986) A light- and electron-microscopic study of mesencephalic neurons projecting to the ganglion of the nervus terminalis in the goldfish. *Cell Tissue Res.*, 246: 63–70.

White, J. and Meredith, M. (1987) Synaptic interactions in the nervus terminalis ganglion of elasmobranchs. *Ann. N.Y. Acad. Sci.*, 519: 33–49.

White, J. and Meredith, M. (1993) Spectral analysis and modelling of ACh and NE effects on shark nervus terminalis activity. *Brain Res. Bull.*, 31: 369–374.

White, J. and Meredith, M. (1995) Nervus terminalis ganglion of the bonnethead shark (*Sphyrna tiburo*): evidence for cholinergic and catecholaminergic influence on two cell types distin-

58

guished by peptide immunocytochemistry. *J. Comp. Neurol.*, 351: 385–403.

Whitlock, K., Stephens, E. and Sanders, L. (2000) Olfactory placode development in zebrafish. *Soc. Neurosci. Abstr.*, 26: 1612.

Wirsig, C.R. and Getchell, T.V. (1986) Amphibian terminal nerve: distribution revealed by LHRH and AChE markers. *Brain Res.*, 385: 10–21.

Wirsig, C.R. and Leonard, C.M. (1986) Acetylcholinesterase and luteinizing hormone-releasing hormone distinguish separate populations of terminal nerve neurons. *Neuroscience*, 19: 719–740.

Wirsig, C.R. and Leonard, C.M. (1987) Terminal nerve damage impairs the mating behavior of the male hamster. *Brain Res.*, 417: 293–303.

Wirsig-Wiechmann, C.R. (1990) The nervus terminalis in the chick: a FMRFamide-immunoreactive and Ache-positive nerve. *Brain Res.*, 523: 175–179.

Wirsig-Wiechmann, C.R. (1993) Peripheral projections of nervus terminalis LHRH-containing neurons in the tiger salamander, *Ambystoma tigrinum. Cell Tiss. Res.*, 273: 31–40.

Wirsig-Wiechmann, C.R. (1997) Nervus terminalis lesions: II. enhancement of lordosis induced by tactile stimulation in the hamster. *Physiol. Behav.*, 61: 867–871.

Wirsig-Wiechmann, C.R. and Basinger, S.F. (1988) FMRFamide-immunoreactive retinopetal fibers in the frog, *Rana pipiens*: demonstration by lesion and immunocytochemical techniques. *Brain Res.*, 449: 116–134.

Wirsig-Wiechmann, C.R. and Ebadifar, B. (2002) The naris muscles in tiger salamander II. Innervation as revealed by enzyme histochemistry and immunocytochemistry. *Anat. Embryol.*, 205: 181–186.

Wirsig-Wiechmann, C.R. and Holliday, K.R. (2002) The naris muscles in tiger salamander I. Potential functions and innervation as revealed by biocytin tracing. *Anat. Embryol.*, 205: 169–179.

Wirsig-Wiechmann, C.R. and Jennes, L. (1993) Gonadotropin-releasing hormone agonist binding in tiger salamander nasal cavity. *Neurosci. Lett.*, 160: 201–204.

Wirsig-Wiechmann, C.R. and Lee, C.E. (1999) Estrogen regulates gonadotropin-releasing hormone in the nervus terminalis of *Xenopus laevis. Gen. Comp. Endocrinol.*, 115: 301–308.

Wirsig-Wiechmann, C.R. and Lepri, J.J. (1991) LHRH-immunoreactive neurons in the pterygopalatine ganglion of voles: a component of the nervus terminalis? *Brain Res.*, 568: 289–293.

Wirsig-Wiechmann, C.R. and Oka, Y. (2002) The terminal nerve ganglion cells project to the olfactory mucosa in the dwarf gourami. *Neurosci. Res.*, 44: 337–341.

Wirsig-Wiechmann, C.R. and Wiechmann, A.F. (2001) The prairie vole vomeronasal organ is a target for gonadotropin-releasing hormone. *Chem. Senses*, 26: 1193–1202.

Wirsig-Wiechmann, C.R. and Wiechmann, A.F. (2002) The vole retina is a target for gonadotropin-releasing hormone. *Brain Res.*, 950: 210–217.

Wirsig-Wiechmann, C.R., Wiechmann, A.F. and Delay, R.J. (2000) GnRH modulates rodent chemosensory neuron responses to odors. *Soc. Neurosci. Abstr.*, 26: 2199.

Witkin, J.W. (1987) Immunocytochemical demonstration of luteinizing hormone-releasing hormone in optic nerve and nasal region of fetal rhesus macaque. *Neurosci. Lett.*, 79: 73–77.

Witkin, J.W. and Silverman, A.J. (1983) Luteinizing hormone-releasing hormone (LHRH) in rat olfactory systems. *J. Comp. Neurol.*, 218: 426–432.

Wray, S., Grant, P. and Gainer, H. (1989) Evidence that cells expressing luteinizing hormone-releasing hormone mRNA in the mouse are derived from progenitor cells in the olfactory placode. *Proc. Natl. Acad. Sci. USA*, 86: 8132–8136.

Wysocki, C.J. (1979) Neurobehavioral evidence for the involvement of the vomeronasal system in mammalian reproduction. *Neurosci. Biobehav. Rev.*, 3: 301–341.

Yamamoto, N. and Ito, H. (2000) Afferent sources to the ganglion of the terminal nerve in teleosts. *J. Comp. Neurol.*, 428: 355–375.

Yamamoto, N., Oka, Y. and Kawashima, S. (1997) Lesions of gonadotropin-releasing hormone-immunoreactive terminal nerve cells: effects on the reproductive behavior of male dwarf gouramis. *Endocrinology*, 65: 403–412.

Yamamoto, N., Ito, H. and Oka, Y. (2001) Glutamate may be a co-transmitter of terminal nerve GnRH neurons. In: I. Parhar and Y. Oka (Organizers), *Second International Symposium on the Comparative Biology of GnRH: Molecular Forms and Receptors.*

I.S. Parhar (Ed.)
Progress in Brain Research, Vol. 141
© 2002 Published by Elsevier Science B.V.

CHAPTER 5

Angiogenesis in association with the migration of gonadotropic hormone-releasing hormone (GnRH) systems in embryonic mice, early human embryos and in a fetus with Kallmann's syndrome

Marlene Schwanzel-Fukuda *, Donald W. Pfaff

Laboratory of Neurobiology and Behavior, The Rockefeller University, 1230 York Avenue, Box 275, New York, NY 10021, USA

Introduction

The purpose of this review is to examine the theoretical notion that development of the vasculature in the head, and in particular, in the nasal mesenchyme, may play a role in the migration and ultimately, in the function of the gonadotropin-releasing hormone (GnRH), or luteinizing hormone-releasing hormone (LHRH) systems in mammals. From early embryogenesis, fetal development and birth, through puberty and generation of the reproductive cycles in both sexes, a close association is consistently found between the GnRH cells and the blood vessels (Silverman, 1988). The GnRH neurosecretory cells, in response to blood borne secretions of estrogen by the ovaries or of testosterone by the testes, release their hormone into the vasculature of the primary portal plexus of the median eminence. Circulating through the sinusoids of the anterior pituitary gland, the GnRH is taken up by receptors on the gonadotrops which evoke the secretion of luteinizing hormone (LH) or of follicle-stimulating hormone (FSH) into

the bloodstream. Every cell along this vascular route is exposed to these hormones, but only those cells in the ovaries and the testes, which contain the required receptors, are activated. In response, steroid hormones are released into the blood. In turn, the steroid hormones are bound by receptors in neuroendocrine cells and directly by receptors and in the gonadotrops of the anterior pituitary gland. The vasculature is thus an integral and essential part of the GnRH systems, from the nascent formation of the migration route and the origin of GnRH cells in the olfactory placode through the actual mediation of reproductive functions (Pfaff and Schwanzel-Fukuda, 1995).

Observations in mice

(a) Formation of the cellulovascular bridge and aggregate below the rostral tip of the forebrain visualized by N-CAM immunoreactive axons and cell bodies from the epithelium of the olfactory placode. (b) The initial appearance of GnRH-immunoreactive neurons in the epithelium of the olfactory pit. (c) Organization of the GnRH cell migration route along the N-CAM-immunoreactive axons in contact with the rostral forebrain is concomitant with the formation of the olfactory bulb.

In embryonic mice, on days 9 through 10 of gestation (22 to 26 pairs of somites are present) the ol-

---
* Correspondence to: M. Schwanzel-Fukuda, Laboratory of Neurobiology and Behavior, The Rockefeller University, 1230 York Avenue, Box 275, New York, NY 10021, USA. Tel.: +1 (212) 327-8661; Fax: +1 (212) 327-8664; E-mail: schwanm@mail.rockefeller.edu

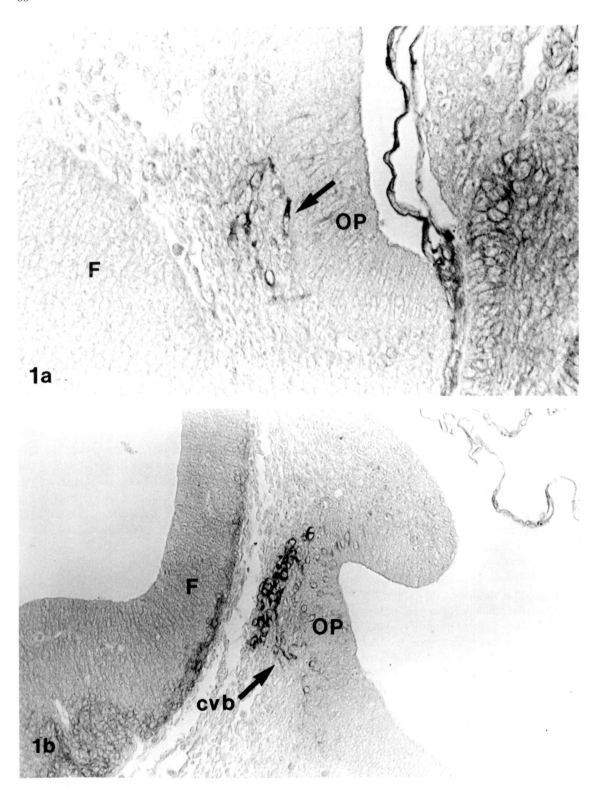

1a

1b

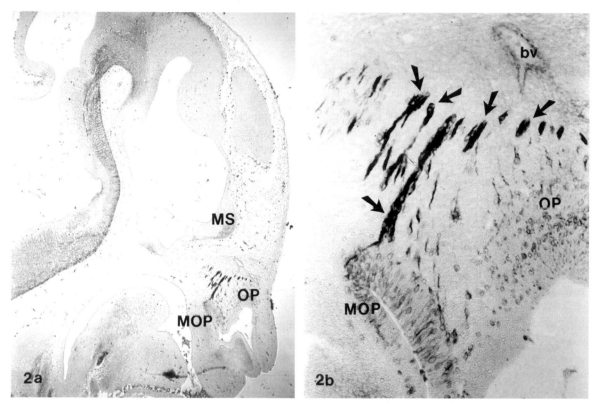

Fig. 2. (a) Photomicrograph of an 8 μm sagittal section through the head of a 12-day-old embryonic mouse at low (a) and higher (b) magnification. This section is just on midline, between the forebrain vesicles and includes a part of the developing medial septal region of the brain (MS), the olfactory pit (OP) and the medial olfactory pit (MOP), the anlage of the vomeronasal organ. (b) The cellulovascular bridge is seen here in its most medial aspect, and N-CAM-immunoreactive axons and cell bodies form a part of the migration route for GnRH neurons (arrow).

factory placodes (thickenings of the ectoderm on the ventrolateral sides of the developing head) invaginate to form simple olfactory pits either side of midline (Fig. 1a, Schwanzel-Fukuda et al., 1992). Soon after, a secondary recess, the anlage of the vomeronasal organ, forms in the medial wall of either olfactory pit. Within the luminal, intermediate and basal layers of the olfactory epithelium N-CAM immunoreactive cells are visible. The N-CAM-immunoreactive ax-

ons, soon followed by N-CAM-immunoreactive cell bodies emerge from the basal layer of this epithelium into the nasal mesenchyme where they aggregate around the numerous small blood vessels (Fig. 1a, b) which are present in this part of the nasal mesenchyme between the olfactory pit and the developing forebrain (Schwanzel-Fukuda et al., 1992, 1996). These N-CAM-immunoreactive axons and cell bodies and the blood vessels make up the cellulovas-

Fig. 1. (a) Photomicrograph of an 8 μm sagittal section through the head of an early 10-day-old embryonic mouse. The olfactory placode has invaginated to form a simple olfactory pit (OP). A fusiform N-CAM-immunoreactive cell (arrow) is seen at the basal surface of the epithelium and in this plane of section five lightly stained N-CAM-immunoreactive cells are seen nearby. At this age the forebrain (F) is comparatively close to the developing olfactory pit. (b) Photomicrograph of an 8 μm section through the head of a 10-day-old embryonic mouse, a little older than that seen in Fig. 1a. The olfactory pit (OP) is more deeply invaginated and several lightly stained-NCAM-immunoreactive cells are seen in the intermediate and basal strata of its epithelium. Darkly stained N-CAM-immunoreactive cells form a cellular aggregate in the nasal mesenchyme. A few N-CAM-immunoreactive fibers (arrow) extend from the epithelium of the olfactory pit into the cellular aggregate, the beginning of the formation of the cellulovascular bridge (CVB).

cular bridge, 'strand' (Bossy, 1980) or 'blastema' (Lejour, 1967) seen in the nasal mesenchyme, coursing from the epithelium of the medial olfactory pit, across the developing nasal septum, into the forebrain (Fig. 2a, b). This cellulovascular bridge (CVB), the medial part of which ultimately forms the migration route for the GnRH neurons, is made up (from medial to lateral) of N-CAM-immunoreactive axons of the nervus terminalis, the vomeronasal and the olfactory nerves (Schwanzel-Fukuda and Pfaff, 1989; Schwanzel-Fukuda et al., 1992). The N-CAM-immunoreactive cell bodies form the cellulovascular aggregate, actually a part of the cellulovascular bridge, seen just below the anlage of the forebrain (Fig. 3). This aggregate of olfactory placode-derived cells includes the N-CAM immunoreactive axons of the olfactory, vomeronasal and terminalis nerves, and the cell bodies of the Schwann cells which accompany them throughout their peripheral distribution. A population of N-CAM immunoreactive ganglion cell bodies associated with the terminalis nerve are seen in the *ganglion terminale* (Huber and Guild, 1913) the largest ganglion of this nerve. It is seen on the most medial side of the cellulovascular aggregate, either side of midline, usually above the anlage of the cribriform plate at later stages (from days 14) of gestation. Smaller ganglia are seen at nodal points along the course of the nervus terminalis across the nasal septum, through the cribriform plate and by three or four short central roots into the medial forebrain and preoptic areas.

GnRH-immunoreactive neurons, in mice, are first detected by immunocyotchemical procedures and antibodies to GnRH (Fig. 4) at about 11 to 11.5 days of embryogenesis (more than 45 pairs of somites are present at this age). The GnRH-immunoreactive neurons are found in the epithelium of the medial part of the olfactory pit, either side of midline (Schwanzel-Fukuda and Pfaff, 1989; Wray et al., 1989). Widely separated from each other throughout the placodal epithelium, as the GnRH neurons emerge into the nasal mesenchyme they form cords along, or surrounded by, N-CAM-immunoreactive axons of the terminalis and vomeronasal nerves and perforating branches of the nasal arteries and veins (Fig. 2a, b). The GnRH cells are never found independent of these axons in the nasal mesenchyme. Thus, they begin their migration across the nasal

septum. In transverse section they can be seen on the most medial parts of the cellular aggregates either side of midline (Fig. 3) and can be followed into the medial sides of the cerebral hemispheres where they enter the brain. In sagittal section the GnRH-immunoreactive cells enter the ventromedial forebrain, caudal to the anlage of the olfactory bulb, and course along the medial surfaces of the cerebral hemispheres to enter the septal and preoptic areas of the brain (Fig. 5a and b). While on the nasal septum, the majority of the nervus terminalis ganglion cells, including those of the large ganglion terminale are N-CAM immunoreactive and appear to outnumber the GnRH cells. By day 12 of embryogenesis in mice (48–52 pairs of somites are present), olfactory axons have made contact with the rostral forebrain and the anlage of the olfactory bulb can be recognized. Soon after, from about day 13.5 (60 pairs of somites present at this age) through day 16, the part of the N-CAM-immunoreactive cellulovascular aggregate that receives the olfactory nerves *en route* to the olfactory bulb develops into the olfactory nerve layer of the olfactory bulb (Schwanzel-Fukuda et al., 1992). Schwann cells are seen among the axons of the olfactory nerves in this layer. The distinctive cellular morphology of the olfactory bulb soon develops, and the accessory olfactory bulb receives the incoming axons of the vomeronasal nerves and a recurrent branch of the nervus terminalis. Three or four central roots of the nervus terminalis, the most medially placed of the olfactory placode-derived axons, enter the septal and preoptic areas of the ventromedial forebrain, caudal to the olfactory bulbs (Fig. 5a, b). In mice, the migration of GnRH cells into the brain is largely over by 16 days of gestation (Schwanzel-Fukuda and Pfaff, 1989).

Formation of the meninges in mice and development of the vasculature

(a) The leptomeningeal (pia and arachnoid matres) plexus of blood vessels along the ventromedial surface of the brain. (b) The pachymeninx (dura mater) and formation of the dural venous sinuses. (c) The initial appearance of nucleated red blood cells.

In 10-day-old embryonic mice (30–34 somites are present at this age), the anterior and posterior neuropores have closed and the neural tube is now

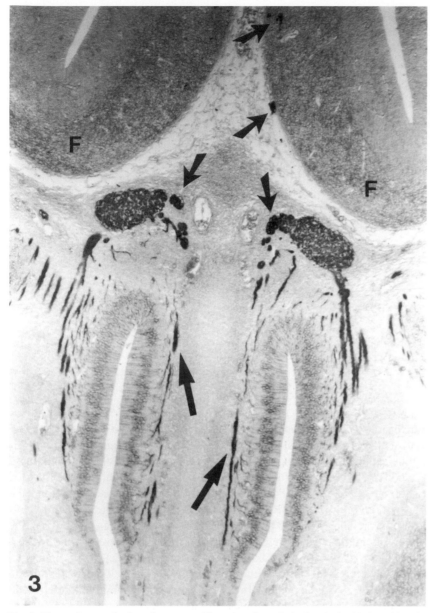

Fig. 3. Photomicrograph of an 8 μm coronal section through the head of a 13-day-old embryonic mouse. At this age the nuclei of the N-CAM immunoreactive cells are seen in the middle or intermediate zone of the epithelia of the olfactory pits, in contrast the location of these nuclei in the 11- to 12-day-old embryonic mouse. The N-CAM-immunoreactive axons and cell bodies are coursing from the lateral and dorsal parts of the placodal epithelium into the N-CAM-immunoreactive cellular aggregates seen just below the anlagen of the forebrain (F). GnRH-immunoreactive cells migrate toward the brain along the medial part of the cellulovascular bridge and aggregate (arrows) and some cell have reached the medial surfaces of the cerebral hemispheres (arrows).

completely closed (Kaufman, 1998). The brain then undergoes a rapid development which is paralleled by the development of the cranial blood vessels, which soon "invest the central nervous system with an extensive vascular plexus" (Theiler, 1989). As early as 8.5 days of gestation in mice (8 somites are present) a vascular anastomosis is present between the two dorsal aortae and the unpaired omphalo-

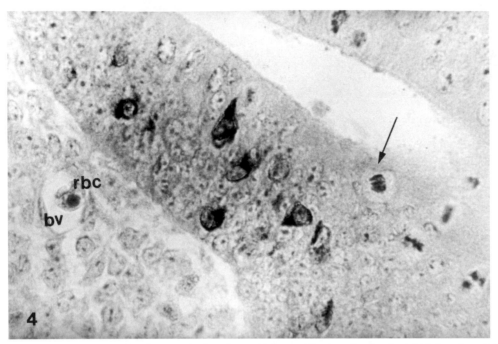

Fig. 4. Photomicrograph of an 8 μm sagittal section, at high magnification, showing the initial detection of GnRH-immunoreactive cells (black) in the epithelium of the medial olfactory pit of an 11.5-day-old embryonic mouse. This section has been lightly stained with cresyl violet. Near the luminal surface, a cell is seen in mitosis (arrow). Small blood vessels (bv) are seen along the basal surface, one of which contains a nucleated red blood cell (rbc).

mesenteric trunk which formed from fusion of the paired vitelline arteries. The embryonic and yolk sac circulations are now in direct communication and a small number of nucleated red blood cells are seen in the embryonic vasculature (Kaufman, 1998). Remarkably, this vascular anastomosis will go on to form the unpaired abdominal aorta, and the singular celiac and superior and inferior mesenteric arteries which branch extensively from it (Theiler, 1989; Kaufman, 1998) to supply blood to the abdominal viscera of the digestive system.

The blood vessels which proliferate in the nasal mesenchyme in mice, from days 9.5 (15 to 20 pairs of somites) on, appear to result from an outgrowth of this investing vascular plexus including branches of the internal carotid arteries which developed from the first and third aortic arch arteries (Fig. 6a and b and see Sadler, 1985). *Thus, angiogenesis rather than de novo synthesis or vasculogenesis in the nasal mesenchyme probably accounts for the proliferation of small vessels in precisely that area of the nasal mesenchyme which receives the emerging N-CAM immunoreactive axons and cell bodies and gives origin to (1) the cellulovascular bridge (CVB) and (2) the cellular aggregate.* We hypothesize that these structures are crucial for dispatching the olfactory, the vomeronasal and the terminalis nerves (and the GnRH neurons) to their appropriate destinations in the brain. Lejour (Lejour, 1967) in a study of the development of the olfactory nerves in the rat, using alkaline phosphatase histochemistry, opines "... it is a mass of neuroblast-like cells to which all the neural fibres originating in the nasal epithelium and the Jacobson organ converge ... this blastema thus appears as a large ganglionic relay in the course of the olfactory fibres."

The loose mesenchyme which surrounds the developing brain and spinal cord, a forerunner of the leptomeninges (the pia and arachnoid maters) is believed by most developmental anatomists to be derived from the neural crest (O'Rahilly and Muller, 1986). The blood vessels which lie directly adjacent to the brain are considered to be the first pial vessels (Figs. 1a, b and 6a, b). With further growth

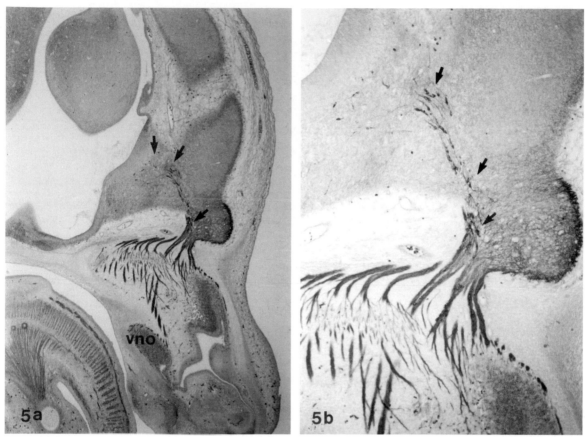

Fig. 5. (a) Photomicrograph of an 8 μm sagittal section through the head of a 14-day-old embryonic mouse. N CAM-immunoreactive cells are seen in the epithelium of the vomeronasal organ (vno), in cords on the nasal septum and entering the ventromedial forebrain with axons of the nervus terminalis, caudal to the olfactory bulbs. (b) On the nasal septum and in arc through the forebrain, GnRH-immunoreactive cells (arrows) accompany the N-CAM-immunoreactive axons. Please see the color plate, Fig. 16 (p. 76), for greater detail of this figure.

a membrane or cellular sheet, the pia mater, develops between the blood vessels and the cerebral wall. The leptomeningeal meshwork develops in the mesenchyme immediately adjacent to the brain, giving rise to a pia-vascular layer and an overlying fluid-filled subarachnoid space beneath the arachnoid layer of the meninges (Weed, 1917). This part of the cranial leptomeningeal meshwork is considered the "central part of the primitive or primary meninx" (O'Rahilly and Muller, 1986). The third layer of the meninges, the pachymeninx or dura mater, is not derived from the neural crest but from the mesoderm. This is the outer most layer of the meninges and it is present in the 14-day-old embryonic mouse (Kaufman, 1998). It lines the cranial vault and dou-

ble layers of the dura mater form the falx cerebri, the tentorium cerebelli and the dural sinuses including the superior, the inferior and the transverse venous sinuses (Theiler, 1989; Kaufman, 1998) which receive the venous blood from the brain and the cerebral spinal fluid from the subarachnoid spaces.

Observations of the development of the GnRH systems and meninges in human embryos. Comparisons with those seen in mice

In human embryos, the development of the N-CAM-immunoreactive migration route: the formation of the cellulovascular bridge (CVB), the cellular aggregate, and the origin and migration of the GnRH

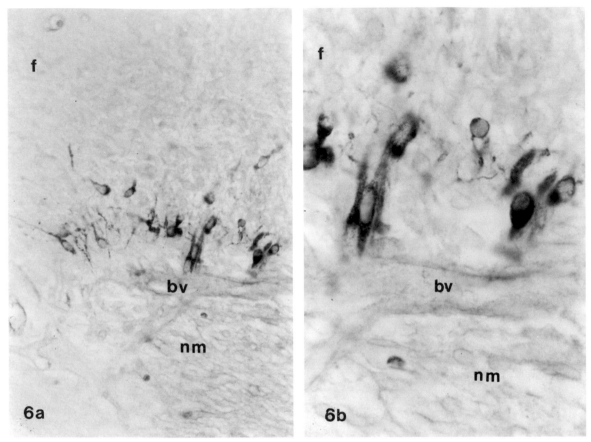

Fig. 6. (a) Photomicrograph of an 8 μm sagittal section through the base of the ventromedial forebrain (f) of a 16-day-old embryonic mouse. At low magnification, a number of GnRH-immunoreactive cells (darkly stained with DAB chromogen, arrows) are seen emerging from the nasal mesenchyme in close association with small blood vessels (bv) containing vascular endothelial growth factor (VEGF)-labelled endothelial cells. (b) At higher magnification, these blood vessels appear to be opening into the parenchyma of the brain (arrows).

neurons into the brain is very similar to that seen in mice and other mammalian embryos. Note, that the GnRH-immunoreactive cells are seen by immunocytochemical procedures entering the ventromedial surface of the brain, caudal to the medial sides of the olfactory bulbs, either side of midline. They enter the brain with the central processes of the terminal (and to some extent the vomeronasal) nerves. While there are species specific differences in the gestational ages of these events, the stages at which particular developments take place in mice and in human embryos are generally equivalent.

Fig. 7. (a) In an 8 μm sagittal section through a whole 28-day-old human embryo, N-CAM-immunoreactivity is seen in all strata of the epithelium of the simple olfactory pit (OP) and a few N-CAM cells have emerged into the nasal mesenchyme (NM). At this age Rathke's pouch is present and open. Numerous small blood vessels are seen in the nasal mesenchyme between the olfactory pit and the developing forebrain (f). At higher magnification of this section (b and c) we trace the association of N-CAM-immunoreactive cells with the developing blood vessels (bv) in the nasal mesenchyme (NM) and the arrangement of the N-CAM cells in the epithelium of the olfactory pit. Note that these N-CAM immunoreactive cells are only present in the deepest part (arrow) of this olfactory pit. In (c) the nucleated red blood cells can be seen in the walls of the small blood vessels (bv) in the nasal mesenchyme. The N-CAM-immunoreactive cells consistently appear to aggregate around these vessels. No GnRH-immunoreactive cells are present in the 28-day-old human embryo.

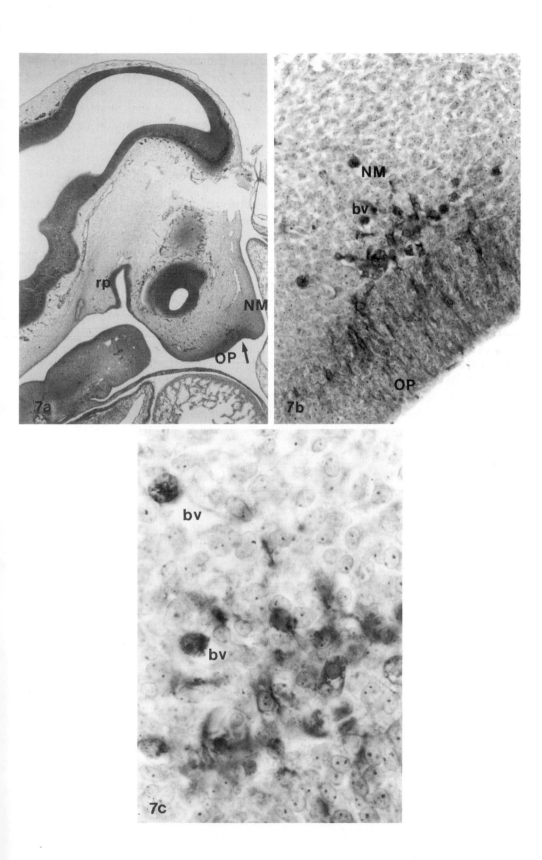

68

Briefly, in human embryos, at about 28 days of gestation (30 or more pairs of somites), both the anterior and the posterior neuropores are closed and the olfactory placode invaginates to form a simple olfactory pit (Fig. 7a–c; and see Pearson, 1941; Schwanzel-Fukuda et al., 1996). As the brain now begins to rapidly differentiate, a vascular meningeal plexus forms along the surface of the brain. At the same time, blood vessels elaborate in the nasal mesenchyme (Weed, 1917) from primitive branches of the first and third aortic arches. These vessels are the precursors of the sphenopalatine branch of the maxillary artery and the anterior and posterior ethmoidal branches of the ophthalmic artery. Branches of these arteries and their corresponding veins form dense arteriovenous anastomoses, actually "... a very fine network in the mucous membrane ... and a rich cavernous plexus well-marked over the inferior concha and lower part of the {nasal} septum." (Lockhart et al., 1969). The arterial blood is thus shunted rapidly into the venous circulation of the head.

N-CAM-immunoreactivity is visible in the cells in the luminal, intermediate and basal parts of the epithelium of the medial part of the olfactory pit (Fig. 8a–c), a structure which will later form the vomeronasal organ. As in mice, N-CAM-immunoreactive axons, followed by the N-CAM-immunoreactive cell bodies from the epithelium of the olfactory pit, migrate into the nasal mesenchyme (Fig. 9a, b). Clustering around the nasal blood vessels, these axons and cell bodies form a distinctive cellulovascular bridge below the anlage of the rostral forebrain, the medial part of which will eventually serve as the migration route for GnRH cells into the brain in older embryos. The structure of the cellulovascular bridge in older human embryos 32, 42, and 46 days, as studied by light microscopy in this lab (Schwanzel-Fukuda et al., 1996), is larger in proportion to the nasal mesenchyme and appears to be more compact than that seen in mice (compare Fig. 5 and Figs. 9–11). At 42 (but not 28 or 36 days of gestation, GnRH-immunoreactive cells are detected in the epithelium of the medial olfactory pit, with N-CAM-immunoreactive axons on the nasal septum and in ganglia of the nervus terminalis, along with a greater number of N-CAM-immunoreactive cell bodies (Figs. 10–12). However, it's interesting to note that the majority, if not all, of the cells that traverse the cribriform plate and actually enter the medial basal forebrain with the central roots of the nervus terminalis are GnRH-immunoreactive (Boehm et al., 1994; and personal unpublished observations in human embryos). In some species, including the rat, the guinea pig, and the opossum, the GnRH immunoreactive cells are rarely seen in the epithelium of the olfactory pit (Schwanzel-Fukuda and Silverman, 1980; Schwanzel-Fukuda et al., 1985, 1988). Rather, the cords of cells emerge from the epithelium and begin to show GnRH-immunoreactivity on the nasal septum, or as they traverse the cribriform plate and enter the forebrain. It may be that some of the N-CAM-immunoreactive cells are destined to begin synthesis and secretion of GnRH on the nasal septum or as they traverse the cribriform plate and enter the brain with roots of the nervus terminalis (Schwanzel-Fukuda et al., 1996).

One notable difference seen in the 42- and 46-day-old human embryos compared to mice is the distinctive separation of the N-CAM immunoreactive cellulovascular bridge into lateral and medial nerve fascicles or 'plexus-like laminae' (Bossy, 1980) by primitive branches of the anterior cerebral arteries to the nasal regions, either side of midline (Fig. 11a, b). The lateral nerve fascicles or laminae are made up of the axons of the olfactory nerves and the Schwann (ensheathing) cells, which accompany them to the olfactory bulbs (Figs. 9a, b; 10a, b; 11a, b; and see Pearson, 1941). These cells are smaller and the axon

Fig. 8. (a) At 29-30 days of embryogenesis in an 8 μm sagittal section of a human embryo, a larger, more elaborate vascular field is seen in the nasal mesenchyme (NM) between the olfactory pit (OP) and the developing forebrain (f). (b, c) At higher magnifications (compare with the same as seen in Fig. 7b and c) smaller, now diamond shaped N-CAM-immunoreactive cells are seen the epithelium of the olfactory placode (OP). In the vascular field, the blood vessels (bv) are larger and contain many more red blood cells (rbc). The N-CAM immunoreactive cells still aggregate around the blood vessels and at this age fine processes appear to sprout from these cells and form a kind of plexus in the nasal mesenchyme. Fig. 8c shows details of the association between the blood vessels and the N-CAM cell bodies and processes. No GnRH-immunoreactive cells are present in human embryos at these ages.

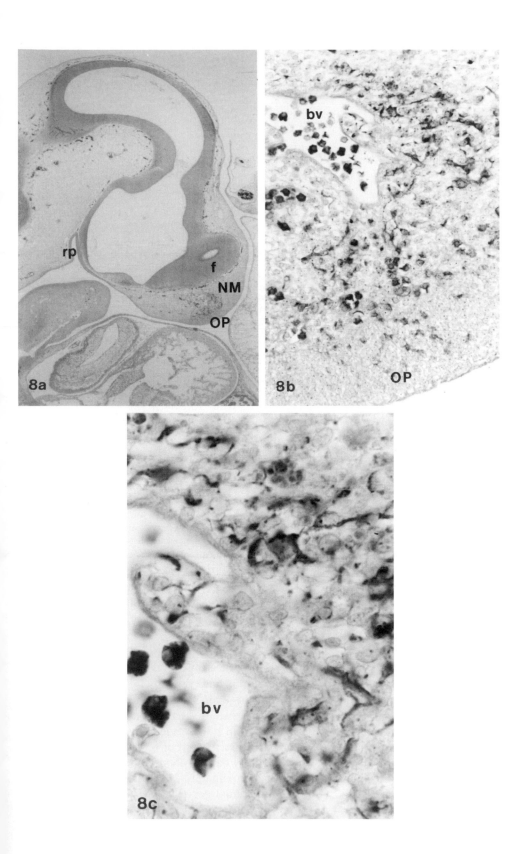

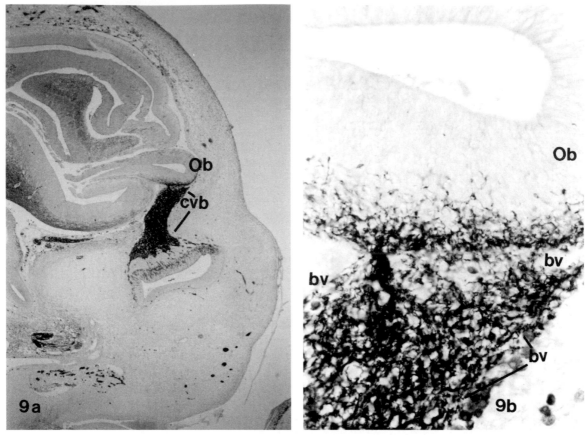

Fig. 9. (a) Low power photomicrograph of an 8 μm sagittal section through a 42-day-old human embryo. A broad, thick band of N-CAM immunoreactive cell bodies and axons is seen in the more lateral parts of the cellulovascular bridge (cvb) and forms a distinctive, dense part of the migration route from the olfactory epithelium, through the nasal mesenchyme. The rostral part of this bridge is in contact with the developing olfactory bulb (Ob). These ingrowing axons are the central processes of the olfactory nerves, and they are accompanied by Schwann cells, which originate in the olfactory placode. These axons form an N-CAM-immunoreactive net-like cap over the anlage of the olfactory bulb. GnRH cells are never seen in these parts of the migration route. (b) At higher magnification, the contact of the N-CAM-immunoreactive axons can be seen in proximity to the surrounding blood vessels (bv).

bundles form a compact course from the epithelium of the olfactory pit into the developing olfactory nerve layer and the anlage of the olfactory bulb. The nerve fascicles or laminae from the medial part of the olfactory epithelium consist of the central processes of the vomeronasal and terminalis nerves (Pearson, 1941). In contrast to the lateral cellulovascular bridge and aggregate, the GnRH- and N-CAM-immunoreactive cells are seen on N-CAM-immunoreactive axons emerging into the nasal mesenchyme leading into the ganglia of the terminalis nerve along the medial part of the nasal septum (Figs. 3, 11a, b and 12). The GnRH and the N-CAM-immunoreactive cells are distinctly larger than the

N-CAM-immunoreactive Schwann cells seen among the axons of the olfactory nerves (Schwanzel-Fukuda et al., 1996).

The basement membrane along the ventral surface of the is breached in several places along the medial basal forebrain and cords of GnRH cells have begun pass into the brain (Fig. 11a, b). There is also evidence that capillaries have begun to vascularize the basal forebrain at these ages (Figs. 7a, b; 8a, b; 11a, b and 13a, b). By the end of the embryonic period, 57 postovulatory days or about 8 weeks, the human embryo is approximately 27–31 mm in length. At this time, the cranial nerves are visible and the choroid plexus is vascularized by deep branches of the an-

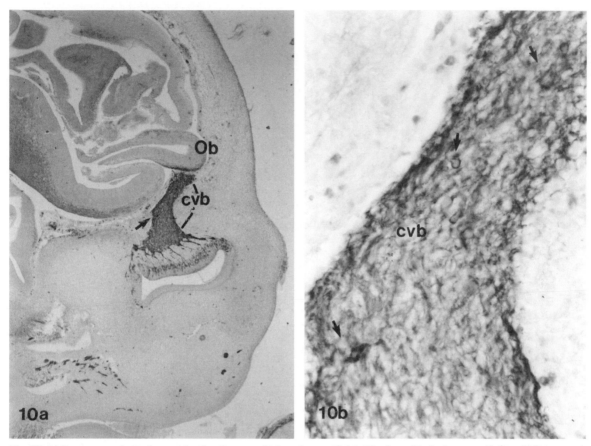

Fig. 10. (a) Low power photomicrograph of an 8 μm sagittal section, medial to Fig. 9a and b, in the same 42-day-old human embryo. In this section, closer to midline, a few GnRH-immunoreactive cells (arrow) are seen among the N-CAM-immunoreactive cell bodies and axons of the cellulovascular bridge (cvb). In this section and in Fig. 9a and b, the connections between the olfactory placodal epithelium and the developing olfactory bulb can be seen clearly. (b) At higher magnification, the GnRH cells (arrows) can be seen along the migration route of the cellulovascular bridge. Please see this figure reproduced in the color plate, Fig. 16 (p. 76), for greater detail.

terior cerebral arteries (Theiler, 1989). The circulus arteriosus or 'circle of Willis' has been present on the ventral surface of the brain since approximately 37 postovulatory days and by 57 days closely resembles the adult pattern of vascularization (O'Rahilly and Muller, 1994).

The search for a single molecular or chemical cue which provides the magical 'Open Sesame' for the movement of GnRH-immunoreactive cells from the nose into the brain has thus far (at least at this writing) shown several promising yet still elusive clues. Schwarting and co-workers (Schwarting et al., 2001) showed that 'Deleted in colorectal cancer (DCC)' a vertebrate receptor for the guidance molecule netrin-1, during development of the olfac-

tory system is expressed in cells migrating from the olfactory epithelium and vomeronasal organ from embryonic days 11 to 14. They also found that DCC is 'downregulated' beginning at day 12 of embryogenesis. Study of GnRH cell migration in transgenic DCC−/− mice showed abnormal trajectories for the GnRH neurons, thus demonstrating that DCC is important in guiding the migration of the GnRH cells to their appropriate destinations. Susan Wray (Wray, 2001) discovered a novel factor which she termed 'nasal embryonic LHRH factor' or NELF which is expressed in both the peripheral and central nervous systems, including olfactory sensory cells and GnRH cells in nasal areas. Results of her study raised the possibility that NELF acts as a common guidance

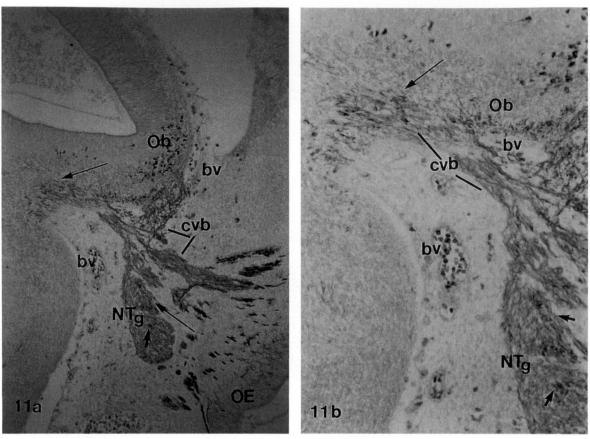

Fig. 11. (a) Low power photomicrograph of an 8 μm sagittal section through the midline brain and nasal regions of a 46-day-old human embryo. In this section, a number of N-CAM-immunoreactive axons and be seen emerging from the olfactory epithelium (OE) onto the nasal septum and extending into contact with the rostral–ventral olfactory bulb. The ganglion terminale of the nervus terminalis (Ntg) is also visible in this section, with its population of GnRH cells (arrow). (b) N-CAM immunoreactive axons accompanied by GnRH cells can be seen entering the ventral–medial forebrain, caudal to the medial surface of the olfactory bulbs. Note that the cellulovascular bridge (cvb) is in continuity with the basal forebrain at this point. The vascular layer (bv) which clearly separates the olfactory from the terminal-vomeronasal complex is visible in these near midline sections.

molecule for olfactory axon projections and the migration of GnRH cells, and thus acts as a migratory signal. All such studies show the importance, as anticipated, of signaling molecules, and suggest that the coordinated interactions of more than one factor may be essential for the complex migration of the GnRH cells into the brain. Part of this complexity may be due to the fact that GnRH is a crucial hormone for development of the reproductive system as well as for essential functions in puberty and adulthood (Silverman, 1988). The trajectories taken by the migrating GnRH cells and the eventual regulated release of GnRH may require a sequence of signals dependent on and changing with the stages of de-

velopment during the embryonic and fetal periods, both morphological and in response to circulating hormones.

Kallmann's syndrome: a genetic defect involving derivatives of the olfactory placode, including GnRH neurons

The genetic basis of Kallmann's syndrome was first described by Kallmann and co-workers (Kallmann et al., 1944). A similar disorder was described by De Morsier in his remarkable study of median cranioencephalic dysraphias and olfactogenital dysplasia (De Morsier, 1974). The puzzle of why this genetic de-

---

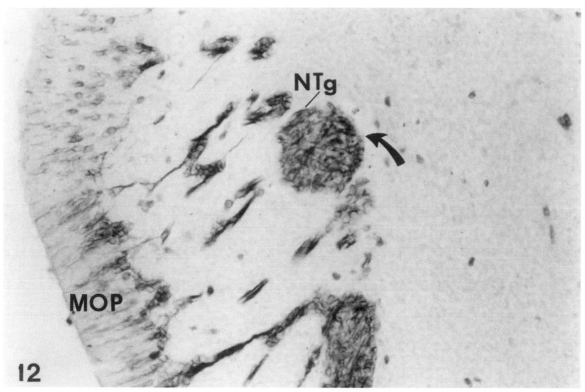

Fig. 12. An 8 μm sagittal section near midline through the nasal mucosa showing the epithelium of the medial olfactory placode (MOP) and the ganglion terminale of the nervus terminalis (NTg). A population of GnRH-immunoreactive cells are visible among the N-CAM-immunoreactive cell bodies and axons in the smaller ganglia seen along the course of the nervus terminalis. Please see the color plate, Fig. 16 (p. 76), for greater detail.

fect involving olfactory placodal derivatives should become evident after they have undergone considerable differentiation is very likely worthy of study. Given our current hypothesis about the role of angiogenesis on development of the GnRH neuronal migration, we are looking back over our material to see if any clues can be found in the patterns of angiogenesis between the Kallmann fetus compared to those of the normal sex and age-matched controls.

In this laboratory, we have had the privilege of examining by immunocytochemical procedures and antibodies to GnRH, a 19-week-old male fetus with Kallmann's syndrome and three normal male fetuses of the same age (Schwanzel-Fukuda et al., 1989). The Kallmann fetus showed an absence of both olfactory bulbs, and no GnRH-immunoreactivity was located in any part of the brain. In striking contrast, thick fascicles of GnRH-immunoreactive axons and

clusters of GnRH neurons were seen deep in the nasal regions, on the nasal septum and along the dorsal surface of the cribriform plate (Fig. 14), and closely applied to either side of the crista galli, deep to the overlying meninges. In the nasal regions, the olfactory, the vomeronasal and the nervus terminalis fibers appeared to have developed normally from the olfactory placode, up to a point. The cribriform plate of the ethmoid bone showed eccentric perforations, and axons of the olfactory nerves ran through the perforations to the dorsal surface of the cribriform plate. Here, either side of midline, the olfactory nerve fibers formed neuromas, tangles of axons on the cribriform plate, deep to the meninges. Importantly, there was no indication of any contact with the forebrain. Deep in the nasal regions, the thick fascicles of GnRH axons and clumps of ganglion cells (Fig. 15) also enforced the impression that development of the vomeronasal and terminalis nerves,

74

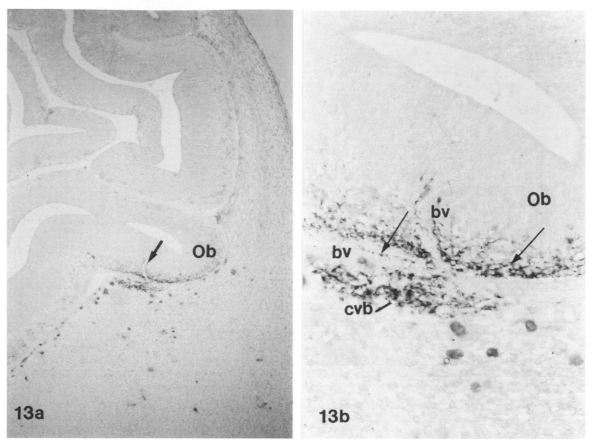

Fig. 13. (a) An 8 μm sagittal section through the brain and nasal regions of a 42-day-old human embryo. N-CAM-immunoreactivity is seen in cells of the leptomeningeal vascular layer along the ventral surface of the developing brain (arrow), and in the walls of a small blood vessel (bv) which has broken through the basement membrane of the ventral surface of the olfactory bulb (OB). (b) At higher magnification a part of the cellulovascular bridge (cvb) and the pial vasculature (arrow) can be distinguished.

and the GnRH-immunoreactive neurons from the olfactory placode, was normal for an initial period of embryonic development.

Over the past months, we have examined Bouin's-fixed, paraffin embedded 8 μm tissue sections of embryonic mice and humans, with antibodies to metalloproteinases (MMP-2) and to endothelial cell markers (Kalebic et al., 1983), including vascular endothelial growth factor (VEGF; Fig. 6a, b) and the tyrosine kinase receptors Flt-1 and Flk-1. Our purpose is to visualize and compare essential patterns in the development of the GnRH systems and the vasculature. One of the strongest coming from these observations is that during early embryogenesis, the areas of the brain undergoing rapid development with neurogenesis and the migration of cells through

a differentiating matrix have very high energy requirements for nutrients and disposal of waste products. Both sets of needs are best met by a rich blood supply.

For 13 years we have hoped to define the sequence of transient events (time-ordered and morphological) which result in the genesis of the GnRH neurons in the olfactory placode and the remarkable phenomenon of their migration along an elaborate 'cellulovascular bridge' from the placodal epithelium across the nasal septum and into the ventromedial forebrain. If the current hypothesis is correct, it will provide a large step toward understanding this intricate and transient system, repeated during embryogenesis in all mammals and most vertebrates.

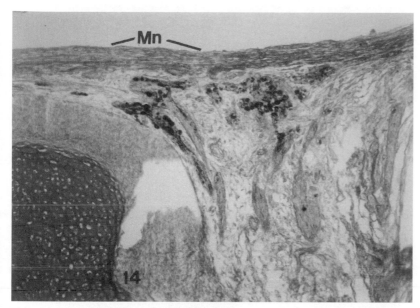

Fig. 14. Nineteen-week-old male fetus with Kallmann's syndrome. An 8 μm sagittal section through the brain and nasal regions of this fetus shows an absence of the olfactory bulb, and an accumulation of GnRH immunoreactive ganglion cell bodies (arrow) and axons on the dorsal surface of the ethmoid bone, deep to meninges (Mn). No GnRH immunoreactivity was detected in any part of the brain. This section was counterstained lightly with cresyl violet.

Fig. 15. Nineteen-week-old male fetus with Kallmann's syndrome. An 8 μm sagittal section shows thick GnRH-immunoreactive axons bundles and cell bodies (arrows) in ganglia of the nervus terminalis in the nasal mucosa. Please see this figure reproduced in the color plate, Fig. 16 (p. 76).

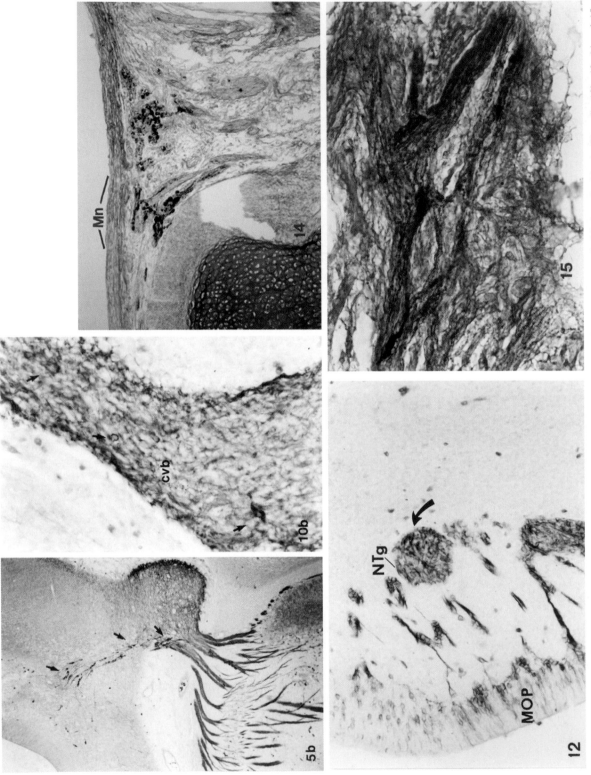

Fig. 16. Color plate to show the double-labeling of GnRH neurons and N-CAM-immunoreactive neurons seen in black and white in Figs. 5b, 10b, 12, 14 and 15. GnRH-immunoreactivity is seen as red–brown reaction product. N-CAM-immunoreactivity is seen as blue–gray reaction product. All sections are lightly counterstained with cresyl violet.

Acknowledgements

This research was supported by a Shannon award from the National Institutes of Health for M.S.-F. All studies in human embryos were carried out with the approval of the Internal Review Board of The Rockefeller University Hospital.

References

Boehm, N., Roos, J. and Gasser, B. (1994) Luteinizing hormone-releasing hormone (LHRH)-expressing cells in the nasal septum of human fetuses. *Dev. Brain Res.*, 82: 175–180.

Bossy, Y. (1980) Development of olfactory and related structures in staged human embryos. *Anat. Embryol.*, 161: 225–236.

De Morsier, G. (1962) Median cranioencephalic dysraphias and olfactogenital dysplasia. *World Neurol.*, 3: 485 504.

Huber, G.C. and Guild, S.R. (1913) Observations on the peripheral distribution of the nervus terminalis in mammalia. *Anat. Rec.*, 7: 253–272.

Kalebic, T., Garbisa, S., Glaser, B. and Liotta, L.A. (1983) Basement membrane collagen: degradation by migrating endothelial cells. *Science*, 221: 281–283.

Kallmann, F., Schoenfeld, W.A. and Barrera, S.E. (1944) The genetic aspects of primary eunuchoidism. *Am. J. Mental Def.*, 48: 203–236.

Kaufman, M.H. (1990) Morphological stages of postimplantation embryonic development. In: A.J. Copp and D.L. Cockroft (Eds.), *Postimplantation Mammalian Embryos, A Practical Approach. The Practical Approach Series*. IRL Press at Oxford University, New York, NY, Chapter 4, pp. 81–91.

Kaufman, M.H. (1998) *The Atlas of Mouse Development*. Academic Press, San Diego, CA.

Lejour, M. (1967) Activite de quatre enzymes dephosphorlants au cours de la morphogenese du palaisprimaire chez le rat (phosphatase alcaline phosphatase acide, ATP-et AMP-ases). *Arch. Biol.*, 78: 389–450.

Lockhart, R.D., Hamilton, G.F. and Fyfe, F.W. (1969) Nasal mucous membrane. In: *Anatomy of the Human Body*, J.P. Lippencott & Co., Philadelphia, PA, pp. 479–480.

O'Rahilly, R. and Muller, F. (1986) The meninges in human development. *J. Neuropathol. Exp. Neurol.*, 45: 588–608.

O'Rahilly, R. and Muller, F. (1994) *The Embryonic Human Brain*. Wiley-Liss Inc., New York, NY.

Pearson, A. (1941) The development of the nervus terminalis in man. *J. Comp. Neurol.*, 75: 39–66.

Pfaff, D.W. and Schwanzel-Fukuda, M. (1995) Development of GnRH neurons: important for the onset of reproductive endocrine and behavioral functions. In: T.M. Plant and P.A. Lee (Eds.), *The Neurobiology of Puberty*, Journal of Endocrinology Limited, Almondsbury, Bristol, pp. 3–13.

Sadler, T.W. (1985) Arterial system, In: *Langman's Medical Embryology, 5th Edition*. Williams and Wilkins, Baltimore, MD.

Schwanzel-Fukuda, M. and Silverman, A.-J. (1980) The nervus terminalis of the guinea pig: a new luteinizing hormone-releasing hormone (LHRH) neuronal system. *J. Comp. Neurol.*, 191: 213–225.

Schwanzel-Fukuda, M., Morell, J.I. and Pfaff, D.W. (1985) Ontogenesis of neurons producing luteinizing hormone-releasing hormone (LHRH) in the rat. *J. Comp. Neurol.*, 238: 348–364.

Schwanzel-Fukuda, M., Fadem, B.H., Garcia, M.S. and Pfaff, D.W. (1988) The Immunocytochemical localization of luteinizing hormone-releasing hormone (LHRH) in the brain and nervus terminalis of the adult and early neonatal gray short-tailed opossum (*Monodelphis domestica*). *J. Comp. Neurol.*, 276: 44–60.

Schwanzel-Fukuda, M., Bick, D. and Pfaff, D.W. (1989) Luteinizing hormone-releasing hormone (LHRH)-expressing cells do not migrate in an inherited hypogonadal (Kallmann) syndrome. *Mol. Brain Res.*, 6: 311 326.

Schwanzel-Fukuda, M., Abraham, S., Crossin, K.L., Edelman, G.M. and Pfaff, D.W. (1992) Immunocytochemical demonstration of neural cell adhesion molecule (N-CAM) along the migration route of luteinizing hormone-releasing hormone (LHRH) neurons in mice. *J. Comp. Neurol.*, 321: 1–18.

Schwanzel-Fukuda, M., Pfaff, D.W., Bouloux, P.M.G., Hardelin, J.-P. and Petit, C. (1996) Migration of luteinizing hormone-releasing hormone in early human embryos. *J. Comp. Neurol.*, 366: 547–557.

Schwanzel-Fukuda, M. and Pfaff, D.W. (1989) Origin of luteinizing hormone-releasing hormone neurons. *Nature (London)*, 338: 161–164.

Schwarting, G.A., Kostek, C., Bless, E., Ahmad, N. and Tobet, S. (2001) Deleted in colorectal cancer (DCC) regulates the migration of luteinizing hormone-releasing hormone neurons to the basal forebrain. *J. Neurosci.*, 21(3): 911–919.

Silverman, A.-J. (1988) The gonadotropin-releasing hormone (GnRH) neuronal systems: Immunocytochemistry, In: E. Knobil and J. Neill (Eds.), *The Physiology of Reproduction*. Raven Press, New York, NY, pp. 1283 1304.

Theiler, K. (1989) *The House Mouse, Atlas of Embryonic Development*. Springer-Verlag Inc., New York, NY.

Weed, L.H. (1917) The cerebro-spinal spaces in pig and man. *Contrib. Embryol.*, 5(14): 7–133.

Wray, S. (2001) Molecular mechanisms for differentiation and migration of placodally-derived GnRH neurons. Assoc. Chemoreception Sciences, Annual Meeting. Abstract published in *Chem. Senses*, Sarasota, FL.

Wray, S., Nieburgs, A. and Elkabes, S. (1989) Spatiotemporal cell expression of luteinizing hormone-releasing hormone in the prenatal mouse: evidence for an embryonic origin in the olfactory placode. *Dev. Brain Res.*, 46: 309–318.

I.S. Parhar (Ed.)
Progress in Brain Research, Vol. 141
© 2002 Elsevier Science B.V. All rights reserved

CHAPTER 6

Recent advances in the pathogenesis of Kallmann's syndrome

Pierre-Marc Bouloux *, Youli Hu, Gavin MacColl

Centre for Neuroendocrinology, Royal Free and University College Medical School, London NW3 2QG, UK

Introduction

Kallmann's syndrome (KS: olfactogenital dysplasia) represents the prototype of a human developmental defect caused by a disorder of axonal pathfinding and defective GnRH neuronal ontogeny. Affecting 1 in 8,000 males, and 1 in 40,000 females, olfacto-genital dysplasia is characterised by absence of olfaction (anosmia), due to olfactory bulb and tract agenesis and hypogonadotrophic hypogonadism (HH, the functional consequence of failed hypothalamo-hypophyseal gonadotrophin releasing hormone (GnRH) secretion. Although a largely sporadic condition, a significant proportion of KS is familial, with pedigrees demonstrating autosomal dominant, recessive and X-linked modes of inheritance (X-KS) (Oliveira et al., 2001; Quinton et al., 2001).

Genetics of KS

Distinct phenotypic differences exist between the autosomal and X-linked modes of transmission: In addition to anosmia and HH, X-KS is characterized by bimanual synkinesis (upper body mirror movements: 85%) and unilateral renal agenesis (33%). In autosomal KS, additional phenotypes include occasional midline developmental defects such as cleft palate,

choanal atresia, and coloboma (Oliveira et al., 2001; Quinton et al., 2001). Pedigree analysis in isolated XKS pedigrees, and positional cloning approaches led to the identification of the *KAL1* gene in 1991 (OMIM: 308700). This encodes a 680 aa modular extracellular membrane associated protein (anosmin-1) comprising an N terminal whey acidic protein (WAP) like domain, followed by four consecutive fibronectin type III like repeats, and a histidine-rich C terminus. *KAL1* is expressed in the mitral and tufted cells of the developing human and chick olfactory bulbs; these secrete anosmin-1 into the surrounding extracellular matrix suggesting a fundamental role in bulb histogenesis (Legouis et al., 1993; Duke et al., 1995; Hardelin et al., 1999). The additional autosomal genes, *KAL2* and *KAL3* (OMIM: 147950), have remained elusive because of infertility in patients, and therefore a lack of suitable pedigrees for study (MacColl et al., 2002).

Pathophysiology of KS

The pathophysiology of X-KS has been the most extensively studied hitherto. Endocrine profiling, histological/histopathological explorations and observations on comparative GnRH ontogeny, and finally, MRI imaging and electrophysiological investigations have given insights into the pathogenesis of this disease (Legouis et al., 1993; Kirk et al., 1994; Lutz et al., 1994; Duke et al., 1995, 1998; Schwanzel-Fukuda et al., 1996; Krams et al., 1997, 1999; Mayston et al., 1997; Quinton et al., 1997b; Hardelin et al., 1999; Schwanzel-Fukuda,

---

* Correspondence to: P.-M. Bouloux, Centre for Neuroendocrinology, Royal Free and University College Medical School, London NW3 2QG, UK. Fax: +44 0207-431-6435; E-mail: pmgb@rfc.ucl.ac.uk

1999; Deeb et al., 2001; Oliveira et al., 2001; Wray, 2001; Pitteloud et al., 2002).

Endocrine profiling demonstrates very low or absent gonadotrophin secretion in KS patients, but preserved pituitary responsiveness to exogenous GnRH, indicative of hypothalamic GnRH deficiency (Oliveira et al., 2001; Pitteloud et al., 2002), the consequence of a critical reduction in the 2,000 or so hypothalamic GnRH neurons, which normally discharge pulsatile GnRH into the median eminence capillary loops. Moreover, detailed immunohistochemical examination of a 19 week *KAL1* deleted foetus revealed that GnRH neurons, and olfactory, n. terminalis and vomeronasal nerves axonal terminals (cranial nerve 1 complex) failed to reach the brain, instead terminating their journey from their olfactory placodal origin at the upper nasal sub-cribriform area. A pathogenetic cranial nerve 1 complex elongation pathway defect was suggested with, a consequent failure of olfactory bulb induction and ascending GnRH neuronal migratory arrest interpreted as representing a secondary event (Schwanzel-Fukuda, 1999). Postnatal nasal biopsies in X-KS demonstrate immature olfactory epithelial neuronal bodies — similar to appearances seen following experimental bulbectomy (Schwob et al., 1993), and these, together with the identification of persistent GnRH neurons in olfactory epithelial biopsies confirm the normal ontogeny of the early olfactory placodal structures in X-KS (Quinton et al., 1997a).

GnRH neuronal ontogeny

GnRH neuronal ontogeny and its interrelationship with peripheral olfactory pathways is remarkably preserved across a broad range of vertebrate species, from: amphibians, (Muske and Moore, 1990), fish (Chiba et al., 1994) to chick, quail (Murakami et al., 1991), rats and primates (Schwanzel-Fukuda et al., 1996; Terasawa et al., 2001). Both GnRH and olfactory receptor neurons originate in the olfactory placodes, and olfactory bulb genesis is induced by pioneer olfactory axonal contact with telencephalic vesicles (Gong and Shipley, 1995). Disruption of this common pathway, not unexpectedly disturbs both olfactory and GnRH neuronal development. This is reinforced by brain MRI of patients confirming defective olfactory bulb formation and hypoplasia of secondary processing areas including the entorhinal cortex (Quinton et al., 1996; MacColl et al., 2002).

XKS and synkinesis ('mirror movements')

Persistent mirror movements are seen in about 85% of X-KS patients. Focal magnetic stimulation of the motor cortex hand area in patients with synkinesis has revealed fast conducting bilateral corticospinal projections from each motor cortex, and cross correlation analysis of multi-unit EMGs recorded during simultaneous voluntary sustained activation of homologous left and right pairs of distal upper limb muscles confirms the presence of a common drive to left and right homologous motor neuron pools, demonstrating a novel ipsilateral corticospinal tract (Krams et al., 1997, 1999; Mayston et al., 1997). Statistical analysis of pooled white matter data from structural T1 weighted MR images using SPM-96 software in autosomal vs. X-linked KS patients also demonstrated a hypertrophied corticospinal tract in the X-KS cohort, consistent with the abnormal development of ipsilateral corticospinal tract fibres (Krams et al., 1999). Changes in regional cerebral blood flow with $H_2^{15}O$-PET during an externally paced finger voluntary thumb/finger opposition task in the same individuals revealed ipsilateral M1 cortical activation in XKS supporting in addition the presence of an abnormal transcallosal pathway. However, it is also possible that sensory feedback from the involuntary mirroring hand may have contributed to these findings (Krams et al., 1997).

The above investigations suggest a defect in axonal guidance with aberrant decussation of the pyramidal corticospinal pathways KS — a hypothesis consistent with demonstration of the *KAL1* transcript in the 45 post-fertilization developing human spinal cord (Duke et al., 1995).

In summary, loss of *KAL1* function in X-KS leads to two separate neurological defects, in the developing olfactory system and in the descending corticospinal tracts. Both appear to be caused by axon targeting defects.

KAL1 and kidney development

Anosmin-1 demonstrates pleiotropism during embryonic development, and *KAL1* and anosmin-1 are also expressed during nephrogenesis at the ureteric bud/metanephric blastema interface, possibly participating in reciprocal mesenchymal–epithelial cell interactions necessary for organogenesis (Duke et al., 1998; Hardelin et al., 1999; Woolf, 2001). In 67% of males with X-KS, this defect is compensatable, with apparent preservation of normal renal function (Kirk et al., 1994); this is not invariant however, as we have recently provided evidence that the solitary kidney is frequently dysplastic leading to with proteinuria, hypertension (Duke et al., 1998) and early renal failure (Deeb et al., 2001). Adhesive interactions between the developing mesenchymal–epithelial cell boundaries are important for glomerular development and several of the missense mutations which cause X-KS alter the adhesion of kidney epithelial cells to an immobilised recombinant anosmin-1 (Soussi-Yanicostas et al., 1998; Robertson et al., 2001), supporting a putative 'adhesive–inductive' role for anosmin-1 in these interactions.

Cross-species 'anosmins'

While cellular adhesion and axonal guidance appear to represent intrinsic properties of anosmin-1, akin to other molecules that influence development of the olfactory and GnRH systems, the developing corticospinal tracts and the kidneys (Dellovade et al., 1998a,b; Allen et al., 1999, 2000; Kramer and Wray, 2000; Schwarting et al., 2001; Woolf, 2001; Yokoyama et al., 2001) the absence of a murine model of KS has limited experimental progress in defining the biological actions of anosmin-1 and the characterisation of interacting ligands.

KAL-1 is of fundamental importance in the ontogeny of these systems in humans, reflected by the effects of loss of function mutations/deletions (Oliveira et al., 2001). *KAL-1* homologues have also been identified in *Drosophila* to *Caenorhabtidis,* zebra fish, chick, and macaque (Schultz et al., 1998, 2000). All of these cross-species 'anosmins' contain a WAP domain, followed by one or more Fn3 domains — a domain combination that has been conserved during vertebrate and invertebrate evolution.

The *C. elegans* orthologue of *KAL-1* (*CeKal-1*) is expressed in a subset of neurons, as well as the excretory canal and uterine lumen (O. Hobert, pers. comm). Loss of function of the *CeKal1* generates a

number of low to medium penetrance phenotypes, including: ventral enclosure defects, ray (tail) abnormalities in males and neurite outgrowth/axon targeting defects (Rugarli et al., 2002). Neurite outgrowth defects include an extra branching dendrite in EF3 neurons that terminate prematurely and do not appear to make contact with their targets in male nematodes (Rugarli et al., 2002; Bulow personal communication). A similar effect is seen in nematodes overexpressing *CeKal1*, where neuron-specific overexpression induced dosage-dependent branching defects and axon misrouting (Bulow, personal communication). Studies in *C. elegans* also support a role for *KAL1* in cell adhesion, as a loss of function of *CeKal1* causes malformations in adherent junctions between epithelial cells leading to ventral enclosure defects (Rugarli et al., 2002).

What determines localisation of anosmin-1, and what genetic pathways influence its actions?

Heparan sulphate proteoglycans (HSPG) localise anosmin-1 to the surface of cells expressing protein (Soussi-Yanicostas et al., 1996) and binding of anosmin to heparan sulphate side chains are also important for its biological activity, as the phenotypic abnormalities resulting from overexpression of the *CeKal-1* homologue in *C. elegans* are nullified in genetic backgrounds deficient in heparan side chain synthesis and modification (Bulow, personal communication; Hobert, personal communication).

Molecular modelling studies of the four Fn3 domains from human anosmin-1 molecule reveal several basic regions, which may interact with heparan sulphate side chains (Robertson et al., 2001). Consistent with these observations, a deletion mutant of the Fn3 domains abolishes the biological activity of *CeKal1*, as does a point mutant in a region of the first Fn3 domain conserved between Human, Chick and *C. elegans* (Bulow et al., 2002). The physical interactions between anosmin-1 and heparan sulphate are clearly important, and currently being investigated in our laboratory.

Conclusion

In summary, at least three genes cause the X-linked and autosomal forms of KS. The study of X-KS pedigrees indicate that loss of *KAL1* function produces neurological defects caused by abnormal axonal targeting in the olfactory and corticospinal regions, and morphogenetic abnormalities in the developing kidneys. The conservation of domain structure in cross-species 'anosmins' indicate that protein function is also conserved at the molecular and cellular level. The *C. elegans* studies support this hypothesis, with loss of function and overexpression of mutants causing abnormalities in neurite outgrowth, axon targeting and cell–cell contacts during gastrulation/ventral enclosure. Future studies in our laboratory aim to investigate the function of *D. melanogaster Kal-1* (*DKal1*) by engineering a loss of function mutant, identify the specific sites of *DKal1* expression, overexpress the protein in these regions and identify interacting proteins. We believe this may subsequently enable us to identify interactants in the human genome data base thereby revealing potential autosomal loci for KS. Ultimately such genes would be validated as candidates for KS by mutational screening of autosomal KS pedigrees.

References

Allen, M.P., Zeng, C., Schneider, K., Xiong, X., Meintzer, M.K., Bellosta, P., Basilico, C., Varnum, B., Heidenreich, K.A. and Wierman, M.E. (1999) Growth arrest-specific gene 6 (Gas6)/adhesion related kinase (Ark) signaling promotes gonadotropin-releasing hormone neuronal survival via extracellular signal-regulated kinase (ERK) and Akt. *Mol. Endocrinol.*, 13: 191–201.

Allen, M.P., Xu, M., Zeng, C., Tobet, S.A. and Wierman, M.E. (2000) Myocyte enhancer factors-2B and -2C are required for adhesion related kinase repression of neuronal gonadotropin releasing hormone gene expression. *J. Biol. Chem.*, 275: 39662–39670.

Bulow H.E., Berry, K.L., Topper, L.H., Peles, H. and Hobert, O. (2002) Heparan sulphate proteoglycan-dependant induction of axon branching and axon misrouting by the Kallmann syndrome gene *Kal-1*. *Proc. Natl. Acad. Sci.*, 99(9): 6346–6351.

Chiba, A., Oka, S. and Honma, Y. (1994) Ontogenetic development of gonadotropin-releasing hormone-like immunoreactive neurons in the brain of the chum salmon, Oncorhynchus keta. *Neurosci. Lett.*, 178: 51–54.

Deeb, A., Robertson, A., MacColl, G., Bouloux, P.M., Gibson, M., Winyard, P.J., Woolf, A.S., Moghal, N.E. and Cheetham, T.D. (2001) Multicystic dysplastic kidney and Kallmann's syndrome: a new association? *Nephrol. Dial. Transplant.*, 16: 1170–1175.

Dellovade, T.L., Pfaff, D.W. and Schwanzel-Fukuda, M. (1998a) Olfactory bulb development is altered in small-eye (Sey) mice. *J. Comp. Neurol.*, 402: 402–418.

Dellovade, T.L., Pfaff, D.W. and Schwanzel-Fukuda, M. (1998b) The gonadotropin-releasing hormone system does not develop in Small-Eye (Sey) mouse phenotype. *Dev. Brain Res.*, 107: 233–240.

Duke, V.M., Winyard, P.J., Thorogood, P., Soothill, P., Bouloux, P.M. and Woolf, A.S. (1995) KAL, a gene mutated in Kallmann's syndrome, is expressed in the first trimester of human development. *Mol. Cell Endocrinol.*, 110: 73–79.

Duke, V., Quinton, R., Gordon, I., Bouloux, P.M. and Woolf, A.S. (1998) Proteinuria, hypertension and chronic renal failure in X-linked Kallmann's syndrome, a defined genetic cause of solitary functioning kidney. *Nephrol. Dial. Transplant.*, 13: 1998–2003.

Gong, Q. and Shipley, M.T. (1995) Evidence that pioneer olfactory axons regulate telencephalon cell cycle kinetics to induce the formation of the olfactory bulb. *Neuron*, 14: 91–101.

Hardelin, J.P., Julliard, A.K., Moniot, B., Soussi-Yanicostas, N., Verney, C., Schwanzel-Fukuda, M., Ayer-Le Lievre, C. and Petit, C. (1999) Anosmin-1 is a regionally restricted component of basement membranes and interstitial matrices during organogenesis: implications for the developmental anomalies of X chromosome-linked Kallmann syndrome. *Dev. Dyn.*, 215: 26–44.

Kirk, J.M., Grant, D.B., Besser, G.M., Shalet, S., Quinton, R., Smith, C.S., White, M., Edwards, O. and Bouloux, P.M. (1994) Unilateral renal aplasia in X-linked Kallmann's syndrome. *Clin. Genet.*, 46: 260–262.

Kramer, P.R. and Wray, S. (2000) Novel gene expressed in nasal region influences outgrowth of olfactory axons and migration of luteinizing hormone-releasing hormone (LHRH) neurons. *Genes Dev.*, 14: 1824–1834.

Krams, M., Quinton, R., Mayston, M.J., Harrison, L.M., Dolan, R.J., Bouloux, P.M., Stephens, J.A., Frackowiak, R.S. and Passingham, R.E. (1997) Mirror movements in X-linked Kallmann's syndrome. II. A PET study. *Brain*, 120(7): 1217–1228.

Krams, M., Quinton, R., Ashburner, J., Friston, K.J., Frackowiak, R.S., Bouloux, P.M. and Passingham, R.E. (1999) Kallmann's syndrome: mirror movements associated with bilateral corticospinal tract hypertrophy. *Neurology*, 52: 816–822.

Legouis, R., Lievre, C.A., Leibovici, M., Lapointe, F. and Petit, C. (1993) Expression of the KAL gene in multiple neuronal sites during chicken development. *Proc. Natl. Acad. Sci. USA*, 90: 2461–2465.

Lutz, B., Kuratani, S., Rugarli, E.I., Wawersik, S., Wong, C., Bieber, F.R., Ballabio, A. and Eichele, G. (1994) Expression of the Kallmann syndrome gene in human fetal brain and in the manipulated chick embryo. *Hum. Mol. Genet.*, 3: 1717–1723.

MacColl, G., Quinton, R. and Bouloux, P.M. (2002) GnRH neuronal development: insights into hypogonadotrophic hypogonadism. *Trends Endocrinol. Metab.*, 13: 112–118.

Mayston, M.J., Harrison, L.M., Quinton, R., Stephens, J.A., Krams, M. and Bouloux, P.M. (1997) Mirror movements in X-linked Kallmann's syndrome. I. A neurophysiological study. *Brain*, 120(Pt 7): 1199–1216.

Murakami, S., Seki, T., Wakabayashi, K. and Arai, Y. (1991) The ontogeny of luteinizing hormone-releasing hormone (LHRH) producing neurons in the chick embryo: possible evidence for migrating LHRH neurons from the olfactory epithelium expressing a highly polysialylated neural cell adhesion molecule. *Neurosci. Res.*, 12: 421–431.

Muske, L.E. and Moore, F.L. (1990). Ontogeny of immunoreactive gonadotropin-releasing hormone neuronal systems in amphibians. *Brain Res.*, 534: 177–187.

Oliveira, L.M., Seminara, S.B., Beranova, M., Hayes, F.J., Valkenburgh, S.B., Schipani, E., Costa, E.M., Latronico, A.C., Crowley, W.F. Jr. and Vallejo, M. (2001) The importance of autosomal genes in Kallmann syndrome: genotype-phenotype correlations and neuroendocrine characteristics. *J. Clin. Endocrinol. Metab.*, 86: 1532–1538.

Pitteloud, N., Hayes, F.J., Boepple, P.A., DeCruz, S., Seminara, S.B., MacLaughlin, D.T. and Crowley, W.F. Jr. (2002) The role of prior pubertal development, biochemical markers of testicular maturation, and genetics in elucidating the phenotypic heterogeneity of idiopathic hypogonadotropic hypogonadism. *J. Clin. Endocrinol. Metab.*, 87: 152–160.

Quinton, R., Duke, V.M., de Zoysa, P.A., Platts, A.D., Valentine, A., Kendall, B., Pickman, S., Kirk, J.M., Besser, G.M., Jacobs, H.S. and Bouloux, P.M. (1996) The neuroradiology of Kallmann's syndrome: a genotypic and phenotypic analysis. *J. Clin. Endocrinol. Metab.*, 81: 3010–3017.

Quinton, R., Hasan, W., Grant, W., Thrasivoulou, C., Quiney, R.E., Besser, G.M. and Bouloux, P.M. (1997a) Gonadotropin-releasing hormone immunoreactivity in the nasal epithelia of adults with Kallmann's syndrome and isolated hypogonadotropic hypogonadism and in the early midtrimester human fetus. *J. Clin. Endocrinol. Metab.*, 82: 309–314.

Quinton, R., Hasan, W., Grant, W., Thrasivoulou, C., Quiney, R.E., Besser, G.M. and Bouloux, P.M. (1997b) Gonadotropin-releasing hormone immunoreactivity in the nasal epithelia of adults with Kallmann's syndrome and isolated hypogonadotropic hypogonadism and in the early midtrimester human fetus. *J. Clin. Endocrinol. Metab.*, 82: 309–314.

Quinton, R., Duke, V.M., Robertson, A., Kirk, J.M., Matfin, G., de Zoysa, P.A., Azcona, C., MacColl, G.S., Jacobs, H.S., Conway, G.S., Besser, M., Stanhope, R.G. and Bouloux, P.M. (2001) Idiopathic gonadotrophin deficiency: genetic questions addressed through phenotypic characterization. *Clin. Endocrinol. (Oxford)*, 55: 163–174.

Robertson, A., MacColl, G.S., Nash, J.A., Boehm, M.K., Perkins, S.J. and Bouloux, P.M. (2001) Molecular modelling and experimental studies of mutation and cell-adhesion sites in the fibronectin type III and whey acidic protein domains of human anosmin-1. *Biochem. J.*, 357: 647–659.

Rugarli, E.I., Di Schiavi, E., Hilliard, M.A., Arbucci, S., Ghezzi, C., Facciolli, A., Coppola, G., Ballabio, A. and Bazzicalupo, P. (2002) The Kallmann syndrome gene homolog in C. elegans is involved in epidermal morphogenesis and neurite branching. *Development*, 129: 1283–1294.

Schultz, J., Milpetz, F., Bork, P. and Ponting, C.P. (1998) SMART, a simple modular architecture research tool: identification of signaling domains. *Proc. Natl. Acad. Sci. USA*, 95: 5857–5864.

Schultz, J., Copley, R.R., Doerks, T., Ponting, C.P. and Bork, P. (2000) SMART: a web-based tool for the study of genetically mobile domains. *Nucleic Acids Res.*, 28: 231–234.

Schwanzel-Fukuda, M. (1999) Origin and migration of luteinizing hormone-releasing hormone neurons in mammals. *Microsc. Res. Tech.*, 44: 2–10.

Schwanzel-Fukuda, M., Crossin, K.L., Pfaff, D.W., Bouloux, P.M., Hardelin, J.P. and Petit, C. (1996) Migration of luteinizing hormone-releasing hormone (LHRH) neurons in early human embryos. *J. Comp. Neurol.*, 366: 547–557.

Schwarting, G.A., Kostek, C., Bless, E.P., Ahmad, N. and Tobet, S.A. (2001) Deleted in colorectal cancer (DCC) regulates the migration of luteinizing hormone-releasing hormone neurons to the basal forebrain. *J. Neurosci.*, 21: 911–919.

Schwob, J.E., Szumowski, K.E., Leopold, D.A. and Emko, P. (1993) Histopathology of olfactory mucosa in Kallmann's syndrome. *Ann. Otol. Rhinol. Laryngol.*, 102: 117–122.

Soussi-Yanicostas, N., Hardelin, J.P., Arroyo-Jimenez, M.M., Ardouin, O., Legouis, R., Levilliers, J., Traincard, F., Betton, J.M., Cabanie, L. and Petit, C. (1996) Initial characterization of anosmin-1, a putative extracellular matrix protein synthesized by definite neuronal cell populations in the central nervous system. *J. Cell Sci.*, 109(Pt 7): 1749–1757.

Soussi-Yanicostas, N., Faivre-Sarrailh, C., Hardelin, J.P., Levilliers, J., Rougon, G. and Petit, C. (1998) Anosmin-1 underlying the X chromosome-linked Kallmann syndrome is an adhesion molecule that can modulate neurite growth in a cell-type specific manner. *J. Cell Sci.*, 111(Pt 19): 2953–2965.

Terasawa, E., Busser, B.W., Luchansky, L.L., Sherwood, N.M., Jennes, L., Millar, R.P., Glucksman, M.J. and Roberts, J.L. (2001) Presence of luteinizing hormone-releasing hormone fragments in the rhesus monkey forebrain. *J. Comp. Neurol.*, 439: 491–504.

Woolf, A.S. (2001) The life of the human kidney before birth: its secrets unfold. *Pediatr. Res.*, 49: 8–10.

Wray, S. (2001) Development of luteinizing hormone releasing hormone neurones. *J. Neuroendocrinol.*, 13: 3–11.

Yokoyama, N., Romero, M.I., Cowan, C.A., Galvan, P., Helmbacher, F., Charnay, P., Parada, L.F. and Henkemeyer, M. (2001) Forward signaling mediated by ephrin-B3 prevents contralateral corticospinal axons from recrossing the spinal cord midline. *Neuron*, 29: 85–97.

SECTION II

GnRH receptors

I.S. Parhar (Ed.)
Progress in Brain Research, Vol. 141
© 2002 Elsevier Science B.V. All rights reserved

CHAPTER 7

Differences in structure–function relations between nonmammalian and mammalian GnRH receptors: what we have learnt from the African catfish GnRH receptor

Marion Blomenröhr [1,*], Jan Bogerd [1], Rob Leurs [2], Henk Goos [1]

[1] *Department Experimental Zoology, Utrecht University, Padualaan 8, 3584 CH, Utrecht, The Netherlands*
[2] *Department Pharmacochemistry, Free University, De Boelelaan 1083, 1081 HV, Amsterdam, The Netherlands*

Introduction

Mammalian GnRH receptors differ from other G-protein coupled receptors (GPCRs) in lacking the intracellular C-terminal tail and in showing an exchange of two otherwise highly conserved Asp and Asn residues in TM 2 and 7, respectively. However, the first GnRH receptor characterized from a nonmammalian vertebrate, the African catfish, contains an intracellular C-terminal tail and has Asp residues in TM 2 and 7 (Fig. 1 and Tensen et al., 1997). Subsequently, the cloning of chicken (Troskie et al., 1997), goldfish (Illing et al., 1999), bullfrog (Wang et al., 2001), *Xenopus laevis* (Troskie et al., 2000), *Seriola dumerilii* (manuscript in preparation) and red seabream (manuscript in preparation) GnRH receptors revealed that all nonmammalian GnRH receptors have a C-terminal tail and the Asp/Asp motif in TM 2 and 7. Thus, these unique features described for mammalian GnRH receptors are not found in their nonmammalian counterparts.

Next to structural variations, differences between mammalian and nonmammalian GnRH receptors were found in their pharmacology and their regulation. In the following sections, the results of studies on catfish and mammalian GnRH receptor expression, regulation and activation, and ligand binding will be summarized. To this end, mammalian–nonmammalian chimeric GnRH receptors, site-directed mutagenesis of GnRH receptors, various GnRH analogs and a three-dimensional model of the receptor–ligand complex were used.

The carboxyl-terminal tail

In many GPCRs, the intracellular C-terminal tail has been demonstrated to be important for receptor expression (Iida-Klein et al., 1995; Oksche et al., 1998) and for regulatory processes, e.g. agonist-induced phosphorylation and subsequent receptor desensitization and internalization (Ferguson et al., 1996; Lefkowitz, 1998). The presence of the C-terminal tail in nonmammalian GnRH receptors enabled studies on the role of this C-terminal tail in these receptors as well as the functional consequences of its absence in mammalian GnRH receptors.

Progressive truncations of the C-terminal tail decreased cell surface expression of the catfish GnRH receptor (Blomenröhr et al., 1999), whereas the addition of the catfish C-terminal tail to the naturally tailless rat GnRH receptor resulted in elevated lev-

* Correspondence to: M. Blomenröhr, Department Experimental Zoology, Utrecht University, Tel.: +31 30 253-3084; Fax: +31 30 253-2837;
E-mail: m.blomenrohr@med.uu.nl

88

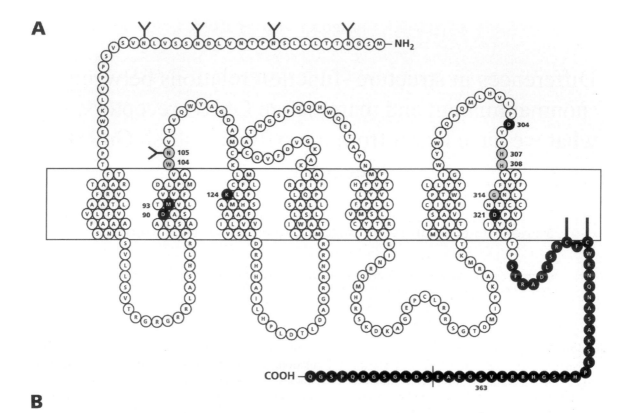

els of receptor expression at the cell surface (Lin et al., 1998). The enhanced expression of mammalian GnRH receptors, fused to the C-terminal tail of the catfish GnRH receptor, in transfected cells could be of value in screening for therapeutically useful GnRH analogs. However, in mammalian gonadotropes, the expression level of the naturally tailless GnRH receptor seems to be sufficient, since it has been demonstrated that occupancy of 20% of GnRH receptors can evoke 80% of the biological response (Naor et al., 1980). In addition to enhancing mammalian GnRH receptor expression, the C-terminal tail of the catfish GnRH receptor has also been used as a linker between the rat GnRH receptor and the green fluorescence protein in order to visualize receptor localization in living cells (Cornea et al., 1999).

Furthermore, it has been demonstrated that the catfish GnRH receptor is susceptible to agonist-induced phosphorylation and that Ser^{363} in the C-terminal tail is the major phospho-acceptor site in this process (Blomenröhr et al., 1999). Mammalian GnRH receptors, on the contrary, are resistant to agonist-dependent phosphorylation due to the lack of a C-terminal tail (Willars et al., 1999). This was substantiated by comparing wild-type mammalian GnRH receptors with chimeras of the mammalian GnRH receptor fused to the C-terminal tail of either the mammalian TRH receptor or the catfish GnRH receptor: only the latter two are phosphorylated in an agonist-dependent manner (Willars et al., 1999). Thus, the presence of a functional C-terminal tail confers agonist-induced GnRH receptor phosphorylation.

Agonist-induced phosphorylation seems to be a prerequisite for desensitization of the catfish GnRH receptor. It undergoes rapid desensitization of the IP response in HEK 293 and COS-7 cells (Heding et al., 1998; Willars et al., 1999), whereas the tailless rat GnRH receptor showed no desensitization of the IP response within seconds to minutes in αT3-1, HEK 293 or COS-7 cells (Anderson et al., 1995; Heding et al., 1998; Willars et al., 1998, 1999). Addition of the C-terminal tail of either the catfish GnRH receptor or the rat TRH receptor to the rat GnRH receptor resulted in rapid desensitization of the IP response (Heding et al., 1998; Willars et al., 1999). Possibly the pattern of GnRH release demands GnRH receptors that are susceptible to desensitization in the catfish pituitary, where GnRH is locally secreted in the vicinity of the gonadotropes, whereas in mammals GnRH is released into the portal system in a pulsatile fashion. Still, mammalian gonadotropin secretion is desensitized upon continous GnRH stimulation, because the absence of mammalian GnRH receptor desensitization is compensated by a desensitizable postreceptor mechanism, i.e. the down regulation of IP_3 receptors and consequent desensitization of GnRH effects on cytosolic calcium (McArdle et al., 1999). On the other hand, it has not been investigated properly whether catfish gonadotropes show rapid desensitization of the gonadotropin secretion because catfish GnRH receptors are susceptible to desensitization within seconds to minutes. It has been demonstrated that continous administration of 10 μM mGnRH analog to perifused catfish pituitaries resulted in a constant amount of released gonadotropins measured

Fig. 1. Schematic side view of the catfish GnRH receptor in the plasma membrane (A). Black circles indicate residues that have been investigated by site-directed mutagenesis, grey circles indicate residues that have been identified by three-dimensional molecular models for the binding of cGnRH-II or cGnRH-II-R8 to the catfish GnRH receptor. Schematic top view of the catfish GnRH receptor as seen from the extracellular side with each helix represented by a circle and the extracellular interconnecting loops (ELs) and the N-terminal domain as solid lines (B). The helices are organized sequentially in a counterclockwise fashion. Extracellular loops 1 and 2 are joined by a disulfide bridge. D^{321} (TM 7) is supposed to be protonated, thereby allowing an interaction with D^{90} in TM 2. Moreover, K^{124} and G^{314} interact with each other in a way that TM 3 and 7 come into close proximity. M^{93} (TM 2) sterically hinders an interaction of D^{90} (TM 2) and K^{124} (TM 3). N^{105} (EL 1) interacts with the carboxyl-terminal $G^{10}-NH_2$, K^{124} (TM 3) with the amino-terminal pE^1, and D^{304} (EL 3) with R^8 of mGnRH as well as with Y^8 of cGnRH-II. The aromatic ring of Y^8 of cGnRH-II is embedded in a pocket formed by the aromatic side chains of residues W^{104}, H^{107} and H^{108} in the receptor. Agonist binding to the receptor triggers phosphorylation of S^{363} in the intracellular carboxyl-terminal tail. S^{363} and the last twelve amino acid residues of the C-terminal tail are important for β-arrestin-dependent receptor internalization. In addition, the C-terminal tail plays a role in cell surface expression of the receptor and ligand binding.

in 10-min interval fractions (De Leeuw et al., 1986). This finding might indicate that catfish gonadotropes are not susceptible to desensitization. However, rapid desensitization occurs within seconds to minutes and a study monitoring gonadotropin secretion during the first 10 minutes has never been performed on catfish gonadotropes.

The process of agonist-induced internalization of the catfish GnRH receptor also depends on receptor phosphorylation, since the removal of the phosphorylation site resulted in impaired receptor internalization compared to the wild-type receptor (Blomenröhr et al., 1999). However, truncated catfish GnRH receptors, still containing the phosphorylation site, also showed slower internalization kinetics than the wild-type receptor. The last 12 amino acids of the C-terminal tail seemed to be important for interaction with accessory proteins like β-arrestin and as such influence agonist-dependent receptor internalization (Blomenröhr et al., 1999). The rat GnRH receptor, on the contrary, has been demonstrated to internalize in a β-arrestin-independent, but dynamin-dependent, manner (Vrecl et al., 1998; Heding et al., 2000). Rat GnRH receptor-internalization kinetics are slower than those of the rat TRH receptor and of the catfish GnRH receptor, but an enhanced internalization rate was achieved under conditions of high β-arrestin when the rat TRH receptor C-terminal tail was added to the rat GnRH receptor (Heding et al., 1998). The β-arrestin-independent part of the catfish GnRH receptor internalization is similar to that of the rat GnRH receptor (Willars et al., 1999). Apart from the C-terminal tail, that is important for β-arrestin-dependent internalization, other residues in the catfish GnRH receptor protein may be involved in the β-arrestin-independent part of receptor internalization, since studies on mutant mammalian GnRH receptors have provided evidence for the importance in the internalization of the mammalian GnRH receptor of conserved amino acids in the DRY/S triplet in the second intracellular loop, a conserved apolar amino acid in the third intracellular loop, and of aromatic amino acids in TM 7 (Arora et al., 1995, 1996, 1997; Chung et al., 1999). Homologous residues in the catfish GnRH receptor may also be involved in the β-arrestin-independent part of catfish GnRH receptor internalization.

Residues in TM 2 and TM 7

The apparent interchange of otherwise highly conserved Asp and Asn residues in TM 2 and 7 of the mammalian GnRH receptors raised the possibility that these two residues interact with each other bringing TM 2 and 7 into close proximity. Studies on mutant mouse GnRH receptors indeed indicated a functional and spatial relationship of these two side chains (Zhou et al., 1994). Data from all other GPCRs studied are in accordance with a spatial proximity of the conserved TM 2 and TM 7 side chains (reviewed in Flanagan et al., 1999). The nonmammalian GnRH receptors, however, have two Asp residues at the homologous positions (Fig. 1 and Tensen et al., 1997). Possibly, the presence of the two Asp residues in nonmammalian GnRH receptors represents an evolutionary intermediate between the conserved Asp/Asn arrangement found in most GPCRs and the Asn/Asp motif of mammalian GnRH receptors. Indeed, crystallographic studies have demonstrated that Asp side chains can occur in spatial proximity within proteins and can form hydrogen bonds when one of the Asp side chains is protonated (Fig. 1) (Davies, 1990; Harrison and Weber, 1994).

It has been described that Asp[90] in TM 2 of the catfish GnRH receptor is implicated in receptor protein expression at the cell surface (Blomenröhr et al., 2001). This is in accordance with findings on the mouse GnRH receptor that also requires the Asn residue at the homologous position in TM 2 for receptor expression (Flanagan et al., 1999). Furthermore, it has been demonstrated for the catfish GnRH receptor that a negatively charged residue in this part of TM 2 is important for ligand binding and signaling (Blomenröhr et al., 2001). It is more likely that Asp[90] is necessary for a proper receptor conformation than for direct interaction with the ligand. Based on a model for the human GnRH receptor it was speculated that Asp[90] in TM 2 interacts with Lys[124] in TM 3 and as such stabilizes the inactive state of the catfish GnRH receptor. However, the role of these two residues in catfish GnRH receptor binding and activation is rather independent (Blomenröhr et al., 2001).

The acidic side chain of Asp[321] in TM 7 does not seem to play a role in either ligand binding or efficient receptor coupling to G proteins (Blomen-röhr et al., 1997). The homologous Asp[318] residue in the mouse GnRH receptor, on the contrary, has been implicated in efficient PLC coupling (Flanagan et al., 1999). Mammalian GnRH receptors have obviously fixed the requirement for an acidic side chain in TM 7, in contrast to most other GPCRs that require an acidic side chain in TM 2 for receptor activation (Flanagan et al., 1999).

In conclusion, it seems that the acidic side chain of Asp[90] in TM 2 of the catfish GnRH receptor is important for catfish receptor functioning, whereas Asp[321] in TM 7 is likely to be protonated. As such, TM 2 and TM 7 of the catfish GnRH receptor are capable of coming into close proximity, as has been postulated for mammalian GnRH receptors as well as for other GPCRs.

Receptor–ligand interactions

Mammalian GnRH receptors have a high affinity for GnRH peptides with a positively charged Arg on position 8 (Millar et al., 1989). It has been demonstrated that the negatively charged Glu[301] of the mouse GnRH receptor plays a role in the recognition of Arg[8] in the ligand (Flanagan et al., 1994). The catfish GnRH receptor contains an Asp residue at the position homologous to Glu[301] in the mouse GnRH receptor (Tensen et al., 1997). However, this nonmammalian GnRH receptor does not show a high affinity for mGnRH that contains an Arg on position 8 (Blomenröhr et al., 1997). To our surprise, mutagenesis studies and a three-dimensional molecular model of the catfish GnRH receptor having cGnRH-II-R8 docked in the binding pocket revealed that Asp[304] of the catfish GnRH receptor nevertheless recognizes Arg[8] in mGnRH and in chimeric GnRH analogs (cfGnRH-R8, cGnRH-II-R8, cGnRH-II-dW6,R8; Blomenröhr et al., 2002; Fig. 1). It is likely that the low affinity of the catfish GnRH receptor for mGnRH is rather due to an unfavorable fit of residue 5 (Tyr) and/or residue 7 (Leu) of this ligand into the binding pocket of the catfish GnRH receptor than to the inability of this receptor to specifically recognize Arg[8] in mGnRH.

Instead of a preference for mGnRH, the catfish GnRH receptor, like other nonmammalian GnRH receptors, shows a high affinity for cGnRH-II (Seal-

fon et al., 1997). Using native and chimeric GnRH analogs varying at positions 5, 7 and 8, we identified His[5], Trp[7] and Tyr[8] as features of cGnRH-II conferring specificity for the catfish GnRH receptor (Blomenröhr et al., 2002). Results of binding studies with the native and chimeric peptides on D304A and D304N mutant catfish GnRH receptors suggested that Asp[304] is not important for the recognition of Tyr[8] in cGnRH-II (Blomenröhr et al., 2002). However, the molecular model for binding of cGnRH-II to the catfish GnRH receptor indicated that Tyr[8] is able to hydrogen bond with Asp[304] in the wild-type receptor (Blomenröhr et al., 2002; Fig. 1). But in the absence of Asp[304] Tyr[8] can also make a hydrogen bond with Ala[195], and the aromatic ring of Tyr[8] is embedded in a pocket formed by the aromatic rings of residues Trp[104] in TM 2 and His[307] and His[308] in TM 7 (Blomenröhr et al., 2002; Fig. 1). Therefore Asp[304] is less important for the recognition of Tyr[8] in cGnRH-II than of Arg[8] in mGnRH.

Molecular dynamics simulations further illustrate that His[5] of cGnRH-II is in close proximity with Gln[295] and His[307] of the catfish GnRH receptor (Blomenröhr et al., 2002), whereas Tyr[5] of mGnRH interacts with Asp[293] (homologous to Gln[295] of the catfish GnRH receptor) in the human GnRH receptor (Hoffmann et al., 2000). The Trp[7] residue of cGnRH-II is embedded in a hydrophobic pocket formed by residues of TM 3, 4 and 7 together with residues of the second extracellular loop (Blomenröhr et al., 2002).

In addition, binding studies on a Lys[124]-mutant catfish GnRH receptor (Blomenröhr et al., 2001) and molecular models for the binding of cGnRH-II to the catfish GnRH receptor (Blomenröhr et al., 2002) implicated Lys[124] (TM 3) as contact site for the amino-terminal pGlu[1] and Asn[105] (EL 1) for the carboxyl-terminal Gly[10]-NH$_2$ (Fig. 1). Moreover, cGnRH-II binds to the catfish GnRH receptor in a constrained β-turn conformation, with Gly[5] facing to the entrance of the binding pocket (Blomenröhr et al., 2002). These results are in accordance with findings on mGnRH binding to mammalian GnRH receptors (Zhou et al., 1995; Davidson et al., 1996; Hoffmann et al., 2000).

Conclusion

The studies on the catfish GnRH receptor are an example as how comparative studies between evolutionary distant animal species may reveal similarities and differences in molecular forms of hormones and their cognate receptors, and of mechanisms of action. That is important to understand the functioning of hormone-controlled processes and to formulate general principles.

References

Anderson, L., McGregor, A., Cook, J.V., Chivers, E. and Eidne, K.A. (1995) Rapid desensitization of GnRH-stimulated intracellular signalling events in αT3-1 and HEK-293 cells expressing the GnRH receptor. *Endocrinology*, 136: 5228–5231.

Arora, K.K., Sakai, A. and Catt, K.J. (1995) Effect of second intracellular loop mutations on signal transduction and internalization of the gonadetropin-releasing hormone receptor. *J. Biol. Chem.*, 270: 22820–22826.

Arora, K.K., Cheng, Z.Y. and Catt, K.J. (1996) Dependence of agonist activation on an aromatic moiety in the DPLIY motif of the gonadotropin-releasing hormone receptor. *Mol. Endocrinol.*, 10: 979–986.

Arora, K.K., Cheng, Z.Y. and Catt, K.J. (1997) Mutations of the conserved DRS motif in the second intracellular loop of the gonadotropin-releasing hormone receptor affect expression, activation, and internalization. *Mol. Endocrinol.*, 11: 1203–1212.

Blomenröhr, M., Bogerd, J., Leurs, R., Schulz, R.W., Tensen, C.P., Zandbergen, M.A. and Goos, H.J.Th. (1997) Differences in structure–function relations between nonmammalian and mammalian gonadotropin-releasing hormone receptors. *Biochem. Biophys. Res. Com.*, 238: 517–522.

Blomenröhr, M., Heding, A., Sellar, R., Eidne, K.A., Leurs, R., Bogerd, J. and Willars, G.B. (1999) Pivotal role for the cytoplasmic carboxyl-terminal tail of a non-mammalian gonadotropin-releasing hormone receptor in cell surface expression, ligand binding, and receptor phosphorylation and internalization. *Mol. Pharmacol.*, 56: 1229–1237.

Blomenröhr, M., Kühne, R., Hund, E., Leurs, R., Bogerd, J. and ter Laak, T. (2001) Proper receptor signalling in a mutant catfish gonadotropin-releasing hormone receptor lacking the highly conserved Asp[90] residue. *FEBS Lett.*, 25059: 1–4.

Blomenröhr, M., ter Laak, T., Kühne, R., Beyermann, M., Hund, E., Bogerd, J. and Leurs, R. (2002) Chimeric gonadotropin-releasing hormone (GnRH) peptides with improved affinity for the catfish GnRH receptor. *Biochem. J.*, in press.

Chung, H.-O., Yang, Q., Catt, K.J. and Arora, K.K. (1999) Expression and function of the gonadotropin-releasing hormone receptor are dependent on a conserved apolar amino acid in the third intracellular loop. *J. Biol. Chem.*, 274: 35756–35762.

Cornea, A., Janovick, J.A., Lin, X. and Conn, P.M. (1999) Simultaneous and independent visualization of the gonadotropin-releasing hormone receptor and its ligand: evidence for independent processing and recycling in living cells. *Endocrinology*, 140: 4272–4280.

Davidson, J.S., McArdle, C.A., Davies, P., Elario, R., Flanagan, C.A. and Millar, R.P. (1996) Asn(102) of the gonadotropin-releasing hormone receptor is a critical determinant of potency for agonists containing C-terminal glycinamide. *J. Biol. Chem.*, 271: 15510–15514.

Davies, D.R. (1990) The structure and function of the aspartic proteinases. *Annu. Rev. Biophys. Biohys. Chem.*, 19: 189–215.

De Leeuw, R., Goos, H.J.Th. and Van Oordt, P.G.W.J. (1986) The dopaminergic inhibition of the gonadotropin-releasing hormone-induced gonadotropin release: an in vitro study with fragments and cell suspensions from pituitaries of the African catfish, *Clarias gariepinus*. *Gen. Comp. Endocrinol.*, 63: 171–177.

Ferguson, S.S.G., Barak, L.S., Zhang, J. and Caron, M.G. (1996) G-protein-coupled receptor regulation: role of G protein-coupled receptor kinases and arrestins. *Can. J. Physiol. Pharmacol.*, 74: 1095–1110.

Flanagan, C.A., Becker, I.I., Davidson, J.S., Wakefield, I.K., Zhou, W., Sealfon, S.C. and Millar, R.P. (1994) Glutamate 301 of the mouse gonadotropin-releasing hormone receptor confers specificity for arginine 8 of mammalian gonadotropin-releasing hormone. *J. Biol. Chem.*, 36: 22636–22641.

Flanagan, C.A., Zhou, W., Chi, L., Yuen, T., Rodic, V., Robertson, D., Johnson, M., Holland, P., Millar, R.P., Weinstein, H., Mitchell, R. and Sealfon, S.C. (1999) The functional microdomain in transmembrane helices 2 and 7 regulates expression, activation, and coupling pathways of the gonadotropin-releasing hormone receptor. *J. Biol. Chem.*, 274: 28880–28886.

Harrison, R.W. and Weber, I.T. (1994) Molecular dynamics simulations of HIV-1 protease with the peptide substrate. *Protein Eng.*, 7: 1353–1363.

Heding, A., Vrecl, M., Bogerd, J., McGregor, A., Sellar, R., Taylor, P.L. and Eidne, K.A. (1998) Gonadotropin-releasing hormone receptors with intracellular carboxyl-terminal tails undergo acute desensitization of total inositol phosphate production and exhibit accelerated internalization kinetics. *J. Biol. Chem.*, 273: 11472–11477.

Heding, A., Vrecl, M., Hanyaloglu, A.C., Sellar, R., Taylor, P.L. and Eidne, K.A. (2000) The rat gonadotropin-releasing hormone receptor internalizes via a β-arrestin-independent, but dynamin-dependent, pathway: addition of a carboxyl-terminal tail confers β-arrestin dependency. *Endocrinology*, 141: 299–306.

Hoffmann, S.H., Ter Laak, A.M., Kühne, R., Beckers, T. and Beckers, T. (2000) Residues lacated in transmembrane helices 2 and 5 of the human gonadotropin-releasing hormone receptor contribute to the binding of agonistic and antagonistic peptides. *Mol. Endocrinol.*, 14: 1099–1115.

Iida-Klein, A., Guo, J., Xie, L.Y., Jüppner, H., Potts, J.T., Kronenberg, H.M., Brinhurts, F.R., Abou-Samra, A.B. and Segre, G.V. (1995) Truncation of the carboxyl-terminal region of the rat parathyroid hormone (PTH)/PTH-related peptide receptor enhances PTH stimulation of adenylyl cyclase but not phospholipase C. *J. Biol. Chem.*, 270: 8458–8465.

Illing, N., Troskie, B.E., Nahorniak, C.S., Hapgood, J.P., Peter, R.E. and Millar, R.P. (1999) Two gonadotropin-releasing hormone receptor subtypes with distinct ligand selctivity and differential distribution in brain and pituitayr in the goldfish (*Carassius auratus*). *Proc. Natl. Acad. Sci. USA*, 96: 2526–2531.

Lefkowitz, R.J. (1998) G protein-coupled receptors: III. New roles for receptor kinases and beta-arrestins in receptor signaling and desensitization. *J. Biol. Chem.*, 273: 18677–18680.

Lin, X., Janovick, J.A., Brothers, S., Blomenröhr, M., Bogerd, J. and Conn, P.M. (1998) Addition of catfish gonadotropin-releasing hormone (GnRH) receptor intracellular carboxyl-terminal tail to rat GnRH receptor alters receptor expression and regulation. *Mol. Endocrinol.*, 12: 161–171.

McArdle, C.A., Davidson, J.S. and Willars, G.B. (1999) The tail of the gonadotrophin-releasing hormone receptor: desensitization at, and distal to, G protein-coupled receptors. *Mol. Cell. Endocrinol.*, 151: 129–136.

Millar, R.P., Flanagan, C., Milton, R.C. and King, J.A. (1989) Chimeric analogues of vertebrate gonadotropin-releasing hormones comprising substitutions of the variant amino acids in position 5, 7, and 8. *J. Biol. Chem.*, 264: 21007–21013.

Naor, Z., Clayton, R.N. and Catt, K.J. (1980) Characterization of gonadotropin-releasing hormone receptors in cultured rat pituitary cells. *Endocrinology*, 107: 1144–1152.

Oksche, A., Dehe, M., Schuelein, R., Wiesner, B. and Rosenthal, W. (1998) Folding and cell surface expression of the vasopressin V_2 receptor: requirement of the intracellular C-terminus. *FEBS Lett.*, 424: 57–62.

Sealfon, S.C., Weinstein, H. and Millar, R.P. (1997) Molecular mechanisms of ligand interaction with the gonadotropin-releasing hormone receptor. *Endocr. Rev.*, 18: 180–205.

Tensen, C.P., Okuzawa, K., Blomenröhr, M., Rebers, F., Leurs, R., Bogerd, J., Schulz, R. and Goos, H. (1997) Distinct efficacies for two endogenous ligands on a single cognate gonadoliberin receptor. *Eur. J. Biochem.*, 243: 134–140.

Troskie, B.E., Sun, Y., Hapgood, J., Sealfon, S.C., Illing, N. and Millar, R.P. (1997) Mammalian GnRH receptor functional features revealed by comparative sequences of goldfish, frog and chicken receptors. *Program of the 79th Annual Meeting of the Endocrine Society (Abstract)*.

Troskie, B.E., Hapgood, J.P., Millar, R.P. and Illing, N. (2000) Complementary deoxyribonucleic acid cloning, gene expression, and ligand selectivity of a novel gonadotropin-releasing hormone receptor expressed in the pituitary and midbrain of *Xenopus laevis*. *Endocrinology*, 141: 1764–1771.

Vrecl, M., Anderson, L., Hanyaloglu, A., Mcgregor, A.M., Groarke, A.D., Milligan, G., Taylor, P.L. and Eidne, K.A. (1998) Agonist-induced endocytosis and recycling of the gonadotropin-releasing hormone receptor: effect of β-arrestin on internalization kinetics. *Mol. Endocrinol.*, 12: 1818–1829.

Wang, L., Bogerd, J., Choi, H.S., Soh, J.M., Chun, S.Y., Seong, J.Y., Blomenröhr, M., Troskie, B.E., Millar, R.P. and Kwon, H.B. (2001) Three distinct types of gonadotropin-releasing hormone receptors characterized in a single diploid species. *Proc. Natl. Acad. Sci. USA*, 98: 361–366.

Willars, G.B., McArdle, C.A. and Nahorski, S.R. (1998) Acute desensitization of phospholipase C-coupled muscarinic M3 receptors but not gonadotropin-releasing hormone receptors co-expressed in αT3-1 cells: implication for mechanisms of rapid desensitization. *Biochem. J.*, 333: 301–308.

Willars, G.B., Heding, A., Vrecl, M., Sellar, R., Blomenröhr, M., Nahorski, S.R. and Eidne, K.A. (1999) Lack of a C-terminal tail in mammalian gonadotropin-releasing hormone receptor confers resistance to agonist-dependent phosphorylation and rapid desensitization. *J. Biol. Chem.*, 274: 30146–30153.

Zhou, W., Flanagan, C., Ballesteros, J.A., Konvicka, K., Davidson, J.S., Weinstein, H., Millar, R.P. and Sealfon, S.C. (1994) A reciprocal mutation supports helix 2 and helix 7 proximity in the gonadotropin-releasing hormone receptor. *Mol. Pharmacol.*, 45: 165–170.

Zhou, W., Rodic, V., Kitanovic, S., Flanagan, C.A., Chi, L., Weinstein, H., Maayani, S., Millar, R.P. and Sealfon, S.C. (1995) A locus of the gonadotropin-releasing hormone receptor that differentiates agonist and antagonist binding sites. *J. Biol. Chem.*, 270: 18853–18857.

I.S. Parhar (Ed.)
Progress in Brain Research, Vol. 141

CHAPTER 8

Regulation of GnRH and its receptor in a teleost, red seabream

Koichi Okuzawa [1,*], Naoki Kumakura [2], Akiko Mori [2], Koichiro Gen [1],
Sonoko Yamaguchi [3], Hirohiko Kagawa [4]

[1] *Inland Station, National Research Institute of Aquaculture, Fisheries Research Agency, 224-1, Hiruta, Tamaki, Watarai, Mie 519-0423, Japan*
[2] *Department of Aquatic Biosciences, Tokyo University of Fisheries, Tokyo 108-8477, Japan*
[3] *Department of Animal and Marine Bioresources Science, Faculty of Bioresources and Bioenvironmental Sciences, Kyushu University, Fukuoka, Fukuoka 812-8581, Japan*
[4] *National Research Institute of Aquaculture, Fisheries Agency, 422-1, Nakatsuhamaura, Nansei, Watarai, Mie 516-0913, Japan*

Introduction

As well as in mammalian species, gonadotropin-releasing hormone (GnRH) is well known as a primary factor which stimulates the synthesis and release of gonadotropin (GTH) from the pituitary in teleosts (see reviews by Peter, 1983; Okuzawa and Kobayashi, 1999). In these two decades, studies on GnRH of teleosts have significantly progressed since Sherwood et al. (1983) discovered a teleost specific GnRH, namely salmon GnRH (sGnRH); molecular forms and localizations of GnRHs in teleosts brain have been intensely studied (see reviews by Amano et al., 1997; Sherwood et al., 1997; Okuzawa and Kobayashi, 1999). However, the regulatory mechanism of the synthesis and release of GnRH in teleosts are not well understood compared to mammals. Therefore, we aimed to reveal the mechanisms underlying the regulation of the synthesis and release of GnRH, in order to help solve several basic and prac-

tical problems in fish reproductive physiology and aquaculture, such as the control of puberty in fish.

Red seabream, *Pagrus* (*Chrysophrys*) *major*, belongs to the order perciformes, the largest and evolutionally the most advanced fish group. It spawns almost everyday even in captivity during its spawning season, thus red seabream is an excellent experimental model for the reproductive biology of fish. In addition, it is one of the most important species for aquaculture in Japan. In this chapter, we will describe our recent achievements in studies on GnRH and GnRH receptor (GnRH-R) in red seabream, and propose possible models for the regulatory mechanisms of the onset of puberty and the seasonal reproduction of this species.

Molecular forms of GnRH and distribution in brain

To date, 14 different forms of GnRH have been identified based on the primary structure or complementary DNA (cDNA) in vertebrates (Carolsfeld et al., 2000; Okubo et al., 2000a; Yoo et al., 2000; Montaner et al., 2001; Adams et al., 2002). Among the 14 forms, 11 GnRHs and 8 GnRHs have been detected in fish and teleosts, respectively. In teleosts, two or three molecular forms of GnRH were found

* Correspondence to: K. Okuzawa, Inland Station, National Research Institute of Aquaculture, Fisheries Research Agency, 224-1, Hiruta, Tamaki, Watarai, Mie 519-0423, Japan. Tel.: +81-596-58-6411; Fax: +81-596-58-6413; E-mail: kokuzawa@fra.affrc.go.jp

in the brain of all species examined, with chicken (c) GnRH-II as a common form (Okuzawa and Kobayashi, 1999). Combined studies using high-performance liquid chromatography (HPLC) and radioimmunoassay (RIA) revealed that three Gn-RHs; namely, sGnRH, cGnRH-II, and seabream (sb) GnRH exist in the brain of red seabream (Fig. 1). Molecular cloning of cDNAs encoding corresponding GnRHs (Okuzawa et al., 1994a, 1997; Kumakura et al., unpublished results) confirmed the existence of these three forms in the red seabream brain. On the other hand, two GnRHs, namely sbGnRH and sGnRH have been detected in the pituitary of red seabream by HPLC/RIA study (Mori et al., unpublished results). Quantification of GnRH with specific RIA revealed that the concentration of sbGnRH in the pituitary is about 2,000 fold higher than sGnRH (Senthilkumaran et al., 1999), suggesting that sbGnRH is more important in terms of GTH secretion from the pituitary.

We have studied the localization of the three GnRHs in the brain of red seabream with immunocytochemistry and *in situ* hybridization (Okuzawa et al., 1997; Okuzawa et al., unpublished results), and found three distinct GnRH neuronal systems in the brain of red seabream, namely the terminal nerve (TN), preoptic, and midbrain GnRH systems, producing sGnRH, sbGnRH, and cGnRH-II, respectively (Fig. 2). The sbGnRH-producing cell bodies in the preoptic area (POA) project their axons mainly to the pituitary, while sGnRH- and cGnRH-II-producing cells project their axons widely in the brain, but not to the pituitary. These morphological characteristics of the GnRH neuronal systems are common to other perciform and pleuronectiform fish species, such as the African cichlid, *Haplochromis burtoni* (White et al., 1995), dwarf gourami, *Colisa lalia* (Yamamoto et al., 1995), tilapia, *Oreochromis mossambicus* (Parhar, 1997), gilthead seabream, *Sparus aurata* (Gothilf et al., 1996), European sea bass, *Dicentrarchus labrax* (Gonzalez-Martinez et al., 2001) and barfin flounder, *Verasper moseri* (Amano et al., 2002a). These results clearly indicate that sbGnRH produced in the preoptic GnRH neuronal system functions as hypophysiotropic form of GnRH (GnRH in its true meaning). While sGnRH and cGnRH-II have neuromodulator or neurotransmitter functions as suggested by Oka (1997).

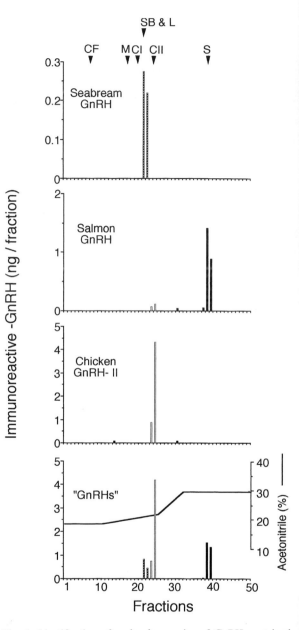

Fig. 1. Identification of molecular species of GnRH contained in the brain of red seabream, *Pagrus major*, by a combined study of high performance liquid chromatography (HPLC) and radioimmunoassays (RIAs). Extract of brains was fractioned by reverse phase HPLC, and each fraction was tested by four kinds of RIA, namely three specific RIAs that are specific to seabream GnRH, salmon GnRH and chicken GnRH-II, and a RIA that can detect several GnRH molecules ('GnRHs'). The arrow heads indicate retention times of synthetic GnRH peptides, CF, catfish GnRH; M, mammalian GnRH; CI, chicken GnRH-I; SB, seabream GnRH; L, lamprey GnRH-I; CII, chicken GnRH-II; S, salmon GnRH.

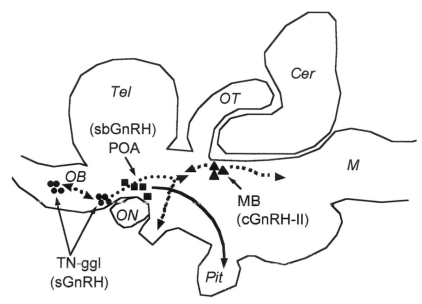

Fig. 2. Schematic illustration of a para-sagittal section of the red seabream, *Pagrus major*, showing the distribution of cell bodies of salmon (s) GnRH (circles), seabream (sb) GnRH (squares) and chicken (c) GnRH-II (triangles), and the projections (arrows). Cer, cerebellum; M, medulla oblongata; MB, midbrain; OB, olfactory bulb; ON, optic nerve; OT, optic tectum; Pit, pituitary; POA, preoptic area; Tel, telencephalon; TN-ggl, ganglions of the terminal nerve. Modified from Okuzawa et al. (1997).

Development of GnRH neuronal systems

At the present time it is well established that GnRH neurons originate in the olfactory organ, and then migrate into the forebrain during the development in mammals, birds and amphibians (see Chapter 3 by Murakami et al. in this volume). Also in teleosts, several reports (Chiba et al., 1994; Parhar et al., 1995; Parhar, 1997; White and Fernald, 1998) have demonstrated that terminal nerve GnRH neurons are derived from the olfactory placode. However, the origin of the three distinct GnRH systems in advanced teleosts, such as perciform fish is still uncertain. We, therefore, studied development of the three distinct GnRH neuronal systems in red seabream by immunocytochemistry and *in situ* hybridization (Ookura et al., 1999). sGnRH immunoreactive (ir) cells were first detected one day after hatching (day 1) in the olfactory pit. Then, sGnRH-ir cells disappeared from the olfactory organ, and they were observed along the olfactory nerve and in the telencephalon (Fig. 3A). cGnRH-II-ir axons and cell bodies were first observed in the midbrain tegmentum on day 3 and day 6, respectively, and the number

of cell bodies and fibers increased with development (Fig. 3B). Immunoreactivity to sbGnRH first emerged in cell bodies in the POA, in axons elongated from the POA to the pituitary, and in the neurohypophysis around day 37 (Fig. 3C). These results suggest that sGnRH, sbGnRH, and cGnRH-II systems have distinct ontogenetic origins, namely the olfactory placode, POA, and midbrain, respectively. Parhar (1997) reported similar results for tilapia. On the other hand, in a perciform African cichlid, it is demonstrated that both sGnRH and sbGnRH neurons migrate from the olfactory placode to their appropriate adult location in the brain, namely the ganglion of the terminal nerve and the POA, respectively, whereas cGnRH-II neurons arise directly from the midbrain ventricle (White and Fernald, 1998). Recently, Amano and co-workers conducted olfactory epithelia lesions with fry of masu salmon, *Oncorhynchus masou*, just after hatching. Fish were sampled 212 days after the operation, and it was found that neurons expressing sGnRH mRNA (a pivotal hypophysiotropic form in salmonoids) were detected in the ventral telencephalon and POA in the control group while there were no such neurons

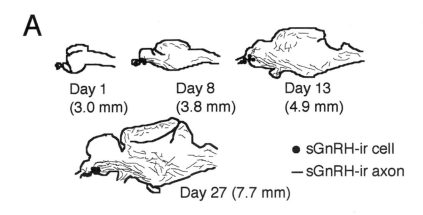

A

Day 1
(3.0 mm)

Day 8
(3.8 mm)

Day 13
(4.9 mm)

● sGnRH-ir cell
— sGnRH-ir axon

Day 27 (7.7 mm)

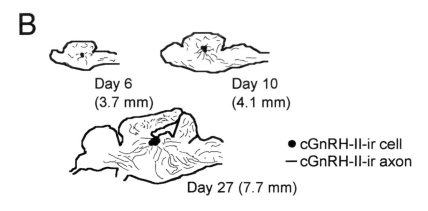

B

Day 6
(3.7 mm)

Day 10
(4.1 mm)

● cGnRH-II-ir cell
— cGnRH-II-ir axon

Day 27 (7.7 mm)

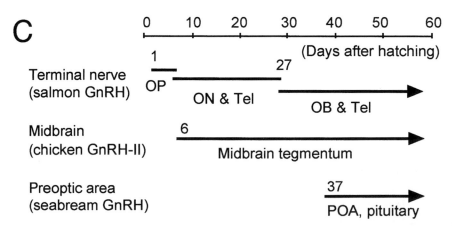

C

0   10   20   30   40   50   60

(Days after hatching)

1   27

Terminal nerve
(salmon GnRH)

OP

ON & Tel

OB & Tel

Midbrain
(chicken GnRH-II)

6

Midbrain tegmentum

Preoptic area
(seabream GnRH)

37

POA, pituitary

Fig. 3. Schematic illustrations of sagittal sections of red seabream, *Pagrus major*, showing the ontogeny of salmon (s) GnRH (A) and chicken (c) GnRH-II (B) neurons, and a summary of the ontogeny of the three GnRH neuronal systems in red seabream (C). Number in parentheses indicates the body length of each specimen. OB, olfactory bulb; ON, olfactory nerve; OP, olfactory pit; POA, preoptic area; Tel, telencephalon.

in those areas in the olfactory epithelium lesioned group (Amano et al., 2002b). This clearly indicates that the origin of the sGnRH neurons in the ventral telencephalon and POA in masu salmon is the olfactory epithelium. A similar experimental approach is necessary to conclude the precise origin of sbGnRH neurons in POA in perciforms, despite it being very difficult because of the tiny size of the fry of perciform fish compared to salmonids.

Seasonal changes in GnRH and GnRH mRNA

Two serious difficulties in the quantification of GnRH peptides or mRNAs occur; one is the multiplicity of GnRH molecular species within the brain of one organism, and the other is the multiple functions of one specific GnRH molecule in distinct neuronal systems (see Okuzawa and Kobayashi, 1999). It is necessary to employ the appropriate quantitative method specific for each GnRH peptide or mRNA species, and to distinguish the same GnRH molecular species belonging to the distinct GnRH neuronal systems in one organism, for example sGnRH in TN and preoptic GnRH neuronal systems in salmonid and cyprinid fish. In these fishes, histological techniques, such as immunocytochemistry or *in situ* hybridization, are the only way to overcome the problem so far, and dividing the brain into several parts is not appropriate. In this section, I will refer to our recent work on red seabream and recent studies in which the two above mentioned difficulties were overcome.

Red seabream is a seasonal spawner; in female red seabream vitellogenesis commences in February and the ovaries attain full maturity in April. Their spawning season begins in April and continues until May, and fish spawn almost every day during the spawning season. A seasonal change in the gonadosomatic index (GSI, gonad weight/body weight × 100) of 2-year-old female red seabream is shown in Fig. 4. Fish were immature and mean GSI was low in October. They had ovaries containing oocytes at the perinucleolus stage. In December, the mean GSI slightly increased and oocytes at the oil stage were observed in their ovaries. In February, the mean GSI further increased, and the most advanced oocytes attained the primary yolk globule stage. In March, the GSI increased rapidly, and the stage of the oocytes proceeded to the secondary yolk globule stage. In

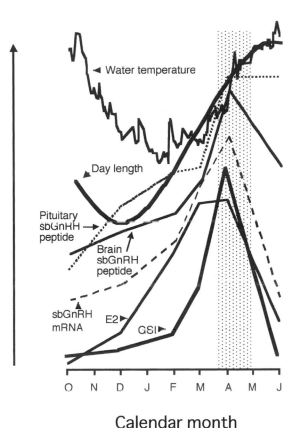

Calendar month

Fig. 4. Schematic illustration of the seasonal fluctuations of environmental factors (water temperature and day length), and gonadosomatic index (GSI), serum levels of estradiol-17β (E$_2$), brain and pituitary levels of seabream (sb) GnRH peptide and brain levels of sbGnRH mRNA in female red seabream, *Pagrus major*. The shaded area indicates the spawning season. Modified from Senthilkumaran et al., 1999 and Okuzawa et al., unpublished results.

April, the spawning season, the mean GSI reached the maximum and all sampled fish had oocytes at various developmental stages from the perinucleolus to the mature stage. In June, after spawning season, the GSI decreased abruptly. Ovaries of all fish were regressed and only oocytes at the perinucleolus stage were observed in the ovaries. Seasonal fluctuations of serum sex steroids, estradiol-17β (E$_2$) (Fig. 4) and testosterone (T) (data not shown) were closely associated with changes in GSI.

In order to obtain basic information for understanding the regulatory mechanism of GnRH, we have studied the seasonal changes in the three native GnRHs (sGnRH, cGnRH-II and sbGnRH) in the brain and pituitary using a specific RIA for each form of GnRH (Senthilkumaran et al., 1999), and the levels of messenger RNA (mRNA) encoding the three GnRHs of female red seabream (Okuzawa et al., unpublished results) using ribonuclease (RNase) protection assay (Fig. 4). The three GnRH neuronal systems in red seabream have distinct localizations and functions, and the three distinct GnRH molecular species are produced in different GnRH neuronal systems (Fig. 2), therefore we can observe fluctuations of activity of each GnRH neuronal system, even if we use the whole brain sample for estimation of GnRH peptide or mRNA.

Both brain and pituitary sbGnRH levels increased from October (immature phase) and reached peaks in April (spawning phase) in correlation with the increase in GSI and vitellogenesis. Brain levels of sbGnRH decreased in June (regressed phase) but the levels were still rather high and similar to levels in March (vitellogenic phase). However, the pituitary levels of sbGnRH remained high in completely regressed fish in June. On the other hand, levels of sGnRH and cGnRH-II in the brain were high in immature (October) and regressed (June) phases, but remained low during the spawning phase (Senthilkumaran et al., 1999). Pituitary levels of sGnRH were much lower than those of sbGnRH and showed no seasonal change, and pituitary cGnRH-II was always below the limit of detection (Senthilkumaran et al., 1999).

The levels of sbGnRH mRNA were correlated with the ovarian development (Fig. 4); the levels of sbGnRH mRNA of immature fish in October and December were low, and increased in February and March in association with active vitellogenesis. The sbGnRH mRNA levels reached maximum levels in April, and after which they rapidly decreased with the ovarian regression in June. In contrast, the levels of sGnRH mRNA showed no variation throughout the year and those of cGnRH-II mRNA displayed only slightly elevated levels in March and April (data not shown). Taken together, the levels of both brain sbGnRH peptide and mRNA were associated with the gonadal status, while seasonal fluctuations of peptides and mRNAs of the other two GnRH forms were rather independent of the gonadal status. Although several reports about the changes in sbGnRH have been published after the discovery

of this third form of GnRH in advanced teleost (Powell et al., 1994), to our knowledge, our work on red seabream is the only example that shows a clear correlation between the brain GnRH peptide or mRNA contents and gonadal maturation. Therefore, together with the morphological observations based on immunocytochemistry and *in situ* hybridization analyses mentioned in the former section (Fig. 2), it is strongly suggested that sbGnRH is the pivotal hypophysiotropic form of GnRH in red seabream. To sum up the evidence showing that sbGnRH is the 'real' GnRH in red seabream: (1) sbGnRH is dominant in the pituitary; (2) solely sbGnRH neurons in the POA project to the pituitary; (3) only sbGnRH peptide in the brain and pituitary increased with gonadal maturation; (4) only sbGnRH mRNA in the brain increased with the gonadal maturation, and decreased with the gonadal regression.

In many studies, the pituitary content of GnRH peptide has been shown to be associated with gonadal maturation; sGnRH in the pituitary increased in parallel with gonadal maturation in rainbow trout, *Oncorhynchus mykiss* (Okuzawa et al., 1990) and masu salmon (Amano et al., 1992, 1993). In a sex changing fish, gilthead seabream (*Sparus aurata*), mature males and females had higher sbGnRH concentrations in the pituitary than bisexual recrudescence fish (Holland et al., 1998). sbGnRH and cGnRH-II levels in the pituitary of maturing striped bass correlated to changes in oocyte diameter and GSI (Holland et al., 2001). In a pleuronectiform fish turbot, *Scophthalmus maximus*, sbGnRH content in pituitary increased in association with the increase in oocyte diameter (Andersson et al., 2001). In our experiments, although sbGnRH levels in the pituitary were correlated with gonadal maturation, the levels remained high in completely regressed fish in June. Similar results were reported in turbot (Andersson et al., 2001) and a viviparous teleost, the grass rockfish (*Sebastes rastrelliger*) (Collins et al., 2001); in post-spawning turbot sbGnRH content in pituitary was high and comparable to mature fish. In grass rockfish, sbGnRH content in brain and pituitary were markedly higher in post-spawn females and regressed males than in other reproductive conditions. Although Collins et al. (2001) explained this phenomenon as a result of a decline in the sbGnRH secretion at the end of reproductive cycle, further studies with different species of teleosts, including viviparous species, are required to explain the physiological meaning of the high sbGnRH levels in the pituitary of regressed fish.

Although the level of mRNA is considered to be a good indicator of the biosynthesis of peptides, only a few studies have been performed about GnRH mRNA levels in teleosts. Amano et al. (1995a) reported that sGnRH mRNA expression in the ventral telencephalon and POA of the brain, detected by *in situ* hybridization, significantly increased during gonadal maturation in masu salmon. Gothilf et al. (1997) reported the preovulatory changes in the levels of three GnRH (sGnRH, sbGnRH and cGnRH-II) mRNAs in female gilthead seabream using RNase protection assay, and found that all three mRNAs fluctuate throughout the day, reaching highest levels 8 h before spawning, concurrent with the preovulatory GTH-II (LH) preovulatory surge. This suggests the involvement of the three GnRHs in the control of ovulation and spawning. Further studies of GnRH mRNA levels will elucidate the mechanism controlling the annual fluctuations of the activities in GnRH neuronal systems.

Regulation of GnRH activities

Regulation of GnRH neuronal activities in teleosts was clearly documented by Yu et al. (1997), therefore in this section, we will concentrate on our recent study in red seabream and recent studies of other investigators. Red seabream is a good experimental model for the study of the regulation of the GnRH neuronal system, because the sbGnRH system of red seabream is localized only in POA and functions only as a hypophysiotropic hormone (namely, GnRH in its true meaning) and does not function as a neurotransmitter within the brain. We will discuss the effect of environmental and physiological factors on the activities of preoptic sbGnRH system.

Environmental factors

It is well accepted that environmental factors, such as photoperiod and water temperature, strongly affect seasonal reproductive cycles in many teleosts (see review by Lam, 1983). We consider that red seabream, a seasonal spawner is not an exception; it is very likely that photoperiod and/or water temperature are critical factors for the control of annual reproductive cycle of this species as well as other spring spawners such as cyprinid fishes (Hanyu et al., 1983).

The abrupt increase in brain sbGnRH peptide and mRNA levels of female red seabream from February to April (Fig. 4) is closely correlated with the rapid increase in day length at the same period (Fig. 4), thus we can hypothesized that the long photoperiod up-regulates the sbGnRH synthesis and in turn enhances GTH secretion from pituitary and gonadal development in red seabream. To test this hypothesis, we have conducted preliminary experiments about the effect of photoperiod on sbGnRH mRNA expression of female red seabream in November, and found no difference in both the gonadal development and the sbGnRH mRNA levels in brain between long and short photoperiod groups (Okuzawa et al., unpublished result). Therefore, photoperiod seems not to be an important factor for gonadal maturation of red seabream at least in females in this season. However, further experiments, for example, conducted in other seasons, are required to conclude that photoperiod is not important. There is no report concerning the relationship between photoperiod and GnRH synthesis except studies on masu salmon (Amano et al., 1995b, 1999), a typical autumn (short photoperiod) spawner, similar to other salmonid fishes. Immature male masu salmon were reared under long (16 hours L: 8 hours D) or short photoperiod (8 hours L: 16 hours D) regimes. After two months, precocious maturation occurred and the number of sGnRH mRNA expressing cells in the ventral telencephalon and POA, determined by *in situ* hybridization, had increased in the short photoperiod group, while fish remained immature and the number of sGnRH expressing cells was unchanged in the long photoperiod group. Furthermore, in castrated masu salmon (Amano et al., 1999), the number of sGnRH mRNA-expressing cells in POA had increased under short photoperiod but not under long photoperiod. Therefore, photoperiod affects sGnRH synthesis in the brain of masu salmon, and this effect is not mediated by gonadal factors such as sex steroids.

Another environmental factor for the reproduction of red sea bream is water temperature. Most likely, similar to spring spawning cyprinids, such as bittering, *Acheilognathus tabira* (Shimizu and Hanyu, 1982) and honmoroko, *Gnathopogon caerulescens* (Okuzawa et al., 1989), increase of water temperature (above 20°C) in late May may inactivate the hypothalamo–hypophysial–gonadal axis, maybe through down-regulation of sbGnRH synthesis, and in turn terminates its spawning season. Probably, high water temperature may keep the red seabream reproductively inactive during summer season. An artificial abrupt decrease of water temperature to the optimum water temperature (from 28°C to 18°C) combined with a long photoperiod (15 hours L: 9 hours D) induced gonadal maturation and spawning in winter, four months earlier than the natural spawning season in red seabream (Kato, personal communication). Therefore, initiation factor for gonadal recrudescence in red seabream is considered to be a decrease of the water temperature to optimum. On the other hand, an increase of water temperature to optimum in combination with serum steroid levels (see below) in spring may function as an accelerator of sbGnRH synthesis (Fig. 4) and in turn pituitary and gonadal activities in red seabream. Red seabream, kept in a higher temperature (16°C) than the ambient water temperature during winter, initiated spawning 20 days earlier than the natural population (Fukusho et al., 1986). To our knowledge, reports on the effect of water temperature on GnRH activities are very scarce. Okuzawa et al. (1994b) reported that the optimum (15°C) water temperature increased sGnRH content in the brain and pituitary of a cyprinid fish honmoroko. This study, however, did not distinguish sGnRH in the preoptic system from sGnRH in the terminal nerve system, which is not essential for gonadal maturation or spawning of cyprinid (Kobayashi et al., 1992, 1994). Further studies are required to elucidate the effect of water temperature on GnRH activities.

Social interactions also affect the activity of the preoptic GnRH neurons; in territorial males of the African cichlid, which are characterized as aggressive and reproductively active, have significantly larger hypothalamic GnRH neurons than non-territorial males. Furthermore, a switch in the social status of an adult male causes a corresponding change in the GnRH neuron size (Francis et al., 1993). Similar situations have been described in a labrid fish, bluehead wrasse (*Thalassoma bifas-*

ciatum); terminal phase males have 2–3-fold more GnRH preoptic cells than primary phase males and females (Grober and Bass, 1991). Quantification of brain GnRH peptide and/or GnRH mRNA will enable a clearer picture to be drawn of the regulation of GnRH by social cues.

Steroids

Besides the environmental factors mentioned above, physiological or internal factors are also important. Amongst physiological factors, gonadal steroids are one of the most probable candidates for the regulator of GnRH neuronal system in teleosts. There are several reports indicating that sex steroids affect the GnRH system in the brain, especially articles on positive feedback in male fish are common. In male and female bluehead wrasse, a sex changing fish, implants of 11-ketotestosterone (11KT), the major androgen of teleosts, induced increases in the GnRH preoptic cell number (Grober et al., 1991). In male masu salmon, administration of 17α-methyltestosterone increased the number of sGnRH-expressing neurons in POA (Amano et al., 1994). In male African catfish, *Clarias gariepinus*, T and E_2 exerted a positive influence on the amount of catfish GnRH (Dubois et al., 1998, 2001). In male tilapia, *Oreochromis niloticus*, E_2 increased the number of preoptic sbGnRH neurons (Parhar et al., 2000). On the other hand, studies in female fish are rather scarce. In female eels, *Anguilla anguilla*, E_2 induced an increase in the levels of brain and pituitary mammalian GnRH, which is the preoptic GnRH of eels (Montero et al., 1995).

In female red seabream, seasonal fluctuations of the levels of E_2 (Fig. 4) and T (data not shown) in serum were parallel to those of the sbGnRH mRNA levels, suggesting that sex steroids affect sbGnRH synthesis via a positive feedback mechanism. To examine this hypothesis, we have conducted several experiments of *in vivo* steroid implants using silastic capsules. Prepubertal (16-months-old) female red seabream were implanted with silastic capsules containing either E_2, 11KT, T, or a solvent (control) in May, and reared under natural environmental conditions for three weeks. Implantation of 11KT increased the brain sbGnRH mRNA levels, estimated by RNase protection assay, while E_2 and T had no

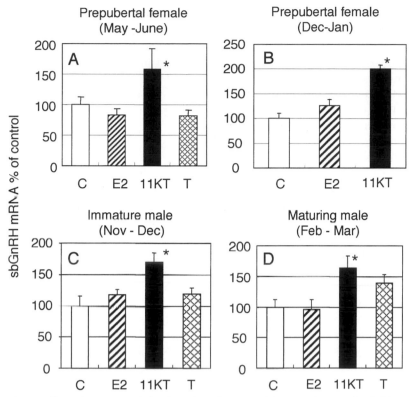

Fig. 5. Effect of a one month implantation of sex steroids on seabream (sb) GnRH messenger RNA levels in the brains of female and male red seabream, *Pagrus major*. C, control; E2, estradiol-17β; 11KT, 11-ketotestosterone; T, testosterone. Bars represent mean ± SEM of 4–6 fish. Asterisks indicate significant differences compared with the control ($P < 0.05$; One-way ANOVA after logarithmic transformation and Duncan's multiple range test).

effect (Fig. 5A). We obtained similar results in another experiment, conducted in December, using the same age females or hermaphrodite red seabream (Fig. 5B). In addition, female red seabream which were artificially matured by the GnRH agonist exhibited no increase in brain sbGnRH mRNA levels despite both serum E_2 and T concentrations being markedly elevated (Kumakura et al., unpublished results). These data suggest that in prepubertal female red seabream E_2 and T, two major sex steroids of females, do not affect the brain sbGnRH mRNA expression, and that 11KT which is contained in female red seabream serum with a low concentration (less than 100 pg/ml, Yamaguchi et al., unpublished results) exerts a positive effect. The physiological role of the positive feedback effect of 11KT on the brain sbGnRH neuronal system in females is still unclear. In immature (Fig. 5C) or maturing (Fig. 5D) male red seabream, on the other hand, implantation

of 11KT for one-month significantly increased the brain sbGnRH mRNA levels, while implants of E_2 or T exerted no effect. 11KT is considered to be the major androgen of male red seabream, therefore, a positive feedback of 11KT on brain sbGnRH mRNA plays an important physiological role in the annual reproductive cycle of males.

Negative feedback effects of sex steroids on the GnRH system have been reported in only one species; in territorial male African cichlids, castration enlarged cell size of POA GnRH neurons (Francis et al., 1992) and 11KT and T prevented the increase of cell size of castrated males (Soma et al., 1996). It is still under debate if the same steroid (11KT) exerts both positive and negative effects on the GnRH activity. It is likely that the status of fish, such as its gonadal maturity, determines the direction (negative or positive) of the feedback effects of sex steroids.

Neurotransmitters

Although, it is still unclear how these environmental and physiological factors influence the reproductive endocrine system, several neurotransmitters, such as monoamines, amino acids or neuropeptides contained in fish brain (see review by Yu et al., 1997) are candidates that mediate information from these factors. Several reviews are available about the regulation of GnRH activities by neurotransmitters (Peter et al., 1991; Yu et al., 1997), therefore we are going to mention only about our recent work on red seabream. As the first step to understand the central regulation of the preoptic sbGnRH neuronal system of red sea bream, we have conducted experiments concerning the effects of serotonin (5-HT), GABA and neuropeptide Y (NPY) on the *in vitro* release of sbGnRH from slices of the preoptic–anterior hypothalamus (P-AH) and pituitary (Senthilkumaran et al., 2001). 5-HT, GABA and NPY all stimulated the release of sbGnRH from the P-AH but not from the pituitary of immature red seabream. They also stimulated sbGnRH release from the P-AH with a similar potency during the course of gonadal development. Specific agonists and/or antagonists of 5-HT, GABA and NPY showed that 5-HT and GABA utilize 5-HT_2 and GABA_A receptor subtypes, respectively, to mediate their action, and that NPY employs at least NPY_{Y1} and NPY_{Y2} receptor subtypes to stimulate sbGnRH release. Furthermore, combinations of different antagonists for 5-HT, GABA and noradrenaline/adrenaline did not block the stimulatory influence of NPY on the release of sbGnRH, and the brain distribution of NPY-ir neurons overlapped with sbGnRH neurons in red seabream (Okuzawa et al., unpublished result). These data indicate that the action of NPY on the sbGnRH neuronal system is probably direct. Positive effects of 5-HT and NPY on *in vitro* GnRH release from P-AH have also been reported in goldfish (Yu et al., 1991; Peng et al., 1993).

Cloning and characterization of the GnRH receptor gene

GnRH exerts its efficacies through binding to the GnRH receptor (GnRH-R), a plasma membrane-bound receptor belonging to the family of G protein-coupled receptors (GPC-Rs). To date, cDNAs encoding GnRH-R have been cloned from several vertebrates including some teleosts. However, still limited information is available as regards the regulatory mechanisms of the expression of GnRH-R in teleosts. Therefore, we cloned a cDNA encoding GnRH-R from the pituitary of the red seabream using a RACE method, and performed some experiments in order to clarify the regulatory mechanisms of GnRH-R, which is important not only for understanding the basic reproductive physiology of teleosts but also for artificial breeding of fishes in aquaculture.

The cloned GnRH-R of the red seabream (PmGnRH-R) cDNA contained an open reading frame of 1245 bps encoding 415 amino acid residues (Okuzawa et al., unpublished results). Hydrophobicity analysis of the deduced amino acid sequence indicated the presence of seven transmembrane-spanning domains, a characteristic of GPC-Rs. Unlike mammalian GnRH-Rs but similar to other non-mammalian GnRH-Rs, the PmGnRH-R contains an intracellular carboxy-terminal tail, indicating that signal transduction mechanisms may in part differ between piscine and mammalian GnRH-Rs. When the PmGnRH-R deduced amino acid sequence was compared to GnRH-Rs isolated from the pituitary of other animals, it has the highest identity (87%) with striped bass GnRH-R (Alok et al., 2000) followed by medaka GnRH-R1 (73%) (Okubo et al., 2001), bullfrog GnRH-R1 (57%) (Wang et al., 2001), *Xenopus* (42%) (Troskie et al., 2000), Japanese eel GnRH-R (42%) (Okubo et al., 2000b), catfish GnRH-R (41%) (Tensen et al., 1997), goldfish GfB (40%), goldfish GfA (37%) (Illing et al., 1999), mouse (34%) (Tsutsumi et al., 1992), and other mammalian GnRH-Rs. A phylogenetic tree, generated by the UPGMA method, revealed that PmGnRH-R was clustered with the GnRH-Rs of evolutionarily advanced fishes, striped bass and medaka, and an amphibian, the bullfrog. On the other hand, GnRH-Rs of phylogenetically older species, catfish, goldfish and eel, were included in another lineage.

Exon/intron organizations of the transcribed regions were determined by sequence analyses of the PmGnRH-R gene. The PmGnRH-R gene is composed of four exons and three introns that are located in the 5′ untranslated region, transmembrane domain

IV and the third cytoplasmic loop. Interestingly, this exon/intron structure is the same as the GnRH-R gene of *Xenopus laevis* (Troskie et al., 2000), and different from medaka GnRH-R1 that possessed only two introns (Okubo et al., 2001).

We characterized PmGnRH-R pharmacologically using a mammalian cell line (293 cells) (Okuzawa et al., unpublished results). A colorimetric assay for measuring the activation of G protein-coupled signaling pathways (Chen et al., 1995) was applied for the characterization of PmGnRH-R. A PmGnRH-R construct was prepared using a mammalian expression vector (pcDNA 3.1), which possessed the CMV promoter, and co-transfected to the 293 cells with a reporter plasmid (pCRE/β-gal), in which the bacterial β-galactosidase (β-gal) gene was under the control of a promoter containing five cyclic AMP response elements (CREs). Transfected cells were incubated for 48 hours and stimulated by several GnRHs including the three native GnRHs of red sea bream and other neuropeptides (isotocin and arginine vasotosin) for 6 hours. Then, the production of β-gal was estimated. The three native GnRHs of red seabream, sGnRH, cGnRH-II, and sbGnRH increased β-gal production in a dose dependent manner. On the other hand, isotocin and arginine vasotosin showed very weak or no stimulation. These data clearly showed that the cloned PmGnRH-R is functional and specific to GnRH. In this assay, sGnRH and cGnRH-II were about 10-fold more effective than sbGnRH, a putative endogenous ligand. Similarly, cGnRH-II was more effective than endogenous preoptic GnRH in a similar pharmacological experiments in *Xenopus* (Troskie et al., 2000), catfish (Tensen et al., 1997), goldfish (Illing et al., 1999), African cichlid (Robison et al., 2001), and medaka (Okubo et al., 2001).

We studied the expression of the PmGnRH-R mRNA in several tissues of red seabream including pituitary, brain, ovary, testis, liver, kidney, heat, stomach, and muscle using RNase protection assay, and found that PmGnRH-R mRNA was expressed only in the pituitary. Hence, it is likely that the endogenous ligand of PmGnRH-R is sbGnRH, a hypophysiotropic form of GnRH among the three native GnRHs in red seabream, which is dominantly contained in the pituitary. However, it cannot be ruled out that small amounts of mRNA of GnRH-R, below the detection limit of the RNase protection assay, were expressed in extrahypophyseal tissues as reported in other teleosts (Illing et al., 1999; Alok et al., 2000; Madigou et al., 2000; Okubo et al., 2000b; Robison et al., 2001).

Seasonal fluctuation of the expression of red sea bream GnRH receptor

As a basis for understanding the regulatory mechanism of GnRH-R expression, we determined the seasonal fluctuation of the PmGnRH-R mRNA levels in the pituitary of red seabream by a quantitative PCR method (Mori et al., unpublished results). PmGnRH-R mRNA levels were maintained at high levels during winter and early spring and reached a peak one (females) or two months earlier (males) than the onset of the spawning season, then the levels suddenly decreased from March to April (spawning season) and reached the lowest levels. After the spawning season, the levels gradually increased from May to August. It is an interesting finding that the expression of mRNA of GnRH-R is the lowest during the spawning season. A negative feedback mechanism may take part in this phenomenon. We will discuss this issue in the next section.

Regulation of the GnRH receptor

How is the mRNA expression of GnRH-R in red seabream controlled? To answer this question, we have conducted experiments about the effect of three possible environmental or physiological factors, namely water temperature, GnRH and sex steroids.

GnRH

Immature juvenile red seabream of one-year-old were implanted intramuscularly with a cholesterol pellet containing GnRH agonist (200 μg/kg body weight). Control fish received a cholesterol pellet without the GnRH agonist. Ten days or 20 days later, the fish were killed and the PmGnRH-R mRNA levels in the pituitaries were estimated using the RNase protection assay (Kumakura et al., unpublished results). The GnRH agonist treated groups showed significantly higher GnRH-R mRNA levels compared to the control groups. This result suggests that GnRH itself is a positive regulator of its own receptor. A study of GnRH-R binding assay in goldfish support this notion; *in vivo* injection of GnRH agonist increased binding capacity of GnRH-R in the pituitary (Omeljaniuk et al., 1989). However, the involvement of gonadal factors, such as steroids or activin is also possible in this positive effect, because the experiments were conducted *in vivo*. Further *in vitro* and/or gonadectomy studies are required to determine if the positive effect of GnRH on GnRH-R expression is direct or indirect.

Sex steroids and water temperature

In December, one-year-old juvenile fish of mixed sex were implanted intraperitoneally with silastic capsules containing E_2, 11KT or solvent. Then, the rearing temperature was increased from 15°C to 20°C, and fish were reared for one month. At the end of the experiment, the mRNA levels of PmGnRH-R in the pituitary were estimated by RNase protection assay (Kumakura et al., unpublished results). The PmGnRH-R mRNA levels of the control group which were implanted with silastic capsule containing solvent was significantly higher than the initial control fish. This suggests that the elevation of water temperature increases the transcription of the PmGnRH-R gene. On the other hand, the levels of PmGnRH-R mRNA of sex steroids (E_2 or 11KT) treated fish were significantly lower than the control fish. This suggests that steroids exert a negative effect on the GnRH-R mRNA expression. Similar positive and negative effects were reported in the three-spined stickleback, *Gasterosteus aculeatus*, and African catfish, respectively, using GnRH-R binding assay. In the three-spined stickleback, an exposure to warm temperature (20°C) in winter increased the binding capacity of GnRH-R in the pituitary irrespective of the photoperiod (Andersson et al., 1992). In African catfish, castration in males resulted in a two-fold increase in the pituitary GnRH receptor binding capacity, and replacement with androstendion reversed the increase of the binding capacity (Habibi et al., 1989).

Besides sex steroids, insulin-like growth factor I (IGF-I) or leptin are other candidates as regulators of GnRH-R, because IGF-I (Huang et al., 1998; Weil et al., 1999) and leptin (Peyon et al., 2001) affect GTH release *in vitro* in teleosts.

Possible model for the regulatory mechanism of the onset of puberty and seasonal reproductive cycle of red seabream

Onset of puberty

Normally, red seabream spawns when they become two-year-old (2+), and underyearling (0+) or one-year-old (1+) red seabream do not spawn even if they are subjected to suitable environmental conditions in the spawning season. In other words, prepubertal 0+ or 1+ fish do not respond to suitable environmental cues for gonadal maturation. Implant of GnRH agonist into 1+ fish, however, induces gonadal maturation and ovulation (Kumakura et al., unpublished results). Therefore, the pituitary–gonad axis is already functional in 1+ fish, while the GnRH neuronal system and/or higher central nervous system (CNS) controlling reproduction is not developed or functional yet in 0+ and 1+ red seabream. Why the reproductive brain, namely the GnRH system and higher CNS, of 0+ to 1+ prepubertal fish does not respond to environmental cues? There are two possibilities; (1) the system of the reproductive brain is not completely developed. (2) an unknown inhibitor mechanism exists in prepubertal fish. Morphologically, the sbGnRH neuronal system of juvenile fish of 37 days after hatching is similar to adult fish, therefore it is likely that the sbGnRH neuronal system of 1+ fish is completed. However, we do not know about the higher CNS controlling sbGnRH system. Also, no evidence is available about any inhibitory factor. If an inhibitor mechanism exists under the regulation of puberty, somatic growth may be a required condition in order to overcome this inhibition. The mediators of somatic growth, such as IGF-I or leptin, are candidates that could feedback centrally and inactivate the inhibitor mechanism.

Seasonal reproductive cycle

After two years from hatching, red seabream reaches the size capable of puberty; 2+ fish can respond to suitable environmental conditions. In late autumn, a decrease of water temperature to the optimum induces gonadal recrudescence through activating the CNS–GnRH system. Increase in photoperiod and/or water temperature from winter to spring enhances gonadal maturation. The increase of sbGnRH secretion underlies the gonadal maturation in this season; stimuli of long photoperiod and/or optimum temperature are accepted by sensory organs, and activate the CNS regulating sbGnRH secretion. In turn, secreted sbGnRH not only stimulates GTH release but also increases the expression of its own receptor, therefore GTH secretion is largely enhanced. In addition, the increase in water temperature in early spring also increases the GnRH-R expression. In accordance with the increase in GTH levels, production of sex steroids is activated and the steroids feedback centrally and up-regulate sbGnRH synthesis. On the other hand, steroids negatively feedback to GnRH-R synthesis and maintain GTH levels at adequate concentrations. Further increase in the water temperature in early summer inactivates sbGnRH synthesis and/or release and in turn reduced the activity of the entire brain–pituitary–gonad axis, and terminates the spawning season in red seabream.

Still a lot of speculation is contained in this model, thus the following studies are required in order to validate this model; (1) to test if environmental factors do affect sbGnRH synthesis and release, (2) to clarify the CNS mediating the effect of environmental factors and internal factors such as sex steroids, (3) to determine whether the effect of water temperature on GnRH-R expression is direct or indirect, namely mediated via sbGnRH secretion.

Acknowledgements

The research from our laboratory described in this chapter was supported in part by a Bio-Design Program (BDP-02-IV-2-3) from the Ministry of Agriculture, Forestry and Fisheries of Japan.

References

Adams, B.A., Vickers, E.D., Warby, C., Park, M., Fischer, W. H., Craig, A.G., Rivier, J.E. and Sherwood, N.M. (2002) Three forms of gonadotropin-releasing hormone, including a novel form, in a basal salmonid, *Coregonus clupeaformis. Biol. Reprod.*, 67: 232–239.

Alok, D., Hassin, S., Sampath Kumar, R., Trant, J.M., Yu, K. and Zohar, Y. (2000) Characterization of a pituitary GnRH-receptor from a perciform fish, *Morone saxatilis*: functional expression in a fish cell line. *Mol. Cell. Endocrinol.*, 168: 65–75.

Amano, M., Aida, K., Okumoto, N. and Hasegawa, Y. (1992) Changes in salmon GnRH and chicken GnRH-II contents in the brain and pituitary, and GTH contents in the pituitary in female masu salmon, *Oncorhynchus masou*, from hatching through ovulation. *Zool. Sci.*, 9: 375–386.

Amano, M., Aida, K., Okumoto, N. and Hasegawa, Y. (1993) Changes in levels of GnRH in the brain and pituitary and GTH in the pituitary in male masu salmon, *Oncorhynchus masou*, from hatching to maturation. *Fish Physiol. Biochem.*, 11: 233–240.

Amano, M., Hyodo, S., Urano, A., Okumoto, N., Kitamura, S., Ikuta, K., Suzuki, Y. and Aida, K. (1994) Activation of salmon gonadotropin-releasing hormone synthesis by 17α-methyltestosterone administration in yearling masu salmon, *Oncorhynchus masou. Gen. Comp. Endocrinol.*, 95: 374–380.

Amano, M., Hyodo, S., Kitamura, S., Ikuta, K., Suzuki, Y., Urano, A. and Aida, K. (1995a) Salmon GnRH synthesis in the preoptic area and the ventral telencephalon is activated during gonadal maturation in female masu salmon. *Gen. Comp. Endocrinol.*, 99: 13–21.

Amano, M., Hyodo, S., Kitamura, S., Ikuta, K., Suzuki, Y., Urano, A. and Aida, K. (1995b) Short photoperiod accelerates preoptic and ventral telencephalic salmon GnRH synthesis and precocious maturation in underyearling male masu salmon. *Gen. Comp. Endocrinol.*, 99: 22–27.

Amano, M., Urano, A. and Aida, K. (1997) Distribution and function of gonadotropin-releasing hormone (GnRH) in the teleost brain. *Zool. Sci.*, 14: 1–11.

Amano, M., Ikuta, K., Kitamura, S. and Aida, K. (1999) Effects of photoperiod on salmon GnRH mRNA levels in brain of castrated underyearling precocious male masu salmon. *Gen. Comp. Endocrinol.*, 115: 70–75.

Amano, M., Oka, Y., Yamanome, T., Okuzawa, K. and Yamamori, K. (2002a) Three GnRH systems in the brain and pituitary of a pleuronectiform fish, the barfin flounder *Verasper moseri. Cell Tissue Res.*, 309: 323–329.

Amano, M., Okubo, K., Ikuta, K., Kitamura, S., Okuzawa, K., Yamada, H., Aida, K. and Yamamori, K. (2002b) Ontogenic origin of salmon GnRH neurons in the ventral telencephalon and the preoptic area in masu salmon. *Gen. Comp. Endocrinol.*, 127: 256–262.

Andersson, E., Borg, B. and Goos, H.J.T. (1992) Temperature, but not photoperiod, influences gonadotropin-releasing hormone binding in the pituitary of the three-spined stickleback, *Gasterosteus aculeatus. Gen. Comp. Endocrinol.*, 88: 111–116.

Andersson, E., Fjelldal, P.G., Klenke, U., Vikingstad, E., Taranger, G.L., Zohar, Y. and Stefansson, S.O. (2001) Three forms of GnRH in the brain and pituitary of the turbot, *Scophthalmus maximus*: immunological characterization and seasonal variation. *Comp. Biochem. Physiol. B*, 129: 551–558.

Carolsfeld, J., Powell, J.F.F., Park, M., Fischer, W.H., Craig, A.G., Chang, J.P., Rivier, J.E. and Sherwood, N.M. (2000) Primary structure and function of three gonadotropin-releasing

108

hormones, including a novel form, from an ancient teleost, herring. *Endocrinology*, 141: 505–512.

Chen, W., Shields, T.S., Stork, P.J. and Cone, R.D. (1995) A colorimetric assay for measuring activation of Gs- and Gq-coupled signaling pathways. *Anal. Biochem.*, 226: 349–354.

Chiba, A., Oka, S. and Honma, Y. (1994) Ontogenic development of gonadotropin-releasing hormone-like immunoreactive neurons in the brain of the chum salmon, *Oncorhynchus keta*. *Neurosci. Lett.*, 178: 51–54.

Collins, P.M., O'Neill, D.F., Barron, B.R., Moore, R.K. and Sherwood, N.M. (2001) Gonadotropin-releasing hormone content in the brain and pituitary of male and female grass rockfish (*Sebastes rastrelliger*) in relation to seasonal changes in reproductive status. *Biol. Reprod.*, 65: 173–179.

Dubois, E.A., Florijn, M.A., Zandbergen, M.A., Peute, J. and Goos, H.J.T. (1998) Testosterone accelerates the development of the catfish GnRH system in the brain of immature African catfish (*Clarias gariepinus*). *Gen. Comp. Endocrinol.*, 112: 383–393.

Dubois, E.A., Slob, S., Zandbergen, M.A., Peute, J. and Goos, H.J.T. (2001) Gonadal steroids and the maturation of the species-specific gonadotropin-releasing hormone system in brain and pituitary of the male African catfish (*Clarias gariepinus*). *Comp. Biochem. Physiol. B*, 129: 381–387.

Francis, R.C., Jacobson, B., Wingfield, J.C. and Fernald, R.D. (1992) Hypertrophy of gonadotropin releasing hormone-containing neurons after castration in the teleost, *Haplochromis burtoni*. *J. Neurobiol.*, 23: 1084–1093.

Francis, R.C., Soma, K. and Fernald, R.D. (1993) Social regulation of the brain–pituitary–gonadal axis. *Proc. Natl. Acad. Sci. USA*, 90: 7794–7798.

Fukusho, K., Fujimura, T. and Yamamoto, T. (1986) Broodstock and advanced spawning of the red seabream in an indoor tank with manipulation of water temperature. *Suisan Zoshoku*, 34: 69–75.

González-Martínez, D., Madigou, T., Zmora, N., Anglade, I., Zanuy, S., Zohar, Y., Elizur, A., Muñoz-Cueto, J.A. and Kah, O. (2001) Differential expression of three different prepro-GnRH (gonadotrophin-releasing hormone) messengers in the brain of the European sea bass (*Dicentrarchus labrax*). *J. Comp. Neurol.*, 429: 144–155.

Gothilf, Y., Muñoz-Cueto, J.A., Sagrillo, C.A., Selmanoff, M., Chen, T.T., Kah, O., Elizur, A. and Zohar, Y. (1996) Three formes of gonadotropin-releasing hormone in a perciform fish (*Sparus aurata*): complementary deoxyribonucleic acid characterization and brain localization. *Biol. Reprod.*, 55: 636–645.

Gothilf, Y., Meiri, I., Elizur, A. and Zohar, Y. (1997) Preovulatory changes in the levels of three gonadotropin-releasing hormone-encoding messenger ribonucleic acids (mRNAs), gonadotropin β-subunit mRNAs, plasma gonadotropin, and steroids in the female gilthead seabream, *Sparus aurata*. *Biol. Reprod.*, 57: 1145–1154.

Grober, M.S. and Bass, A.H. (1991) Neuronal correlates of sex/role change in labrid fishes: LHRH-like immunoreactivity. *Brain Behav. Evol.*, 38: 302–312.

Grober, M.S., Jackson, I.M.D. and Bass, A.H. (1991) Gonadal

steroids affect LHRH preoptic cell number in a sex/role changing fish. *J. Neurobiol.*, 22: 734–741.

Habibi, H.R., De Leeuw, R., Nahorniak, C.S., Goos, H.J.T. and Peter, R.E. (1989) Pituitary gonadotropin-releasing hormone (GnRH) receptor activity in goldfish and catfish: seasonal and gonadal effects. *Fish Physiol. Biochem.*, 7: 109–118.

Hanyu, I., Asahina, K., Shimizu, A., Razani, H. and Kaneko, T. (1983) Environmental regulation of reproductive cycles in teleosts. In: *Proceedings of 2nd North Pacific Aquaculture Symposium*. Shimizu and Tokyo, pp. 173–188.

Holland, M.C., Hassin, S. and Zohar, Y. (2001) Seasonal fluctuations in pituitary levels of the three forms of gonadotropin-releasing hormone in striped bass, *Morone saxatilis* (Teleostei), during juvenile and pubertal development. *J. Endocrinol.*, 169: 527–538.

Holland, M.C.H., Gothilf, Y., Meiri, I., King, J.A., Okuzawa, K., Elizur, A. and Zohar, Y. (1998) Levels of the native forms of GnRH in the pituitary of the gilthead seabream, *Sparus aurata*, at several characteristic stages of the gonadal cycle. *Gen. Comp. Endocrinol.*, 112: 394–405.

Huang, Y.S., Rousseau, K., Le Belle, N., Vidal, B., Burzawa-Gérard, E., Marchelidon, J. and Dufour, S. (1998) Insulin-like growth factor-I stimulates gonadotrophin production from eel pituitary cells: a possible metabolic signal for induction of puberty. *J. Endocrinol.*, 159: 43–52.

Illing, N., Troskie, B.E., Nahorniak, C.S., Hapgood, J.P., Peter, R.E. and Millar, R.P. (1999) Two gonadotropin-releasing hormone receptor subtypes with distinct ligand selectivity and differential distribution in brain and pituitary in the goldfish (*Carassius auratus*). *Proc. Natl. Acad. Sci. USA*, 96: 2526–2531.

Kobayashi, M., Amano, M., Hasegawa, Y., Okuzawa, K. and Aida, K. (1992) Effects of olfactory tract section on brain GnRH distribution, plasma gonadotropin levels, and gonadal stage in goldfish. *Zool. Sci.*, 9: 765–773.

Kobayashi, M., Amano, M., Kim, M.-H., Furukawa, K., Hasegawa, Y. and Aida, K. (1994) Gonadotropin-releasing hormones of terminal nerve origin are not essential to ovarian development and ovulation in goldfish. *Gen. Comp. Endocrinol.*, 95: 192–200.

Lam, T.J. (1983) Environmental influences on gonadal activity in fish. In: W.S. Hoar and D.J. Randall (Eds.), *Fish Physiology, Vol. IX. Reproduction, Part B*. Academic Press, New York, NY, pp. 65–116.

Madigou, T., Mañanos-Sanchez, E., Hulshof, S., Anglade, I., Zanuy, S. and Kah, O. (2000) Cloning, tissue distribution, and central expression of the gonadotropin-releasing hormone receptor in the rainbow trout (*Oncorhynchus mykiss*). *Biol. Reprod.*, 63: 1857–1866.

Montaner, A.D., Park, M.K., Fischer, W.H., Craig, A.G., Chang, J.P., Somoza, G.M., Rivier, J.E. and Sherwood, N.M. (2001) Primary structure of a novel gonadotropin-releasing hormone in the brain of a teleost, Pejerrey. *Endocrinology*, 142: 1453–1460.

Montero, M., Le Belle, N., King, J.A., Millar, R.P. and Dufour, S. (1995) Differential regulation of the two forms of gonadotropin-releasing hormone (mGnRH and cGnRH-II) by sex steroids in the European female silver eel (*Anguilla anguilla*). *Neuroendocrinology*, 61: 525–535.

Oka, Y. (1997) GnRH neuronal system of fish brain as a model system for the study of peptidergic neuromodulation. In: I.S. Parhar and Y. Sakuma (Eds.), *GnRH Neurons, Gene to Behavior*. Brain Shuppan, Tokyo, pp. 245–276.

Okubo, K., Amano, M., Yoshiura, Y., Suetake, H. and Aida, K. (2000a) A novel form of gonadotropin-releasing hormone in the medaka, *Oryzias latipes*. *Biochem. Biophys. Res. Comm.*, 276: 298–303.

Okubo, K., Suetake, H., Usami, T. and Aida, K. (2000b) Molecular cloning and tissue-specific expression of a gonadotropin-releasing hormone receptor in the Japanese eel. *Gen. Comp. Endocrinol.*, 119: 181–192.

Okubo, K., Nagata, S., Ko, R., Kataoka, H., Yoshiura, Y., Mitani, H., Kondo, M., Naruse, K., Shima, A. and Aida, K. (2001) Identification and characterization of two distinct GnRH receptor subtypes in a teleost, the medaka *Oryzias latipes*. *Endocrinology*, 142: 4729–4739.

Okuzawa, K. and Kobayashi, M. (1999) Gonadotropin-releasing hormone neuronal systems in the teleostean brain and functional significance. In: P.D. Prasada Rao and R.E. Peter (Eds.), *Neural Regulation in the Vertebrate Endocrine System*. Kluwer Academic/Plenum Publishers, New York, NY, pp. 85–100.

Okuzawa, K., Furukawa, K., Aida, K. and Hanyu, I. (1989) Effects of photoperiod and temperature on gonadal maturation, and plasma steroid and gonadotropin levels in a cyprinid fish, the honmoroko *Gnathopogon caerulescens*. *Gen. Comp. Endocrinol.*, 75: 139–147.

Okuzawa, K., Amano, M., Kobayashi, M., Aida, K., Hanyu, I., Hasegawa, Y. and Miyamoto, K. (1990) Differences in salmon GnRH and Chicken GnRH-II contents in discrete brain areas of male and female rainbow trout according to age and stage of maturity. *Gen. Comp. Endocrinol.*, 80: 116–126.

Okuzawa, K., Araki, K., Tanaka, H., Kagawa, H. and Hirose, K. (1994a) Molecular cloning of a cDNA encoding the preprosalmon gonadotropin-releasing hormone of the red seabream. *Gen. Comp. Endocrinol.*, 96: 234–242.

Okuzawa, K., Furukawa, K., Aida, K. and Hanyu, I. (1994b) The effects of water temperature on gonadotropin-releasing hormone contents in the discrete brain areas and pituitary of male honmoroko *Gnathopogon caerulescens*. *Fish. Sci.*, 60: 155–158.

Okuzawa, K., Granneman, J., Bogerd, J., Goos, H.J.T., Zohar, Y. and Kagawa, H. (1997) Distinct expression of GnRH genes in the red seabream brain. *Fish Physiol. Biochem.*, 17: 71–79.

Omeljaniuk, R.J., Habibi, H.R. and Peter, R.E. (1989) Alterations in pituitary GnRH and dopamine receptors associated with the seasonal variation and regulation of gonadotropin release in the goldfish (*Carassius auratus*). *Gen. Comp. Endocrinol.*, 74: 392–399.

Ookura, T., Okuzawa, K., Tanaka, H., Gen, K. and Kagawa, H. (1999) The ontogeny of gonadotropin-releasing hormone neurons in the red seabream *Pagrus major*. In: B. Norberg, O.S. Kjesbu, G.L. Taranger, E. Andersson and S.O. Stefansson (Eds.), *Proceedings of 6th International Symposium on the Reproductive Physiology of Fish*. University of Bergen, Bergen, Norway, pp. 47–49.

Parhar, I.S. (1997) GnRH in tilapia: three genes, three origins and their roles. In: I.S. Parhar and Y. Sakuma (Eds.), *GnRH Neurons, Gene to Behavior*. Brain Shuppan, Tokyo, pp. 99–122.

Parhar, I.S., Iwata, M., Pfaff, D.W. and Schwanzelfukuda, M. (1995) Embryonic development of gonadotropin-releasing hormone neurons in the sockeye salmon. *J. Comp. Neurol.*, 362: 256–270.

Parhar, I.S., Soga, T. and Sakuma, Y. (2000) Thyroid hormone and estrogen regulate brain region-specific messenger ribonucleic acids encoding three gonadotropin-releasing hormone genes in sexually immature male fish, *Oreochromis niloticus*. *Endocrinology*, 141: 1618–1626.

Peng, C., Chang, J.P., Yu, K.L., Wong, A.O.-L., van Goor, F., Peter, R.E. and Rivier, J.E. (1993) Neuropeptide-Y stimulates growth hormone and gonadotropin-II secretion in the goldfish pituitary: involvement of both presynaptic and pituitary cell action. *Endocrinology*, 132: 1820–1829.

Peter, R.E. (1983) The brain and neurohormones in teleost reproduction. In: W.S. Hoar, D.J. Randall and E.M. Donaldson (Eds.), *Fish Physiology, Vol. IX. Reproduction, Part A*. Academic Press, New York, NY, pp. 97–135.

Peter, R.E., Trudeau, V.L. and Sloley, B.D. (1991) Brain regulation of reproduction in teleosts. *Bull. Inst. Zool., Acad. Sinica, Monogr.*, 16: 89–118.

Peyon, P., Zanuy, S. and Carrillo, M. (2001) Action of leptin on *in vitro* luteinizing hormone lelease in the European sea bass (*Dicentrarchus labrax*). *Biol. Reprod.*, 65: 1573–1578.

Powell, J.F.F., Zohar, Y., Elizur, A., Park, M., Fischer, W.H., Craig, A.G., Rivier, J.E., Lovejoy, D.A. and Sherwood, N.M. (1994) Three forms of gonadotropin-releasing hormone characterized from brains of one species. *Proc. Natl. Acad. Sci. USA*, 91: 12081–12085.

Robison, R.R., White, R.B., Illing, N., Troskie, B.E., Morley, M., Millar, R.P. and Fernald, R.D. (2001) Gonadotropin-releasing hormone receptor in the teleost *Haplochromis burtoni*: structure, location, and function. *Endocrinology*, 142: 1737–1743.

Senthilkumaran, B., Okuzawa, K., Gen, K., Ookura, T. and Kagawa, H. (1999) Distribution and seasonal variations in levels of three native GnRHs in the brain and pituitary of perciform fish. *J. Neuroendocrinol.*, 11: 181–186.

Senthilkumaran, B., Okuzawa, K., Gen, K. and Kagawa, H. (2001) Effects of serotonin, GABA and neuropeptide Y on seabream gonadotropin releasing hormone release *in vitro* from preoptic–anterior hypothalamus and pituitary of red seabream, *Pagrus major*. *J. Neuroendocrinol.*, 13: 395–400.

Sherwood, N., Eiden, L., Brownstein, M., Spiess, J., Rivier, J. and Vale, W. (1983) Characterization of a teleost gonadotropin-releasing hormone. *Proc. Natl. Acad. Sci. USA*, 80: 2794–2798.

Sherwood, N.M., Schalburg, K. and Lescheid, D.W. (1997) Origin and evolution of GnRH in vertebrates and invertebrates. In: I.S. Parhar and Y. Sakuma (Eds.), *GnRH Neurons, Gene to Behavior*. Brain Shuppan, Tokyo, pp. 3–25.

Shimizu, A. and Hanyu, I. (1982) Environmental regulation

of annual reproductive cycle in a spring-spawning bitterling *Acheilognathus tabira. Nippon Suisan Gakkaishi*, 48: 1563–1568.

Soma, K.K., Francis, R.C., Wingfield, J.C. and Fernald, R.D. (1996) Androgen regulation of hypothalamic neurons containing gonadotropin-releasing hormone in a cichlid fish: integration with social cues. *Horm. Behav.*, 30: 216–226.

Tensen, C., Okuzawa, K., Blomenröhr, M., Rebers, F., Leurs, R., Bogerd, J., Schulz, R. and Goos, H. (1997) Distinct efficacies for two endogenous ligands on a single cognate gonadoliberin receptor. *Eur. J. Biochem.*, 243: 134–140.

Troskie, B.E., Hapgood, J.P., Millar, R.P. and Illing, N. (2000) Complementary deoxyribonucleic acid cloning, gene expression, and ligand selectivity of a novel gonadotropin-releasing hormone receptor expressed in the pituitary and midbrain of *Xenopus laevis. Endocrinology*, 141: 1764–1771.

Tsutsumi, M., Zhou, W., Millar, R.P., Mellon, P.L., Roberts, J.L., Flanagan, C.A., Dong, K., Gillo, B. and Sealfon, S.C. (1992) Cloning and functional expression of a mouse gonadotropin-releasing hormone receptor. *Mol. Endocrinol.*, 6: 1163–1169.

Wang, L., Bogerd, J., Choi, H.S., Seong, J.Y., Soh, J.M., Chun, S.Y., Blomenröhr, M., Troskie, B.E., Millar, R.P., Yu, W.H., McCann, S.M. and Kwon, H.B. (2001) Three distinct types of GnRH receptor characterized in the bullfrog. *Proc. Natl. Acad. Sci. USA*, 98: 361–366.

Weil, C., Carré, F., Blaise, O., Breton, B. and Le Bail, P.Y. (1999) Differential effect of insulin-like growth factor I on *in vitro* gonadotropin (I and II) and growth hormone secretions in rainbow trout (*Oncorhynchus mykiss*) at different stages of the reproductive cycle. *Endocrinology*, 140: 2054–2062.

White, R.B. and Fernald, R.D. (1998) Ontogeny of gonadotropin-releasing hormone (GnRH) gene expression reveals a distinct origin for GnRH-containing neurons in the midbrain. *Gen. Comp. Endocrinol.*, 112: 322–329.

White, S.A., Kasten, T.L., Bond, C.T., Adelman, J.P. and Fernald, R.D. (1995) Three gonadotropin-releasing hormone genes in one organism suggest novel roles for an ancient peptide. *Proc. Natl. Acad. Sci. USA*, 92: 8363–8367.

Yamamoto, N., Oka, Y., Amano, M., Aida, K., Hasegawa, Y. and Kawashima, S. (1995) Multiple gonadotropin-releasing hormone (GnRH)-immunoreactive systems in the brain of the dwarf gourami, *Colisa lalia*: immunohistochemistry and radioimmunoassay. *J. Comp. Neurol.*, 355: 354–368.

Yoo, M.S., Kang, H.M., Choi, H.S., Kim, J.W., Troskie, B.E., Millar, R.P. and Kwon, H.B. (2000) Molecular cloning, distribution and pharmacological characterization of a novel gonadotropin-releasing hormone ([Trp[8]] GnRH) in frog brain. *Mol. Cell. Endocrinol.*, 164: 197–204.

Yu, K.L., Rosenblum, P.M. and Peter, R.E. (1991) *In vitro* release of gonadotropin-releasing hormone from the brain preoptic–anterior hypothalamic region and pituitary of female goldfish. *Gen. Comp. Endocrinol.*, 81: 256–267.

Yu, K.L., Lin, X.W., Bastos, J.C. and Peter, R.E. (1997) Neural regulation of GnRH in teleost fishes. In: I.S. Parhar and Y. Sakuma (Eds.), *GnRH Neurons, Gene to Behavior*. Brain Shuppan, Tokyo, pp. 277–312.

I.S. Parhar (Ed.)
Progress in Brain Research, Vol. 141
© 2002 Published by Elsevier Science B.V.

CHAPTER 9

Multiplicity of gonadotropin-releasing hormone signaling: a comparative perspective

Christian Klausen [1], John P. Chang [2], Hamid R. Habibi [1,*]

[1] *Department of Biological Sciences, University of Calgary, 2500 University Drive N.W., Calgary, AB T2N 1N4, Canada*
[2] *Department of Biological Sciences, University of Alberta, CW 405 Biological Sciences Centre, Edmonton, AB T6G 2E9, Canada*

Introduction

Gonadotropin-releasing hormone (GnRH) is best known for its regulation of the synthesis and secretion of pituitary gonadotropin hormones (GtH), which include follicle-stimulating hormone (FSH) and luteinizing hormone (LH). The production of LH and FSH is also regulated by other neuromodulatory factors including, steroids, catecholamines, amino acids and gonadal peptides (Chang and Jobin, 1994; Shupnik, 1996; Van Der Kraak et al., 1998). In addition to its effects on the pituitary, GnRH exerts effects in other peripheral tissues including the brain, gonads and placenta (King and Millar, 1995). The signal transduction of GnRH-induced GtH synthesis and secretion has been studied extensively in mammals (Naor, 1990; Stojilkovic et al., 1994; Stojilkovic and Catt, 1995; Naor et al., 1998; Shacham et al., 1999) and in non-mammalian vertebrates, in particular, GnRH-induced control of GtH release in fishes (Chang and Jobin, 1994; Chang et al., 1996b, 2000). The objective of this review is to summarize the information on the mechanisms of GnRH-induced GtH release and subunit gene expression from a comparative perspective with emphasis on the studies carried out on goldfish (Tables 1 and 2). We will also examine the evidence for the functional significance of GnRH and GnRH receptor multiplicity in vertebrates.

Multiplicity of GnRH and its receptor

The primary sequences of 14 molecular forms of GnRH have been characterized in various vertebrate and protochordate species (Montaner et al., 2001). All vertebrate classes were found to have two or more forms of GnRH, including one or two molecular forms of GnRH as well as chicken (c)GnRH-II which appears to be ubiquitous in the vertebrate classes studied (King and Millar, 1995; Millar et al., 1997). There is evidence for differential distribution and functional diversity of GnRH variants in the brain–pituitary axis, as well as, neural and peripheral tissues in vertebrates (King and Millar, 1995; Pati and Habibi, 1998; Habibi and Matsoukas, 1999). For example, goldfish brain and pituitary contains salmon (s)GnRH and cGnRH-II (Yu et al., 1988). Both GnRH forms act at the level of the pituitary to stimulate the synthesis and release of GtHs in the goldfish (Chang et al., 1990; Khakoo et al., 1994; Klausen et al., 2001). The discovery of cGnRH-II in the brains of mammals (including humans), and a cGnRH-II specific receptor in the pituitary suggest that cGnRH-II may also have hypophysiotrophic function in mammals (Lescheid et al., 1997; Chen

---

* Correspondence to: H.R. Habibi, Department of Biological Sciences, University of Calgary, 2500 University Drive N.W., Calgary, AB T2N 1N4, Canada. Tel.: +1-403-220-5270; Fax: +1-403-282-0048;
E-mail: habibi@ucalgary.ca

TABLE 1

Signal transduction of GnRH-induced FSH and LH synthesis and secretion [a]

	Release				Gene expression				
	PKC	Ca^{2+}	AA	PKA	PKC	Ca^{2+}	AA	ERK	PKA
FSH									
Rat pit. cells	Y[1]	Y[2]	Y[2]	–	Y[3]	N[3]	–	Y[4]	–
	N[5]	–	–	–	–	–	–	–	–
Rat hemipit.	Y[6]	–	–	–	–	–	–	–	–
Rat-Luc (GGH₃)					Y[7]	N[7]	–	–	–
Ovine pit. cells	–	Y[8]	–	–	–	–	–	–	–
Ovine-Luc (HeLa)					Y[9]	–	–	–	–
Frog hemipit.	–	Y[10]	–	–	–	–	–	–	–
Goldfish pit. cells									
sGnRH	–	–	–	–	N[11]	–	–	–	–
cGnRH-II	–	–	–	–	N[11]	–	–	–	–
Tilapia pit. cells	–	–	–	–	N[12]	–	–	N[13]	Y[12]
LH									
Rat pit. cells	Y[5,14]	Y[2,15]	Y[2]	–	Y[3]	Y[3,16]	–	N[4]	–
	N[17]	–	–	–	Y[17]	–	–	–	–
Rat hemipit.	Y[6]	–	–	–	–	–	–	–	–
Rat-Luc (GGH₃)					Y[7]	Y[7]	–	–	–
Rat-Luc (αT3-1)					–	Y[16]	–	N[16]	–
Equine-Luc (αT3-1)					Y[18]	N[18]	–	Y[18]	–
Ovine pit. cells	–	Y[8]	–	–	–	–	–	–	–
Chicken pit. cells	Y[19]	Y[20,21]	–	–	–	–	–	–	–
Frog hemipit.	–	Y[10]	–	–	–	–	–	–	–
Goldfish pit. cells									
sGnRH	Y[22]	Y[22]	Y[22]	N[22]	N[11]	–	–	–	–
cGnRH-II	Y[22]	Y[22]	N[22]	N[22]	N[11]	–	–	–	–
Tilapia pit. cells	Y[23]	–	–	Y[23]	Y[23]	–	–	Y[13]	Y[23]
Tilapia pit. fragments	–	Y[24]	–	–	–	–	–	–	–
Murrel pit. cells	–	Y[25]	–	–	–	–	–	–	–

[a] Superscript numbers refer to references and (–) denotes unknown or not determined. [1] Wiebe et al., 1994; [2] Chang et al., 1988; [3] Ben-Menahem and Naor, 1994; [4] Haisenleder et al., 1998; [5] Audy et al., 1990; [6] Johnson et al., 1992; [7] Saunders et al., 1998; [8] Kile and Nett, 1994; [9] Strahl et al., 1998; [10] Porter and Licht, 1986; [11] C. Klausen, J.P. Chang, H.R. Habibi, unpublished; [12] Yaron et al., 2001; [13] Gur et al., 2001a; [14] Stojilkovic et al., 1988b; [15] Stojilkovic et al., 1988a; [16] Weck et al., 1998; [17] Andrews et al., 1988; [18] Call and Wolfe, 1999; [19] Johnson and Tilly, 1991; [20] Smith et al., 1987; [21] Davidson et al., 1988; [22] Chang et al., 1996b; [23] Melamed et al., 1996; [24] Levavi-Sivan and Yaron, 1989; [25] Jamaluddin et al., 1989.

et al., 1998; Millar et al., 2001; Neill et al., 2001) similar to that described previously for goldfish (Yu et al., 1988; Lin and Peter, 1996; Peter and Yu, 1997). Thus, understanding the functional aspects of multiple GnRH forms in a single target tissue is of increasing interest and urgency. To date, information on functional aspects of multiple forms of GnRH in a single target tissue has been largely derived from non-mammalian models, and especially from studies in teleost fishes. Among these, studies on the goldfish in particular have generated novel ideas on the functional significance of multiple GnRH forms.

The existence of multiple forms of GnRH represents more than 500 million years of evolution and apparently involves parallel evolution of GnRH receptors. To date, the existence of three types of GnRH receptors (Type I, Type II and Type III) with different structures and ligand selectivity have been demonstrated in the pituitary and extrapituitary tissues in a number of vertebrates. In mammals, including human, there is evidence for the existence of two GnRH receptor subtypes, Type I and Type II, with different structure and ligand selectivity (Millar et al., 2001; Neill et al., 2001). The type II GnRH

TABLE 2

Signal transduction of GnRH-induced GtH-α and GH synthesis and secretion [a]

	Release				Gene expression				
	PKC	Ca^{2+}	AA	PKA	PKC	Ca^{2+}	AA	ERK	PKA
GtH-α									
Human-Luc (GGH$_3$)					Y[1]	Y[1]	–	–	–
Human-Luc (αT3-1)					Y[2]	Y[3,4]	–	Y[2]	–
Human-Luc (rat pit.)					–	–	–	Y[2]	–
Human-Luc (HeLa)					Y[5]	–	–	–	–
Rat pit. cells	–	–	–	–	Y[6]	Y[6]	–	Y[7]	–
	–	–	–	–	–	N[8]	–	–	–
Rat-Luc (αT3-1)					–	N[8]	–	Y[8]	–
Mouse (αT3-1)	–	Y[3]	–	–	Y[9]	Y[9]	Y[10]	–	–
Mouse-Luc (αT3-1)					–	–	–	Y[11]	–
Goldfish pit. cells									
sGnRH	–	–	–	–	N[12]	–	–	–	–
cGnRH-II	–	–	–	–	N[12]	–	–	–	–
Tilapia pit. cells	–	–	–	–	Y[13]	–	–	Y[14]	Y[13]
GH									
Goldfish pit. cells									
sGnRH	Y[15]	Y[15]	N[15]	N[15]	N[12]	–	–	–	–
cGnRH-II	Y[15]	Y[15]	N[15]	N[15]	N[12]	–	–	–	–
Tilapia pit. cells	Y[16]	–	–	N[16]	–	–	–	–	–

[a] Superscript numbers refer to references and (–) denotes unknown or not determined. [1] Saunders et al., 1998; [2] Sundaresan et al., 1996; [3] Holdstock et al., 1996; [4] Call and Wolfe, 1999; [5] Strahl et al., 1998; [6] Ben-Menahem and Naor, 1994; [7] Haisenleder et al., 1998; [8] Weck et al., 1998; [9] Ben-Menahem et al., 1995; [10] Ben-Menahem et al., 1994; [11] Roberson et al., 1995; [12] C. Klausen, J.P. Chang, H.R. Habibi, unpublished; [13] Gur et al., 2001b; [14] Gur et al., 2001a; [15] Chang et al., 1996b; [16] Melamed et al., 1996.

receptor was found to be selective for cGnRH-II with somewhat different signal transduction mechanisms (Millar et al., 2001; Neill et al., 2001). Two subtypes of GnRH receptor (GfA and GfB) have been identified in the goldfish pituitary and these are expressed also in the gonads (Illing et al., 1999). Experimental evidence suggests that both GfA and GfB have similar ligand selectivity and are isoforms of the Type I GnRH receptor family (Troskie et al., 1998; Wang et al., 2001). Three pure antagonists of the human GnRH receptor acted as full or partial agonists for a recently cloned chicken pituitary GnRH receptor (Sun et al., 2001). In addition, recent findings on the physiological role of GnRH in the autocrine/paracrine control of ovarian function suggests that there may be additional forms of GnRH receptors with different ligand selectivity and coupling to signal transduction mechanisms in goldfish (Pati and Habibi, 2000, 2002; Habibi et al., 2001). Indeed, evidence is increasing in support of a functional role for GnRH and GnRH receptor plurality.

Protein kinase C (PKC)

GnRH-induced GtH release

Membrane bound phosphatidylinositol-specific phospholipase C (PLC) is highly selective for phosphatidylinositol 4,5-bisphosphate and catalyzes its cleavage yielding cytosolic inositol 1,4,5-trisphosphate (IP$_3$) and membrane-soluble diacylglycerol (DAG). IP$_3$ is involved in the release of calcium (Ca^{2+}) from intracellular stores via specific IP$_3$ receptors. The primary role of DAG is to stimulate the activity of certain isoforms of PKC. DAG can also be generated by pathways involving phosphatidylcholine-specific PLC and phospholipase D (PLD) (Berridge, 1987; Rasmussen et al., 1995).

In rat pituitary gonadotropes, GnRH treatment results in a rapid rise in the levels of DAG, IP$_3$ and other inositol phosphates (IPs) (Andrews and Conn, 1986; Morgan et al., 1987; Naor, 1990; Horn et al., 1991; Stojilkovic et al., 1994). This action requires a pertussis toxin-insensitive G-protein (G_q and/or G_{11}) which couples receptor activation to the stimulation of PLCβ (Hsieh and Martin, 1992; Grosse et al., 2000). GnRH receptors may also couple to other G-proteins including, G_i and G_s (Stanislaus et al., 1998). A delayed activation of PLD by GnRH, possibly via PKC, is also observed in αT3-1 cells and may contribute additional DAG for the activation of PKC (Stojilkovic et al., 1994; Stojilkovic and Catt, 1995; Naor et al., 1998). Johnson et al. (1992) reported that the PKC inhibitors, staurosporine and H7, reduced GnRH-stimulated FSH or LH release from rat hemipituitaries. In another report, PKC depletion did not affect GnRH-induced LH release in primary cultures of rat pituitary cells (Andrews et al., 1988). Also in rat pituitary cells, Audy et al. (1990) found that PKC depletion reduced GnRH-stimulated LH but not FSH release. While there are contradictory reports regarding the involvement of PKC in GnRH-induced GtH release, there is sufficient evidence for the involvement of PKC in the GnRH-induced response in rat and in a number of other vertebrates. GnRH stimulates the redistribution of PKC and the effects of GnRH on FSH and LH release can be reduced by inhibition and depletion of PKC (Chang et al., 1987; Stojilkovic et al., 1988b; Naor, 1990; Wiebe et al., 1994). PKC depletion effectively reduced chicken GnRH-I-induced LH secretion in primary cultures of dispersed chicken pituitary cells (Johnson and Tilly, 1991). In goldfish, the use of PKC inhibitors and cells depleted of PKC has provided sound evidence for the involvement of PKC in acute and prolonged sGnRH- and cGnRH-II-stimulated GtH release (Chang et al., 1991a; Jobin et al., 1993). Similarly, PKC is also involved in sGnRH-induced GtH secretion in primary cultures of dispersed tilapia pituitary cells (Melamed et al., 1996). Interestingly, the cGnRH-II-stimulated GtH response in goldfish was more sensitive to inhibition by a PKC inhibitor than that to sGnRH (Chang et al., 1991a). In this context, there is increasing evidence for the existence of different PKC-sensitive pathways mediating the action of GnRH variants. Some of these possibilities are discussed in the following section on PKC multiplicity and signaling diversity.

GnRH-induced GtH gene expression

Available data suggest that PKC also mediates GnRH effects at the level of GtH gene expression. In cultured rat pituitary cells, PKC depletion and treatment with GF109203X (PKC inhibitor) inhibited the stimulation of GtH-α, FSH-β and LH-β mRNA levels by GnRH (Andrews et al., 1988; Ben-Menahem and Naor, 1994). Selective deletions of the GtH-α promoter linked to a luciferase reporter gene in transfected αT3-1 cells demonstrated co-localization of responsiveness to GnRH and the phorbol ester (PMA) (Schoderbek et al., 1993). Endogenous GtH-α mRNA levels were found to be increased following treatment with GnRH and phorbol esters in αT3-1 cells. The effect of GnRH could be inhibited by GF109203X and by phorbol ester-induced depletion of PKC. The responses to GnRH and PMA or the Ca^{2+} ionophore, ionomycin, were not additive; neither were the responses to PMA and ionomycin. These data indicate that PKC and Ca^{2+} probably act sequentially or converge in the regulation of GnRH-induced GtH-α gene expression. This idea is further reinforced by the observation that PKC depletion and incubation in Ca^{2+}-free medium abolished the responses to ionomycin and PMA, respectively (Horn et al., 1991; Ben-Menahem et al., 1995). PKC depletion reduced GnRH-induced human GtH-α-luciferase activity in αT3-1 cells (Sundaresan et al., 1996). In GGH_3 cells (GH_3 cells stably expressing the GnRH receptor) transfected with luciferase constructs containing promoters for human GtH-α, rat FSH-β and rat LH-β, GF109203X reduced GnRH-induced luciferase activity for all three constructs (Saunders et al., 1998). In αT3-1 cells, GnRH-stimulated expression of a luciferase construct containing the equine LH-β promoter was reduced by PKC depletion and GF109203X (Call and Wolfe, 1999). Similarly, GF109203X reduced GnRH-induced expression of an ovine FSH-β-luciferase or a human GtH-α-luciferase construct in HeLa cells transfected with the mouse GnRH receptor (Strahl et al., 1998). The mechanism by which GnRH-induced PKC activity regulates gene expression is not well understood. Recent studies using phorbol esters have demonstrated that the early growth response gene 1 product may be involved in mediating PKC-induced effects of GnRH on LH-β gene expression (Halvorson et al., 1999; Tremblay and Drouin, 1999). Currently, the investigation of the regulatory elements contained within the promoter regions of GtH subunit genes is an area of intense study.

There is also a limited amount of information in non-mammalian vertebrates. In dispersed tilapia pituitary cells, GF109203X effectively reduced sGnRH-induced GtH-α and LH-β, but not FSH-β mRNA levels (Melamed et al., 1996; Gur et al., 2001b; Yaron et al., 2001). In goldfish, PKC is involved in the negative regulation of basal GtH subunit gene expression and appears not to be involved in GnRH-induced increases in GtH-α, FSH-β or LH-β mRNA levels (Klausen et al., 2000; C. Klausen, J.P. Chang, H.R. Habibi, unpublished).

When the results discussed above are viewed together, it is clear that the involvement of PKC in the regulation of LH release by GnRH has been confirmed in a variety of cell types and species. However, the role of PKC in GnRH-induced FSH release is not as well understood, in part, due to the lack of a specific antibody for FSH. The involvement of PKC in GnRH-induced GtH subunit gene expression appears to be common in a number of species. This is not surprising since GnRH is coupled to the PLC-activating $G_{q/11}$ G-proteins. In addition, the dual (PKC/Ca^{2+}) nature of PLC-mediated signaling provides GnRH with capacity to regulate many physiological processes. In tilapia, the lack of PKC involvement in sGnRH-induced FSH-β gene expression may reflect differences in the 5′ regulatory regions of the various GtH genes in this species. However, the possibility that GnRH or GnRH receptor multiplicity may be responsible has not been investigated. As discussed later, PKC also plays a central role in the activation of mitogen-activated protein kinases by GnRH in gonadotropes.

PKC multiplicity and signaling diversity

Structural properties and co-factor requirements have led to the classification of PKCs into three groups. Conventional PKCs include the α, βI, βII and γ isozymes which are activated by Ca^{2+}, DAG and phosphatidylserine. Novel PKC isozymes (δ, ε, η and θ) lack the characteristic Ca^{2+}-binding domain and are activated by DAG, phosphatidylserine and unsaturated fatty acids. The ζ and λ(ι) isozymes

comprise the third group referred to as the atypical PKCs. They may have some constitutive activity but are insensitive to Ca^{2+} or DAG and are activated by phosphatidylserine, phosphatidylinositides or unsaturated fatty acids (Liu and Heckman, 1998). In mammalian cells, GnRH-induced activation of PLC is followed closely by the activation of PLD and phospholipase A_2 (PLA$_2$). It has been hypothesized that DAG and Ca^{2+}, derived from the initial activation of PLC, may be responsible for the rapid activation of conventional PKCs involved in secretion. The subsequent activation of PLD and PLA$_2$ would generate signaling molecules for the activation of novel PKCs or atypical PKCs that could be implicated in prolonged release and gene expression (Naor et al., 1998). As such, the diversity in GnRH signaling may be the result of differential activation of PKC isozymes. The report that inhibition of DAG lipase and subsequent increase in DAG levels did not elevate LH and FSH secretion supports this hypothesis (Chang et al., 1988). It indicates that a specific PKC isozyme requiring simultaneous signals from both DAG and Ca^{2+} or some other signaling molecules may be involved in the regulation of hormone secretion. While pituitary cells have been shown to express PKCα, βII, δ, ε and ζ, little is known about the roles of specific PKC isozymes in GnRH-induced GtH release or synthesis (Naor et al., 1998; Shacham et al., 1999). Isozyme specificity has been demonstrated in digitonin permeabilized rat pituitary cells where the addition of α or β, but not γ PKC, was able to recover PMA-induced GtH secretion (Naor et al., 1989). Together these results provide evidence for the role of distinct PKC isozymes in the functional specificity of GnRH signaling. The differential recruitment of PKCs may provide the cell with the ability to differentiate between stimuli that may or may not require the up-regulation of gene expression. In this context, certain isoforms or sets of isoforms may be responsible for a diverse array of GnRH-mediated events. It is possible that distinct GnRH receptor subtypes with differential ligand selectivity, generating dissimilar second messengers may independently regulate cellular processes by activating different PKC isozymes. Although largely untested, there is evidence for GnRH receptor subtypes with differential signaling in goldfish gonadotropes (discussed below).

The role of calcium

GnRH-induced GtH release

In mammals, treatment of pituitary cells with GnRH results in the mobilization of Ca^{2+} from IP_3-sensitive stores followed by the influx of extracellular Ca^{2+} via L-type voltage-sensitive Ca^{2+} channels (VSCCs). Indeed, Ca^{2+} mobilization from IP_3-sensitive stores was found to be necessary for initial rapid phase of GtH secretion whereas sustained GnRH-induced LH release relies in part on the influx of Ca^{2+} from the extracellular environment via VSCCs (Naor, 1990; Naor et al., 1998). In rat pituitary cells, the stimulation of FSH and LH release by GnRH is reduced in Ca^{2+}-free medium and in the presence of VSCC blockers (Chang et al., 1986; Chang et al., 1988; Stojilkovic et al., 1988a; Blotner et al., 1990). In αT3-1 cells, thapsigargin but not nifedipine was able to reduce GnRH-stimulated secretion of GtH-α subunit indicating a role for intracellular but not extracellular Ca^{2+} (Holdstock et al., 1996). Incubation of ovine pituitary cells in Ca^{2+}-free medium abolished the secretion of LH and FSH in response to GnRH (Kile and Nett, 1994). VSCC blockers and Ca^{2+}-free medium have been used to demonstrate the involvement of Ca^{2+} in GnRH-stimulated LH release from dispersed chicken pituitary cells (Smith et al., 1987; Davidson et al., 1988). Incubation of frog hemipituitaries in Ca^{2+}-free medium or in the presence of $CoCl_2$ abolished GnRH-stimulated FSH and LH secretion (Porter and Licht, 1986). In goldfish, the involvement of L-type VSCCs in acute and prolonged sGnRH- and cGnRH-II-induced LH release from primary pituitary cells has also been demonstrated pharmacologically (Chang et al., 2000). Interestingly, sGnRH and cGnRH-II differ in their abilities to generate IPs and their requirements for extracellular Ca^{2+} influx and intracellular Ca^{2+} stores are distinct. sGnRH and cGnRH-II also generate Ca^{2+} signals that differ in their temporal features in identified goldfish gonadotropes (Johnson et al., 1999). Treatment of [^{3}H]inositol-prelabeled goldfish pituitary cells with sGnRH was found to increase levels of IP_1, IP_2, IP_3 and other higher IPs while cGnRH-II only elevated levels of IP_2. The lack of IP_3 production by cGnRH-II agrees with evidence indicating that the effect of cGnRH-II on LH re-

lease is more dependent on influx of extracellular Ca^{2+} than that of sGnRH (Jobin and Chang, 1992; Chang et al., 1995). These data are also consistent with the findings that, unlike the response to sGnRH, cGnRH-II stimulation of LH release is independent of mobilization of Ca^{2+} from Xestospongin C-sensitive IP_3 receptor channels (Jobin and Chang, 1992; Chang et al., 1995; Johnson et al., 2000). On the other hand, the LH release response to both GnRH forms requires ryanodine-sensitive Ca^{2+} stores (Johnson et al., 2000). Surprisingly, sGnRH-induced, but not cGnRH-II-elicited, LH release was abolished by prior exposure to caffeine (a common ryanodine receptor agonist) and caffeine-stimulated LH secretion was mediated by mobilization of Ca^{2+} from ryanodine- and dantrolene-insensitive stores (Johnson et al., 2000, 2002b). These data indicate that pharmacologically distinct IP_3/Xestospongin C-, caffeine- and ryanodine-sensitive intracellular Ca^{2+} stores participate differentially in sGnRH and cGnRH-II stimulation of gonadotropes. These differences may underlie known potency and efficacy characteristics of these two GnRHs in eliciting LH secretion (Chang et al., 1990). In addition, differences in Ca^{2+} signaling may provide the substrate upon which reported seasonal differences in GnRH-stimulated LH synthesis/release and direct gonadal steroid action on these responses may be manifested (Habibi and Huggard, 1998; Lo and Chang, 1998b; Huggard-Nelson et al., 2002). Consistent with this hypothesis, the relative dependence on ryanodine-sensitive signaling in sGnRH-induced LH release varies according to the reproductive status (Johnson and Chang, 2002) and that testosterone positive-feedback action targets PKC-dependent signaling pathways leading to GnRH-stimulated LH release (Lo and Chang, 1998a).

The participation of extracellular Ca^{2+} influx and intracellular Ca^{2+} mobilization has similarly been demonstrated in several other teleosts. GnRH analog-induced secretion of GtH from tilapia pituitary fragments was found to be dependent on extracellular Ca^{2+} (Levavi-Sivan and Yaron, 1989). Similarly, in the murrel (*Channa punctatus*), stimulation of GtH release by mammalian GnRH was found to be dependent on extracellular Ca^{2+} (Jamaluddin et al., 1989). In the African catfish, treatment with catfish GnRH or cGnRH-II increased IPs and intracellular

Ca^{2+} levels (Rebers et al., 2000). However, in these teleost species, it is not known if different GnRH forms elicit different Ca^{2+} responses and the roles of these responses in GnRH-induced LH release requires further investigation.

GnRH-induced GtH gene expression

In addition to its role in hormone secretion, Ca^{2+} is also involved in mediating GnRH-induced increases in GtH subunit mRNA levels. In cultured rat pituitary cells, ionomycin was able to mimic the GnRH-induced changes in LH-β mRNA levels, although GtH-α mRNA level was not affected after 24 hours of incubation (Ben-Menahem and Naor, 1994). In this experiment, FSH-β mRNA level was decreased up to 12 hours after treatment with a subsequent increase occurring at 24 hours. Moreover, incubation of the cells in Ca^{2+}-free medium abolished the response of GtH-α and LH-β, but not FSH-β, to GnRH. The combined effects of phorbol ester and ionomycin were distinct for the different GtH subunits (Ben-Menahem and Naor, 1994). Treatment with the VSCC blocker, nimodipine, was found to reduce GnRH-stimulated LH-β without affecting GtH-α mRNA levels and luciferase activity in cultured rat pituitary cells and transfected αT3-1 cells, respectively (Weck et al., 1998). Ca^{2+} mobilization and influx were both demonstrated to be involved in the regulation of GtH-α mRNA levels by GnRH in αT3-1 cells in experiments using ionomycin, Ca^{2+}-free medium and the cell-permeant Ca^{2+} chelator BAPTA/AM (Ben-Menahem et al., 1995). In αT3-1 cells transfected with a human GtH-α-luciferase construct, treatment with VSCC blockers and incubation in Ca^{2+}-free medium reduced GnRH-induced GtH-α promoter activation; in contrast, thapsigargin selectively affected GnRH-induced GtH-α release but not the increase in promoter activity (Holdstock et al., 1996; Call and Wolfe, 1999). In αT3-1 cells stably expressing an equine LH-β-luciferase construct, GnRH-induced expression was not dependent on extracellular Ca^{2+} (Call and Wolfe, 1999). Saunders et al. (1998) demonstrated that the VSCC agonist, BayK8644, increased only GtH-α-luciferase activity and nimodipine reduced GnRH-induced human GtH-α- and rat LH-β-, but not rat FSH-β-luciferase activity in GGH$_3$ cells.

The evidence presented above clearly indicates that Ca^{2+} derived from extracellular and intracellular sources may differentially mediate GnRH-induced GtH-α gene expression and secretion, respectively (Holdstock et al., 1996). On the other hand, more research will be needed to investigate the role of different sources of Ca^{2+} in GnRH-induced GtH gene expression in order to clarify the differences observed. It is possible that cell type specific differences in promoter elements may explain some of the observed variations regarding the involvement of Ca^{2+} in GtH subunit gene expression. In this context, vertebrate GtH subunit genes contain 5' regulatory regions with different complements of transcription factor binding sites (reviewed in Ando et al., 2001). It is becoming increasingly evident that the multiplicity of Ca^{2+} stores provides an intricate system of spatially and temporally distinct Ca^{2+} signals important for the differential regulation of diverse physiological processes (Berridge et al., 2000; Johnson and Chang, 2000). As such, the paucity of data regarding the relative roles of various Ca^{2+} stores in the regulation of GtH subunit gene expression needs to be addressed. In goldfish pituitary cells, intracellular and extracellular Ca^{2+} stores appear to be differentially involved in the regulation of basal LH-β subunit gene expression (J.D. Johnson, C. Klausen, H.R. Habibi and J.P. Chang, unpublished).

Arachidonic acid metabolism

GnRH-induced GtH release and gene expression

Arachidonic acid (AA) can be derived directly from phospholipids by the action of PLA$_2$ and indirectly through the action of phospholipases A$_1$, C and D in concert with a host of secondary and tertiary enzymes. To date, the rat and goldfish models are the only models in which the role of AA in GnRH-stimulated GtH release has been rigorously examined. The liberation of AA via the activation of PLA$_2$ by GnRH occurs in αT3-1 cells (Naor et al., 1998). GnRH-induced LH and FSH secretion involves AA liberated through the action of PLA$_2$ and DAG lipase (Chang et al., 1988). Furthermore, treatment with AA was able to mimic the initial phase of the secretory response to GnRH (Chang et al., 1987). In rat pituitary cells, GnRH stimulates the metabolism of AA to form leukotrienes LTC$_4$, LTD$_4$ and LTE$_4$ as well as 5- and 15-hydroxyeicosatetraenoic acids (Dan-Cohen et al., 1992). Moreover, lipoxygenase inhibitors and a peptidoleukotriene receptor antagonist were able to reduce GnRH-induced LH release suggesting the involvement of AA metabolites in GnRH action (Chang et al., 1987; Dan-Cohen et al., 1992).

A major difference in the involvement of AA metabolism in GnRH-induced LH release has been identified in goldfish. Mixed action and specific lipoxygenase but not cyclooxygenase inhibitors reduced sGnRH- and abolished AA-induced LH release. In contrast, none of these inhibitors reduced the response to cGnRH-II indicating that sGnRH but not cGnRH-II, in part through activation of lipoxygenase metabolites, stimulates acute and prolonged LH release (Chang et al., 1989, 1991b, 1995). The enzyme responsible for generating the AA involved in GnRH-stimulated GtH release is PLA$_2$. PLA$_2$ inhibitors reduced the sGnRH-stimulated response whereas, the DAG lipase inhibitor, U-57908, was ineffective (Chang et al., 1994a).

AA and its lipoxygenase metabolites have been implicated in the regulation of GtH subunit mRNA levels. The effect of GnRH was shown to be mimicked by AA and the 5-lipoxygenase products 5-hydroxyeicosatetraenoic acid and LTC$_4$ in αT3-1 cells (Ben-Menahem et al., 1994). Treatments with inhibitors of PLA$_2$ and 5-lipoxygenase reduced GnRH-induced elevations in GtH-α mRNA levels in αT3-1 cells, although a cyclooxygenase inhibitor was without effect (Ben-Menahem et al., 1994). In cultured rat pituitary cells, AA and LTC$_4$ stimulated GtH-α, FSH-β and LH-β mRNA levels; whereas, 5-hydroxyeicosatetraenoic acid stimulated only FSH-β (Ben-Menahem et al., 1994).

So far, the role of AA and its metabolites in GnRH-induced GtH release and gene expression has only been examined in the rats among mammals. In goldfish, however, major differences were observed in the involvement of AA in the sGnRH- and cGnRH-II-induced responses (Chang et al., 1996b). These observations are now very relevant to the understanding of the control of GtH synthesis and release in mammals in view of the recent identification of a GnRH receptor specific for cGnRH-II in the pituitary. The role of AA in the regulation of GtH sub-

unit gene expression by GnRH is poorly understood. The observation that 5-hydroxyeicosatetraenoic acid increases GtH-α mRNA levels in αT3-1 cells but not in cultured pituitary cells indicates that a difference may exist between the clonal and cultured pituitary cell systems. Moreover, the ability of 5-hydroxyeicosatetraenoic acid to increase only FSH-β mRNA levels suggests the existence of different signaling pathways coupling GnRH receptors to GtH release and subunit gene expression. In this context, AA represents a rather poorly understood pathway with potential functional implications in the multiplicity of GnRH signaling.

Mitogen-activated protein kinases (MAPKs)

MAPK pathways can be subdivided into three subfamilies; the c-jun amino-terminal kinase/stress-activated protein kinase (JNK/SAPK) pathway, the p38 MAPK pathway and the extracellular signal-regulated kinase (ERK) pathway. MAPK pathways consist of three sequentially activated protein kinases. A serine/threonine MAPK kinase kinase phosphorylates and activates a dual specificity threonine/tyrosine MAPK kinase which in turn phosphorylates and activates a serine/threonine MAPK. Active MAPK will then phosphorylate transcription factors or further regulatory kinases. In the ERK pathway the central MAPK kinase kinase is Raf-1, the MAPK kinase is MEK and the MAPK is ERK (Lewis et al., 1998; Kolch, 2000; Pearson et al., 2001). Evidence to date indicates that MAPK signal transduction pathways are activated in response to GnRH in both primary pituitary and αT3-1 cells. Indeed, many G-protein-coupled receptors are able to activate MAPKs through a wide array of mechanisms (Van Biesen et al., 1996; Lopez-Ilasaca, 1998; Naor et al., 2000; Pierce et al., 2001).

In rats, GnRH stimulates ERK activity in vivo and in vitro (Sundaresan et al., 1996; Haisenleder et al., 1998). Treatment of αT3-1 cells with GnRH has been found to activate ERK1 and ERK2 in a MEK- and PKC-dependent manner although other Ca^{2+}- and tyrosine phosphorylation-dependent components may participate in this process (Sundaresan et al., 1996; Reiss et al., 1997; Call and Wolfe, 1999; Mulvaney et al., 1999; Benard et al., 2001). PKC-dependent activation of ERKs by GnRH has also been demonstrated in GGH$_3$ cells (Han and Conn, 1999). GnRH is thought to activate ERKs by two distinct pathways converging at the level of Raf-1. One pathway involves the activation of Raf-1 by PKC and another involves Src and Ras (Benard et al., 2001). The ERK pathway appears to be involved in GnRH-stimulated promoter activity of human GtH-α in primary cultures of rat pituitary cells and αT3-1 cells (Sundaresan et al., 1996). Similarly, stimulation of mouse GtH-α promoter activity by GnRH was found to involve the ERK pathway (Roberson et al., 1995). In rat pituitary cells, ERKs are involved in GnRH-induced increases in GtH-α and FSH-β but not LH-β mRNA levels (Haisenleder et al., 1998). Likewise, in αT3-1 cells transfected with rat GtH-α- or LH-β-luciferase constructs, a MEK inhibitor reduced GnRH-stimulated GtH-α but not LH-β promoter activity (Weck et al., 1998). In contrast, GnRH-induced expression of an equine LH-β-luciferase construct was found to be reduced by the MEK inhibitor PD98059 (Call and Wolfe, 1999). In addition to activating ERKs, GnRH also stimulates the activity of JNK, p38 MAPK and big MAPK (Levi et al., 1998; Roberson et al., 1999; Naor et al., 2000). Although activated by GnRH, p38 MAPK does not appear to regulate the mouse GtH-α gene expression (Roberson et al., 1999).

sGnRH was found to activate both ERK1 and ERK2 in a MEK- and PKC-dependent manner in primary cultures of dispersed tilapia pituitary cells. PD98059 inhibited sGnRH-stimulated increases in GtII-α and LH-β but not FSH-β mRNA levels (Gur et al., 2001a). The lack of PKC-dependent ERK involvement in the regulation of FSH-β mRNA levels agrees with the report that sGnRH-induced increases in FSH-β mRNA levels are unaffected by PKC inhibition (Yaron et al., 2001).

Together, these results demonstrate the differential involvement of MAPK in GnRH-induced GtH subunit gene expression between and within species. These differences may reflect variations in the elements contained within the promoter regions that govern the transcription of the different GtH subunits or other species specific differences. To date, there have been no studies examining the possibility of differential regulation by multiple GnRHs. The recently cloned mammalian Type II GnRH receptor differs from the Type I receptor in its ability to

activate p38 MAPK, although the functional significance of such a difference has not been determined (Millar et al., 2001). Currently there is no information regarding the involvement of MAPK signaling in the control of GtH release. Given that PLA$_2$ can be activated by ERK, a role for ERK in GnRH-induced GtH secretion could exist and needs to be investigated.

PKA and GnRH?

The production of cyclic adenosine 3′,5′-monophosphate (cAMP) by adenylyl cyclase in response to receptor activation and the subsequent activation of cAMP-dependent protein kinase A (PKA) is involved in the intracellular signaling of many hormones (Beebe, 1994). In primary pituitary and αT3-1 cells, cAMP levels are not affected by GnRH treatment and as such are considered not to participate directly in GnRH-induced changes in GtH secretion or gene expression (Conn et al., 1979; Horn et al., 1991). However, there are reports of GnRH-induced cAMP production in rat hemipituitaries (Bourne, 1988), monkey COS-7 cells transfected with the mouse GnRH receptor (Arora et al., 1998) and rat GGH$_3$ cells (Kaiser et al., 1997). In goldfish, treatment with sGnRH or cGnRH-II did not elevate levels of cAMP, and administration of the PKA inhibitor, H89, was without effect on GnRH-induced GtH secretion (Chang et al., 1992; Jobin et al., 1996). Similarly, cAMP levels were not affected by treatment with either catfish GnRH or cGnRH-II in primary cultures of African catfish pituitary cells (Rebers et al., 2000). In tilapia, however, there is evidence for the involvement of PKA in GnRH-induced GtH synthesis and release. The PKA inhibitor H89 was found to reduce sGnRH-stimulated GtH release and subunit mRNA levels (Melamed et al., 1996; Gur et al., 2001b; Yaron et al., 2001). Although PKA is not involved directly in GnRH action in most systems, this pathway may be of importance as a modulator potentially interacting with GnRH-mediated pathways. In rat pituitary cells, forskolin, cAMP analogues and flufenamic acid have been used to demonstrate that the PKA pathway regulates GtH-α, FSH and LH secretion as well as subunit mRNA levels (Ishizaka et al., 1993; Holdstock and Burrin, 1994). In goldfish, elevation of cAMP levels or treatment with forskolin can potentiate the GtH response to native GnRHs via interactions with downstream PKC- and Ca^{2+}-dependent signaling mechanisms (Chang et al., 1992; Chang et al., 2001).

GnRH and growth hormone (GH)

In goldfish, GnRH receptors are present in somatotropes and GnRH stimulates the synthesis and release of GH (Marchant et al., 1989; Cook et al., 1991; Habibi et al., 1992; Mahmoud et al., 1996; Klausen et al., 2001; C. Klausen, J.P. Chang and H.R. Habibi, unpublished). Although not universally observed (Bosma et al., 1997), GnRH binding sites and GnRH involvement in GH secretion has been reported in several other fish species (Lin et al., 1993; Melamed et al., 1996; Stefano et al., 1999), as well as in certain clinical conditions in humans (Marchant et al., 1989). GH interacts with the reproductive axis and potentiates the effects of GtH on steroidogenesis (Le Gac et al., 1993). Like LH release, PKC and extracellular Ca^{2+} are required for both acute and prolonged sGnRH- and cGnRH-II-induced GH release in goldfish (Chang and De Leeuw, 1990; Chang et al., 1991a, 1994b). In contrast, AA and its metabolites are not involved in either sGnRH- or cGnRH-II-induced GH release since inhibitors of PLA_2, cyclooxygenase and lipoxygenase did not influence GnRH activity (Chang and De Leeuw, 1990; Chang et al., 1994b, 1996a). It should be noted that AA pathway is active in somatotropes and is involved in dopamine-induced GH release in the goldfish pituitary (Chang et al., 1996a). There are also differences in the mechanisms mediating the effects of sGnRH and cGnRH-II in the goldfish somatotropes. Only cGnRH-II action involves a ryanodine-sensitive intracellular Ca^{2+} store, although both sGnRH and cGnRH-II stimulation of GH release involve caffeine- and TMB-8-sensitive intracellular Ca^{2+}-mobilization events. Conversely, sGnRH-induced GH release, but not cGnRH-II, is dependent on Xestospongin C (IP_3)-sensitive mechanism (Johnson and Chang, 2002). Thus, it is evident that the action of both sGnRH and cGnRH-II are somehow different between somatotropes and gonadotropes, supporting the hypothesis for the existence of different GnRH receptor subtypes (or suites of receptor subtypes) in somatotropes and

gonadotropes in goldfish. In this context, amino acid substitutions in GnRH were shown to reduce its GtH-releasing potency without affecting GH-releasing activity (Habibi et al., 1992).

In tilapia, the PKC pathway is directly involved in GnRH-stimulated GH release as in the case for GtH secretion. However, unlike the GtH response, GnRH stimulation of GH secretion is unaffected by H89, a PKA inhibitor which is effective against dopamine-induced GH release (Melamed et al., 1996). Thus, the complement of GnRH receptors in gonadotropes and somatotropes appears to differ in tilapia, as in the case of the goldfish.

GnRH stimulates GH release but not mRNA levels in tilapia, demonstrating that the control of secretion and gene expression by GnRH can be completely uncoupled in this species (Melamed et al., 1996). In contrast, in goldfish, GH mRNA levels are increased following treatment with both sGnRH and cGnRH-II, in vivo and in vitro (Mahmoud et al., 1996; Klausen et al., 2001; C. Klausen, J.P. Chang and H.R. Habibi, unpublished). Currently there is no information regarding the mechanisms of GnRH-induced GH gene expression. Recent work in our labs indicates that Ca^{2+} is involved in the regulation of basal GH mRNA levels in goldfish pituitary cells. The effects of increased intracellular Ca^{2+} on GH mRNA levels were different depending on the origin of the Ca^{2+} involved, suggesting that functional specificity in GnRH signaling may involve multiplicity of Ca^{2+} mobilization (Johnson et al., 2002a).

GnRH signaling in the ovary

In addition to its hypophysiotrophic activity, GnRH is known to function as a paracrine/autocrine regulator of gonadal function. The demonstration of ovarian GnRH and GnRH receptor gene expression and their regulation by various factors in the ovary provide clear evidence that GnRH peptides are important autocrine/paracrine factors involved in the regulation of ovarian function. In mammals, GnRH directly affects follicular steroidogenesis and resumption of meiosis. While treatment with GnRH was found to stimulate basal steroidogenesis, it inhibited gonadotropin-induced cAMP and steroid production in a maturational stage-dependant manner (Hsueh and Jones, 1981; Hsueh and Schaefer, 1985; Knecht et al., 1985; Leung and Steele, 1992). GnRH exerts direct action in the ovary through specific GnRH receptors coupled to G_q and PLC (Leung and Steele, 1992). Recent studies on isolated human granulosa cells demonstrated that GnRH actions on steroidogenesis are significantly enhanced in the presence of prostaglandin F2alpha (Vaananen et al., 1997). There is a significant degree of similarity in intracellular signaling cascades mediating GnRH actions in the pituitary and ovary, including involvement of PLA_2, PLC and PLD as well as the activation of MAPK (Leung and Steele, 1992; Steele and Leung, 1993; Kang et al., 2001). Direct ovarian action of GnRH has also been investigated in fish and amphibians (Chieffi et al., 1991; Habibi, 1999; Habibi et al., 2001). In goldfish, GnRH exerts both stimulatory and inhibitory actions on oocyte meiosis and follicular steroidogenesis, depending upon the presence or absence of GtH. In follicle-enclosed goldfish oocytes, a number of GnRH variants including cGnRH-II, sGnRH and sGnRH analogs were found to individually stimulate oocyte meiosis in vitro as well as histone H1 kinase activity, which is an indicator of maturation promoting factor (MPF) activity (Habibi et al., 1988; Pati and Habibi, 2000; Pati et al., 2000). However, in the presence of GtH, sGnRH and a sGnRH analog were found to inhibit GtH-induced germinal vesicle breakdown, while cGnRH-II had no effect on GtH-induced responses (Habibi et al., 1988; Pati and Habibi, 2000). Interestingly, addition of a GnRH antagonist was found to effectively block the stimulatory effect of both sGnRH and cGnRH-II on oocyte meiosis, without affecting the inhibitory actions of sGnRH on GtH-induced response, suggesting the involvement of different receptors/signaling pathways mediating the stimulatory and inhibitory actions of sGnRH (Habibi and Pati, 1993; Habibi, 1999; Habibi et al., 2001). Further studies carried out in goldfish demonstrated a difference in the postreceptor mechanisms involving the stimulatory and inhibitory actions of sGnRH on reinitiation of oocyte meiosis and steroidogenesis. The stimulatory effect of both sGnRH and cGnRH-II on the reinitiation of oocyte meiosis was completely blocked by PKC inhibitors (H7 and GF109203X), suggesting the involvement of a PLC/DAG pathway in the mechanisms of GnRH-induced meiosis (Pati and Habibi, 2002). Administration of ETYA, an inhibitor of AA metabolism only inhibited the stimulatory effect of sGnRH and had no effect on cGnRH-II-induced meiosis. Furthermore, the inhibitory effect of sGnRH on GtH-induced meiosis and steroidogenesis was completely reversed by addition an AA metabolism inhibitor, while PKC inhibitors had no effect. There appear to be differences between the mechanisms of GnRH stimulation and the inhibition of gonadal functions in goldfish and rat (Leung and Steele, 1992). In the rat ovary, the PKC pathway and AA metabolism have been suggested to mediate the inhibitory and stimulatory effects of GnRH, respectively. Furthermore, an inhibitor of lipoxygenase, but not cyclooxygenase, was found to block GnRH and AA-induced progesterone production in the rat ovary (Wang and Leung, 1988). In contrast to the rat, cyclooxygenase metabolites have been reported to be important in the AA-induced testosterone production in the goldfish ovary (Van Der Kraak and Chang, 1990). The use of a general inhibitor of AA metabolism, ETYA reduces the products of cyclooxygenase and lipoxygenase metabolites, and could potentially increase the concentration of AA itself. Therefore, the possibility that AA might have played a direct role in reversing the inhibitory actions of sGnRH could not be ruled out.

These findings provide functional evidence in support of the hypothesis that goldfish ovarian follicles contain GnRH receptor subtypes with different ligand selectivity that mediate the stimulatory and inhibitory actions of sGnRH in the goldfish ovary. Similarly, there is evidence for the presence of different GnRH receptors, coupling independently to PLC and PLA_2 in the rat luteal cells (Watanabe et al., 1990).

Summary

GnRH regulation of GtH synthesis and release involves PKC- and Ca^{2+}-dependent pathways. There are differential signaling mechanisms in different cells, tissues and species. Signaling mechanisms involved in GnRH-mediated GtH release appear to be more conserved compared to that of GnRH-induced GtH gene expression. This may in part be due to different 5' regulatory regions on the GtH-subunit genes. Cell type specific expression of various signaling and/or exocytotic components may also be responsible for the observed differences in signaling between gonadotropes and somatotropes in the goldfish and tilapia pituitaries. However, this can not explain the observed differences in post receptor mechanisms for sGnRH and cGnRH-II in gonadotropes which is more likely to result from the existence of GnRH receptor subtypes. Support for this hypothesis is also provided by observations on mechanisms of autocrine/paracrine regulation of ovarian function by sGnRH and cGnRH-II in the goldfish ovary in which GnRH antagonists only block GnRH stimulation of oocyte meiosis and do not affect inhibitory effects of sGnRH. It should be easier to explain observed variations concerning GnRH-induced responses as more information becomes available on different types of GnRH receptors, and their distribution and function in mammals and non-mammalian vertebrates.

Abbreviations

AA	arachidonic acid
BAPTA/AM	1,2-bis(o-aminophenoxy) ethane-N,N,N′,N′-tetraacetic acid tetra (acetoxymethyl) ester
BayK8644	1,4-dihydro-2,6-dimethyl-5-nitro-4-[2′-(trifluoromethyl)phenyl]-3-pyridinecarboxylic acid methyl ester
Ca^{2+}	calcium
cAMP	cyclic adenosine 3′,5′-monophosphate
cGnRH-II	chicken GnRH-II
DAG	diacylglycerol
ERK	extracellular signal-regulated kinase
ETYA	5,8,11,14-Eicosatetraynoic acid
FSH	follicle-stimulating hormone
GF109203X	2-[1-(3-dimethylaminopropyl)-1H-indol-3-yl]-3-(1H-indol-3-yl)-maleimide
GfA	goldfish GnRH receptor A
GfB	goldfish GnRH receptor B
GH	growth hormone
GnRH	gonadotropin-releasing hormone
GtH	gonadotropin hormone
H7	1-(5-isoquinolinesulfonyl)-2-methylpiperazine
H89	N-[2-((p-bromocinnamyl)amino) ethyl]-5-isoquinolinesulfonamide
IP	inositol phosphate

IP_1	inositol 1-phosphate
IP_2	inositol 1, 4-bisphosphate
IP_3	inositol 1, 4, 5-trisphosphate
JNK	c-jun amino-terminal kinase
LH	luteinizing hormone
LTC_4	leukotriene C4₄
LTD_4	leukotriene D_4
LTE_4	leukotriene E_4
MAPK	mitogen-activated protein kinase
MEK	MAPK/ERK kinase
MPF	maturation promoting factor
PD98059	2′-Amino-3′-methoxyflavone
PKA	protein kinase A
PKC	protein kinase C
PLA_2	phospholipase A_2
PLC	phospholipase C
PLD	phospholipase D
PMA	phorbol-12-myristate-13-acetate
SAPK	stress-activated protein kinase
sGnRH	salmon GnRH
TMB-8	8-(N,N-diethylamino)-octyl-3,4,5-trimethoxybenzoate
U-57908	1,6-bis(cyclohexyloximinocarbonyl-amino)hexane
VSCC	voltage-sensitive calcium channel

Acknowledgements

The authors acknowledge grant support from the Natural Sciences and Engineering Research Council of Canada. C.K. is supported by an Alberta Heritage Foundation for Medical Research studentship.

References

Ando, H., Hew, C.L. and Urano, A. (2001) Signal transduction pathways and transcription factors involved in the gonadotropin-releasing hormone-stimulated gonadotropin subunit gene expression. *Comp. Biochem. Physiol. B*, 129: 525–532.

Andrews, W.V. and Conn, P.M. (1986) Gonadotropin-releasing hormone stimulates mass changes in phosphoinositides and diacylglycerol accumulation in purified gonadotropes cell cultures. *Endocrinology*, 118: 1148–1158.

Andrews, W.V., Maurer, R.A. and Conn, P.M. (1988) Stimulation of rat luteinizing hormone-β messenger RNA levels by gonadotropin releasing hormone. Apparent role for protein kinase C. *J. Biol. Chem.*, 263: 13755–13761.

Arora, K.K., Krsmanovic, L.Z., Mores, N., O'Farrell, H. and Catt, K.J. (1998) Mediation of cyclic AMP signaling by the first intracellular loop of the gonadotropin-releasing hormone receptor. *J. Biol. Chem.*, 273: 25581–25586.

Audy, M.C., Boucher, Y. and Bonnin, M. (1990) Estrogen modulated gonadotropin release in relation to gonadotropin-releasing-hormone (GnRH) and phorbol ester (PMA) actions in superfused rat pituitary cells. *Endocrinology*, 126: 1396–1402.

Beebe, S.J. (1994) The cAMP-dependent protein kinase and cAMP signal transduction. *Semin. Cancer Biol.*, 5: 285–294.

Benard, O., Naor, Z. and Seger, R. (2001) Role of dynamin, Src, and Ras in the protein kinase C-mediated activation of ERK by gonadotropin-releasing hormone. *J. Biol. Chem.*, 276: 4554–4563.

Ben-Menaham, D. and Naor, Z. (1994) Regulation of gonadotropin mRNA levels in cultured rat pituitary cells by gonadotropin-releasing hormone (GnRH): role for Ca^{2+} and protein kinase C. *Biochemistry*, 33: 3698–3704.

Ben-Menaham, D., Shraga-Levine, Z., Limor, R. and Naor, Z. (1994) Arachidonic acid and lipoxygenase products stimulate gonadotropin α-subunit mRNA levels in pituitary αT3-1 cell line: role in gonadotropin-releasing hormone action. *Biochemistry*, 33. 12795–12799.

Ben-Menaham, D., Shraga-Levine, Z., Mellon, P.L. and Naor, Z. (1995) Mechanism of action of gonadotropin-releasing hormone upon gonadotropin α-subunit mRNA levels in the αT3-1 cell line: role of Ca^{2+} and protein kinase C. *Biochem. J.*, 309: 325–329.

Berridge, M.J. (1987) Inositol triphosphate and diacylglycerol: two interacting second messengers. *Annu. Rev. Biochem.*, 56: 159–193.

Berridge, M.J., Lipp, P. and Bootman, M.D. (2000) The versatility and universality of calcium signaling. *Nat. Rev. Mol. Cell. Biol.*, 1: 11–21.

Blotner, M., Shangold, G.A., Lee, E.Y., Murphy, S.N. and Miller, R.J. (1990) Nitrendipine and Ω-conotoxin modulate gonadotropin release and gonadotrope $[Ca^{2+}]_i$. *Mol. Cell. Endocrinol.*, 71: 205–216.

Bosma, P.T., Kolk, S.M., Rebers, F.E.M., Lescroart, O., Roelants, I., Willems, P.H.G.M. and Schulz, R.W. (1997) Gonadotrophs but not somatotrophs carry gonadotropin-releasing hormone receptors: receptor localisation, intracellular calcium, and gonadotrophin and GH release. *J. Endocrinol.*, 152: 437–446.

Bourne, G.A. (1988) Cyclic AMP indirectly mediates the extracellular Ca^{2+}-independent release of LH. *Mol. Cell. Endocrinol.*, 58: 155–160.

Call, G.B. and Wolfe, M.W. (1999) Gonadotropin-releasing hormone activates the equine luteinizing hormone β promoter through a protein kinase C/mitogen-activated protein kinase pathway. *Biol. Reprod.*, 61: 715–723.

Chang, J.P. and De Leeuw, R. (1990) In vitro goldfish growth hormone responses to gonadotropin-releasing hormone: possible roles of extracellular calcium and arachidonic acid metabolism? *Gen. Comp. Endocrinol.*, 80: 155–164.

Chang, J.P. and Jobin, R.M. (1994) Regulation of gonadotropin release in vertebrates: a comparison of GnRH mechanisms of action. In: K. Davey, R. Peter and S. Tobe (Eds.), *Perspectives in Comparative Endocrinology*. National Research Council of Canada, Ottawa, ON, pp. 41–51.

Chang, J.P., McCoy, E.E., Graeter, J., Tasaka, K. and Catt, K.J. (1986) Participation of voltage-dependent calcium channels in the action of gonadotropin-releasing hormone. *J. Biol. Chem.*, 261: 9105–9108.

Chang, J.P., Graeter, J. and Catt, K.J. (1987) Dynamic actions of arachidonic acid and protein kinase C in pituitary stimulation by gonadotropin-releasing hormone. *Endocrinology*, 120: 1837–1845.

Chang, J.P., Morgan, R.O. and Catt, K.J. (1988) Dependence of secretory responses to gonadotropin-releasing hormone on diacylglycerol metabolism. Studies with a diacylglycerol lipase inhibitor, RHC 80267. *J. Biol. Chem.*, 263: 18614–18620.

Chang, J.P., Freedman, G.L. and De Leeuw, R. (1989) Participation of arachidonic acid metabolism in gonadotropin-releasing hormone stimulation of goldfish gonadotropin release. *Gen. Comp. Endocrinol.*, 76: 2–11.

Chang, J.P., Cook, H., Freedman, G.L., Wiggs, A.J., Somoza, G.M., De Leeuw, R. and Peter, R.E. (1990) Use of a pituitary cell dispersion method and primary culture system for the studies of gonadotropin-releasing hormone action in the goldfish, *Carassius auratus*. *Gen. Comp. Endocrinol.*, 77: 256–273.

Chang, J.P., Jobin, R.M. and De Leeuw, R. (1991a) Possible involvement of protein kinase C in gonadotropin and growth hormone release from dispersed goldfish pituitary cells. *Gen. Comp. Endocrinol.*, 81: 447–463.

Chang, J.P., Wildman, B. and Van Goor, F. (1991b) Lack of involvement of arachidonic acid metabolism in chicken gonadotropin-releasing hormone II (cGnRH-II) stimulation of gonadotropin secretion in dispersed pituitary cells of goldfish, *Carassius auratus*. Identification of a major difference in salmon GnRH and cGnRH-II mechanisms of action. *Mol. Cell. Endocrinol.*, 79: 75–83.

Chang, J.P., Wong, A.O.L., Van Der Kraak, G. and Van Goor, F. (1992) Relationship between cyclic AMP-stimulated and native gonadotropin-releasing hormone-stimulated gonadotropin release in the goldfish. *Gen. Comp. Endocrinol.*, 86: 359–377.

Chang, J.P., Van Goor, F. and Neumann, C.M. (1994a) Interactions between protein kinase C and arachidonic acid in the gonadotropin response to salmon and chicken gonadotropin-releasing hormone-II in goldfish. *Gen. Comp. Endocrinol.*, 93: 304–320.

Chang, J.P., Van Goor, F., Wong, A.O.L., Jobin, R.M. and Neumann, C.M. (1994b) Signal transduction pathways in GnRH- and dopamine D1-stimulated growth hormone secretion in the goldfish. *Chin. J. Physiol.*, 37: 111–127.

Chang, J.P., Garofalo, R. and Neumann, C.M. (1995) Differences in the acute actions of sGnRH and cGnRH-II on gonadotropin release in goldfish pituitary cells. *Gen. Comp. Endocrinol.*, 100: 339–354.

Chang, J.P., Abele, J.T., Van Goor, F., Wong, A.O.L. and Neumann, C.M. (1996a) Role of arachidonic acid and calmodulin in mediating dopamine D1- and GnRH-stimulated growth hormone release in goldfish pituitary cells. *Gen. Comp. Endocrinol.*, 102: 88–101.

Chang, J.P., Van Goor, F., Jobin, R.M. and Lo, A. (1996b) GnRH signaling in goldfish pituitary cells. *Biol. Signals*, 5: 70–80.

Chang, J.P., Johnson, J.D., Van Goor, F., Wong, C.J.H., Yunker, W.K., Uretsky, A.D., Taylor, D., Jobin, R.M., Wong, A.O.L. and Goldberg, J.I. (2000) Signal transduction mechanisms mediating secretion in goldfish gonadotropes and somatotropes. *Biochem. Cell Biol.*, 78: 139–153.

Chang, J.P., Wirachowsky, N.R., Kwong, P. and Johnson, J.D. (2001) PACAP stimulation of gonadotropin-II secretion in goldfish pituitary cells: mechanisms of action and interaction with gonadotropin-releasing hormone. *J. Neuroendocrinol.*, 13: 540–550.

Chen, A., Yahalom, D., Ben-Aroya, N., Kaganovsky, E., Okon, E. and Koch, Y. (1998) A second isoform of gonadotropin-releasing hormone is present in the brain of human and rodents. *FEBS Lett.*, 435: 199–203.

Chieffi, G., Pierantoni, R. and Fasano, S. (1991) Immunoreactive GnRH in hypothalamic and extrahypothalamic areas. *Int. Rev. Cytol.*, 127: 1–55.

Conn, P.M., Morrell, D.V., Dufau, M.L. and Catt, K.J. (1979) Gonadotropin-releasing hormone action in cultured pituicytes: independence of luteinizing hormone release and adenosine 3′,5′-monophosphate production. *Endocrinology*, 104: 448–453.

Cook, H., Berkenbosch, J.W., Fernhout, M.J., Yu, K.L., Peter, R.E., Chang, J.P. and Rivier, J.E. (1991) Demonstration of gonadotropin releasing-hormone receptors on gonadotrophs and somatotrophs of the goldfish: an electron microscope study. *Regul. Pept.*, 36: 369–378.

Dan-Cohen, H., Sofer, Y., Schwartzman, M.L., Natarajan, R.D., Nadler, J.L. and Naor, Z. (1992) Gonadotropin releasing hormone activates the lipoxygenase pathway in cultured pituitary cells: role in gonadotropin secretion and evidence for a novel autocrine/paracrine loop. *Biochemistry*, 31: 5442–5448.

Davidson, J.S., Wakefield, I.K., King, J.A., Mulligan, G.P. and Millar, R.P. (1988) Dual pathways of calcium entry in spike and plateau phases of luteinizing hormone release from chicken pituitary cells: sequential activation of receptor-operated and voltage-sensitive calcium channels by gonadotropin-releasing hormone. *Mol. Endocrinol.*, 2: 382–390.

Grosse, R., Schmid, A., Schoneberg, T., Herrlich, A., Muhn, P., Schultz, G. and Gudermann, T. (2000) Gonadotropin-releasing hormone receptor initiates multiple signaling pathways by exclusively coupling to $G_{q/11}$ proteins. *J. Biol. Chem.*, 275: 9193–9200.

Gur, G., Bonfil, D., Safarian, H., Naor, Z. and Yaron, Z. (2001a) GnRH receptor signaling in tilapia pituitary cells: role of mitogen-activated protein kinase (MAPK). *Comp. Biochem. Physiol. B*, 129: 517–524.

Gur, G., Rosenfeld, H., Melamed, P., Meiri, I., Elizur, A. and Yaron, Z. (2001b) Tilapia glycoprotein hormone α subunit: cDNA cloning and hypothalamic regulation. *Mol. Cell. Endocrinol.*, 182: 49–60.

Habibi, H.R. (1999) Gonadotropin-releasing hormone as a paracrine regulator of ovarian function. In: P. Rao and P. Kluwer (Eds.), *Neural Regulation in the Vertebrate Endocrine System*. Academic/Plenum Publishers, New York, NY, pp. 101–110.

Habibi, H.R. and Huggard, D.L. (1998) Testosterone regulation of gonadotropin production in goldfish. *Comp. Biochem. Physiol. C*, 119: 339–344.

Habibi, H.R. and Matsoukas, J.M. (1999) Gonadotropin-releasing hormone: structural and functional diversity. In: J. Matsoukas and T. Mavromoustakos (Eds.), *Bioactive Peptides in Drug Discovery and Design: Medical Aspects Volume 22 in Biomedical and Health Research*. IOS Press, Amsterdam, pp. 247–255.

Habibi, H.R. and Pati, D.P. (1993) Endocrine and paracrine control of ovarian function: role of compounds with GnRH-like activity in goldfish. In: F. Facchinetti, I.W. Henderson, R. Pierantoni and A.M. Polzonetti-Magni (Eds.), *Cellular Communication in Reproduction. J. Endocrinol.*, pp. 59–70.

Habibi, H.R., Van Der Kraak, G., Bulanski, E. and Peter, R.E. (1988) Effects of teleost GnRH on reinitiation of oocyte meiosis in goldfish in vitro. *Am. J. Physiol.*, 255: R268–R273.

Habibi, H.R., Van Der Kraak, G., Fraser, R. and Peter, R.E. (1989) Effect of a teleost GnRH analog on steroidogenesis by the follicle-enclosed goldfish oocytes. *Gen. Comp. Endocrinol.*, 76: 95–105.

Habibi, H.R., Peter, R.E., Nahorniak, C.S., Milton, R.C.L. and Millar, R.P. (1992) Activity of vertebrate gonadotropin-releasing hormones and analogs with variant amino acid residues in positions 5, 7 and 8 in the goldfish pituitary. *Regul. Pept.*, 37: 271–284.

Habibi, H.R., Andreu-Vieyra, C. and Mirhadi, E. (2001) Functional significance of gonadal gonadotropin-releasing hormone. In: H.J.T.H. Goos, R.K. Rastogi, H. Vaudry and R. Pierantoni (Eds.), *Perspective in Comparative Endocrinology: Unity and Diversity*. Monduzzi Editore, pp. 959–968.

Haisenleder, D.J., Cox, M.E., Parsons, S.J. and Marshall, J.C. (1998) Gonadotropin-releasing hormone pulses are required to maintain activation of mitogen-activated protein kinase: role in stimulation of gonadotrope gene expression. *Endocrinology*, 138: 3104–3111.

Halvorson, L.M., Kaiser, U.B. and Chin, W.W. (1999) The protein kinase C system acts through the early growth response protein 1 to increase LHβ gene expression in synergy with steroidogenic factor-1. *Mol. Endocrinol.*, 13: 106–116.

Han, X.B. and Conn, P.M. (1999) The role of protein kinases A and C pathways in the regulation of mitogen-activated protein kinase activation in response to gonadotropin-releasing hormone receptor activation. *Endocrinology*, 140: 2241–2251.

Holdstock, J.G. and Burrin, J.M. (1994) Regulation of glycoprotein hormone free α-subunit secretion and intracellular α-subunit content in primary pituitary cells. *Endocrinology*, 134: 685–694.

Holdstock, J.G., Aylwin, S.J.B. and Burrin, J.M. (1996) Calcium and glycoprotein hormone α-subunit gene expression and secretion in αT3-1 gonadotropes. *Mol. Endocrinol.*, 10: 1308–1317.

Horn, F., Bilezikjian, L.M., Perrin, M.H., Bosma, M.M., Windle, J.J., Huber, K.S., Blount, A.L., Hille, B., Vale, W. and Mellon, P.L. (1991) Intracellular responses to gonadotropin-releasing hormone in a clonal cell line of the gonadotrope lineage. *Mol. Endocrinol.*, 5: 347–355.

Hsieh, K.P. and Martin, T.F.J. (1992) Thyrotropin-releasing hormone and gonadotropin-releasing hormone receptors activate phospholipase C by coupling to the guanosine triphosphate-binding proteins G_q and G_{11}. *Mol. Endocrinol.*, 6: 1673–1681.

Hsueh, A.J.W. and Jones, P.B.C. (1981) Extrapituitary actions of gonadotropin-releasing hormone. *Endocr. Rev.*, 2: 437–461.

Hsueh, A.J.W. and Schaefer, J.M. (1985) Gonadotropin-releasing hormone as a paracrine hormone and neurotransmitter in extrapituitary sites. *J. Steroid Biochem.*, 23: 757–764.

Huggard-Nelson, D.L., Nathwani, P.S., Kermouni, A. and Habibi, H.R. (2002) Molecular characterization of LH-β and FSH-β subunits and their regulation by estrogen in the goldfish pituitary. *Mol. Cell. Endocrinol.*, 188: 171–193.

Illing, N., Troskie, B.E., Nahorniak, C.S., Hapgood, J.P., Peter, R.E. and Millar, R.P. (1999) Two gonadotropin-releasing hormone receptor subtypes with distinct ligand selectivity and differential distribution in brain and pituitary in the goldfish (*Carassius auratus*). *Proc. Natl. Acad. Sci. USA*, 96: 2526–2531.

Ishizaka, K., Tsujii, T. and Winters, S.J. (1993) Evidence for a role for the cyclic adenosine $3',5'$-monophosphate/protein kinase-A pathway in regulation of the gonadotropin subunit messenger ribonucleic acids. *Endocrinology*, 133: 2040–2048.

Jamaluddin, M.D., Banerjee, P.P., Manna, P.R. and Bhattacharya, S. (1989) Requirement of extracellular calcium in fish pituitary gonadotropin release by gonadotropin hormone-releasing hormone. *Gen. Comp. Endocrinol.*, 74: 190–198.

Jobin, R.M. and Chang, J.P. (1992) Actions of two native GnRHs and protein kinase C modulators on goldfish pituitary cells. Studies on intracellular calcium levels and gonadotropin release. *Cell Calcium*, 13: 531–540.

Jobin, R.M., Ginsberg, J., Matowe, W.C. and Chang, J.P. (1993) Downregulation of protein kinase C levels leads to inhibition of GnRH-stimulated gonadotropin secretion from dispersed pituitary cells of goldfish. *Neuroendocrinology*, 58: 2–10.

Jobin, R.M., Van Goor, F., Neumann, C.M. and Chang, J.P. (1996) Interactions between signaling pathways in mediating GnRH-stimulated GTH release from goldfish pituitary cells: protein kinase C, but not cyclic AMP is an important mediator of GnRH-stimulated gonadotropin secretion in goldfish. *Gen. Comp. Endocrinol.*, 102: 327–341.

Johnson, J.D. and Chang, J.P. (2000) Function- and agonist-specific Ca^{2+} signaling: the requirement for and mechanism of spatial and temporal complexity in Ca^{2+} signals. *Biochem. Cell Biol.*, 78: 217–240.

Johnson, J.D. and Chang, J.P. (2002) Agonist-specific and sexual stage-dependent inhibition of GnRH-stimulated gonadotropin and growth hormone release by ryanodine: relationship to sexual stage-dependent caffeine-sensitive hormone release. *J. Neuroendocrinol.*, 14: 144–155.

Johnson, A.L. and Tilly, J.L. (1991) Second messenger pathways mediating chicken luteinizing hormone secretion from dispersed pituitary cells. *Biol. Reprod.*, 45: 64–72.

Johnson, M.S., Mitchell, R. and Thomson, F.J. (1992) The priming effect of luteinizing hormone-releasing hormone (LHRH) but not LHRH-induced gonadotropin release, can be prevented by certain protein kinase C inhibitors. *Mol. Cell. Endocrinol.*, 85: 183–193.

Johnson, J.D., Van Goor, F., Wong, C.J.H., Goldberg, J.I. and Chang, J.P. (1999) Two endogenous gonadotropin-releasing hormones generate dissimilar Ca^{2+} signals in identified goldfish gonadotropes. *Gen. Comp. Endocrinol.*, 116: 178–191.

Johnson, J.D., Van Goor, F., Jobin, R.M., Wong, C.J.H., Goldberg, J.I. and Chang, J.P. (2000) Agonist-specific Ca^{2+} signaling systems, composed of multiple intracellular Ca^{2+} stores, regulate gonadotropin secretion. *Mol. Cell. Endocrinol.*, 170: 15–29.

Johnson, J.D., Klausen, C., Habibi, H.R. and Chang, J.P. (2002a) Function-specific calcium stores selectively regulate growth hormone secretion, storage, and mRNA level. *Am. J. Physiol. Endocrinol. Metab.*, 282: E810–E819.

Johnson, J.D., Wong, C.J.H., Yunker, W.K. and Chang, J.P. (2002b) Caffeine-stimulated GTH-II release involves Ca^{2+} stores with novel properties. *Am. J. Physiol. Cell Physiol.*, 282: C635–C645.

Kaiser, U.B., Conn, P.M. and Chin, W.W. (1997) Studies of gonadotropin-releasing hormone (GnRH) action using GnRH receptor-expressing pituitary cell lines. *Endocr. Rev.*, 18: 46–70.

Kang, S.K., Tai, C.J., Nathwani, P.S., Choi, K.C. and Leung, P.C. (2001) Stimulation of mitogen-activated protein kinase by gonadotropin-releasing hormone in human granulosa-luteal cells. *Endocrinology*, 142: 671–679.

Khakoo, Z., Bhatia, A., Gedamu, L. and Habibi, H.R. (1994) Functional specificity for salmon gonadotropin-releasing hormone (GnRH) and chicken GnRH-II coupled to the gonadotropin release and subunit messenger ribonucleic acid level in the goldfish pituitary. *Endocrinology*, 134: 838–847.

Kile, J.P. and Nett, T.M. (1994) Differential secretion of follicle-stimulating hormone and luteinizing hormone from ovine pituitary cells following activation of protein kinase A, protein kinase C, or increased intracellular calcium. *Biol. Reprod.*, 50: 49–54.

King, J.A. and Millar, R.P. (1995) Evolutionary aspects of gonadotropin-releasing hormone and its receptor. *Cell. Mol. Neurobiol.*, 15: 5–23.

Klausen, C., Chang, J.P. and Habibi, H.R. (2000) Involvement of protein kinase C in growth hormone and gonadotropin subunit gene expression in the goldfish pituitary. In: *Program and Abstracts 4th Int. Symp. of Fish Endocrinol.*, Seattle, WA, p. 40 (abstract O-79).

Klausen, C., Chang, J.P. and Habibi, H.R. (2001) The effect of gonadotropin-releasing hormone on growth hormone and gonadotropin subunit gene expression in the pituitary of goldfish, *Carassius auratus*. *Comp. Biochem. Physiol. B*, 129: 511–516.

Knecht, M., Ranta, T., Feng, P., Shinohara, O. and Catt, K.J. (1985) Gonadotropin-releasing hormone as a modulator of ovarian function. *J. Steroid Biochem.*, 23: 771–777.

Kolch, W. (2000) Meaningful relationships: the regulation of the Ras/Raf/MEK/ERK pathway by protein interactions. *Biochem. J.*, 351: 289–305.

Le Gac, F., Blaise, O., Fostier, A., Le Bail, P.Y., Loir, M.,

Mourot, B. and Weil, C. (1993) Growth hormone (GH) and reproduction: a review. *Fish Physiol. Biochem.*, 11: 219–232.

Lescheid, D.W., Teresawa, E., Abler, L.A., Urbanski, H.F., Warby, C.M., Millar, R.P. and Sherwood, N.M. (1997) A second form of gonadotropin-releasing hormone (GnRH) with characteristics of chicken GnRH-II is present in the primate brain. *Endocrinology*, 138: 5618–5629.

Leung, P.C.K. and Steele, G.L. (1992) Intracellular signaling in gonads. *Endocr. Rev.*, 13: 476–498.

Levavi-Sivan, B. and Yaron, Z. (1989) Gonadotropin secretion from perifused tilapia pituitary in relation to gonadotropin-releasing hormone, extracellular calcium, and activation of protein kinase C. *Gen. Comp. Endocrinol.*, 75: 187–194.

Levi, N.L., Hanoch, T., Benard, O., Rozenblat, M., Harris, D., Reiss, N., Naor, Z. and Seger, R. (1998) Stimulation of Jun N-terminal kinase (JNK) by gonadotropin-releasing hormone in pituitary αT3-1 cell line is mediated by protein kinase C, c-Src, and CDC42. *Mol. Endocrinol.*, 12: 815–824.

Lewis, T.S., Shapiro, P.S. and Ahn, N.G. (1998) Signal transduction through MAP kinase cascades. *Adv. Cancer Res.*, 74: 49–139.

Lin, X.W. and Peter, R.E. (1996) Expression of salmon gonadotropin-releasing hormone (GnRH) and chicken GnRH-II precursor messenger ribonucleicacids in the brain and ovary of goldfish. *Gen. Comp. Endocrinol.*, 101: 282–296.

Lin, X.W., Lin, H.R. and Peter, R.E. (1993) Growth hormone and gonadotropin secretion in the common carp (*Cyprinus carpio* L.): in vitro interactions of gonadotropin-releasing hormone, somatostatin, and the dopamine agonist apomorphine. *Gen. Comp. Endocrinol.*, 89: 62–71.

Liu, W.S. and Heckman, C.A. (1998) The sevenfold way of PKC regulation. *Cell Signal.*, 10: 529–542.

Lo, A. and Chang, J.P. (1998a) In vitro action of testosterone in potentiating gonadotropin-releasing hormone-stimulated gonadotropin-II secretion in goldfish pituitary cells: involvement of protein kinase C, calcium and testosterone metabolites. *Gen. Comp. Endocrinol.*, 111: 318–333.

Lo, A. and Chang, J.P. (1998b) In vitro application of testosterone potentiates gonadotropin-releasing hormone-stimulated gonadotropin-II secretion from cultured goldfish pituitary cells. *Gen. Comp. Endocrinol.*, 111: 334–346.

Lopez-Ilasaca, M. (1998) Signaling from G-protein-coupled receptors to mitogen-activated protein (MAP)-kinase cascades. *Biochem. Pharmacol.*, 56: 269–277.

Mahmoud, S.S., Moloney, M.M. and Habibi, H.R. (1996) Cloning and sequencing of the goldfish growth hormone cDNA. *Gen. Comp. Endocrinol.*, 101: 139–144.

Marchant, T.A., Chang, J.P., Nahorniak, C.S. and Peter, R.E. (1989) Evidence that gonadotropin-releasing hormone also functions as a growth hormone-releasing factor in the goldfish. *Endocrinology*, 124: 2509–2518.

Melamed, P., Gur, G., Elizur, A., Rosenfeld, H., Sivan, B., Rentier-Delrue, F. and Yaron, Z. (1996) Differential effects of gonadotropin-releasing hormone, dopamine and somatostatin and their second messengers on the mRNA levels of gonadotropin IIβ subunit and growth hormone in the teleost fish, Tilapia. *Neuroendocrinology*, 64: 320–328.

Millar, R.P., Troskie, B., Sun, Y.M., Ott, T., Wakefield, I., Myburgh, D., Pawson, A., Davidson, J.S., Flanagan, C., Katz, A., Hapgood, J., Illing, N., Weinstein, H., Sealfon, S.C., Peter, R.E., Terasawa, E. and King, J.A. (1997) Plasticity in the structural and functional evolution of GnRH: a peptide for all seasons. In: S. Kawashima and S. Kikuyama (Eds.), *Proceedings of the XIIIth International Congress of Comparative Endocrinology*, Monduzzi Editore, pp. 15–27.

Millar, R., Lowe, S., Conklin, D., Pawson, A., Maudsley, S., Troskie, B., Ott, T., Millar, M., Lincoln, G., Sellar, R., Faurholm, B., Scobie, G., Kuestner, R., Terasawa, E. and Katz, A. (2001) A novel mammalian receptor for the evolutionarily conserved type II GnRH. *Proc. Natl. Acad. Sci. USA*, 98: 9636–9641.

Montaner, A.D., Park, M.K., Fischer, W.H., Craig, A.G., Chang, J.P., Somoza, G.M., Rivier, J.E. and Sherwood, N.M. (2001) Primary structure of a novel gonadotropin-releasing hormone in the brain of a teleost, pejerrey. *Endocrinology*, 142: 1453–1460.

Morgan, R.O., Chang, J.P. and Catt, K.J. (1987) Novel aspects of gonadotropin-releasing hormone action on inositol polyphosphate metabolism in cultured pituitary gonadotrophs. *J. Biol. Chem.*, 262: 1166–1171.

Mulvaney, J.M., Zhang, T., Fewtrell, C. and Roberson, M.S. (1999) Calcium influx through L-type channels is required for selective activation of extracellular signal-regulated kinase by gonadotropin-releasing hormone. *J. Biol. Chem.*, 274: 29796–29804.

Naor, Z. (1990) Signal transduction mechanisms of Ca^{2+} mobilizing hormones: the case of gonadotropin-releasing hormone. *Endocr. Rev.*, 11: 326–353.

Naor, Z., Dan-Cohen, H., Hermon, J. and Limor, R. (1989) Induction of exocytosis in permeabilized pituitary cells by α- and β-type protein kinase C. *Proc. Natl. Acad. Sci. USA*, 86: 4501–4504.

Naor, Z., Harris, D. and Shacham, S. (1998) Mechanism of GnRH receptor signaling: combinatorial cross-talk of Ca^{2+} and protein kinase C. *Front. Neuroendocrinol.*, 19: 1–19.

Naor, Z., Benard, O. and Seger, R. (2000) Activation of MAPK cascades by G-protein-coupled receptors: The case of gonadotropin-releasing hormone receptor. *Trends Endocrinol. Metab.*, 11: 91–99.

Neill, J.D., Duck, L.W., Sellers, J.C. and Musgrove, L.C. (2001) A gonadotropin-releasing hormone (GnRH) receptor specific for GnRH II in primates. *Biochem. Biophys. Res. Commun.*, 282: 1012–1018.

Pati, D. and Habibi, H.R. (1998) Presence of salmon gonadotropin-releasing hormone (GnRH) and compounds with GnRH-like activity in the ovary of goldfish. *Endocrinology*, 139: 2015–2024.

Pati, D. and Habibi, H.R. (2000) Direct action of GnRH variants on goldfish oocyte meiosis and follicular steroidogenesis. *Mol. Cell. Endocrinol.*, 160: 75–88.

Pati, D. and Habibi, H.R. (2002) The involvement of protein kinase C and arachidonic acid pathways in the gonadotropin-releasing hormone regulation of oocyte meiosis and follicular steroidogenesis in the goldfish ovary. *Biol. Reprod.*, 66: 813–822.

Pati, D., Lohka, M.J. and Habibi, H.R. (2000) Time-related effect of GnRH on histone H1 kinase activity in the goldfish follicle-enclosed oocyte. *Can. J. Physiol. Pharmacol.*, 78: 1067–1071.

Pearson, G., Robinson, F., Gibson, T.B., Xu, B.E., Karandikar, M., Berman, K. and Cobb, M.H. (2001) Mitogen-activated protein (MAP) kinase pathways: regulation and physiological functions. *Endocr. Rev.*, 22: 153–183.

Peter, R.E. and Yu, K.L. (1997) Neuroendocrine regulation of ovulation in fishes: basic and applied aspects. *Rev. Fish Biol. Fish.*, 7: 173–197.

Pierce, K.L., Luttrell, L.M. and Lefkowitz, R.J. (2001) New mechanisms in heptahelical receptor signaling to mitogen activated protein kinase cascades. *Oncogene*, 20: 1532–1539.

Porter, D.A. and Licht, P. (1986) Dependence of GnRH action on Na^+, K^+, and Ca^{2+} in the frog, *Rana pipiens*, pituitary. *J. Exp. Zool.*, 239: 379–391.

Rasmussen, H., Isales, C.M., Calle, R., Throckmorton, D., Anderson, M., Gasalla-Herraiz, J. and McCarthy, R. (1995) Diacylglycerol production, Ca^{2+} influx, and protein kinase C activation in sustained cellular responses. *Endocr. Rev.*, 16: 649–681.

Rebers, F.E., Bosma, P.T., Van Dijk, W., Goos, H.J.T. and Schulz, R.W. (2000) GnRH stimulates LH release directly via inositol phosphate and indirectly via cAMP in African catfish. *Am. J. Physiol. Regul. Integr. Comp. Physiol.*, 278: R1572–R1578.

Reiss, N., Llevi, L.N., Shacham, S., Harris, D., Seger, R. and Naor, Z. (1997) Mechanism of mitogen-activated protein kinase activation by gonadotropin-releasing hormone in the pituitary αT3-1 cell line: differential roles of calcium an protein kinase C. *Endocrinology*, 138: 1673–1682.

Roberson, M.S., Misra-Press, A., Laurance, M.E., Stork, P.J.S. and Maurer, R.A. (1995) A role for mitogen-activated protein kinase in mediating activation of the glycoprotein hormone α-subunit promoter by gonadotropin-releasing hormone. *Mol. Cell. Biol.*, 15: 3531–3539.

Roberson, M.S., Zhang, T., Li, H.L. and Mulvaney, J.M. (1999) Activation of the p38 mitogen-activated protein kinase pathway by gonadotropin-releasing hormone. *Endocrinology*, 140: 1310–1318.

Saunders, B.D., Sabbagh, E., Chin, W.W. and Kaiser, U.B. (1998) Differential use of signal transduction pathways in the gonadotropin-releasing hormone-mediated regulation of gonadotropin subunit gene expression. *Endocrinology*, 139: 1835–1843.

Schoderbek, W.E., Roberson, M.S. and Maurer, R.A. (1993) Two different DNA elements mediate gonadotropin releasing hormone effects on expression of the glycoprotein hormone α-subunit gene. *J. Biol. Chem.*, 268: 3903–3910.

Shacham, S., Cheifetz, M.N., Lewy, H., Ashkenazi, I.E., Becker, O.M., Seger, R. and Naor, Z. (1999) Mechanism of GnRH receptor signaling: from the membrane to the nucleus. *Ann. Endocrinol. (Paris)*, 60: 79–88.

Shupnik, M.A. (1996) Gonadotropin gene modulation by steroids and gonadotropin-releasing hormone. *Biol. Reprod.*, 54: 279–286.

Smith, C.E., Wakefield, I., King, J.A., Naor, Z., Millar, R.P. and Davidson, J.S. (1987) The initial phase of GnRH-stimulated LH release from pituitary cells is independent of calcium entry through voltage-gated channels. *FEBS Lett.*, 225: 247–250.

Stanislaus, D., Pinter, J.H., Janovick, J.A. and Conn, P.M. (1998) Mechanisms mediating multiple physiological responses to gonadotropin-releasing hormone. *Mol. Cell. Endocrinol.*, 144: 1–10.

Steele, G.L. and Leung, P.C.K. (1993) Signal transduction mechanisms in ovarian cells. In: E.Y. Adashi and P.C.K. Leung (Eds.), *The Ovary*. Raven Press, New York, NY, pp. 113–127.

Stefano, A.V., Vissio, P.G., Paz, D.A., Somoza, G.M., Maggese, M.C. and Barrantes, G.E. (1999) Colocalization of GnRH binding sites with gonadotropin-, somatotropin-, somatolactin-, and prolactin-expressing pituitary cells of the pejerrey, *Odontesthes bonariensis*, in vitro. *Gen. Comp. Endocrinol.*, 116: 133–139.

Stojilkovic, S.S. and Catt, K.J. (1995) Novel aspects of GnRH-induced intracellular signaling and secretion in pituitary gonadotrophs. *J. Neuroendocrinol.*, 7: 739–757.

Stojilkovic, S.S., Chang, J.P., Izumi, S.I., Tasaka, K. and Catt, K.J. (1988a) Mechanisms of secretory responses to gonadotropin-releasing hormone and phorbol esters in cultured pituitary cells. Participation of protein kinase C and extracellular calcium mobilization. *J. Biol. Chem.*, 263: 17301–17306.

Stojilkovic, S.S., Chang, J.P., Ngo, D. and Catt, K.J. (1988b) Evidence for a role of protein kinase C in luteinizing hormone synthesis and secretion. Impaired responses to gonadotropin-releasing hormone in protein kinase C-depleted pituitary cells. *J. Biol. Chem.*, 263: 17307–17311.

Stojilkovic, S.S., Reinhart, J. and Catt, K.J. (1994) Gonadotropin-releasing hormone receptors: structure and signal transduction pathways. *Endocr. Rev.*, 15: 462–499.

Strahl, B.D., Huang, H.J., Sebastian, J., Ghosh, B.R. and Miller, W.L. (1998) Transcriptional activation of the ovine follicle-stimulating hormone β-subunit gene by gonadotropin-releasing hormone: involvement of two activating protein-1-binding sites and protein kinase C. *Endocrinology*, 139: 4455–4465.

Sun, Y.M., Flanagan, C.A., Illing, N., Ott, T.R., Sellar, R., Fromme, B.J., Hapgood, J., Sharp, P., Sealfon, S.C. and Millar, R.P. (2001) A chicken gonadotropin-releasing hormone receptor that confers agonist activity to mammalian antagonists. Identification of D-Lys[6] in the ligand and extracellular loop two of the receptor as determinants. *J. Biol. Chem.*, 276: 7754–7761.

Sundaresan, S., Colin, I.M., Pestell, R.G. and Jameson, J.L. (1996) Stimulation of mitogen-activated protein kinase by gonadotropin-releasing hormone: evidence for the involvement of protein kinase C. *Endocrinology*, 137: 304–311.

Tremblay, J.J. and Drouin, J. (1999) Egr-1 is a downstream effector of GnRH and synergizes by direct interaction with Ptx1 and SF-1 to enhance luteinizing hormone β gene transcription. *Mol. Cell. Biol.*, 19: 2567–2576.

Troskie, B., Illing, N., Rumbak, E., Sun, Y.M., Hapgood, J., Sealfon, S., Conklin, D. and Millar, R. (1998) Identification of three putative GnRH receptor subtypes in vertebrates. *Gen. Comp. Endocrinol.*, 112: 296–302.

128

Vaananen, J.E., Tong, B.L., Vaananen, C.M., Chan, I.H., Yuen, B.H. and Leung, P.C. (1997) Interaction of prostaglandin F2alpha and gonadotropin-releasing hormone on progesterone and estradiol production in human granulosa-luteal cells. *Biol. Reprod.*, 57: 1346–1353.

Van Biesen, T., Luttrell, L.M., Hawes, B.E. and Lefkowitz, R.J. (1996) Mitogenic signaling via G protein-coupled receptors. *Endocr. Rev.*, 17: 698–714.

Van Der Kraak, G. and Chang, J.P. (1990) Arachidonic acid stimulates steroidogenesis in goldfish preovulatory ovarian follicles. *Gen. Comp. Endocrinol.*, 77: 221–228.

Van Der Kraak, G., Chang, J.P. and Janz, D.M. (1998) Reproduction. In: D.H. Evans (Ed.), *The Physiology of Fishes*. CRC Press, New York, NY, pp. 465–488.

Wang, J. and Leung, P.C.K. (1988) Role of arachidonic acid in luteinizing hormone-releasing hormone action: stimulation of progesterone production in rat granulosa cells. *Endocrinology*, 122: 906–911.

Wang, L., Bogerd, J., Choi, H.S., Seong, J.Y., Soh, J.M., Chun, S.Y., Blomenrohr, M., Troskie, B.E., Millar, R.P., Yu, W.H., McCann, S.M. and Kwon, H.B. (2001) Three distinct types of GnRH receptor characterized in the bullfrog. *Proc. Natl. Acad. Sci. USA*, 98: 361–366.

Watanabe, H., Tanaka, S., Akino, T. and Hasegawa-Sasaki, H. (1990) Evidence for coupling of different receptors for gonadotropin-releasing hormone to phospholipase C and A2 in cultured rat luteal cells. *Biochem. Biophys. Res. Commun.*, 168: 328–334.

Weck, J., Fallest, P.C., Pitt, L.K. and Shupnik, M.A. (1998) Differential gonadotropin-releasing hormone stimulation of rat luteinizing hormone subunit gene transcription by calcium influx and mitogen-activated protein kinase-signaling pathways. *Mol. Endocrinol.*, 12:451–457.

Wiebe, J.P., Dhanvantari, S., Watson, P.H. and Huang, Y. (1994) Suppression in gonadotropes of gonadotropin-releasing hormone-stimulated follicle-stimulating hormone release by the gonadal- and neurosteroid 3α-hydroxy-4-pregnen-20-one involves cytosolic calcium. *Endocrinology*, 134: 377–382.

Yaron, Z., Gur, G., Melamed, P., Rosenfeld, H., Levavi-Sivan, B. and Elizur, A. (2001) Regulation of gonadotropin subunit genes in tilapia. *Comp. Biochem. Physiol. B*, 129: 489–502.

Yu, K.L., Sherwood, N.M. and Peter, R.E. (1988) Differential distribution of two molecular forms of gonadotropin-releasing hormone in discrete brain areas of goldfish (*Carassius auratus*). *Peptides*, 9: 625–630.

I.S. Parhar (Ed.)
Progress in Brain Research, Vol. 141
© 2002 Elsevier Science B.V. All rights reserved

CHAPTER 10

Gonadotropin-releasing hormone receptor: cloning, expression and transcriptional regulation

Sham S. Kakar*, M. Tariq Malik, Stephen J. Winters

Department of Medicine, University of Louisville, Louisville, KY 40202, USA

Introduction

Gonadotropin-releasing hormone (GnRH), also referred to as LHRH (luteinizing hormone-releasing hormone), is a hypothalamic decapeptide (pGlu–His–Trp–Ser–Tyr–Gly–Leu–Arg–Pro–Gly–NH$_2$), which was isolated and characterized by groups led by Schally and Guillemin, the 1977 Nobel laureates. It is synthesized by hypothalamic neurons and released into the portal blood in a pulsatile fashion. GnRH stimulates the synthesis and secretion of gonadotropins (follicle stimulating hormone, FSH; and luteinizing hormone, LH) from gonadotropes in the anterior pituitary. Gonadotropins, in turn, regulate both the gametogenic and steroidogenic functions of the gonads (Fink, 1988). These actions of GnRH are achieved through its high affinity receptors that are expressed on the plasma cell membranes of gonadotropes. The expression of mRNA for GnRH and its receptor in pituitary as well as extra-pituitary tissues, including the placenta, ovary, myometrium, endometrium, breast, prostate, and blood mononuclear cells, has been documented. The binding of GnRH to its specific receptors initiates activation. Ligand binding allows coupling of receptor to the Gαq$_{11}$ protein with generation of several second messengers, most notable diacylglycerol

and inositol triphosphate (IP$_3$). The increased level of diacyglycerol leads to activation of protein kinase C, and the increased level of inositol triphosphate leads to the production of cAMP and the release of calcium from intracellular pools, with both of these downstream events leading to secretion of FSH and LH. Following prolonged, continuous activation of the GnRH receptor by GnRH, however, the release of gonadotropins is followed by receptor desensitization and a cessation of LH secretion. Our understanding of the functioning of GnRH receptor, the molecular mechanisms that mediate its signal transduction, and its transcriptional regulation has been advanced by the cloning of the GnRH receptor cDNA and its gene. In this review, we present a summary of the cloning of the GnRH receptor cDNA and its gene, and recent insights into its regulation.

Molecular structure of GnRH receptors

The cloning of the GnRH receptor has greatly improved our understanding of the mechanisms that underlie GnRH regulation of reproductive functions, as well as other extra-pituitary functions. Using the *Xenopus* oocyte as an expression system, Tsutsumi and colleagues (Tsutsumi et al., 1992) successfully isolated a cDNA clone for the GnRH receptor from a mouse gonadotrope cell line (αT31). As anticipated, the GnRH receptor contains seven transmembrane domains, belongs to the G-protein-coupled receptor family, and utilizes Ca^{2+} as a second messenger. Subsequently, Kakar and others isolated the cDNAs for the GnRH receptors of humans (Kakar et al.,

---

* Correspondence to: S.S. Kakar, Department of Medicine, University of Louisville, Louisville, KY 40202, USA. Tel.: +1 (502)-852-0812; Fax: +1 (502)-852-2356; E-mail: sskaka01@louisville.edu

1992; Chi et al., 1993), as well as the rat (Kakar et al., 1994a), cow (Kakar et al., 1993), ovine (Brooks et al., 1993), porcine (Weesner and Matteri, 1994), and marmoset (Byrne et al., 1999). Like the mouse receptor, the GnRH receptors from all the species analyzed to date contain seven transmembrane domains and belong to the family of G-protein-coupled receptors.

The deduced amino acid sequence of the human GnRH receptor is shown in Fig. 1. The human, cow, ovine, porcine and marmoset monkey receptor cDNAs encode a protein of 328 amino acids, whereas the mouse and rat receptor cDNAs encode 327 amino acid proteins, as illustrated in Fig. 2. The predicted amino acid sequences for the GnRH receptors of all of the mammalian species are 85% identical. The binding affinity of the cloned receptors after their transfection into heterologous systems is high, and the pharmacologic properties are similar to those of the native receptors (Brown and Reeves, 1983; Wormald et al., 1985; Horn et al., 1991; Tsutsumi et al., 1992). The cloning of GnRH receptors from human breast and ovarian tumors, gonads and placenta demonstrated that the sequences of these receptors are identical to those of the pituitary receptor (Kakar et al., 1994b; Moumni et al., 1994; Boyle et al., 1998).

Analysis of the amino acid sequences of the GnRH receptors revealed homology with other G-protein-coupled receptors, with some amino acid residues in the transmembrane domains 2 (TM2), 3 (TM3), 5 (TM5), 6 (TM6), and 7 (TM7), and the first extracellular domain being highly conserved among this group of proteins (Davidson et al., 1994). Notably, there is a conserved proline residue in each of the TM2, TM4, TM5, TM6 and TM7 domains, and residues Asn[53] (TM1), Trp[164] (TM4), Ser[167] (TM4), Asn[135] (TM7) and Tyr[323] (TM7) are conserved. There is, however, little sequence homology with other G-protein-coupled receptors in the intra- and extracellular domains of GnRH receptors. The human, cow, ovine and porcine receptor sequences contain two potential sites for N-linked gylcosylation (N-X-S/T), one in the amino terminus and one in the first extracellular domain (Fig. 2), whereas the rodent receptors contain three potential N-linked gylcosylation sites. Although mutation of the N-linked gylcosylation site in the N-terminal domain to Gln caused

a change in apparent molecular weight, the ligand binding affinity for the receptor was not altered. Moreover, alteration of the potential gylcosylation site (Asn) in the first extracellular domain to Gln did not affect the mobility of the receptor nor its binding affinity. These results suggest that although gylcosylation of GnRH receptors may be essential for receptor expression and stability, it probably is not required for GnRH binding (Davidson et al., 1995). As in many of the G-protein-coupled seven transmembrane receptors, cysteine residues are present in the first and second extracellular domains of GnRH receptors, suggesting the presence of disulfide bonds, which have been reported to stabilize ligand binding (Dohlman et al., 1990). Using site-directed mutagenesis and photoaffinity cross-linking, Davidson et al. (1997) identified two cystine disulfide bridges between Cys[14]–Cys[200], and Cys[114]–Cys[196] respectively. Consistent with findings for the human GnRH receptor, Cook and Eidne (1997) reported the presence of two disulfide bridges between Cys[14]–Cys[199], and Cys[114]–Cys[195] for the rat GnRH receptor. These disulfide bridges were proposed to maintain the structure and function of the GnRH receptor (Cook and Eidne, 1997; Davidson et al., 1997). The GnRH receptors also contain five potential sites for phosphorylation. These sites are located in the intracellular loops and may be involved in receptor desensitization and signal transduction (Fig. 1).

There are several features of the GnRH receptor that are unusual in the G-protein-coupled receptor family. These include lack of the entire intracellular C-terminal domain that is required for desensitization of many G-protein-coupled receptors (Attwood et al., 1991). In addition, there are unusual substitutions at loci that are highly conserved, such as modification of the more common DRY sequence, known as the 'signature' sequence, that is localized at the junction of the third transmembrane domain and the second intracellular loop. DRY is replaced by DRS in GnRH receptors (Fig. 1). Mutation of DRS to DRY in the mammalian receptors resulted in a small increase in agonist affinity with no discernible change in signal transduction (Arora et al., 1995). Finally, there is a reciprocal interchange of Asn[87] and Asp[318] residues in transmembrane domains 2 and 7 as compared to the G-protein-coupled receptors.

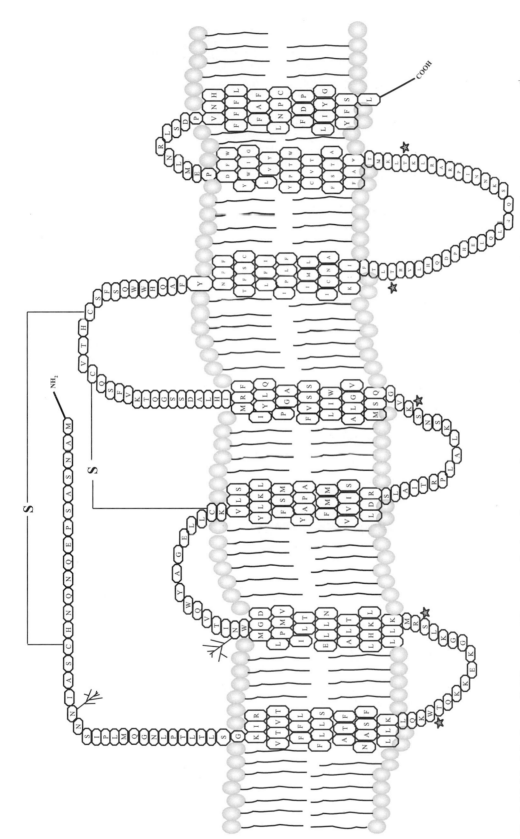

Fig. 1. Model of the human GnRH receptor. The receptor is shown in proposed seven transmembrane topology. The potential sites for N-linked gylcosylation ($\curlyvee$), phosphorylation by protein kinase C ($\bigstar$) and disulfide bridges between cysteine residues ($-$S$-$) are indicated.

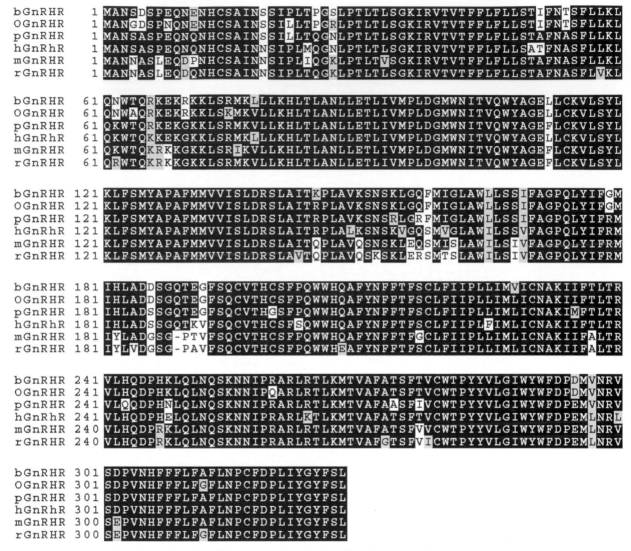

Fig. 2. Comparison of the deduced amino acid sequence of the bovine (b), ovine (o), porcine (p), human (h), mouse (m) and rat (r) GnRH receptors. Amino acids are represented by single letter codes. Identical amino acids are boxed. Amino acid present in bovine, ovine, porcine and human GnRH receptors but absent in mouse and rat GnRH receptors is indicated by hyphen (—).

Cloning of GnRH receptors from non-mammalian species

To date at least 12 forms of GnRH have been identified in vertebrates (Sherwood et al., 1993, 1997; Craig et al., 1997; King and Millar, 1997) and 2 variants of GnRH in an invertebrate, the tunicate (Powell et al., 1996). Two or more forms of GnRH are present in most vertebrate species (Sealfon et al., 1997). One form is represented by mammalian

GnRH and its non-mammalian counterparts, which have a predominant function as neuropeptides regulating the pituitary. The second form of GnRH, first identified in chicken brain (His[5]Trp[7]Tyr[8]GnRH), is the most ubiquitous form in vertebrates, and most species have this form together with one or two other GnRHs. The mammalian GnRH (GnRH-I) exhibits high affinity binding to mammalian GnRH receptors but only low affinity binding to the non-mammalian receptors. On the other hand, non-mammalian GnRH (GnRH-II, also called chicken GnRH-II) binds non-mammalian GnRH receptors with high affinity and binds mammalian receptors with low affinity. This preferential binding raised the question as to whether a distinct GnRH receptor is expressed in non-mammalian vertebrates. To address this issue, Tensen et al. (1997) cloned the GnRH receptor from African catfish, and subsequently others reported the cloning of receptors from chicken (Troskie et al., 1997), frog (Wang et al., 2001), goldfish (Illing et al., 1999), and Xenopus (Troskie et al., 2000). In addition, existence of more than one GnRH receptor subtype has been reported in goldfish (Illing et al., 1999) and frog (Wang et al., 2001). These studies revealed that the GnRH receptors from non-mammalian species encode proteins of 368 to 412 amino acids, which, like mammalian GnRH receptors, contain seven transmembrane domains and belong to the family of G-protein-coupled receptors. The non-mammalian GnRH receptors were found to be only 38% to 41% homologous with the mammalian GnRH receptors at the amino acid level. Unlike mammalian GnRH receptors, non-mammalian GnRH receptors contain the C-terminal domain of 51 amino acids. The significance of this C-terminal domain of the non-mammalian GnRH receptors in the desensitization and internalization of the receptors was demonstrated by Pawson et al. (1998), who showed that deletion of the C-terminal domain from the chicken GnRH receptor resulted in a reduction in the internalization of the receptor. Furthermore, fusion of the C-terminal tail of the catfish GnRH receptor to the rat receptor resulted in an increase in receptor desensitization and internalization (Lin et al., 1998), suggesting that the C-terminal tail is essential for rapid receptor desensitization and internalization.

Cloning of mammalian GnRH receptor II

As discussed above, the existence of GnRH-II (chicken GnRH-II) in non-mammalian species is well established. More recently, several groups have identified a second form of GnRH (GnRH-II) in the brain of mammalian species (Kasten et al., 1996; Lescheid et al., 1997; Chen et al., 1998; White et al., 1998; Urbanski et al., 1999; Latimer et al., 2000), and the GnRH-II gene was cloned from human

(White et al., 1998) and monkey brain (Urbanski et al., 1999). As GnRH-II receptors exist in non-mammalian species, the search for similar receptors in mammalian species was undertaken, and recently, Neill et al. (2001) and Millar et al. (2001) reported the cloning of GnRH-II receptors from rhesus monkey pituitary, African green monkey kidney cell line (COS-7), and marmoset pituitary. The GnRH-II receptors (GnRH receptor II) that have been cloned from these species encode a protein of 379 amino acids and, as anticipated, belong to the family of G-protein-coupled receptors. However, the GnRH receptor II exhibits only 41% identity with the human type I receptor, and unlike the type I receptor, has a C-terminal tail (Fig. 3). Moreover, the type II receptor is highly selective for GnRH-II. In common with the type I receptor, it couples to $G\alpha q_{11}$ and activates extracellular signal-regulated kinase (ERK1/2) but differs in that it activates p38 mitogen-activated protein (MAP) kinase.

The expression of this GnRH receptor II in humans remains controversial, however. It has been reported that exon 3 transcripts can be found in various human tissues (Millar et al., 2001; Neill et al., 2001), but the existence of a full-length transcript have not been described to date. A sequence homologous to the monkey GnRH receptor II has been localized on human chromosome 1, and a truncated complementary sequence homologous to ribosomal protein RBM8A has been localized on human chromosome 8 (Millar et al., 1999; Faurholm et al., 2001). Based on the genomic sequence, Faurholm et al. (2001) predicted that the human GnRH receptor II does not possess a classical methionine as the start codon but does contain a stop codon in exon 2, suggesting that the human GnRH receptor II may be truncated and non-functional.

Cloning of the human GnRH receptor gene and its molecular structure

Using PCR analysis of the somatic hybrid cell lines and chromosomal in situ hybridization, we localized the human GnRH receptor gene to chromosome 4 (4q13.1) (Kakar and Neill, 1994). The mouse gene has been mapped by linkage analysis to be within 1.2 ± 1.2 centimorgans of the chromosome 5 marker Pmv-11 (Kaiser et al., 1994), and the ovine gene

134

```
hGnRHRI    -MANSASPEQNQNHCSAINNSIPLMQGN-LPTLTLSGKIRVTVTFFLFLLSATFNASFLL  58
mGnRHRII   MSAGNGTPWGSAAGEESWAASGVAVEGSELPTFSAAAKVRVGVTIVLFVSSAGGNLAVLW  60

hGnRHRI    KLQKWTQKKEKGKKLSRMKLLLLKHLTLANLLETLIVMPLDGMWNITVQWYAGELLCKVLS  118
mGnRHRII   SVTRPQPSQLRP---SPVRTLFAHLAAADLLVTFVVMPLDATWNITVQWLAEDIACRTLM  117

hGnRHRI    YLKLFSMYAPAFMMVVISLDRSLAITRPLALKSNSKVGQSMVGLAWILSSVFAGPQLYIF  178
mGnRHRII   FLKLMAMYSAAFLPVVIGLDRQAAVLNPLGSRSGVRK---LLGAAWGLSFLLALPQLFLF  174

hGnRHRI    RMIHLADSSGQTKVFSQCVTHCSFSQWWHQAFYNFFTFSCLFIIPLFIMLICNAKIIFTL  238
mGnRHRII   HTVHRA---G-PVPFTQCVTKGSFKARWQETTYNLFTFRCLFLLPLTAMAICYSHIVLSV  230

hGnRHRI    ----TRVL-HQDPHELQLNQSKNNIPRARLKTLKMTVAFATSFTVCWTPYYVLGIWYWFD  293
mGnRHRII   SSPQTRKGSHAPAGEFALCRSFDNCPRVRLWALRLALLILLTFILCWTPYYLLGLWYWFS  290

hGnRHRI    PEMLNRLSDPVNHFFFLFAFLNPCFDPLIYGYFSL*                         328
mGnRHRII   PTMLTEVPPSLSHILFLFGLLNAPLDPLLYGAFTLGCQRGHQELSIDSSNEGSGRMLQQE  350

mGnRHRII   IHALRQQEVQKTVTSRSAGETKDISITSI*                               379
```

Fig. 3. Comparison of the deduced amino acid sequence of the human GnRH receptor type I (hGnRHRI) and monkey (Macaca mulatta) GnRH receptor type II (mGnRHRII). The amino acids are represented by single letter codes. Identical amino acids are boxed. Gaps (-) are generated to obtain maximum alignment.

was localized near the FecB locus on chromosome 6 (Montgomery et al., 1995). The structures of the human, mouse, rat and ovine GnRH receptor genes have been characterized (Albarracin et al., 1994; Fan et al., 1995; Campion et al., 1996; Kakar, 1997; Reinhart et al., 1997). A comparison of the GnRH receptor gene sequences and cDNA sequences reveals that, unlike other G-protein-coupled receptors that are intronless, each of these genes has two introns at the same locations in the reading frame. The first intron (intron 1) is larger in the mouse (15 kb) than in the rat (12 kb), human (4 kb), or sheep (>10 kb). It is located in transmembrane domain 4, with intron 2 being located between transmembrane domains 5 and 6. In the human gene, exon 1 encodes the 5′ untranslated sequence and nucleotides +1 to +522 in the open reading frame; exon 2, encodes nucleotides +523 to +742, and exon 3, encodes nucleotide +743 to +983 in the open reading frame and the 3′ untranslated sequence (Fig. 4). The human exon 2 is three nucleotides longer than the mouse and rat genes, resulting in a single additional amino acid (Lys[191]) (Fig. 2) in the second extracellular loop of the human receptor. There is 42% sequence ho-

mology among the genes of all four species within a 0.9 kb region from the ATG codon. The human GnRH receptor gene is more closely related to the ovine gene with 70% sequence homology, whereas same region of the human gene is 46% homologous with the rat and mouse genes (Campion et al., 1996). One significant difference between the mouse, rat, and ovine genes, and the human gene is the location of the transcriptional start sites. The start sites for the rat and mouse are clustered in a region within 110 bp from the translation codon (ATG) except the distal site described by Clay et al. (1995). Mapping of the transcription start site(s) for the human GnRH receptor using 5′ extension and 5′-RACE analysis of human pituitary RNA revealed multiple transcription start sites in the 5′ flanking region of the gene. Fan et al. (1995) using human brain RNA reported five transcriptional start sites, whereas, using human pituitary RNA, we found 18 transcriptional start sites (Figs. 5 and 6) (Kakar, 1997). The nearest start sites in the human and ovine genes are at least 600 bp upstream from the ATG start codon. The human GnRH receptor gene contains multiple TATA sequences (six compared with none in mouse) and

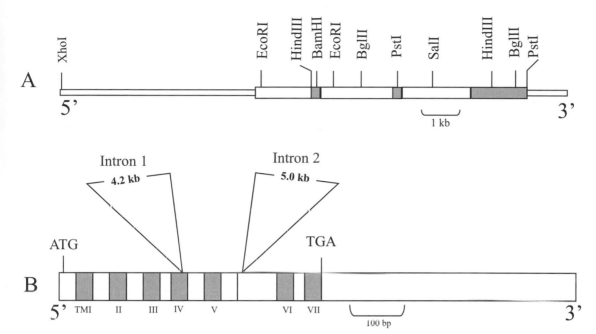

Fig. 4. Genomic organization of the human GnRH receptor gene. Panel A: The solid boxes indicate the exons. The relevant restriction enzymes sites are shown. Panel B: The structure of the human GnRH receptor cDNA. Open box indicate the coding region. Solid boxes indicate the putative transmembrane domains. The location and sizes of the introns are shown. Reproduced from Kakar (1997) with permission.

CAAT sequences (six) distributed over a 669 bp region from ATG (Fan et al., 1995; Kakar, 1997). The longest transcript has been confirmed by PCR and represents a transcript of 1393 bp of 5'-untranslated sequence (Fan et al., 1995). The existence of multiple transcription sites and the differences in these sites in the human pituitary and brain suggest the possibility of alternate transcriptional start sites in the pituitary and differences in the regulatory mechanisms in the pituitary and extra-pituitary tissues. In the human GnRH receptor gene, with regard to transcriptional regulation, the cAMP response element (CRE) has been identified at −2191 (relative to transcription start codon; ATG), and a putative glucocorticoid/progesterone response element (GRE/PRE) with one base pair replacement differentiating it from a functional GRE/PRE is found at −793 (relative to translation start codon). Additionally, AP-1 and AP-2 binding sites that implicate the effect of kinase C on the regulation of the receptor gene have been identified at −1517 and −4570 (relative to translation start site), respectively (Figs. 5 and 6). Pit-1, an anterior pituitary-specific transcription factor (Imagawa et al., 1987), which has been reported to be essential to the expression of growth hormone and prolactin (Bodner et al., 1988; Ingraham et al., 1990), and a steroidgenic factor-1 (SF-1) binding site were identified at −942 and −134 (relative to translation start site), respectively (Fan et al., 1995; Kakar, 1997; Ngan et al., 1999). At the 3' end of the gene, five polyadenylation signals are found, which are distributed over 800 bp (Fan et al., 1995).

Determination of the putative promoter for the GnRH receptor gene

To identify the promoter sequence for the human GnRH receptor gene, we analyzed a 4600 bp sequence upstream of the translation start site (ATG), and prepared a number of pGL3-luc (luciferase) chimeric constructs containing the gene sequences (−4615 to +1), (−4615 to −1018), (−4115 to −1018), (−3571 to −1018), (−2284 to −1018), (−2284 to −1657), (−2106 to −1657), (−2003 to −1657), (−1350 to +1), (−936 to +1), (−936 to −397) and (−811 to −215) (relative to trans-

136

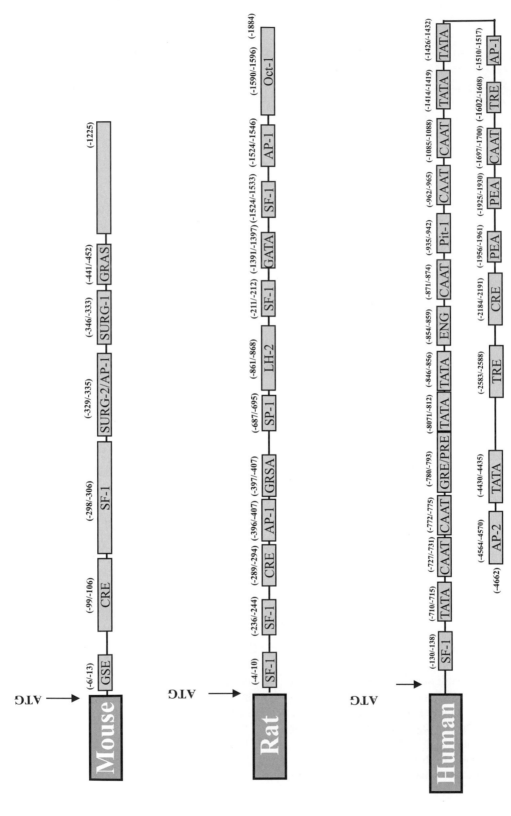

Fig. 5. Representations of the mouse, rat and human GnRH receptor gene 5′ flanking region. The translation start site is indicated by arrow (ATG). The regulatory element sequences are boxed and indicated. The numbering shown is relative to translation start site (ATG; +1).

lation start site). These constructs along with a pSV-gal (galactosidase) plasmid were cotransfected into a mouse pituitary gonadotrope-derived cell line (αT31), a human endometrial tumor cell line (HEC-1A), and an African green monkey kidney cell line (COS-7). After 48 h of transfection, the cells were lysed and luciferase activity was assayed. Based on this activity, we identified two putative promoters and one negative region for the gene. The distal promoter resides between −2003 and −1657 and the proximal promoter resides between −811 to −397, whereas a suppressor resides between −936 to +1 (unpublished results). Consistent with our results is the report of Kang et al. (2000) who, using similar techniques and αT31 and an ovarian tumor cell line (OVCAR-3), identified two putative promoters and a negative regulatory region. They localized one promoter, designated PR1, to a region between −771 to −557, and, a second, PR2 between −1351 to −1022, with a negative control region residing between −1022 to −771 in both the pituitary and ovarian cells. These investigators further reported the utilization of the same promoters to drive the basal promoter activities in both αT31 and OVCAR-3 cell, suggesting that the usage of the human GnRH receptor gene is not cell-specific. Ngan et al. (2000), using 5′ and 3′ deleted constructs of the human GnRH receptor gene flanking region, identified a distal promoter element at −1705/−1674 with reference to start codon (ATG). This promoter was found to be immediately 5′ to a previously identified CAP site at −1673 in the human pituitary gene. In addition, these investigators identified a pyrimidine-rich initiator element (Inr) (−1682) and a CAAT box (−1702) within the promoter sequence, and mutation of these elements abrogated both promoter activity and nuclear protein binding to the promoter sequence, suggesting that these elements may play an important role in regulating promoter activity. On the other hand, Cheng et al. (2000a), using a human placental cell line (JEG-3), reported the existence of promoter sequence between −1737 and −1346, relative to the translation start site. Taken together, these studies suggest the existence of a number of different promoters for the human GnRH receptor gene, and it is tempting to speculate that these differences may be associated with tissue-specific expression of the human GnRH receptor gene. It remains unclear,

however, as to which promoters and transcriptional start sites are utilized in different tissues in particular in pituitary and tumors.

In their analysis of the 1.2 kb 5′ flanking region of the mouse GnRH receptor gene, Albarracin et al. (1994) and Clay et al. (1995) showed a high level of promoter activity in αT31 cells but low levels of promoter activity in a rat somato-gonadotrope cell line (GH3), a human choriocarcinoma cell line (JEG-3), a monkey kidney fibroblast cell line (CV-1), and a monkey kidney cell line (COS-7). These findings support the concept of pituitary-specific promoter activity. In further studies, Clay et al. (1995), Duval et al. (1997a), Norwitz et al. (1999) and White et al. (1999) identified a 500 bp region that supported basal and GnRH-stimulated expression of the mouse GnRH receptor gene. Similarly, a rat 5′ flanking region has been reported to be associated with high promoter activity in gonadotrope αT31 cells but not in somatotrope GH3b, Chinese hamster ovary CHO, and monkey kidney COS-7 cells (Reinhart et al., 1997; Pincas et al., 1998). Again, this cell-specific expression of the GnRH receptor gene indicates the presence of tissue-specific or cell-specific transcriptional regulation of both the rat and mouse GnRH receptor genes. In contrast, however, the human GnRH receptor gene promoter(s) has been found to have equivalent activity in pituitary cells and extra-pituitary cells (unpublished results), a finding which is consistent with our previous finding for the expression of GnRH receptor mRNA in pituitary and extra pituitary tissues (Kakar and Jennes, 1995). The level of human GnRH receptor promoter activity was much lower in αT31 cells when compared to that of the rat and mouse receptor promoters, which may be due to the lack of specific factors in the mouse gonadotrope cell line (αT31) (Cheng et al., 2000a; Ngan et al., 2000).

Transcriptional regulation of the GnRH receptor gene

The concentration of the GnRH receptor protein is regulated tightly and exhibits both up- and down-regulation in response to its cognate ligand, gonadal steroids and peptide hormones (Clayton and Catt, 1981a,b; Lloyd and Childs, 1988; Laws et al., 1990; Gregg et al., 1991; Kaiser et al., 1993; Kakar et al.,

```
-4662 gcctgggcaa tatggtgsss ccctgtctct actaaaaata caaaaattag ctgtgcatgg
                                                AP-2
-4602 tggtgcacac ctgtagtccc agctacttgg gaggctgagg ccgaagaacc acttgaaccc
-4542 aggaaggcga aggttgcagt gagccgagat cgtgccactg cactccagcc tgggcgacag
                                                          TATA
-4482 agcaagactc ttgattgaaa aaaaaattca taacaataca atataagtat aaaatatgtt
-4422 tggaaatatc atatacgagc tttggctagc cttatttttg ·tatccacaa gtatagtgtg
-4362 tagtacataa tgcttaataa gtggttaatt gaattgttga aataaatata attcttatat
-4302 tcctggttta aatcttttta tatattgtta attatactta aaagtaaaga ttactacttt
-4242 tagtatttag tgaacataaa ttttgataag taaacatact tattatgctt ctgcttttct
-4182 gttcatttca tatgttggtg atttatatta tatatgattt ccttaatctg atctccaaaa
-4122 tcgtacatat ttaaccataa tttcccatca tttacctcta agacatctat actgaaataa
-4062 aaattgcaac aatgttttac agtgtgattc tccaaataat gagcacttct tggaataatg
-4002 aaagtgaatc tgccatgtta aatgtatgca taatttaaca tggcttagaa aaatttgtat
-3942 ggcttttaaa atttgttgct ctccttcaca acaattttgt gtatatattt cctcatacta
-3882 cttcatccac agtccttcct caaatgcttc catttaaact taaattttaa aagttcccct
-3822 tttactcatt tcacagttgt cgtagacact atcattcttt gctcaacaaa gtaagtcatg
-3762 ttttcaaaaa ctaaaaaaaa aaaaattctt tttttttctga tattgtgcta cttccatctt
-3702 ctctccttac aggaaacaaa attttttccc aaatacattt gcatacacat tcttcttatt
-3642 tgaaagactg ccttcctcta ttattttatt ctcatgagta aagaagtatg aaaaatattt
-3582 cttgctcccc tgggtcactt ctaaaccatc atcactcttt ccttcagaag ttggacttct
-3522 aattcaaaaa aaaaaaaaaa aactggctag ttaggttcaa gaagatgatg ggaaaaacac
-3462 tatgaagaaa aaaaagatac tataattgaa aaccaactca ggtgatgaat cagcaggagt
-3402 tgctcatcct taaagctgcc tgggctcctg ttggctttct aggtcttgtg tatcatgaaa
-3342 gataactcca ttatctaaaa cagtctccct ttcctgcaac taaaagggcc tgttaacacg
-3282 gcccttcatt cactaaggac tatggcacct ggctcataca cctttccacc ccaactgcta
-3222 ccagaaagta gcattgcata gtagttaaag acaggagctc tgaaggcaga ctgcatgaga
-3162 tgtaacacaa ttctgtggtg tcctaattag ggaaaaggag tcaggctggt gggagcaggg
-3102 gaaagcaaaa agaaaaagca gataagctac aaggacacat agccctcctg tgcaaataac
-3042 tctcaatctt cctatgccca actatcacca gacacctgca agttagctca ctgcaacctt
-2982 ggcattatca gtactgcaca aagccctctt cagcatacag cataaacact atcctataaa
-2922 atctccagca agcctttgtt tctttgcagt cagcttccct tccgctgatc ctgcccattg
-2862 tctccctggc aacgtatttt cctcccttct ctaacaaatt tgcctttctt ttacctacaa
-2802 tggtaaaccc ttttgtccct gcacctcggg cctagatagc caccgctccc catgacaaac
-2742 tctaatatgg atcaggtatg taaattctct ctgcttcacc ttcctcaacc ataagacgaa
-2682 aataatagta ctgacttcta gggttgttgt aaggtctaaa tgtgttaata cgtgtaacat
                                                TRE
-2622 gcttagaaaa gtgctgccac acaaagtagt acccaataaa tattaactat gagtaaatat
-2562 tatcagtatt gacagcacct tcaaaatcta tatagttgac tcatccaaca acttagattc
-2502 aacaacttag agctcctcaa atgcgtcaaa catcattcgc cctctacttc ctctaccgac
-2442 tttcatagcc acaccctgca tttgagggct ttataccttt catgttgtgt tgatagcaag
-2382 ggttttccaa atagattaac ctccctgaat tttgattgcc cactatttcc caaatccctt
-2322 caaggcattc tctgtatgct acacaataag ctttctgaag cataaatctg gccatacctta
-2262 ccgtatatta ctccattctt tatataggta aagcctaaac tccttttctt ggaatatagg
                       CRE
-2202 ctctaaagat ctgaagtctg cctaattatt tacacttttg ctttcacata ccctttgaac
-2142 tttctcacat tgtcttcgtg tttgcatgtg ctgctccagc ttctaagcat gccctccctg
-2082 tcctcatacc ccattcccca gccacttatt acgctatatg ctgtagtccg attcagctcg
-2022 gttgcaacct cttccctaat gaatcagtcc atcattaaca aagaaaggga gggagggagg
      PAE-3                                      PAE-3
-1962 gaggaagaga ggaatgaaag gaggaaaagg gaaggaaggg gaagggaagg ggaagggaag
-1902 gggaagggaa gggaagggga agggaaggaa tgggaggaaa agggacaaat aatgaatgat
-1842 atgctctaat ctttttcccc tagatataga agacsaaaga gcaaaatata cttcactaaa
                                                ⇓
-1782 ttgattttta cataaatttt ctttcctttg tttttttggtt gctggtccac ttacaaacac
                       CAAT                                          ⇓
-1722 ttttcatatt tgtatgtctt tccaatggtt atcctgtttt gttcatttca ggcatatggc
```

```
                                                      ⇓                              TRE
-1662  cctgatcaga  ttaactgaca  tgatgtatat  gcaaagcctt  ttgagttctt  cagaaaaata
                   ⇓
-1602  aattatctta  ttcaagactg  attgcttata  aggaacttat  tatagctaat  atagtaggca
                                          AP-1
-1542  caattttttt  tgtaattctc  ctagatgagt  cagaacttag  ttttgatgta  ggtaaaaatt
                   ⇓                                                    TATA
-1482  ttatggtcac  aaatctcagg  tgtgagaaaa  tctctttcct  tgatactcta  tataaataga
          TATA                                                     ⇓
-1422  ggatataaat  atttcaagtc  tggaagtagt  gagagaagct  ggtaattctg  gacatatagt
                   ⇓
-1362  gacagtcaaa  aaggagctca  ggtacaggac  tggtctaagc  tgrtcaagat  tcaggagaca
-1302  gccagtacac  agagaagctg  aggaaataat  acagatatat  ctaa acact  tatctaacct
-1242  tctgtggtaa  caagctcctt  aaaggggctg  gatgatgttg  tgttcacttt  ttatcaccag
-1182  caaaggctaa  gataatgtat  atagtaaata  tttagtaacc  atttattaaa  taaataaata
                                          CAAT
-1122  tttaagacag  aataaacaag  tataataaat  gaaccaataa  gaatgcacca  tctaagtcaa
-1062  aatagccact  tttatcctta  acattgtacc  tgctttggct  gctgcagaag  caaacttgtt
                                          CAAT
 1002  ggcattagac  aaatcaagct  ggtgatttaa  taaattccaa  tgtaagtctt  accagtattg
          Pit-1
 -942  atgaataact  atccagcact  caccatgaaa  gttaaagaag  caacacagaa  aaagttccta
          CA AT                ENG         TATA
 -882  agtggtccca  atttgaaatg  atcagataac  ctataaaaga  acatattcat  attatactaa
       ⇓           TATA ⇓                  GRE/PRE             CAAT
 -822  cataaacaca  tataaatgca  cttacagcag  ttacacagta  ttctcttcaa  taactagttt
                   ⇓                       CAAT         ⇓         TATA
 -762  ccttatgcat  taatgtgtaa  taacagcaac  tacaatattt  agataattat  aaaaaccaag
       ⇓                      ⇓
 -702  gcaataattt  aaaaactgat  taaccgtttt  actctaactt  aagcatggat  tggatcagta
          ⇓          ⇓          ⇓                       ⇓
 -642  agattgatta  ataaatttga  atgcagtcag  ttggattgat  tctaatttaa  agttttaatt
          ⇓
 -582  tgttgtagaa  taattttaag  tgaatatatt  tgtccagtgt  tcgagtgctc  aacagtgtgt
 -522  ttgaaaagga  aaacaaagaa  tgttttgaga  atgtgttaat  tccttaagac  aatggatttt
 -462  aattggatct  gttgttttca  tttttcttca  ttatcattat  acatctgtat  gttggacaga
 -402  acactaacac  taaatagttt  ttagaaagtg  ttttttgaag  ttatttaaat  cataatatca
 -342  tgactgactt  ttgaattcaa  aattaggctg  tgactatcct  tcttcactta  ggaagagtgt
 -282  tgtgaaagcc  agaccatctg  ctgaggtgct  acagttacat  gtggccctca  gaatgcgttt
 -222  ggcctgctct  gttttagcac  tctgttggat  taccaataca  caaaacaagt  taacctttga
 -162  tctttcacat  taagtatctc  agggacaaaa  tttgacatac  gtctaaacct  gtgacgtttc
 -102  catctaaaga  aggcagaaat  aaaacatgga  ctttagattc  ggttacaata  aaatatcaga
                                                         +1
 - 42  tgcaccagag  acacaaggct  tgaagctctg  tcctgggaaa  atATGGCAAA  CAGTGCCTCT
```

Fig. 6. Nucleotide sequence of the 5′ flanking region of the human GnRH receptor gene. Transcription start sites are shown by arrows. Translation start site (ATG) is capitalized and bolded. The TATA box, CAAT box and other regulatory sequences are shaded. The sequence shown is retrieved from GenBank (accession number AF001950).

1993, 1994a; Fernandez-Vazquez et al., 1996). One mechanism by which the level of the GnRH receptor protein is regulated is through changes in the level of GnRH receptor mRNA (Kaiser et al., 1993; Smith and Reinhart, 1993; Brooks and McNeilly, 1994; Kakar et al., 1994a; Fernandez-Vazquez et al., 1996), suggesting regulation of the protein at the transcrip-

tional level. The half-life of the GnRH receptor mRNA has been estimated to be approximately 21 h (Cheon et al., 2000). However, no studies of its regulation have been performed.

A number of regulatory sequences in the rat, mouse and human receptor genes have been identified (Albarracin et al., 1994; Fan et al., 1995; Cam-

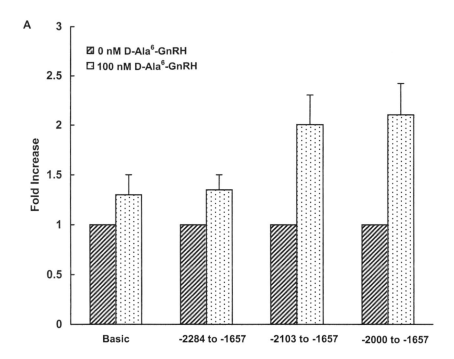

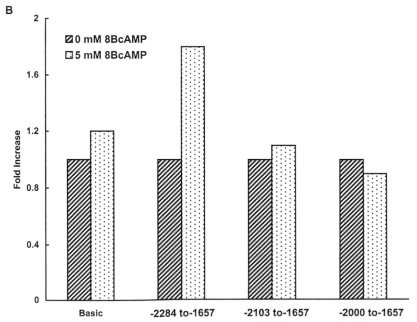

Fig. 7. Effect of 8-bromo cAMP and GnRH agonist (D-Ala⁶-GnRH) on GnRH receptor promoter activity. The luciferase chimeric constructs containing the various 5′ flanking sequence of the human GnRH receptor gene were co-transfected with pSV-β-gal vector into αT31 cells. After 24 h of transfection, the cells were treated with either 8-bromo cAMP (A) or GnRH agonist (B). After 24 h of treatment the cells were lysed and assayed for luciferase and galactosidase activities. Luciferase activity was normalized with galactosidase activity. The results presented are the mean of three independent experiments.

pion et al., 1996; Duval et al., 1997b; Kakar, 1997; Pincas et al., 1998). Using the 5′ flanking region of the human GnRH receptor gene sequence and its 5′ and 3′ deletion constructs, Ngan et al. (1999) reported the importance of the most proximal 173 bp region (relative to translation start site) within the first exon, which is not a promoter itself but instead contains a critical regulatory element(s) essential for the basal expression of the human GnRH receptor gene. These investigators also reported the existence of three putative gonadotrope-specific elements (GSE; consensus 5′-CTGA/TCCTTG-3′) residing at positions −5, −134 and −396. Using the 2300 bp of the flanking region of the human GnRH receptor gene, Cheng et al. (2000a,b) showed a dose-dependent and time-dependent down-regulation of GnRH receptor promoter activity, suggesting the regulation of GnRH receptor gene by GnRH and its analog. By performing the progressive 5′ deletion studies, these investigators also identified a 248-bp DNA fragment (−1018 to −771), relative to the translation start site, that appears to be responsible for the GnRH analog-mediated down-regulation of the human GnRH receptor gene (Cheng et al., 2000a,b). Two AP1 binding sequences also were also identified within the 248 bp sequence, and mutation of these was shown to abolish the GnRH analog-induced response. In contrast, Norwitz et al. (1999) and White et al. (1999) reported stimulation of the mouse GnRH receptor promoter by 100 nM of GnRH agonist (GnRHa). Duval et al. (1997b), using mutation analysis of the 600 bp 5′ flanking region, that has been shown to mediate basal expression of the mouse receptor gene, located a GnRH-responsive fragment, in which a putative AP-1 was subsequently identified (White et al., 1999). This GnRHa-induced increase in promoter activity was mimicked by treatment with phorbol ester, but not by forskolin. Furthermore, pretreatment of the cells with a specific protein kinase C inhibitor (GF109203X) blocked the GnRHa- and TPA-induced increase in promoter activity, suggesting an important role for PKC in regulation of the expression of the mouse GnRH receptor gene. Norwitz et al. (1999), by deleting the 5′ and 3′ sequence of 1.2 kb flanking region of the mouse GnRH receptor gene sequence, identified a 68 bp region located between −300 and −232 (relative to the ATG site) that is responsible for the GnRHa-induced

stimulation. Furthermore, mutation of the AP-1 binding sequence resulted in complete loss of promoter activity, suggesting the importance of the AP-1 binding sequence in GnRH responsiveness. Consistent with these results concerning the mouse GnRH receptor gene, our analysis of a 346 bp sequence in the flanking region of the human GnRH receptor gene (−2003 to −1657) showed activation of promoter activity by GnRH in αT31 cells (Fig. 7A).

In the mouse, the 500 bp sequence upstream of the ATG site was found to be sufficient to drive the promoter activity in αT31 cells. Deletion and mutation analysis of this region allowed the identification of a number of regulatory elements including SF-1 and AP-1 as well as a novel element referred to as the GnRH receptor activating sequence (GRAS: 5′-CTAGTCACAACA-3′). Each of these elements appears to be important in the regulation of expression of the receptor gene as their mutation resulted in a 60% to 80% reduction in promoter activity (Clay et al., 1995; Duval et al., 1997a). In the rat, a 1.2 kb flanking sequence was found to be sufficient to drive the basal promoter activity. However, in this species, various regulatory elements including LH-2 (−869 to −862, relative to ATG site), and SF-1 (−696 to −688, relative to ATG site) were found to be responsible for the regulation of the promoter activity, as mutation of these sequences resulted in a decrease in promoter activity (Duval et al., 1997b).

cAMP may also have a role in regulating GnRH receptor gene expression. The existence of a CRE (cAMP response element) in the flanking regions of the rat, mouse and human GnRH receptor genes has been reported (Albarracin et al., 1994; Fan et al., 1995; Kakar, 1997; Reinhart et al., 1997). To determine if the CRE in the human receptor gene is functional, we tested the 5′ flanking region (−2284 to −1657, containing CRE sequence) of the gene in αT31 cells after its subcloning into the pGL3 vector. As shown in Fig. 7B, treatment of the transfected αT31 cells with the cAMP analog (8-bromo cAMP) resulted in a two-fold increase in luciferase activity. Deletion of the CRE sequence (−2284 to −2104) abolished the cAMP analog effect. In addition, forskolin, an adenylate cyclase activator, mimicked the effect of the cAMP analog (results not shown), suggesting that the CRE sequence in human GnRH receptor gene is functionally active. Similar results were reported by Lin and Conn (1998), Maya-Nunez and Conn (1999) and Maya-Nunez and Conn (2001) for the mouse GnRH receptor gene. These investigators localized the CRE sequence at -107 to -100 (relative to ATG start site) in the mouse gene.

Hormonal regulation of GnRH receptor gene expression

It is well documented that pituitary GnRH receptor is regulated by GnRH. Radioreceptor assays revealed that castration increases GnRH receptor binding, and that testosterone in the male, and estradiol and progesterone in the female prevent this rise (Clayton and Catt, 1981a,b). These effects are likely due to gonadal steroid suppression of GnRH secretion since a GnRH antiserum blocked the castration-associated rise in GnRH receptors (Frager et al., 1981). By contrast, in the mouse pituitary, GnRH receptor concentration was decreased by orchidectomy to 45% below intact values for up to 3 months even though there is a persistent 5-fold rise in serum LH values. Further, treatment of male mice with testosterone propionate prevented the GnRH receptor fall (Naik et al., 1994). A seminal finding was that GnRH up-regulates its receptor only when delivered to gonadotropes in a pulsatile fashion. On the other hand, continuous GnRH causes receptor down-regulation and desensitization. This finding in part lead to the development of long-acting GnRH analogs that produce a chemical gonadectomy and are widely used to treat patients with prostate cancer, precocious puberty, and many other disorders.

Much of the regulation of GnRH receptor protein is through effects on GnRH receptor gene expression. The level of GnRH receptor mRNA varies during the rat estrus cycle, and increases following orchidectomy or ovariectomy in a wide variety of species (Kaiser et al., 1993; Kakar et al., 1994a) including primates (Winters et al., 2001). Experiments using perifused rat pituitary cells indicated that only pulsatile GnRH increases the level of GnRH receptor mRNA (Kaiser et al., 1993), but how this occurs remains poorly understood.

Although estradiol reduces GnRH receptor in vivo in rats (Kaiser et al., 1993; Kakar et al., 1994a), estradiol increases GnRH-R mRNA levels dose-dependently in male and female sheep (Turzillo et al., 1994; Adams et al., 1996). This effect is likely to be directly on the pituitary since it can be mimicked in ovine pituitary cell cultures (Ghosh et al., 1996). Progesterone decreases GnRH binding sites and GnRH receptor mRNA concentrations in ovine pituitary cells (Laws et al., 1990; Sealfon et al., 1990). In primary cultures of female rat pituitary cells, however, neither estradiol nor progesterone modulates the basal expression of GnRH receptor mRNA (Cheon et al., 2000). Kakar et al. (1993) found increased levels of GnRH receptor mRNA in sterile cows. In mouse-derived αT31 cells, estradiol reduces GnRH receptor number as well as the efficiency of coupling of residual GnRH receptors to second messenger generation (McArdle et al., 1992). Approximately 9,100 bp of 5' flanking region from the ovine GnRH receptor (oGnRHR) gene was devoid of transcriptional activity and unresponsive to both estradiol and GnRH in αT31 cells whereas this same 9,100 bp promoter fragment directed tissue-specific expression of luciferase in multiple lines of transgenic mice (Duval et al., 2000). The absence of an estrogen response element (ERE) in the flanking region of the (human) GnRH receptor gene suggests that regulation by estrogen may not be transcriptional.

Using the human GnRH receptor 5' flanking sequence, Cheng et al. (2001) found inhibition of promoter activity by progesterone in αT3-1 cells, but activation in placental JEG-3 cells. Using 5' and 3' deletion analysis, these investigators localized the progesterone response element (PRE) between -535 to -531 (relative to ATG). Furthermore the physiological effects of progesterone were shown to occur through two different isoforms of the progesterone receptor. Over-expression of PR-A reduced promoter activity in both pituitary and placental cells, whereas over-expression of PR-B activated the promoter in placental cells but inhibited the promoter in pituitary cells.

An additional autocine/paracrine mechanism for GnRH receptor mRNA regulation is suggested by the findings that rat GnRH receptors are up-regulated by activin (Braden and Conn, 1992), and decreased by inhibin (Wang et al., 1989) in rat pituitary cell cultures. Activin stimulated transcription of GnRH receptor mRNA in αT3-1 cells (Fernandez-Vazquez et al., 1996), and inhibin suppressed GnRH receptor steady state mRNA levels in male rats (Winters et al., 1996). On the other hand, inhibin increases GnRH receptor mRNA concentrations in ovine pituitary cultures (Sealfon et al., 1990).

Finally, transcription of the GnRH receptor gene is stimulated by PACAP presumably through a cAMP pathway (Pincas et al., 2001). The region between −1727 and −1577 was found to be responsible for PACAP induction (Ngan et al., 2001). This finding expands the interplay between these two hypophysiotropic peptides.

Summary

In summary, isolation of GnRH receptor cDNA, its gene, and identification of regulatory elements in the flanking region of the gene have added to our knowledge regarding the tissue-specific expression of the GnRH receptor gene, and the mechanisms that mediate and influence its transcriptional regulation. However, the interactions of the different regulatory factors (nuclear factors) and the effects of these interactions on the regulation of the GnRH receptor gene remain unclear. Due to existence of multiple promoters and transcriptional start sites in human GnRH receptor gene and the lack of a human gonadotrope cell line, the precise promoter and transcriptional start sites in human pituitary, extra-pituitary tissues and tumors have not yet been identified.

References

Adams, B.M., Sakurai, H. and Adams, T.E. (1996) Concentrations of gonadotropin-releasing hormone (GnRH) receptor messenger ribonucleic acid in pituitary tissue of orchidectomized sheep: effect of estradiol and GnRH. *Biol. Reprod.*, 54: 407–412.

Albarracin, C.T., Kaiser, U.B. and Chinn, W.W. (1994) Isolation and characterization of the 5′-flanking region of the mouse gonadotropin-releasing hormone receptor gene. *Endocrinology*, 135: 2300–2306.

Arora, K.K., Sakai, A. and Catt, K.J. (1995) Effects of second intracellular loop mutations on signal transduction and internalization of the gonadotropin-releasing hormone receptor. *J. Biol. Chem.*, 270: 22820–22826.

Attwood, T.K., Eliopoulos, E.E. and Findlay, J.B. (1991) Multiple sequence alignment of protein families showing low sequence homology: a methodological approach using database pattern-matching discriminators for G-protein linked receptors. *Gene*, 98: 153–159.

Bodner, M., Castrillo, J.L., Theill, L.E., Deerinck, T., Ellisman,

M. and Karin, M. (1988) The pituitary-specific transcription factor GHF-1 is a homeobox-containing protein. *Cell*, 55: 505–518.

Boyle, T.A., Belt-Davis, D.I. and Duello, T.M. (1998) Nucleotide sequence analysis predict that human pituitary and human placental gonadotropin-releasing hormone receptors have identical primary structures. *Endocr. J.*, 9: 281–287.

Braden, T.D. and Conn, P.M. (1992) Activin-A stimulates the synthesis of gonadotropin-releasing hormone receptors. *Endocrinology*, 130: 2101–2105.

Brooks, J. and McNeilly, A.S. (1994) Regulation of gonadotropin-releasing hormone receptor mRNA expression in the sheep. *Endocrinology*, 143: 175–182.

Brooks, J., Taylor, P.L., Saunders, P.T., Eidne, K.A., Struthers, W.J. and McNeilly, A.S. (1993) Cloning and sequencing of the sheep pituitary gonadotropin-releasing hormone receptor and changes in expression of its mRNA during estrous cycle. *Mol. Cell. Endocrinol.*, 94: R23–R27.

Brown, J.L. and Reeves, J.J. (1983) Absence of luteinizing hormone releasing hormone receptors in ovine, bovine and porcine ovaries. *Biol. Reprod.*, 29: 1179–1182.

Byrne, B., McGregor, A., Taylor, P.L., Sellers, R., Rodger, F.E., Fraser, H.M. and Eidne, K.A. (1999) Isolation and characterization of the marmoset gonadotropin releasing hormone receptor: Ser (140) of the DRS motif is substituted by Phe. *J. Endocrinol.*, 163: 447–456.

Campion, C.E., Turzillo, A.M. and Clay, C.M. (1996) The gene encoding the ovine gonadotropin-releasing hormone (GnRH) receptor: cloning and initial characterization. *Gene*, 170: 277–280.

Chen, A., Yahalom, D., Ben-Aroya, N., Kaganovsky, E., Okon, E. and Koch, Y. (1998) A second isoform of gonadotropin-releasing hormone is present in the brain of human and rodents. *FEBS Lett.*, 435: 199–203.

Cheng, K.W., Cheng, C.K. and Leung, P.C.K. (2001) Differential role of PR-A and -B isoforms in transcription regulation of human GnRH receptor gene. *Mol. Endocrinol.*, 15: 2078–2092.

Cheng, K.W., Ngan, E.S.W., Kang, S.K., Chow, B.K.C. and Leung, P.K.C. (2000a) Transcriptional down-regulation of human gonadotropin-releasing hormone (GnRH) receptor gene by GnRH: role of protein kinase C and activating protein 1. *Endocrinology*, 141: 3611–3622.

Cheng, K.W., Nathwani, P.S. and Leung, P.C.K. (2000b) Regulation of human gonadotropin-releasing hormone receptor gene expression in placental cells. *Endocrinology*, 141: 2340–2349.

Cheon, M., Park, D., Park, Y., Kam, K., Park, S.D. and Ryu, K. (2000) Homologous upregulation of gonadotropin-releasing hormone receptor mRNA occurs through transcriptional activation rather than modulation of mRNA stability. *Endocr. J.*, 13: 47–53.

Chi, L., Zhou, W., Prikhozhan, A., Flanagan, C., Davidson, J.S., Golembo, M., Illing, N., Millar, R.P. and Sealfon, S.C. (1993) Isolation and characterization of the human GnRH receptor. *Mol. Cell. Endocrinol.*, 91: R1–R6.

Clay, C.M., Nelson, S.E., DiGreorio, G.B., Campion, C.E., Wiedemann, A.L. and Nett, R.J. (1995) Cell-specific expression of the mouse gonadotropin-releasing hormone (GnRH) gene is conferred by elements residing within 500 bp of proximal 5'-flanking region. *Endocrine*, 3: 615–622.

Clayton, R.N. and Catt, K.J. (1981a) Regulation of pituitary gonadotropin-releasing hormone receptors by gonadal hormones. *Endocrinology*, 108: 887–885.

Clayton, R.N. and Catt, K.J. (1981b) Gonadotropin-releasing hormone receptors: characterization, physiological regulation, and relationship to reproductive function. *Endocr. Rev.*, 2: 186–209.

Cook, J.V. and Eidne, K.A. (1997) An intramolecular disulfide bond between conserved extracellular cysteines in the gonadotorpin-releasing hormone receptor is essential for binding and activation. *Endocrinology*, 138: 2800–2806.

Craig, A.G., Fischer, W.H., Park, M., Rivier, J.E., Musselman, B.D., Powell, J.F.F., Reska-Skinner, S.M., Prakash, M.O., Mackie, G.O. and Sherwood, N.M. (1997) Sequence of two gonadotropin releasing hormone from tunicate suggest an important role of conformation in receptor activation. *FEBS Lett.*, 413: 215–225.

Davidson, J.S., Flanagan, C.A., Becker, I.I., Illing, N., Sealfon, S.C. and Millar, R.P. (1994) Molecular function of the gonadotropin-releasing hormone receptor: insights from site-directed mutagenesis. *Mol. Cell. Endocrinol.*, 100: 9–14.

Davidson, J.S., Flanagan, C.A., Zhou, W., Becker, I.I., Elario, R., Emeran, W., Sealfon, S.C. and Millar, R.P. (1995) Identification of N-gylcosylation sites in the gonadotropin-releasing hormone receptor: role of receptor expression but not ligand binding. *Mol. Cell. Endocrinol.*, 107: 241–245.

Davidson, J.S., Assefa, D., Pawson, A., Davies, P., Hapgood, J., Becker, I., Flanagan, C., Roeske, R. and Millar, R.P. (1997) Irreversible activation of the gonadotropin-releasing hormone receptor by photoaffinity crosslinking: localization of attachment site to cys residue in N-terminal segment. *Biochemistry*, 36: 12881–12889.

Dohlman, H.G., Caron, M.G., DeBlasi, A., Frielle, T. and Lefkowitz, R.J. (1990) Role of extracellular disulfide-bonded cysteines in the ligand function of the beta 2-adrenergic receptor. *Biochemistry*, 29: 2335–2345.

Duval, D.L., Ellsworth, B.S. and Clay, C.M. (1997a) The tripartite basal enhancer of the gonadotropin-releasing hormone (GnRH) receptor gene promoter regulates cell-specific expression through a novel GnRH receptor activating sequence. *Mol. Endocrinol.*, 11: 1814–1821.

Duval, D.L., Nelson, S.E. and Clay, C.M. (1997b) A binding site for steroidogenic factor-1 is part of a complex enhancer that mediates expression of the murine gonadotropin-releasing hormone receptor gene. *Biol. Reprod.*, 56: 160–168.

Duval, D.L., Farris, A.R., Quirk, C.C., Nett, T.M., Hamernik, D.L. and Clay, C.M. (2000) Responsiveness of the ovine gonadotropin-releasing hormone receptor gene to estradiol and gonadotropin-releasing hormone is not detectable in vitro but is revealed in transgenic mice. *Endocrinology*, 141: 1001–1010.

Fan, N.C., Peng, C., Krinsinger, J. and Leung, P.C.K. (1995) The human gonadotropin-releasing hormone receptor gene: complete structure including multiple promoters, transcription initiation sites, and polyadenylation signals. *Mol. Cell. Endocrinol.*, 107: R1–R8.

Faurholm, B., Millar, R.P. and Katz, A.A. (2001) The genes encoding the type II gonadotropin-releasing hormone receptor and the ribonucleoprotein RBM8A in humans overlap in two genomic loci. *Genomics*, 78: 15–18.

Fernandez-Vazquez, G., Kaiser, U.B., Albarracin, C.T. and Chin, W.W. (1996) Transcriptional activation of the gonadotropin-releasing hormone receptor gene by activin A. *Mol. Endocrinol.*, 10: 356–366.

Fink, G. (1988) Gonadotropin secretion and its control. In: E. Knobil and J.D. Neill (Eds.), *The Physiology of Reproduction*. Raven Press, New York, NY, pp. 1349–1377.

Frager, M.S., Pieper, D.R., Tonetta, S.A., Duncan, J.A. and Marshall J.C. (1981) Pituitary gonadotropin-releasing hormone receptors. Effects of castration, steroid replacement, and the role of gonadotropin-releasing hormone in modulating receptors in the rat. *J. Clin. Invest.*, 67: 615–623.

Ghosh, B.R., Wu, J.C., Strahl, B.D., Childs, G.V. and Miller, W.L. (1996) Inhibin and estradiol alter gonadotropes differentially in ovine pituitary cultures: changing gonadotrope numbers and calcium responses to gonadotropin-releasing hormone. *Endocrinology*, 137: 5144–5154.

Gregg, D.W., Schwall, R.H. and Nett, T.M. (1991) Regulation of gonadotropin secretion and number of gonadotropin-releasing hormone receptors by inhibin, activin-A and estradiol. *Biol. Reprod.*, 44: 725–732.

Horn, F., Bilezikjian, L.M., Perrin, M.H., Bosma, M.M., Windle, J.J., Huber, K.S., Blount, A.L., Hille, B., Vale, W. and Mellon, P.L. (1991) Intracellular responses to gonadotropin-releasing hormone in a clonal cell line of the gonadotrope lineage. *Mol. Endocrinol.*, 5: 347–355.

Illing, N., Troskie, B.E., Nahorniak, C.S., Hapgood, J.P., Peter, R.E. and Millar, R.P. (1999) Two gonadotropin-releasing hormone receptor subtypes with distinct ligand selectivity and differential distribution in brain and pituitary in the goldfish (carassius auratus). *Proc. Natl. Acad. Sci. USA*, 96: 2526–2531.

Imagawa, M., Chiu, R. and Karin, M. (1987) Transcription factor AP-2 mediates induction by two different signal-transduction pathways: protein kinase C and cAMP. *Cell*, 51: 251–260.

Ingraham, H.A., Flynn, S.E., Voss, J.W., Albert, V.R., Kapiloff, M.S., Wilson, L. and Rosenfeld, M.G. (1990) The POU-specific domain of Pit-1 is essential for sequence-specific, high affinity DNA binding and DNA-dependent Pit-1–Pit-1 interactions. *Cell*, 61: 1021–1033.

Kaiser, U.B., Jakubowiak, A., Steinberger, A. and Chin, W.W. (1993) Regulation of rat pituitary gonadotropin-releasing hormone mRNA levels in vivo and vitro. *Endocrinology*, 133: 931–934.

Kaiser, U.B., Dushkin, H., Altherr, M.R., Beier, D.R. and Chin, W.W. (1994) Chromosomal localization of the gonadotropin-releasing hormone receptor gene to human chromosome 4q13.1–q21.1 and mouse chromosome 5. *Genomics*, 20: 506–508.

Kakar, S.S. (1997) Molecular structure of the human

gonadotropin-releasing hormone receptor gene. *Eur. J. Endocrinol.*, 137: 183–192.

Kakar, S.S. and Jennes, L. (1995) Expression of gonadotropin-releasing hormone and gonadotropin-releasing hormone receptor mRNAs in various non-reproductive human tissues. *Cancer Lett.*, 98: 57–62.

Kakar, S.S. and Neill, J.D. (1994) The human gonadotropin-releasing hormone receptor gene (GNRHR) maps to chromosome 4q13.1. *Cytogenet. Cell Genet.*, 70: 211–214.

Kakar, S.S., Musgrove, L.C., Devor, D.C., Sellers, J.C. and Neill, J.D. (1992) Cloning, sequencing, and expression of human gonadotorpin-releasing hormone (GnRH) receptor. *Biochem. Biophys. Res. Commun.*, 189: 289–295.

Kakar, S.S., Rahe, C.H. and Neill, J.D. (1993) Molecular cloning, sequencing, and characterizing the bovine receptor for gonadotropin releasing hormone (GnRH). *Domest. Anim. Endocrinol.*, 10: 335–342.

Kakar, S.S., Granthan, K., Musgrove, L.C., Devor, D.C., Sellers, J.C. and Neill, J.D. (1994a) Rat gonadotropin-releasing hormone (GnRH) receptor: tissue expression and hormonal regulation of its mRNA. *Mol. Cell. Endocrinol.*, 1010: 151–157.

Kakar, S.S., Grizzle, W.E. and Neill, J.D. (1994b) The nucleotide sequence of human GnRH receptors in breast and ovarian tumors are identical with that found in pituitary. *Mol. Cell. Endocrinol.*, 106: 145–149.

Kang, S.K., Cheng, K.W., Ngan, E.S.W., Chow, B.K.C., Choi, K. and Leung, P.C.K. (2000) Differential expression of human gonadotropin-releasing hormone receptor gene in pituitary and ovarian cells. *Mol. Cell. Endocrinol.*, 162: 157–166.

Kasten, T.L., White, S.A., Norton, T.T., Bond, C.T., Adelman, J.P. and Fernald, R.D. (1996) Characterization of two new pre-proGnRH mRNAs in the tree shrew: first direct evidence for mesencephalic GnRH gene expression in a placental mammal. *Gen. Comp. Endocrinol.*, 104: 7–19.

King, J.A. and Millar, R.P. (1997) Co-ordinated evolution of GnRH and its receptors. In: I.S. Parhar and Y. Sakuma (Eds.), *GnRH Neurons: Gene to Behaviour*. Brain Shuppan, Tokyo, pp. 51–77.

Latimer, V.S., Rodrigues, S.M., Garyfallou, V.T., Kohama, S.G., White, S.G., Fernald, R.D. and Urbanski, H.F. (2000) Two molecular forms of gonadotropin-releasing hormone (GnRH-I and GnRH-II) are expressed by two separate populations of cells in the rhesus macaque hypothalamus. *Brain. Res. Mol. Brain. Res.*, 75: 287–292.

Laws, S.C., Beggs, M.J., Webster, J.C. and Miller, W.L. (1990) Inhibin increases and progesterone decreases receptors for gonadotropin-releasing hormone in ovine pituitary culture. *Endocrinology*, 127: 373–380.

Lescheid, D.W., Terasawa, E., Abler, L.A., Urbanski, H.F., Warby, C.M., Millar, R.P. and Sherwood, N.M. (1997) A second form of gonadotropin-releasing hormone (GnRH) with characteristics of chicken GnRH II is present in the primate brain. *Endocrinology*, 138: 5618–5629.

Lin, X. and Conn, P.M. (1998) Transcriptional activation of gonadotropin-releasing hormone (GnRH) receptor gene by GnRH and cyclic adenosine monophosphate. *Endocrinology*, 139: 3896–4092.

Lin, X., Janovick, J.A. and Conn, P.M. (1998) Mutations at the consensus phosphorylation sites in the third intracellular loop of the rat gonadotropin-releasing hormone receptor: effects on receptor ligand binding and signal transduction. *Biol. Reprod.*, 59: 1470–1476.

Lloyd, J.M. and Childs, G.V. (1988) Changes in the number of GnRH-receptive cells during the rat estrous cycle: biphasic effects of estradiol. *Neuroendocrinology*, 48: 138–146.

Maya-Nunez, G. and Conn, P.M. (1999) Transcriptional regulation of the gonadotropin-releasing hormone receptor gene is mediated in part by a putative repress or element and by the cyclic adenosine 3',5'-monophosphate response element. *Endocrinology*, 140: 3452–3458.

Maya-Nunez, G. and Conn, P.M. (2001) Cyclic adenosine 3',5'-monophosphate (cAMP) and cAMP responsive element-binding protein are involved in the transcriptional regulation of gonadotropin-releasing hormone (GnRH) receptor by GnRH and mitogen-activated protein kinase signal transduction pathway in GGH(3) cells. *Biol. Reprod.*, 65: 561–567.

McArdle, C.A., Schomerus, E., Groner, I. and Poch, A. (1992) Estradiol regulates gonadotropin-releasing hormone receptor number, growth and inositol phosphate production in alpha T3-1 cells. *Mol. Cell. Endocrinol.*, 87: 95–103.

Millar, R.P., Conkin, D., Lofton-Day, C., Hutchinson, E., Troskie, B., Illing, N., Sealfon, S.C. and Hapgood, J. (1999) A novel human GnRH receptor homolog gene: abundant and wide tissue distribution of the antisense transcript. *J. Endocrinol.*, 162: 117–126.

Millar, R.P., Lowe, S., Conklin, D., Pawson, A., Maudsley, S., Troskie, B., Ott, T., Millar, M., Lincoln, G., Sellar, R., Faurholm, B., Scobie, G., Kuestner, R., Terasawa, E. and Katz, A. (2001) A novel mammalian receptor for the revolutionarily conserved type II GnRH. *Proc. Natl. Acad. Sci. USA*, 98: 9636–9641.

Montgomery, G.W., Penty, J.M., Lord, E.A., Brooks, J. and McNeilly, A.S. (1995) The gonadotropin-releasing hormone receptor maps to sheep chromosome 6 outside of the region of the FecB locus. *Mamm. Genome*, 6: 436–438.

Moumni, M., Kottler, M.L. and Counis, R. (1994) Nucleotide sequence analysis of mRNAs predicts that rat pituitary and gonadal gonadotropin-releasing hormone receptor proteins have identical primary structure. *Biochem. Biophys. Res. Commun.*, 200: 1359–1366.

Naik, S.I., Young, L.S., Charlton, H.M. and Clayton, R.N. (1994) Pituitary gonadotropin-releasing hormone receptor regulation in mice. II: Females. *Endocrinology*, 115: 114–120.

Ngan, E.S.W., Cheng, P.K.W., Leung, P.C.K. and Chow, B.K.C. (1999) Steroidgenic factor-1 interact with gonadotrope-specific element with the first exon of the human gonadotropin-releasing hormone gene to mediate gonadotrope-specific expression. *Endocrinology*, 140: 2452–2462.

Ngan, E.S.W., Leung, P.C.K. and Chow, K.C. (2000) Identification of an upstream promoter in the human gonadotropin-releasing hormone receptor gene. Biochem. *Biophys. Res. Commun.*, 270: 766–772.

146

Ngan, E.S., Leung, P.C. and Chow, B.K. (2001) Interplay of pituitary adenylate cyclase-activating polypeptide with a silencer element to regulate the upstream promoter of the human gonadotropin-releasing hormone receptor gene. *Mol. Cell. Endocrinol.*, 176: 135–144.

Neill, J.D., Duck, L.W., Sellers, J.C. and Musgrove, L.C. (2001) A gonadotropin-releasing hormone (GnRH) receptor specific for GnRH II in primates. *Biochem. Biophys. Res. Commun.*, 282: 1012–1018.

Norwitz, E.R., Cardona, G.R., Jeong, K.H. and Chin, W.W. (1999) Identification and characterization of the gonadotropin-releasing hormone response elements in the mouse gonadotropin-releasing hormone receptor gene. *J. Biol. Chem.*, 274: 2389–2392.

Pawson, A.J., Katz, A., Sun, Y.M., Lopes, J., Illing, N., Millar, R.P. and Davidson, J.S. (1998) Contrasing internalization kinetic of human and chicken gonadotropin-releasing hormone receptors mediated by C-terminal tail. *J. Endocrinol.*, 156: R9–R12.

Pincas, H., Forrai, Z., Chauvin, S., Laverriere, J.N. and Counis, R. (1998) Multiple elements in the distal part of the 1.2 kb 5′ flanking region of the art GnRH receptor gene regulate gonadotrope-specific expression conferred by proximal domain. *Mol. Cell. Endocrinol.*, 144: 95–108.

Pincas, H., Laverriere, J.N. and Counis, R. (2001) Pituitary adenylate cyclase-activating polypeptide and cyclic adenosine 3′,5′-monophosphate stimulate the promoter activity of the rat gonadotropin-releasing hormone receptor gene via a bipartite response element in gonadotrope-derived cells. *J. Biol. Chem.*, 276: 23562–23571.

Powell, J.F., Reska-Skinner, S.M., Prakash, M.O., Fischer, W.H., Park, M., Rivier, J.E., Craig, A.G., Mackie, G.O. and Sherwood, N.M. (1996) Two new forms of gonadotropin-releasing hormone in a protochordate and the evolutionary implications. *Proc. Natl. Acad. Sci. USA*, 93: 10461–10464.

Reinhart, J., Xiao, S., Arora, K.K. and Catt, K.J. (1997) Structural organization and characterization of the promoter region of the rat gonadotropin-releasing hormone receptor gene. *Mol. Cell. Endocrinol.*, 130: 1–12.

Sealfon, S.C., Laws, S.C., Wu, J.C., Gillo, B. and Miller W.L. (1990) Hormonal regulation of gonadotropin-releasing hormone receptors and messenger RNA activity in ovine pituitary culture. *Mol. Endocrinol.*, 4: 1980–1987.

Sealfon, S.C., Weinstein, H. and Millar, R.P. (1997) Molecular mechanisms of ligand interaction with the gonadotropin-releasing hormone receptor. *Endocr. Rev.*, 18: 180–205.

Sherwood, N.M., Lovejoy, D.A. and Coe, I.R. (1993) Origin of mammalian gonadotropin-releasing hormones. *Endocr. Rev.*, 14: 241–254.

Sherwood, N.M., Von Schalburg, K. and Lescheid, D.W. (1997) Origin and evolution of GnRH in vertebrate and invertebrates. In: I.S. Parhar and Y. Sakuma (Eds.), *GnRH Neuro: Gene to Behavior*. Brain Shuppan, Tokyo, pp. 3–25.

Smith, M.S. and Reinhart, J. (1993) Changes in pituitary gonadotropin-releasing hormone receptor messenger ribonucleic acid content during lactation and after pup removal. *Endocrinology*, 133: 2080–2084.

Tensen, C., Okuzawa, K., Blomenrohr, M., Rebers, F., Leurs, R., Bogerd, J., Schulz, R. and Goos, H. (1997) Distinct efficacies for two endogenous ligands on a single cognate gonadoliberin receptor. *Eur. J. Endocrinol.*, 243: 134–140.

Troskie, B.E., Sun, Y., Hapgood, J., Sealfon, S.C., Illing, N. and Millar, R.P. (1997) In: *Program of the 79th Annual Meeting of the Endocrine Society*, Minneapolis, MN, P1–130 (abstract).

Troskie, B.E., Illing, N., Rumbak, E., Sun, Y.M., Sealfon, S.C., Conkin, D. and Millar, R.P. (1998) Identification of three putative GnRH receptor subtypes in vertebrates. *Gen. Comp. Endocrinol.*, 112: 296–302.

Troskie, B.E., Hapgood, J.P., Millar, R.P. and Illing, N. (2000) Complementary deoxyribonucleic acid cloning, gene expression, and ligand selectivity of a novel gonadotropin-releasing hormone receptor expressed in the pituitary and midbrain of Xenopus laevis. *Endocrinology*, 141: 1764–1771.

Tsutsumi, M., Zhou, W., Millar, R.P., Mellon, P.L., Roberts, J.L., Flanagan, C.A., Dong, K.W., Gillo, K. and Sealfon, S.C. (1992) Cloning and functional expression of a mouse gonadotropin-releasing hormone receptor. *Mol. Endocrinol.*, 6: 1163–1169.

Turzillo, A.M., Campion, C.E., Clay, C.M. and Nett, T.M. (1994) Regulation of gonadotropin-releasing hormone (GnRH) receptor messenger ribonucleic acid and GnRH receptors during the early preovulatory period in the ewe. *Endocrinology*, 135: 1353–1358.

Urbanski, H.F., White, R.B., Fernald, R.D., Kohama, S.G. and Densmore, V.S. (1999) Regional expression of mRNA encoding a second form of gonadotropin-releasing hormone in the Macaque brain. *Endocrinology*, 140: 1945–1948.

Wang, L., Bogerd, J., Choi, HS., Seong, J.Y., Soh, J.M., Chun, S.Y., Blomenrohr, M., Troskie, B.E., Millar, R.P., Yu, W.H., McCann, S.M. and Kwon, H.B. (2001) Three distinct types of GnRH receptor characterized in the bullfrog. *Proc. Natl. Acad. Sci. USA*, 98: 361–366.

Wang, Q.F., Farnworth, P.G., Findlay, J.K. and Burger, H.G. (1989) Effect of purified 31K bovine inhibin on the specific binding of gonadotropin-releasing hormone to rat anterior pituitary cells in culture. *Endocrinology*, 123: 2161–2166.

Weesner, G.D. and Matteri, R.L. (1994) Nucleotide sequence of luteinizing hormone-releasing hormone (LHRH) receptor cDNA in the pig pituitary. *J. Anim. Sci.*, 72: 1911.

White, B.R., Duval, D.L., Mulvaney, J.M., Roberson, M.S. and Clay, C.M. (1999) Homologous regulation of the gonadotropin-releasing hormone receptor gene is partially mediated by protein kinase C activation of an activator protein-1 element. *Mol. Endocrinol.*, 13: 566–577.

White, R.B., Eisen, J.A., Kasten, T.L. and Fernald, R.D. (1998) Second gene for gonadotropin-releasing hormone in humans. *Proc. Natl. Acad. Sci. USA*, 95: 305–309.

White, S.A., Bond, C.T., Francis, R.C., Kasten, T.L., Fernald, R.D. and Adelman, J.P. (1994) A second gene for gonadotropin-releasing hormone: cDNA and expression pattern in the brain. *Proc. Natl. Acad. Sci. USA*, 91: 1423–1427.

Winters, S.J., Pohl, C.R., Adedoyin, A. and Marshall, G.R. (1996) Effects of continuous inhibin administration on gonadotropin secretion and subunit gene expression in immature and adult male rats. *Biol. Reprod.*, 55: 1377–1382.

Winters, S.J., Kawakami, S., Sahu, A. and plant T.M. (2001) Plant, T.M. Pituitary follistatin and activin gene expression, and the testicular regulation of FSH in the adult rhesus monkey (Macaca mulatta). *Endocrinology*, 142: 2874–2878.

Wormald, P.J., Eidne, K.A. and Millar, R.P. (1985) Gonadotropin-releasing hormone receptors in human pituitary: ligand structural requirements, molecular size, and cationic effects. *J. Clin. Endocrinol. Metab.*, 61: 1190–1194.

SECTION III

Physiology and regulation

I.S. Parhar (Ed.)
Progress in Brain Research, Vol. 141
© 2002 Elsevier Science B.V. All rights reserved

CHAPTER 11

Hypothalamic control of gonadotropin secretion [‡]

S.M. McCann [1,*], S. Karanth [1], C.A. Mastronardi [1], W.L. Dees [2], G. Childs [3], B. Miller [4], S. Sower [5], W.H. Yu [1]

[1] *Pennington Biomedical Research Center, Louisiana State University, 6400 Perkins Road, Baton Rouge, LA 70808-4124, USA*
[2] *Department of Veterinary Anatomy and Public Health, Texas A&M University, University Drive, College Station, TX 77843, USA*
[3] *Department of Anatomy, University of Arkansas for Medical Sciences, 4301 W. Markham Street, Slot #510, Little Rock, AR 72205, USA*
[4] *Department of Anatomy and Neuroscience, University of Texas Medical Branch, 301 University Boulevard, MRB 10 104, Galveston, TX 77555-1043, USA*
[5] *Department of Biochemistry and Molecular Biology University of New Hampshire, 46 College Road, Room 310, Durham, NH 03824, USA*

Introduction

The control of gonadotropin secretion is extremely complex as revealed by the research of the past 40 years since the discovery of LHRH (McCann et al., 1960), now commonly called gonadotropin-releasing hormone (GnRH) (McCann and Ojeda, 1996). This was the second of the hypothalamic-releasing hormones characterized. It stimulates FSH release, albeit in smaller amounts than LH. For this reason, it was renamed GnRH (Reichlin, 1992; McCann and Ojeda, 1996). Overwhelming evidence indicates that there must be a separate FSHRH because pulsatile release of LH and FSH can be dissociated. In the castrate male rat, roughly half of the FSH pulses occur in the absence of LH pulses and only a small fraction of the pulses of both gonadotropins are co-incident. LHRH antisera or antagonists can suppress

pulsatile release of LH without altering FSH pulses (Culler and Negro-Vilar, 1987; McCann et al., 1993). LH, but not FSH pulses can be suppressed by alcohol (Dees et al., 1985), delta-9-tetrahydrocannabinol, and cytokines, such as interleukin-1 alpha (IL-1α) (Rettori et al., 1991). In addition, a number of peptides inhibit LH, but not FSH release and a few stimulate FSH without affecting LH (McCann and Krulich, 1989; McCann et al., 1993).The hypothalamic areas controlling LH and FSH are separable. Stimulation in the dorsal medial anterior hypothalamic area causes selective FSH release, whereas lesions in this area selectively suppress the pulses of FSH and not LH (Lumpkin et al., 1989). Conversely, stimulations or lesions in the medial preoptic region can augment or suppress LH release respectively without affecting FSH release. Electrical stimulation in the preoptic region releases only LH, whereas lesions in this area inhibit LH release without inhibiting FSH release. The medial preoptic area contains most of the perikarya of LHRH neurons. The axons of these neurons project from the preoptic region to the anterior and mid portions of the median eminence. Extracts of the anterior–mid median eminence contain LH-releasing activity commensurate with the content of immunoassayable LHRH, whereas extracts of the caudal median eminence

* Correspondence to: S.M. McCann, Pennington Biomedical Research Center, Louisiana State University, 6400 Perkins Road, Baton Rouge, LA 70808-4124, USA. Tel.: +1-225-763-3042; Fax: +1-225-763-3030; E-mail: mccannsm@pbrc.edu

[‡] Modified from *Archives of Medical Research*, 32: 476–485, 2001.

and organum vasculosum lamina terminalis (OVLT) contain more FSH-releasing activity than can be accounted for by the content of LHRH (McCann et al., 1993).Lesions confined to the rostral and mid-median eminence can selectively inhibit pulsatile LH release without altering FSH pulsations, whereas lesions that destroy the caudal and mid-median eminence can selectively block FSH pulses in castrated male rats (McCann et al., 1993; Marubayashi et al., 1999). Therefore, the putative FSHRH may be synthesized in neurons with perikarya in the dorsal anterior hypothalamic area with axons that project to the mid and caudal median eminence to control FSH release selectively.

FSHRH

We (followed by several other groups) reported FSH-releasing activity in the stalk-median eminence (Igarashi and McCann, 1964). The activity was purified and separated from the LH-releasing activity in 1965 as measured by in vivo bioassays (Dhariwal et al., 1965, 1967). This separation was confirmed by Schally et al. (1966) and two other laboratories. It has repeatedly been shown that bioactive and radioimmunoassayable FSH-releasing activity can be separated from LHRH by gel filtration through sephadex G-25 on the same column used in the earlier research. FSH- and LH-releasing activity were assayed by the increase in plasma FSH and LH, respectively, in ovariectomized, estrogen–progesterone blocked rats (Lumpkin et al., 1987). The separation of the two activities was also demonstrable by assay of FSH and LH released from hemipituitaries incubated in vitro by bioassay (Mizunuma et al., 1983) and RIA (Yu et al., 2000). In both assay systems, FSHRH emerged from the column just before elution of LHRH.

In the search for FSHRH, we first believed that it might be an analog of LHRH (Dhariwal et al., 1967) and later had many such analogs synthesized. We tested the forms of GnRH that were known to exist in lower species (Yu et al., 1990, 2002). We had not tested lamprey GnRH-III, but when we realized that an antiserum that crossed reacted with lGnRH-I and lGnRH-III, immunostained neural fibers in the arcuate nucleus proceeding to the median eminence of human brains obtained at autopsy, it occurred to us

that lGnRH-III could be the FSHRH since lGnRH-I had little activity to release either LH or FSH (Yu et al., 1997a). Indeed, lGnRH-III is a potent FSH-releasing factor with little or no LH-releasing activity both in vitro when incubated with hemipituitaries of male rats and in vivo when injected into conscious, ovariectomized, estrogen–progesterone-blocked rats (Yu et al., 1997a). The lowest dose tested in that preparation (10 pM) produced a highly significant increase in plasma FSH with no increase in LH, a 10-fold higher dose (100 pM) increased plasma FSH similarly and had no effect on LH release. lGnRH-III 10^{-9} M produced highly significant FSH release in vitro, whereas, LH-releasing activity did not appear until 10^{-6} M (Yu et al., 1997a).

Testing other GnRH analogs in this in vitro system revealed that cGnRH-II had no significant activity to release FSH or LH until a much higher dose was reached and no selective releasing activity. It had slightly selective FSH-releasing activity in vivo in ovariectomized, estrogen–progesterone-blocked rats (Yu et al., 1990). We have tested over 40 natural and synthesized analogs of lGnRH-III and found that the amino acids in positions 5, 6, 7 and 8 of lGnRH-III are important in conveying selective FSH-releasing activity. These 4 amino acids are dissimilar from those in LHRH. From tunicate to man the first 4 amino acids in GnRHs are nearly constant and the last 2 positions, 9 and 10, are constant, leading us to conclude that amino acids 5–8 are crucial for determining whether or not an analog is going to be selective for FSH or LH release (Yu et al., 1999, 2002). These 4 amino acids in lGnRH-III are His_5, Asp_6, Trp_7 and Lys_8; whereas in mLHRH they are Tyr_5, Gly_6, Leu_7, and Arg_8. Thus, lGnRH III has 60% homology with LHRH.

We fractionated 1,000 rat hypothalami by gel filtration on Sephadex G-25 and determined the FSH- and LH-releasing activity of the various fractions by bioassay on male rat hemipituitaries and compared the localization of these activities with that of LHRH and lGnRH determined by RIA. lGnRH was assayed by RIA using a specific antiserum for lGnRH, that recognized lGnRH I and III equally but did not cross-react with LHRH or cGnRH-II. A peak of LHRH immunoreactivity was found as well as 3 peaks of lGnRH immunoreactivity that preceded the peak of LHRH. The first peak eluted was quite small.

The second peak was much larger and the third peak was of greatest magnitude. Only the second peak altered gonadotropin release and it produced selective FSH release.

To determine whether this activity was caused by lGnRH, anterior hemipituitaries were incubated with normal rabbit serum or the lGnRH antiserum (1 : 1,000) and the effect on FSH- and LH-releasing activity of the FSH-releasing fraction and the LH-releasing activity of LHRH was determined. The antiserum had no effect on basal release of either FSH or LH, but eliminated the FSH-releasing activity of the active fraction without altering the LH-releasing activity of LHRH. Since lGnRH-1 has minimal nonselective activity to release FSH or LH, whereas, previous experiments had shown that lGnRH-III highly selectively releases FSH with a potency equal to that of LHRH to release LH, these results support the hypothesis that the FSH-releasing activity observed in the FSHRH fraction was caused by lGnRH-III or a very closely related peptide (Yu et al., 2000).

Localization of lGnRH in the brain of rats by immunocytochemistry

Using the same antiserum that we employed to find the location of lGnRH after gel-filtration of rat hypothalami, we attempted to localize lGnRH neurons by immunocytochemistry. Immunoreactive lGnRH-like cell bodies were found in the ventromedial preoptic area with axons projecting to the rostral wall of the third ventricle (3V) and OVLT. Another population of lGnRH-like cell bodies was located in the dorsomedial preoptic area with axons projecting caudally and ventrally to the external layer of the median eminence. On the other hand, using a highly specific, monoclonal antiserum against mGnRH to localize the mGnRH neurons, so that their localization could be compared with that of lGnRH neurons, we found that there were no mGnRH cells or fibers in the dorsomedial preoptic area that contained perikarya and fibers of lGnRH neurons (Dees et al., 1999).

Furthermore, immunoabsorption studies indicated that the cell bodies of the lGnRH neurons were eliminated by lGnRH-III, but not by mGnRH whereas the axons in the median eminence were eliminated by lGnRH-III but only slightly reduced by absorption with mGnRH. Using an antiserum against cGnRH-II that visualized cGnRH-II neurons in the chicken hypothalamus, no such neurons could be visualized in the rat hypothalamus (Dees et al., 1999).

Since the lGnRH antiserum (#3952) recognized lGnRH-I and lGnRH-III equally, a specific antiserum against lGnRH-III without cross-reactivity with lGnRH-I was needed to prove that the lGnRH neurons visualized in the rat brain were indeed lGnRH-III neurons. Our recent studies indicate that the specific lGnRH-III antiserum (#39-82-78-3) visualizes the same population of neurons seen with the less specific lGnRH antiserum (3952). Also, the staining of cells and fibers could be eliminated by lGnRH-III but was not affected by lGnRH-I or mGnRH. Consequently, the results strongly support the hypothesis that the lGnRH-III neurons are located in the areas of the brain responsible for control of FSH are the FSHRH neurons (Hiney et al., 2002). The lGnRH-III neurons whose cell bodies are located in the caudal dorsal medial preoptic area with axons projecting to the median eminence are in the very regions which have been shown to be selectively involved in the control of FSH release by lesion and stimulation studies described earlier (pages 151 and 152). Lesions in the caudal dorsal medial preoptic and anterior dorsal medial anterior hypothalamic area blocked the castration-induced rise in FSH (Lumpkin and McCann, 1984) and blocked pulsatile FSH release without interfering with pulsatile LH release (Lumpkin et al., 1989). Conversely, implants of prostaglandin E_2 (PGE_2) in this region evoked selective FSH release. These evoked a pattern of FSH selective release from implants of PGE_2 along a path running from the medial dorsal preoptic area to the caudal parts of the median eminence regions (Ojeda et al., 1977) that have been shown to contain more FSH-releasing activity than could be accounted for by LHRH, and that when destroyed can also selectively block FSH and not LH pulses in castrate male rats (Marubayashi et al., 1999).

The other population of lGnRH-III neurons in the ventromedial caudal preoptic area has axons that appear to project to the OVLT and to the wall of the 3V, whereas the more dorsal neurons also project to the ventricle and caudally to the mid-brain central gray along the same pathway as the LHRH neurons (Dees et al., 1999). The projection to the mid-brain central gray suggests the possibility that lGnRH-III may be

154

involved in mating behavior since this is the area which LHRH activates to induce mating behavior.

The function of the caudal ventral medial preoptic neurons projecting to the OVLT is not clear. Possibly, they release lGnRH-III into the ventricular system for actions more caudally after its uptake from the ventricle. Alternatively, their axons may in some manner also reach the median eminence by an unknown trajectory. It is interesting that some of these neurons in this region also contain mLHRH and that there are other neurons there that contain only LHRH. These are located laterally to the positions of the lGnRH-III neurons that are more medial. The fact that some of the neurons contain both peptides suggests that they may interact in an ultra short-loop feedback mechanism to control the release not only of FSHRH, but also LHRH (Hiney et al., 2002). The localization and projection of the lGnRH-III and LHRH neurons are illustrated (Fig. 1).

Evidence for a specific FSHRH receptor

The selectivity of release of FSH by FSHRH and lGnRH-III suggests the probability of a specific FSHRH receptor on pituitary gonadotropes. In an early experiment in collaboration with J. Rivier and Vale (Salk Institute, La Jolla, CA), we tested the FSH-releasing activity of our purified FSHRH using monolayered cultured pituitary cells and were unable to detect any FSH-releasing activity (unpublished data). This led us to the hypothesis that the FSHRH receptors are down regulated during the 4 days in culture in the absence of gonadal steroids. Indeed, in a recent experiment, we have confirmed the fact that lGnRH-III has little activity on monolayered cultured male rat pituitary cells at the same time that LHRH is active at 10^{-10} M. lGnRH-III was also tested in a LHRH receptor assay. The production of inositol phosphate in COS cells transfected with LHRH receptors was not increased by lGnRH-III until a concentration of 10^{-7} M. As the concentration was increased full activity was obtained at 10^{-4} M (Yu et al., 1999, 2002). We hypothesized that we might demonstrate the presence of lGnRH-III receptors on gonadotropes using biotinylated lGnRH-III and mLHRH. To enhance imaging an avidin-based system was used combined with immunocytochemistry for FSH and LH. The lGnRH-III

peptide was extended with one or two spacer arms of aminocaproic acid between the biotin moiety and one amino acid of the peptide. Three biotinylated derivatives of lGnRH-III were tested for bioactivity. The peptide derivative biotinylated on Lys^8 with 2 spacer arms of aminocaproic acid (Bxx–Lys^8–lGnRH-III) had selective FSH-releasing activity at a dose of 10^{-8} M on hemipituitaries incubated in vitro, whereas the analog with one spacer arm Bx–Lys^8) had less activity than that of the above derivative. By contrast, the analog biotinylated on Asp^6 and one spacer arm (Bx–Asp^6) had no selective FSH-releasing activity. Therefore, the two Bxx–lGnRH-III appeared to be satisfactory for locating lGnRH-III binding sites on pituitary gonadotropes. Biotinylated lGnRH-III (10^{-9} M) bound to 80% of FSH gonadotropes and only 50% of LH gonadotropes of acutely dispersed pituitary cells, a finding that indicates that there are receptors on gonadotropes that bind this peptide (Childs et al., 2001). The binding of biotinylated lGnRH-III was not displaced with LHRH that indicates that it is highly specific for the putative FSHRH receptors. It appears that in this situation as with monolayer cultured pituitary cells, the FSHRH receptors disappear with time in culture since with 24 h in culture, the binding of the biotinylated lGnRH-III was significantly decreased. We believe on the basis of our recent studies with the biotinylated lGnRH-III that ultimately a specific receptor for this peptide will be found in the pituitary and we believe it is in all probability the FSHRH.

lGnRH-III binds to the 3 GnRH receptors discovered in the bullfrog, but the affinity to the receptors was less than that of mLHRH or cGnRH-II as determined in the inositol phosphate assay (Wang et al., 2001). Neill et al. (2001) have recently reported the existence of a second GnRH receptor (GnRH-IIr) in the human genome and reported the cloning and characterization of its cDNA from monkeys. The cDNA generated a G protein coupled transmembrane receptor having a C-terminal cytoplasmic tail, whereas GnRH-Ir lacks this tail. GnRH-IIr resembles more closely the type II receptors of amphibians and fish than it does the type I receptor of humans. This receptor is specific for cGnRH-II and is expressed ubiquitously in human tissues (Neill et al., 2001).

Previously, cGnRH-II has been reported in various tissues in monkeys and humans, but the peptide

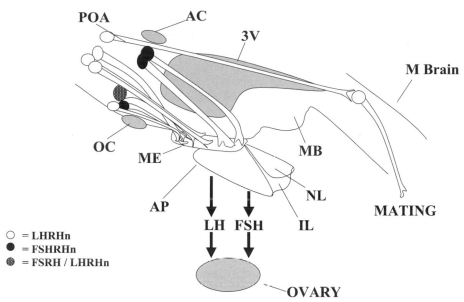

Fig. 1. Parasuggital section of preoptic and hypothalamic region of the rat brain illustrating the distribution of FSHRH and LHRH neurons. Abbreviations: POA = preoptic area, AC = anterior commissure, OC = optic chiasm, 3V = third ventricle, ME = median eminence, MB = mammilliary body, M Brain = midbrain, AP = anterior pituitary, IL = intermediate lobe, NL = neural lobe.

was found in the hypothalamus only of fetal monkeys and also in the mid-brain central gray (Lescheid et al., 1997). In adult monkeys, there were only a few cells and fibers in the caudal hypothalamus suggesting that this peptide would not reach the pituitary via the portal vessels, but must have other actions perhaps on mating behavior or in regulation of cell division. Indeed, as indicated above, we found no cGnRH-II perikarya and only scant fibers in the hypothalamus of the rat in contrast with the readily observed lGnRH-III neurons and terminals in the median eminence (Hiney et al., 2002). Furthermore, cGnRH-II, as indicated above, has little or no selective FSH releasing activity; however, there is little doubt that it is an ancient GnRH existing from fish to mammals.

Mechanism of action of FSHRH, LHRH and leptin on gonadotropin secretion

It is well known that FSH and LH release is controlled by calcium ions (Ca^{2+}) (Wakabayashi et al., 1969; Stojilkovic, 1998) and that interaction of LHRH with its receptor causes an increase in intracellular free calcium and also activates the phos-

phatidyl inositol cycle that mobilizes internal calcium. The resulting increase in intracellular free calcium mediates the releasing action of LHRH (for review, see Stojilkovic, 1998); however, we earlier showed a role for a cGMP and not cyclic adenosine monophosphate (cAMP) in controlling the release of LH and FSH mediated by LHRH (Nakano et al., 1978; Naor et al., 1978; Snyder et al., 1980). This was before it was accepted that NO is a physiologically significant, gaseous transmitter that acts by activation of guanylyl cyclase that converts GTP to cGMP. cGMP activates protein kinase G that causes exocytosis of gonadotropin secretory granules.

To test the hypothesis that the FSH-releasing activity of lGnRH-III (or FSHRH) is mediated by calcium and NO, calcium was removed from the medium and a chelating agent (ethylene glycol-N-N-N'-N'-tetraacetic acid) that would remove any residual Ca^{2+} was added. The action of purified FSHRH and lGnRH was blocked in the absence of Ca^{2+}. N^G-monomethyl-L-arginine (NMMA), a competitive inhibitor of NOS, was added to the medium in other experiments. We found that this competitive inhibitor of NOS, NMMA, completely blocked the FSH-releasing activity of not only purified FSHRH

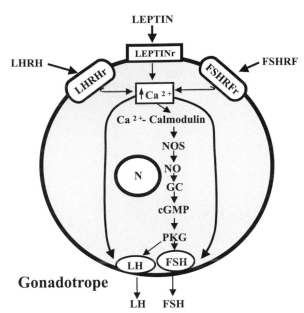

Gonadotrope

Fig. 2. Schematic diagram illustrating the mechanism of the gonadotropin-releasing action of FSHRH, LHRH and leptin, N = nucleus, other abbreviations are in list of abbreviations. The principal pathway is via NO, cGMP and PKG, but Ca²⁺ may have an independent action.

but also of lGnRH-III. Furthermore, sodium nitroprusside (NP), a releaser of NO, stimulated both LH and FSH release and the activity of LHRH to release both LH and FSH was also blocked by NMMA. These data indicate that FSHRH (or lGnRH-III) acts on its putative receptor via a calcium-dependent, nitric oxide pathway to release FSH specifically, whereas LHRH acts on its receptor similarly to increase intracellular Ca²⁺ that activates NOS in the gonadotrophs to cause release of LH and to a lesser extent FSH (Yu et al., 1997c, 1999, 2002) (Fig. 2).

Possible use of lGnRH-III in control of reproduction

Not only does lGnRH-III have selective FSH-releasing activity in vitro but also in vivo in the ovariectomized, estrogen–progesterone-blocked rat (Yu et al., 1997a) and normal male rat (Yu et al., unpublished), but also in the cow (Dees et al., 2001). Treatment of cows with lGnRH-III in the luteal stage of the estrous cycle has produced multiple large follicles in 6 consecutive treatment cycles, in striking

contrast to the normal development of a single large follicle in each cycle. Ovulation of 2 large follicles occurred in 2 cases following injection of human chorionic gonadotropin. On the basis of these findings we believe that lGnRH-III has great potential to alter reproduction in animals and man (Dees et al., 2001).

Role of NO in control of LHRH release

NO is formed in the body by NOS, an enzyme which converts arginine in the presence of oxygen and several cofactors into equimolar quantities of citrulline and NO. There are three isoforms of the enzyme. One of these, nNOS, is found in the cerebellum and various regions of the cerebral cortex and also in various ganglion cells of the autonomic nervous system. Large numbers of nNOS-containing neurons, termed NOergic neurons, were also found in the hypothalamus particularly in the paraventricular and supraoptic nuclei with axons projecting to the median eminence and neural lobe, which also contains large amounts of nNOS. These findings indicated that the enzyme is synthesized at all levels of the neuron from perikarya to axon terminals (McCann and Rettori, 1997).

Because of this distribution in the hypothalamus in regions that contain peptidergic neurons that control pituitary hormone secretion, we decided to determine the role of this soluble gas in the release of LHRH. The approach used was to use sodium NP that spontaneously liberates NO to see if this altered the release of various hypothalamic transmitters. Hemoglobin, which scavenges NO by a reaction with the heme group in the molecule and inhibitors of NOS, such as NMMA, a competitive inhibitor of NOS, were used to determine the effects of decreased NO. Two types of studies were performed. In the first set of experiments, medial basal hypothalamic (MBH) explants were preincubated in vitro and then exposed to neurotransmitters that modify the release of various hypothalamic peptides in the presence or absence of inhibitors of the release of NO. The response to NO itself, provided by sodium NP, was also evaluated. To determine if the results in vitro also held in vivo, substances were microinjected into the 3V of the brain of conscious, freely moving animals to determine the effect on pituitary hormone release (Rettori et al., 1993).

Our most extensive studies were carried out with regard to the release of LHRH. Not only does LHRH act after its secretion into the hypophyseal portal vessels to stimulate LH and to a lesser extent FSH release, but it also induces mating behavior in female rats and penile erection in male rats by hypothalamic action (Mani et al., 1994).

Our experiments showed that release of NO from sodium NP in vitro promoted LHRH release and that the action was blocked by hemoglobin, a scavenger of NO. NP also caused an increased release of PGE_2 from the tissue, which previous experiments showed played an important role in release of LHRH. Furthermore, it caused the biosynthesis and release of prostanoids from ^{14}C arachidonic acid. The effect was most pronounced for PGE_2, but there also was release of lipoxygenase products that have been shown to play a role in LHRH release. Inhibitors of cyclooxygenase, the responsible enzyme for prostanoid synthesis, such as indomethacin and salicylic acid blocked the release of LHRH induced by norepinephrine (NE), providing further evidence for the role of NO in the control of LHRH release via the activation of cyclooxygenase-1. Nedleman's group later showed that NO activates cyclooxygenase-1 and cyclooxygenase-2 in cultured fibroblasts. The action is probably mediated by combination of NO with the heme group of cyclooxygenase altering its conformation. The action on lipoxygenase is similar; although it contains ferrous iron, the actual presence of heme in lipoxygenase has yet to be demonstrated (Rettori et al., 1992, 1993).

As indicated above (page 155), the previously accepted pathway for the physiologic action of NO is by activation of soluble guanylate cyclase by interaction of NO with the heme group of this enzyme, thereby causing conversion of GTP into cGMP, which mediates the effect on smooth muscle by decreasing the intracellular $[Ca^{2+}]$. On the other hand, Muellam's group showed in incubated pancreatic acinar cells that cGMP has opposite effects on intracellular $[Ca^{2+}]$, elevating it at low concentration and lowering it at higher concentrations. We postulate that the NO released from the NOergic neurons, near the LHRH neuronal terminals, increases the intracellular free calcium required to activate phospolypase A_2 (PLA_2). PLA_2 causes the conversion of membrane phospholipids in the LH terminal to arachidonate, which then can be converted to PGE_2 via the activated cyclooxygenase. The released PGE_2 activates adenyl cyclase causing an increase in cAMP release, which activates protein kinase-A, leading to exocytosis of LHRH secretory granules into the hypophyseal portal capillaries for transmission to the anterior pituitary gland (Canteros et al., 1995).

NE has previously been shown to be a releaser of LHRH. It acts by activation of the NOergic neurons because the activation of these neurons and the release of LHRH could be blocked by a competitive inhibitor of NOS, NMMA. NE acts to stimulate the release of NO from the NOergic neurons by α_1 adrenergic receptors since its action can be blocked by phentolamine, an α receptor blocker, and prazosine, a specific α_1 receptor blocker. Activation of the α_1 receptors is postulated to increase intracellular $[Ca^{2+}]$ that combines with calmodulin to activate NOS leading to generation of NO.

We measured the effect of NE on the content of NOS in the MBH explants at the end of the experiments by homogenizing the tissue and adding ^{14}C arginine and measuring its conversion to citrulline on incubation of the homogenate. Because arginine is converted to equimolar quantities of NO and citrulline, measurement of citrulline production provides a convenient estimate of the activity of the enzyme. The NO disappears rapidly making its measurement very difficult. NE caused an increase in the apparent content of the enzyme. That we were actually measuring enzyme content was confirmed, because incubation of the homogenate with L-nitroargine methyl ester, another inhibitor of NOS caused a drastic decline in the conversion of arginine to citrulline. We further confirmed that we actually had increased the content of enzyme by isolating the enzyme according to the method Bredt and Snyder (Rettori et al., 1992) and then measuring the conversion of labeled arginine to citrulline. The conversion was highly significantly increased by NE (Canteros et al., 1996).

Glutamic acid (GA), at least in part by n-methyl-D-aspartate receptors, also plays a physiologically significant role in controlling the release of LHRH. Therefore, we evaluated where GA fits into the picture. It also acted via NO to stimulate LHRH release, but we showed that the effect of GA could be completely obliterated by the α-receptor blocker phentolamine. Consequently, we concluded that GA acted by stimulation of the noradrenergic terminals in the MBH to release NE, which then initiated NO release and stimulation of LHRH release (Kamat et al., 1995).

Oxytocin has actions within the brain to promote mating behavior in the female and penile erection in the male rat. Since LHRH mediates mating behavior, we hypothesized that oxytocin would stimulate the LHRH release that, after secretion into the hypophyseal portal vessels, mediates LH release from the pituitary. Consequently, we incubated MBH explants and demonstrated that oxytocin (10^{-7}–10^{-10} M) induced LHRH release via NE stimulation of nNOS. Therefore, oxytocin may be very important as a stimulator of LHRH release. Furthermore, NO acted as a negative feedback to block oxytocin release (Rettori et al., 1997).

One of the few receptors to be identified on LHRH neurons is the gamma amino butyric acid a (GABAa) receptor. Consequently, we evaluated the role of GABA in LHRH release and the participation of NO in this. The experiments showed that GABA blocked the response of the LHRH neurons to NP that acts directly on the LHRH terminals. We concluded that GABA suppressed LHRH release by blocking response of the LHRH neuronal axons to NO. Additional experiments showed that NO stimulates the release of GABA, providing thereby an inhibitory feed-forward pathway to inhibit the pulsatile release of LHRH initiated by NE. As NE stimulated the release of NO, this would stimulate the release of GABA, which would then block the response of the LHRH neuron to the NO released by NE (Seilicovich et al., 1995a).

Other studies indicated that NO would suppress the release of dopamine and NE. We have already described the ability of NE to stimulate LHRH release and dopamine also acts as a stimulatory transmitter in the pathway. Therefore, there is an ultra short-loop negative feedback mechanism to terminate the pulsatile release of LHRH because the NO released by NE would diffuse to the noradrenergic terminals and inhibit the release of NE, thereby terminating the pulse of NE, LHRH and finally LH (Seilicovich et al., 1995b).

We further examined the possibility that other products from this system might have inhibitory actions. Indeed, we found that as we added increasing amounts of NP, we obtained a bell-shaped dose-response curve of the release of LHRH, such that the release increased with increasing concentrations of NP up to a maximum at around 600 μM and then declined with higher concentrations. When the effect of NP on NOS content at the end of the experiment was measured, we found that high concentrations of NP lowered the NOS content. Furthermore, NP could directly decrease NOS content when incubated with MBH homogenates, results that indicate a direct effect on NOS probably by interaction of NO with the heme group on the enzyme. Thus, when large quantities of NO are released, as could occur after the induction of inducible (i)NOS in the brain during infections, the release of NO would decrease by an inhibitory action on the enzyme at these high concentrations. Furthermore, high concentrations of cGMP released by NO also acted in the explants or even in the homogenates to suppress the activation of NOS. This pathway could also be active in the presence of high concentrations of NO, such as would occur in infection by induction of iNOS by bacterial or viral products (Canteros et al., 1996).

The effect of cytokines (IL-1 and granulocyte macrophage colony-stimulating factor (GMCSF)) on the NOergic control of LHRH release

The cytokines so far tested, for example IL-1 and GMCSF, act within the hypothalamus to suppress the release of LHRH as revealed in both in vivo and in vitro studies. We have examined the mechanism of this effect and found that for IL-1, it occurs by inhibition of cyclooxygenase as shown by the fact that there is blockage of the conversion of labeled arachidonate to prostanoids, particularly PGE$_2$, and the release of PGE$_2$ induced by NE is also blocked (Rettori et al., 1994).

A principle mechanism of action is by suppression of the LHRH release induced by NO donors such as NP. We first believed that there were IL-1 and GMCSF receptors on the LHRH neuron that blocked the response of the neuron to NO. However, because we had also shown that GABA blocks the response to NP and earlier work had shown that GABA receptors are present on the LHRH neurons, we evaluated the possibility that the action of cytokines could be mediated by stimulation of GABAergic neurons in the

MBH. Indeed, in the case of GMCSF, its inhibitory action on LHRH release can be partially reversed by the GABAa receptor blocker, bicuculine, which also blocks the inhibitory action of GABA, itself, on the response of the LHRH terminals to NO. Therefore, we believe that the inhibitory action of cytokines on LHRH release is mediated by stimulation of GABA neurons (Kimura et al., 1997).

Role of NO in mating behavior

LHRH controls lordosis behavior in the female rat and is also involved in mediating male sex behavior. Studies in vivo have shown that NO stimulates the release of LHRH involved in inducing sex behavior. This behavior can be stimulated by third ventricular injection of NP and is blocked by inhibitors of NOS. Apparently, there are two LHRH neuronal systems: one, with axons terminating on the hypophyseal portal vessels, the other with axons terminating on neurons that mediate sex behavior (Mani et al., 1994). NO is also involved in inducing penile erection by the release of NO from NOergic neurons innervating the corpora cavernosa penis. The role of NO in sex behavior in both sexes has led us to change the name of NO to the sexual gas (McCann and Rettori, 1997).

Potential role of leptin in reproduction

The hypothesis that leptin may play an important role in reproduction stems from several findings. First, the Ob/Ob mouse, lacking the leptin gene, is infertile and has atrophic reproductive organs. Gonadotropin secretion is impaired and very sensitive to negative feedback by gonadal steroids as is the case for prepubertal animals (Swerdloff et al., 1976, 1978). It has now been shown that treatment with leptin can recover the reproductive system in the Ob/Ob mouse by leading to growth and function of the reproductive organs and fertility (Chehab et al., 1996) via secretion of gonadotropins (Barash et al., 1996).

The critical weight hypothesis of the development of puberty states that when body fat stores have reached a certain point puberty occurs (Frisch and McArthur, 1974). This hypothesis in its original form does not hold because if animals are underfed, puberty is delayed, but with access to food, rapid weight gain leads to the onset of puberty at weights well below the critical weight under normal nutritional conditions (Ronnekleiv et al., 1978). We hypothesized that during this period of refeeding or at the time of the critical weight in the normally fed animals as the fat stores increase, there is increased release of leptin from the adipocytes into the blood stream and that this acts on the hypothalamus to stimulate the release of LHRH with resultant induction of puberty. Indeed, leptin has been found recently to induce puberty (Chehab et al., 1997).

We initiated studies on its possible effects on hypothalamic–pituitary function. We anticipated that it would also be active in adult rats and therefore studied its effect on the release of FSH and LH from hemipituitaries, and also its possible action to release LHRH from MBH explants in vitro. To determine if it was active in vivo, we used a model that we have often used to evaluate stimulatory effects of peptides on LH release; namely, the ovariectomized, estrogen-primed rat. Because our supply of leptin was limited, we began by microinjecting it into the 3V in conscious animals bearing implanted third ventricular cannulae, and also catheters in the external jugular vein extending to the right atrium, so that we could draw blood samples before and after the injection of leptin and measure the effect on plasma FSH and LH (Yu et al., 1997b).

Effect of leptin on gonadotropin release

We found that under our conditions, leptin had a bell-shaped dose-response curve to release LH from anterior pituitaries incubated in vitro. There was no consistent stimulation of LH release with a concentration of 10^{-5} M. Results became significant with 10^{-7} M and remained on a plateau through 10^{-11} M with reduced release at a concentration of 10^{-12} M that was no longer significant statistically. The release was not significantly less than that achieved with LHRH (4×10^{-8} M). Under these conditions, there was no additional release of LH when leptin (10^{-7} M) was incubated together with LHRH (4×10^{-8} M). In certain other experiments, there was an additive effect when leptin was incubated with LHRH; however, this effect was not uniformly seen. The results indicate that leptin was only slightly less effective to release LH, than LHRH itself (Yu et al., 1997b).

Effect of leptin on FSH release

In the incubates from these same glands, we also measured FSH release and found that it showed a similar pattern as that of LH, except that the sensitivity in terms of FSH release was much less than that for LH. The minimal effective dose for FSH was 10^{-9} M, whereas it was 10^{-11} M for LH. The responses were roughly of the same magnitude at the effective concentrations as obtained with LH and the responses were clearly equivalent to those observed with 4×10^{-9} M LHRH. Combination of LHRH with a concentration of leptin which was just below significance gave a clear additive effect (Yu et al., 1997b). The action of leptin to stimulate both LH and FSH release was mediated by the long form of the leptin receptor that is located on the cell surface of the gonadotropes. The mechanism of action is the same as that of FSHRH and LHRH, the only difference is that the action is mediated by leptin receptors that increase intracellular free Ca^{2+} activating nNOS that generates NO that activates GC followed by PKG leading to extrusion of FSH and LH secretory granulues from the gonadotropes (Fig. 2).

Effect of leptin on LHRH release

There was no significant effect of leptin in a concentration range of 10^{-6}–10^{-12} M on LHRH release during the first 30 min of incubation; however, during the second 30 min, the highest concentration produced a borderline significant decrease in LHRH release with 10^{-6} M, followed by a tendency to increase with lower concentrations and a significant ($P < 0.01$), plateaued increase with the lowest concentrations tested 10^{-10} and 10^{-12} M) (57). Both the FSH and LH-releasing actions of leptin were blocked by NMMA indicating that NO mediates its action (Yu et al., 1997b).

The effect of intraventricularly injected leptin on plasma gonadotropin concentrations in ovariectomized, estrogen-primed rats

The injection of the diluent for leptin into the 3V (Krebs-Ringer Bicarbonate, 5 μl) had no effect on pulsatile FSH or LH release, but the injection of leptin (10 μg) uniformly produced an increase in plasma LH with a variable time-lag ranging from 10 to 50 minutes, so that the maximal increase in LH from the starting value was highly significant $P <$ 0.01) and constituted a mean increase of 60% above the initial concentration. In contrast, leptin inhibited FSH release on comparison with the results with the diluent, but the effect was delayed and occurred mostly in the second hour. Therefore, at this dose of estrogen (10 μg estradiol benzoates, 72 h before experiments), leptin stimulates the release of LHRH and inhibits the release of FSHRH (Walczewska et al., 1999).

Mechanism of action of leptin on the hypothalamic pituitary axis

We have shown that leptin exerts its action at both hypothalamic and pituitary level by activating NOS since its effect to release LHRH, FSH and LH in vitro (Yu et al., 1997b) is blocked by NMMA. Leptin, in essence, is a cytokine secreted by the adipocytes. It, like the cytokines, seems to reach the brain via a transport mechanism mediated by the Ob/Ob_a receptors (Cioffi et al., 1996) in the choroid plexus (Schwartz et al., 1996). These receptors have an extensive extracellular domain, but a greatly truncated intracellular domain (Cioffi et al., 1996) and mediate transport of the cytokine by a saturable mechanism (Banks et al., 1996). Following uptake into the cerebrospinal fluid (CSF) through the choroid plexus, leptin is carried by the flow of CSF to the 3V, where it either diffuses into the hypothalamus through the ependymal layer lining the ventricle or combines with Ob/Ob_a (Cioffi et al., 1996) receptors on terminals of responsive neurons that extend to the ventricular wall.

The Ob/Ob_b receptor has a large intracelluar domain that presumably mediates the action of the protein (Schwartz et al., 1996). These receptors are wide-spread throughout the brain (Schwartz et al., 1996), but particularly localized in the region of the paraventricular (PVN) and arcuate nuclei (AN). Leptin activates stat 3 within 30 min after its intraventricular injection (Vaisse et al., 1996). Stat 3 is a protein that is important in conveying information to the nucleus to initiate DNA-directed messenger ribonucleic acid (mRNA) synthesis. Following injection of bacterial lipopolysaccharide (LPS), it is also activated, but in this case, the time delay is 90 min presumably because LPS has been shown to induce IL-1 beta (β) mRNA in the same areas — namely, the PVN and AN (Schwartz et al., 1996), IL-1β mRNA would then cause production of IL-1β that would activate stat 3. On entrance into the nucleus, stat 3 would activate or inhibit DNA-directed mRNA synthesis. In the case of leptin, it activates corticotropin-releasing hormone (CRH) mRNA in the PVN, whereas in the AN, it inhibits neuropeptide Y (NPY) mRNA resulting in increased CRH synthesis and presumably release in the PVN and decreased NPY synthesis and release in the AN (Schwartz et al., 1996). Presumably, the combination of leptin with these transducing receptors also either increases or decreases the firing rate of that particular neuron. In the case of the AN-median eminence area, leptin may enter the median eminence by diffusion between the tanycytes or alternatively by combining with its receptors on terminals of neurons projecting to the tanycytes. Activation of these neurons would induce LHRH release.

The complete pathway of leptin action in the MBH to stimulate LHRH release is not yet elucidated. Arcuate neurons bearing Ob/Ob receptors may project to the median eminence to the tanycyte/portal capillary junction. Leptin would either combine with its receptors on the terminals that transmit information to the cell bodies in the AN or diffuse to the AN to combine with its receptors on the perikarya of AN neurons. Because leptin decreases NPY mRNA and presumably NPY biosynthesis in NPY neurons in the AN (Chehab et al., 1997), we postulate that leptin causes a decrease in NPY release. Because NPY inhibited LH release in intact and castrated male rats (Reznikov and McCann, 1993), we hypothesize that NPY decreases the release of LHRH by inhibiting the noradrenergic neurons which mediate pulsatile release of LHRH. Therefore, when the release of NPY is inhibited by leptin, noradrenergic impulses are generated, that act on α_1 receptor on the NOergic neurons causing the release of NO which diffuses to the LHRH terminals and activates LHRH release by activating guanylate cyclase and cyclooxygenase$_1$ as shown in our prior experiments reviewed above. Leptin acts to activate NOS as indicated because its release of LHRH is blocked by inhibition of NOS (Yu et al., 1997b).

The LHRH enters the portal vessels and is carried to the anterior pituitary gland where it acts to stimulate FSH and particularly LH release by combining with its receptors on the gonadotropes. The release of LH and to a lesser extent FSH is further increased by the direct action of leptin on its receptors in the pituitary gland (Cioffi et al., 1996; Naivar et al., 1996; Yu et al., 1997b).

We hypothesize that leptin may be a critical factor in induction of puberty as the animal nears the so-called critical weight. Either metabolic signals reaching the adipocytes, or signals related to their content of fat cause the release of leptin, which increases LHRH and gonadotropin release, thereby initiating puberty and finally ovulation and onset of menstrual cycles. In the male, the system would work similarly; however, there is no preovulatory LH surge brought about by the positive feedback of estradiol. Sensitivity to leptin is undoubtedly under steroid control and we are actively working to elucidate this problem.

During fasting, the leptin signal is removed and LH pulsatility and reproductive function decline quite rapidly. In women with anorexia nervosa, this causes a reversion to the prepubertal state, which can be reversed by feeding. Thus, leptin would have a powerful influence on reproduction throughout the reproductive lifespan of the individual. The consequences to gonadotropin secretion of overproduction of leptin, as has already been demonstrated in human obesity, are not clear. There are often reproductive abnormalities in this circumstance and whether they are due to excess leptin production or other factors, remains to be determined. In conclusion, it is now clear that leptin plays an important role in control of reproduction by actions on the hypothalamus and pituitary.

Abbreviations

3V	third ventricle
AN	arcuate nucleus
c	chicken
cAMP	cyclic adenosine monophosphate
cGMP	cyclic guanosine monophosphate
CRH	corticotropin-releasing hormone
CSF	cerebrospinal fluid
FSH	follicle-stimulating hormone
GA	lutamic acid

GABA	gamma amino butyric acid
GMCSF	granulocyte macrophage colony-stimulating factor
GnRH	gonadotropin-releasing hormone
GnRHr	GnRH receptor
GTP	guanosine triphosphate
i	inducible
IL-1α	interleukin-1 alpha
l	lamprey
LPS	lipopolysaccharide
m	mammalian
MBH	medial basal hypothalami
mLHRH	Mammalian luteinizing hormone-releasing hormone
mRNA	messenger ribonucleic acid
NE	norepinephrine
NMMA	N^G-monomethyl-L-arginine
nNOS	neural nitric oxide synthase
NP	nitroprusside
NPY	neuropeptide Y
OVLT	organum vasculosum lamina terminalis
PGE2	prostaglandin E_2
PLA2	phospolipase A_2
PVN	paraventricular nucleus

Acknowledgements

This work was supported by NIH Grant MH51853. We would like to thank Judy Scott and Natasha Hunter for their excellent secretarial support.

References

Banks, W.A., Kastin, A.J., Huang, W., Jaspan, J.B. and Maness, L.M. (1996) Leptin enters the brain by a saturable system independent of insulin. *Peptides*, 17: 305–311.

Barash, I.A., Cheung, C.C., Weigle, D.S., Ren, H., Kabigting, E.B., Kuijper, J.L., Clifton, D.K. and Steiner, R.A. (1996) Leptin is a metabolic signal to the reproductive system. *Endocrinology*, 137: 3144–3147.

Bredt, D.S. and Snyder, S.H. (1989) Nitric oxide mediates glutamate-linked enhancement of cGMP levels in the cerebellum. *Proc. Natl. Acad. Sci.*, 86: 9030–9033.

Canteros, G., Rettori, V., Franchi, A., Genaro, A., Cebral, E., Saletti, A., Gimeno, M. and McCann, S.M. (1995) Ethanol inhibits luteinizing hormone-releasing hormone (LHRH) secretion by blocking the response of LHRH neuronal terminals to nitric oxide. *Proc. Natl. Acad. Sci.*, 92: 3416–3420.

Canteros, G., Rettori, V., Genaro, A., Suburo, A., Gimeno, M. and McCann, S.M. (1996) Nitric oxide synthase (NOS) content of hypothalamic explants: Increase by norepinephrine and

inactivated by NO and cyclic GMP. *Proc. Natl. Acad. Sci.*, 93: 4246–4250.

Chehab, F.F., Lim, M.E. and Ronghua, L. (1996) Correction of the sterility defect in homozygous obese female mice by treatment with the human recombinant leptin. *Nat. Genet.*, 12: 318–320.

Chehab, F.F., Mounzih, K., Lu, R. and Lim, M.E. (1997) Early onset of reproductive function in normal female mice treated with leptin. *Science*, 275: 88–90.

Childs, G.V., Miller, B.T., Chico, D.E., Unabia, G.C., Yu, W.H. and McCann, S.M. (2001) Preferential expression of receptors for lamprey gonadotropin releasing hormone-III (GnRH-III) by FSH cells: support for its function as an FSH-RF. *Endocrine Society's 83rd Annual Meeting*, Denver, CO, June 20–23, pp. 506, Abstract #P3-268.

Cioffi, J., Shafer, A., Zupancic, T., Smith-Gbur, J., Mikhail, A., Platika, D. and Snodgrass, H. (1996) Novel B219/ob receptor isoforms: possible role of leptin in hematopoiesis and reproduction. *Nat. Med.*, 2: 585–588.

Culler, M.D. and Negro-Vilar, A. (1987) Pulsatile follicle-stimulating hormone secretion is independent of luteinizing hormone-releasing hormone (LHRH): Pulsatile replacement of LHRH bioactivity in LHRH-immunoneutralized rats. *Endocrinology*, 120: 2011–2021.

Dees, W.L., Rettori, V., Kozlowski, J.G. and McCann, S.M. (1985) Ethanol and the pulsatile release of luteinizing hormone, follicle stimulating hormone and prolactin in ovariectomized rats. *Alcohol*, 2: 641–646.

Dees, W.L., Hiney, J.K., Sower, S.A., Yu, W.H. and McCann, S.M. (1999) Localization of immunoreactive lamprey gonadotropin-releasing hormone in the rat brain. *Peptides*, 20: 1503–1511.

Dees, W.L., Dearth, R.K., Hooper, R.N., Brinsko, S.P., Romano, J.E., Rahe, C.H., Yu, W.H. and McCann, S.M. (2001) Lamprey gonadotropin-releasing hormone-III selectively releases follicle stimulating hormone in the bovine. *Domest. Anim. Endocrinol.*, 20: 279–288.

Dhariwal, A.P.S., Nallar, R., Batt, M. and McCann, S.M. (1965) Separation of FSH-releasing factor from LH-releasing factor. *Endocrinology*, 76: 290–294.

Dhariwal, A.P.S., Watanabe, S., Antunes-Rodrigues, J. and McCann, S.M. (1967) Chromatographic behavior of follicle stimulating hormone-releasing factor on Sephadex and carboxy methyl cellulose. *Neuroendocrinology*, 2: 294–303.

Frisch, R.E. and McArthur, J.W. (1974) Menstrual cycles: Fatness as a determinant of minimum weight for height necessary for their maintenance or onset. *Science*, 185: 949–951.

Hiney, J.K., Sower, S.A., Yu, W.H., McCann, S.M. and Dees, W.L. (2002) Gonadotropin-releasing hormone neurons in the preoptic-hypothalamic region of the rat contain lamprey gonadotropin-releasing hormone III, mammalian luteinizing hormone-releasing hormone, or both peptides. *Proc. Natl. Acad. Sci.*, 99: 2386–2391.

Igarashi, M. and McCann, S.M. (1964) A hypothalamic follicle stimulating hormone releasing factor. *Endocrinology*, 74: 446–452.

Kamat, A., Yu, W.H., Rettori, V. and McCann, S.M. (1995) Glutamic acid stimulated luteinizing-hormone releasing hormone release is mediated by alpha adrenergic stimulation of nitric oxide release. *Brain Res. Bull.*, 37: 233–235.

Kimura, M., Yu, W.H., Rettori, V. and McCann, S.M. (1997) Granulocyte-macrophage colony stimulating factor suppresses LHRH release by inhibition of nitric oxide synthase and stimulation of gamma-aminobutyric acid release. *Neuroimmunomodulation*, 4: 237–243.

Lescheid, D.W., Terasawa, E., Abler, L.A., Urbanski, H.F., Warby, C.M., Millar, R.P. and Sherwood, N.M. (1997) A second form of gonadotropin-releasing hormone (GnRH) with characteristics of chicken GnRH-II is present in the primate brain. *Endocrinology*, 138: 5618–5629.

Lumpkin, M.D. and McCann, S.M. (1984) Effect of destruction of the dorsal anterior hypothalamus on selective follicle stimulating hormone secretion. *Endocrinology*, 115: 2473–2480.

Lumpkin, M.D., Moltz, J.H., Yu, W.H., Samson, W.K. and McCann, S.M. (1987) Purification of FSH-releasing factor: Its dissimilarity from LHRH of mammalian, avian, and piscian origin. *Brain Res. Bull.*, 18: 175–178.

Lumpkin, M.D., McDonald, J.K., Samson, W.K. and McCann, S.M. (1989) Destruction of the dorsal anterior hypothalamic region suppresses pulsatile release of follicle stimulating hormone but not luteinizing hormone. *Neuroendocrinology*, 50: 220–235.

Mani, S.K., Allen, J.M., Rettori, V., McCann, S.M., O'Malley, B.W. and Clark, J.H. (1994) Nitric oxide mediates sexual behavior in female rats by stimulating LHRH release. *Proc. Natl. Acad. Sci.*, 91: 6468–6472.

Marubayashi, U., Yu, W.H. and McCann, S.M. (1999) Median eminence lesions reveal separate hypothalamic control of pulsatile follicle-stimulating hormone and luteinizing hormone release. *Proc. Soc. Exp. Biol. Med.*, 220: 139–146.

McCann, S.M. and Krulich, L. (1989) Role of neurotransmitters in control of anterior pituitary hormone release. In: L.J. DeGroot (Ed.), *Endocrinology, 2nd ed.* WB Saunders, Philadelphia, PA, pp. 117–130.

McCann, S.M. and Ojeda, S.R. (1996) The anterior pituitary and hypothalamus. In: J. Griffin and S.R. Ojeda (Eds.), *Textbook of Endocrine Physiology, 3rd ed.* Oxford University Press, New York, NY, pp. 101–133.

McCann, S.M. and Rettori, V. (1997) The role of nitric oxide in reproduction. *Proc. Natl. Acad. Sci.*, 94: 2735–2740.

McCann, S.M., Taleisnik, S. and Friedman, K.M. (1960) LH-releasing activity in hypothalamus extracts. *Proc. Soc. Exp. Biol. Med.*, 104: 432–434.

McCann, S.M., Marubayashi, U., Sun, H.-Q. and Yu, W.H. (1993) Control of follicle stimulating hormone and luteinizing hormone release by hypothalamic peptides. In: G.P. Chrousos and T. Tolis (Eds.), *Intraovarian Regulators and Polycystic Ovarian Syndrome: Recent Progress on Clinical and Therapeutic Aspects. Ann. N.Y. Acad. Sci.*, 687: 55–59.

Mizunuma, H., Samson, W.K., Lumpkin, M.D., Moltz, J.H., Fawcett, C.P. and McCann, S.M. (1983) Purification of a bioactive FSH-releasing factor (FSHRH). *Brain Res. Bull.*, 10: 623–629.

Naivar, J.S., Dyer, C.J., Matteri, R.L. and Keisler, D.H. (1996) Expression of leptin and its receptor in sheep tissues. *Proc. Soc. Study Reprod.*, 391: 154 (abstract).

Nakano, H., Fawcett, C.P., Kimura, F. and McCann, S.M. (1978) Evidence for the involvement of guanosine 3',5'-cyclic monophosphate in the regulation of gonadotropin release. *Endocrinology*, 103: 1527–1533.

Naor, Z., Fawcett, C.P. and McCann, S.M. (1978) The involvement of cGMP in LHRH stimulated gonadotropin release. *Am. J. Physiol.*, 235: 586–590.

Neill, J.D., Duck, L.W., Sellers, J.C. and Musgrove, L.C. (2001) A gonadotropin-releasing hormone (GnRH) receptor specific for GnRH II in primates. *Biochem. Biophys. Res. Comm.*, 282: 1012–1018.

Ojeda, S.R., Jameson, H.E. and McCann, S.M. (1977) Hypothalamic areas involved in prostaglandin (PG)-induced gonadotropin release. II: Effect of PGE2 and PGF2alpha implants on follicle stimulating hormone release. *Endocrinology*, 100: 1595–1603.

Reichlin, S. (1992) In: J.D. Wilson and D.W. Foster (Eds.), *Williams' Textbook of Endocrinology, Neuroendocrinology.* WB Saunders, Philadelphia, PA, pp. 135–219.

Rettori, V., Gimeno, M.F., Karara, A., Gonzales, M.C. and McCann, S.M. (1991) Interleukin 1α inhibits prostaglandin E_2 release to suppress pulsatile release of luteinizing hormone but not follicle-stimulating hormone. *Proc. Natl. Acad. Sci.*, 88: 2763–2767.

Rettori, V., Gimeno, M., Lyson, K. and McCann, S.M. (1992) Nitric oxide mediates norepinephrine-induced prostaglandin E_2 release from the hypothalamus. *Proc. Natl. Acad. Sci.*, 89: 11543–11546.

Rettori, V., Belova, N., Dees, W.L., Nyberg, C.L., Gimeno, M. and McCann, S.M. (1993) Role of nitric oxide in the control of luteinizing hormone-releasing hormone release in vivo and in vitro. *Proc. Natl. Acad. Sci.*, 90: 10130–10134.

Rettori, V., Belova, N., Kamat, A., Lyson, K. and McCann, S.M. (1994) Blockade by IL-I-alpha of the nitrocoxidergic control of luteinizing hormone-releasing hormone release in vivo and in vitro. *Neuroimmunomodulation*, 1: 86–91.

Rettori, V., Canteros, G., Renoso, R., Gimeno, M. and McCann, S.M. (1997) Oxytocin stimulates the release of luteinizing hormone-releasing hormone from medial basal hypothalamic explants by releasing nitric oxide. *Proc. Natl. Acad. Sci.*, 94: 2741–2744.

Reznikov, A.G. and McCann, S.M. (1993) Effects of neuropeptide Y on gonadotropin and prolactin release in normal, castrated or flutamide-treated male rats. *Neuroendocrinology*, 57: 1148–1154.

Ronnekleiv, O.K., Ojeda, S.R. and McCann, S.M. (1978) Undernutrition, puberty and the development of estrogen positive feedback in the female rat. *Biol. Reprod.*, 19: 414–424.

Schally, A.V., Saito, T., Arimura, A., Muller, E.E., Bowers, C.Y. and White, W.F. (1966) Purification of follicle-stimulating hormone-releasing factor (FSH-RF) from bovine hypothalamus. *Endocrinology*, 79: 1087–1094.

Schwartz, M.W., Seeley, R.J., Campfield, L.A., Burn, P. and Baskin, D.G. (1996) Identification of targets of leptin action in rat hypothalamus. *J. Clin. Invest.*, 98: 1101–1106.

Seilicovich, A., Duvilanski, B.H., Pisera, D., Thies, S., Gimeno, M., Rettori, V. and McCann, S.M. (1995a) Nitric oxide inhibits hypothalamic luteinizing hormone-releasing hormone release by releasing γ-aminobutyric acid. *Proc. Natl. Acad. Sci.*, 92: 3421–3424.

Seilicovich, A., Lasaga, M., Befurno, M., Duvilanski, B.H., de C Dias, M., Rettori, V. and McCann, S.M. (1995b) Nitric oxide inhibits the release of norepinephrine and dopamine from the medial basal hypothalamus of the rat. *Proc. Natl. Acad. Sci.*, 92: 11299–11302.

Snyder, G., Naor, Z., Fawcett, C.P. and McCann, S.M. (1980) Gonadotropin release and cyclic ucleotides: Evidence for LHRH-induced elevation of cyclic GMP levels in gonadotrophs. *Endocrinology*, 107: 1627–1633.

Stojilkovic, S.S. (1998) Calcium signaling systems. In: M.P. Conn and H.M. Goodman (Eds.), *Handbook of Physiology, Section 7. The Endocrine System, Volume I: Cellular Endocrinology*. Oxford Press, New York, NY, pp. 177–224.

Swerdloff, R., Batt, R. and Bray, G. (1976) Reproductive hormonal function in the genetically obese (ob/ob) mouse. *Endocrinology*, 98: 1359–1364.

Swerdloff, R., Peterson, M., Vera, A., Batt, R., Heber, D. and Bray, G. (1978) The hypothalamic–pituitary axis in genetically obese (ob/ob) mice: response to luteinizing hormone-releasing hormone. *Endocrinology*, 103: 542–547.

Vaisse, C., Halaas, J.L., Horvath, C.M., Darnell, J.E. Jr., Stoffel, M. and Friedman, J.M. (1996) Leptin activation of stat3 in the hypothalamus of wildtype and Ob/Ob mice but not Db/Db mice. *Nat. Genet.*, 14: 95–97.

Wakabayashi, K., Kamberi, I.A. and McCann, S.M. (1969) In vitro responses of the rat pituitary to gonadotrophin-releasing factors and to ions. *Endocrinology*, 85: 1046–1056.

Walczewska, A., Yu, W.H., Karanth, S. and McCann, S.M. (1999) Estrogen and leptin have differential effects on FSH and LH release in female rats. *Proc. Soc. Exp. Biol. Med.*, 222: 70–77.

Wang, L., Bogerd, J., Choi, H.S., Seong, J.Y., Soh, J.M., Chun, S.Y., Blomenrohr, M., Troskie, B.E., Millar, R.P., Yu, W.H., McCann, S.M. and Kwon, H.B. (2001) Three distinct types of GnRH receptor characterized in the bullfrog. *Proc. Natl. Acad. Sci. USA*, 98: 361–366.

Yu, W.H., Millar, R.P., Milton, S.C.F., Del Milton, R.C. and McCann, S.M. (1990) Selective FSH-releasing activity of [D-Trp9] GAP 1-13: comparison with gonadotropin-releasing abilities of analogs of GAP and natural LHRHs. *Brain Res. Bull.*, 25: 867–873.

Yu, W.H., Karanth, S., Walczewska, A., Sower, S.A. and McCann, S.M. (1997a) A hypothalamic follicle-stimulating hormone-releasing decapeptide in the rat. *Proc. Natl. Acad. Sci.*, 94: 9499–9503.

Yu, W.H., Kimura, M., Walczewska, A., Karanth, S. and McCann, S.M. (1997b) Role of leptin in hypothalamic–pituitary function. *Proc. Natl. Acad. Sci.*, 94: 1023–1028.

Yu, W.H., Walczewska, A., Karanth, S. and McCann, S.M. (1997c) Nitric oxide mediates leptin-induced luteinizing hormone-releasing hormone (LHRH) and LHRH and leptin-induced LH release from the pituitary gland. *Endocrinology*, 138: 5055–5058.

Yu, W.H., Karanth, S., Mastronardi, C.A., Sealfon, S., Dees, L. and McCann, S.M. (1999) Follicle stimulating hormone-releasing factor acts on its putative receptor via nitric oxide to specifically release follicle stimulating hormone. *81st Annual Meeting of the Endocrine Society*, San Diego, CA, pp. 2–15, 284.

Yu, W.H., Karanth, S., Sower, S.A., Parlow, A.F. and McCann, S.M. (2000) The similarity of FSH-releasing factor to lamprey gonadotropin-releasing hormone III (lGnRH-III). *Proc. Soc. Exp. Biol. Med.*, 224: 87–92.

Yu, W.H., Karanth, S., Mastronardi, C.A., Sealfon, S., Dean, C., Dees, W.L. and McCann, S.M. (2002) Lamprey GnRH-III acts on its putative receptor via nitric oxide to release follicle-stimulating hormone specifically. *Expl. Biol. Med.*, accepted.

I.S. Parhar (Ed.)
Progress in Brain Research, Vol. 141
© 2002 Elsevier Science B.V. All rights reserved

CHAPTER 12

Gonadotropin-releasing hormone (GnRH) surge generator in female rats

Toshiya Funabashi [1,*], Dai Mitsushima [1], Takahiro J. Nakamura [1], Tsuguo Uemura [2], Fumiki Hirahara [3], Kazuyuki Shinohara [1], Kumiko Suyama [1], Fukuko Kimura [1]

[1] *Department of Physiology, Yokohama City University School of Medicine, 3-9 Fukuura, Kanazawa-ku, Yokohama 236-0004, Japan*
[2] *Department of Gynecology and Obstetrics, Fujisawa City Hospital, 2-6-1 Fujisawa, Fujisawa 251-0052, Japan*
[3] *Department of Gynecology and Obstetrics, Yokohama City University Scool of Medicine, Yokohama 236-0004, Japan*

Introduction

In rats, gonadotropin-releasing hormone (GnRH) expressing neurons, which control luteinizing hormone (LH) and follicle stimulating hormone secretions from the anterior pituitary, are mainly located in the preoptic area (POA) (for review see Silverman et al., 1994), but some GnRH neurons are also found in the mediobasal hypothalamus (MBH) in rats (Kawano and Daikoku, 1981; King et al., 1982; Shivers et al., 1983; Merchenthaler et al., 1984). However, the GnRH neurons found in the MBH are not located in the arcuate nucleus but in the lateral hypothalamic area, close to the dorso-medial surface of the optic tract and the base of the hypothalamus (Funabashi et al., 1997a,b). These GnRH neurons in the MBH are involved in the GnRH pulse generator, which is not described in detail in this chapter. At any rate, this wide and scattered distribution of GnRH neurons in the brain makes study of the physiology of GnRH neurons difficult.

One of most remarkable characteristic of GnRH neurons is that they do not originate from the hy-pothalamus but from the olfactory placode and migrate toward the hypothalamus during developmental stages (Schwanzel-Fukuda et al., 1992). Another interesting characteristic is their secretion patterns. Secretion of LH, which is assumed to reflect GnRH secretion from the hypothalamus into the portal vein, shows two remarkably different patterns. One is pulsatile LH secretion, and another is pre-ovulatory LH surge secretion. The fact that these secretory patterns reflect those of GnRH has now been determined; the pulsatile LH secretion is followed by pulsatile GnRH secretion in rats (Levine and Ramirez, 1982), sheep (Clarke and Cummins, 1982) and monkeys (Woller et al., 1992), and the surge of LH secretion is followed by a surge of GnRH secretion in rats (Sarkar et al., 1976), sheep (Caraty et al., 1989; Moenter et al., 1992), and monkeys (Levine et al., 1985; Pau et al., 1993).

We have hypothesized that these two patterns of GnRH secretion are controlled by different neural mechanisms (for review see Kimura and Funabashi, 1998): the GnRH pulse generator, previously known as the tonic center, and the GnRH surge generator, previously known as the phasic or the cyclic center (Gorski, 1971). Classical neuroendocrinological studies indicate that in female rats, the neuronal component responsible for inducing the LH surge is located in the POA (Everett, 1965; Halász and Gorski, 1967; Gorski, 1971; Kawakami et al., 1971). According to our hypothesis, which includes this tra-

---

* Correspondence to: T. Funabashi, Department of Physiology, Yokohama City University School of Medicine, 3-9 Fukuura, Kanazawa-ku, Yokohama 236-0004, Japan. Tel.: +81-45-787-2579; Fax: +81-45-787-2509; E-mail: toshiya@med.yokohama-cu.ac.jp

ditional concept, the GnRH pulse generator located in the MBH makes intimate contact with opioid neurons that mediate the negative feedback effects of estrogen and the GnRH surge generator located in the POA involves GABA neurons and arginine-vasopressin (AVP) neurons in addition to GnRH neurons. In this chapter, we focus on the putative GnRH surge generator of female rats (Funabashi et al., 2002).

Ultra-short feedback of GnRH

It has now been established, by checking GnRH secretion, that the surge of LH secretion is produced not only by increasing the responsiveness of the anterior pituitary to GnRH but also by increasing GnRH secreted into the portal blood as the result of a positive feedback action of estrogen, which is administered at a certain concentration and duration to the hypothalamus and the pituitary gland (Kalra, 1993; Freeman, 1994). We were first interested in how this pattern of GnRH secretion occurs. Since the bulk of GnRH secretion, the GnRH surge, occurs during a certain period and for a certain duration (Sarkar et al., 1976; Levine et al., 1985), many GnRH neurons need to be excited at the same time. Several lines of morphological evidence indicate that GnRH neurons in the POA make synaptic contacts with each other (Merchenthaler et al., 1984; Leranth et al., 1985; Pelletier, 1987), suggesting that GnRH affects GnRH neurons in the POA. In support of this, electrophysiological studies revealed both excitatory and inhibitory effects of GnRH on neurons in the POA (Moss, 1979). Hence, we thought that an ultra-short feedback mechanism (Hyyppa et al., 1971) is involved in the bulk of GnRH secretion, resulting in a surge of LH secretion. To this end, we injected GnRH into the POA in estrogen-primed ovariectomized (OVX) or proestrous rats. We found that the injection potentiated and advanced the timing of the LH surge (Hiruma et al., 1989), while the injection of GnRH antagonist delayed the LH surge (Funabashi et al., 1994). In accord with these results, it was also reported that the injection of GnRH antagonist into the POA disrupted the estrous cyclicity in rats (Weesner and Pfaff, 1994). However, injection of neither GnRH nor GnRH antagonist into the POA affected the pulsatile LH secretion in OVX rats, sug-

gesting that ultra-short feedback mechanism in the POA is only acting for the GnRH surge generator but not for the GnRH pulse generator (Funabashi et al., 1994). It was recently reported that GnRH receptor mRNA was expressed in the POA (Han et al., 1999). Thus, a stimulatory ultra-short feedback of GnRH in the POA is involved in inducing the GnRH surge.

GABA and LH surge

To understand the mechanism underlying the generation of the GnRH surge, it is necessary to determine the neuronal components of the GnRH surge generator. There is increasing evidence indicating that GABAergic inputs are profoundly involved in regulating the surge of LH secretion (Demling et al., 1985; Adler and Crowley, 1986; Donoso and Banzan, 1986; Flügge et al., 1986; Herbison and Dyer, 1991; Jarry et al., 1995; Mitsushima et al., 1997). Further, it has been shown that GnRH neurons in the POA express subunits of $GABA_A$ receptor (Petersen et al., 1993a; Jung et al., 1998; Sim et al., 2000) and respond to bicuculline, a $GABA_A$ receptor antagonist (Sim et al., 2000). This means that a decrease in inhibitory GABA tone on GnRH neurons results in an activation of GnRH neurons. Consistent with this, we have shown that the intravenous infusion of bicuculline, a $GABA_A$ receptor antagonist, on the morning of proestrus induces a premature surge-like secretion of LH (Kimura and Jinnai, 1994). We then wondered what similarities there were between the spontaneous LH surge and the bicuculline-induced surge-like LH secretion.

Since it was reported that GnRH neurons within the POA express c-Fos protein in association with the LH surge on the day of proestrus (Lee et al., 1990a) and in ovariectomized estrogen- and progesterone-primed female rats (Lee et al., 1990b), we examined whether the infusion of bicuculline on the day of proestrus advances the timing of c-Fos expression in GnRH neurons as well as the surge of LH secretion. We thought that if the premature surge-like LH secretion was due to the advancement of activation of GnRH neurons in the surge generator, GnRH neurons would express c-Fos at an early time when c-Fos expression is usually not observed in proestrous rats. We thus used double immunocyto-

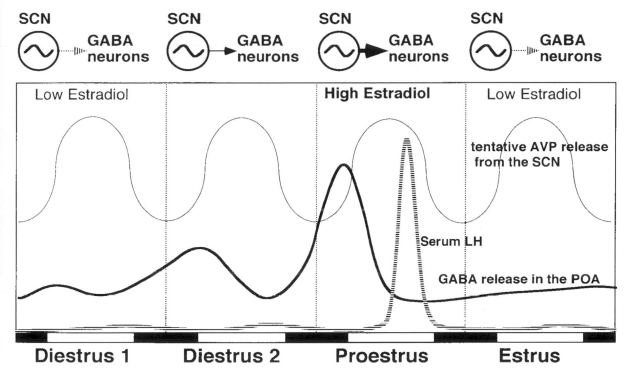

Fig. 1. Schematic illustration showing the relationship among GnRH, GABA and AVP secretions during the estrous cycle in rats. The activity of AVP neurons in the SCN is constantly exhibiting a circadian during the estrous cycle, based on changes in AVP release (Shinohara et al., 1994; Kalsbeek et al., 1995) and its mRNA in the SCN (Krajnak et al., 1998a,b). Since serum estrogen levels are low in diestrus 1 and 2, and thus levels of AVP 1a receptors are low in the POA, circadian signals are not transmitted to GnRH neurons or GABA neurons. Inhibition of GABA neurons on GnRH neurons develops as a function of levels of serum estrogen. An increase in serum estrogen also elicits the expression of AVP 1a receptor in the POA. This makes it possible for GnRH neurons or GABA neurons to communicate with AVP neurons coming from the SCN. Disinhibition of GABA neurons on GnRH neurons that occurs during the critical period is probably permissive for activation of GnRH neurons. Clock information conveyed through AVP neurons may inhibit GABA activity or activate GnRH neurons. Hence, the LH surge is induced as the result of positive feedback effects of estrogen.

chemistry for c-Fos and GnRH (Funabashi et al., 1997a). As expected, c-Fos was expressed in approximately 50% of GnRH neurons in saline-infused rats killed in the late afternoon but not in the saline-infused rats killed in the early afternoon. Similarly, approximately 50% of GnRH neurons in bicuculline-infused rats killed in the early afternoon were found to express c-Fos. The distribution and proportion of GnRH neurons expressing c-Fos in response to bicuculline were identical to those during the spontaneous LH surge as observed in saline-injected rats killed in the late afternoon. These results indicate that bicuculline infusions activate GnRH neurons. This activation of GnRH neurons would account for the observed increase in serum LH, because the intensity of c-Fos expression in GnRH neurons is closely associated with the amount of LH secreted in proestrous rats (Lee et al., 1992). Therefore, we conclude that bicuculline advances the timing of the activation of GnRH neurons. This provides additional evidence that GABAergic neurons are involved in the GnRH surge generator and subsequently the LH surge. It was reported that GABA release in the POA seemed to decline prior to the onset of a surge-like secretion of LH in Ovariectomized estrogen-primed rats (Jarry et al., 1992, 1995). Quite recently, we determined the GABA release in the POA during the estrous cycle by microdialysis (Fig. 1). We found that the GABA release markedly increased from late in the night of diestrus 2 through the morning of proestrus, when it attained its peak, and thereafter it declined sharply until the critical period of proestrus.

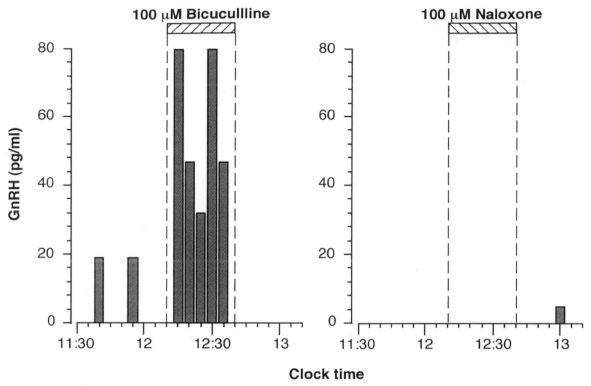

Fig. 2. Effects of bicuculline and naloxone on the GnRH secretion in POA slices derived from proestrous rats. On the morning of proestrus, rats were killed and 3 200-μm-thick POA slices were cut with a brain slicer. After 1 h of incubation in a standard artificial cerebrospinal fluid (aCSF) with 10 mM glucose at a flow rate of 500 μl/min, 50 μl of aCSF were collected at 5-min intervals. Concentrations of GnRH were determined by EIA. As shown here, 100 μM bicuculline infusion markedly increased the GnRH secretion but 100 μM naloxone was without effect. This preliminary result suggests that a decrease in GABA tone in the POA elicits GnRH secretion but that a decrease in opioid tone does not.

This suggests that the preovulatory elevation of the GABA release from the night through to the morning of proestrus, followed by a sharp decline, is closely associated with the onset of the preovulatory LH surge in cyclic female rats (Mitsushima et al., 2002). This also provides strong evidence that a decrease in GABA tone in the POA triggers the activation of GnRH neurons. We further confirmed this concept by checking changes in GnRH release from POA slices. When 100 μM bicuculline was applied for 30 min, GnRH release was markedly increased in 200 μm thick slices obtained from a proestrous rat, but naloxone had no significant effect on GnRH release (Fig. 2). Although this is a preliminary result, it shows that disinhibition of GABA, but not of opioid, on GnRH neurons elicits a surge-like GnRH secretion from the POA, as we have suggested that opioid neurons are not involved in the LH surge (Kimura

and Jinnai, 1994; Funabashi et al., 1997b; Kimura and Funabashi, 1998).

The suprachiasmatic nucleus (SCN), POA, and the GnRH surge generator

A circadian clock located in the SCN (Meijer and Rietveld, 1989) is involved in the timing of the LH surge in rats (Wiegand et al., 1980; van der Beek, 1996). This is demonstrated clearly by the fact that a daily afternoon rise of LH secretion occurs if ovariectomized rats are treated with a certain amount of estrogen (Everett and Sawyer, 1950; Legan and Karsch, 1975), indicating that the neuronal component responsible for the GnRH surge in the POA develops receptiveness to clock information from the SCN (Wiegand et al., 1978) along with the action of estrogen (Petersen et al., 1993b). It is therefore necessary to know which neurons in the SCN send the circadian clock information to the GnRH surge generator in the POA. Since arginine-vasopressin (AVP) and vasoactive intestinal polypeptide (VIP) neurons provide the major output from the SCN (Watts and Swanson, 1987), these two peptides are the most likely candidates (Harney et al., 1996; Krajnak et al., 1998a,b).

To determine whether the SCN can drive the circadian secretion of GnRH from neurons in the POA and which neurons mediate the clock information to the GnRH surge generator in the POA, we measured the secretion of GnRH, AVP and VIP in co-cultures of the POA and the SCN at 2-h intervals over a period of 120 hours (Funabashi et al., 2000a). Co-cultures were treated with antimitotics, because such treatment alters the phase relationship between AVP and VIP rhythms, causing the periods of the two peptide rhythms to diverge from each other (Shinohara et al., 1995).

We found that the secretion of GnRH in co-cultures exhibited a significant circadian rhythm in the presence of estrogen but not in the absence of estrogen. In each co-culture, the period of the GnRH circadian rhythm was the same as that of the AVP circadian rhythm, but different from the VIP circadian rhythm. As mentioned above, the circadian rhythm of the AVP release differed from that of VIP release because of the antimitotic treatment. Furthermore, the peak phase of the GnRH rhythm occurred at the time same as that of the AVP rhythm. However, the peak phase of the GnRH rhythm was not always the same as that of the VIP rhythm. The administration of AVP significantly increased GnRH secretion in individual POA cultures in the presence of estrogen, but the administration of VIP had no effect. These results suggest that, in co-cultures of the SCN and the POA, AVP neurons drive the circadian secretion of GnRH in the presence of estrogen. We therefore conclude that AVP neurons in the SCN mediate circadian clock information to the GnRH surge generator in the POA. In support of this conclusion, the injection of AVP receptor antagonist at 1300 h in proestrous rats resulted in significant decreases in both LH and prolactin secretion (Funabashi et al., 1999). Furthermore, Palm et al. reported that when AVP was injected into the POA, AVP stimulated the LH surge (Palm et al., 1999, 2001). These results

provide further evidence supporting our hypothesis that endogenous AVP plays a stimulatory role in the induction of the surge of LH.

On the basis of these findings, we searched for the site of estrogen action in association with the LH surge. We thought that estrogen may target neurons expressing AVP receptor in the POA. We therefore tried to determine whether estrogen affected the expression of AVP receptor mRNA in the POA of female rats (Funabashi et al., 2000b). By reverse transcription-PCR, we found that all three types of the AVP receptor mRNA, V1a, V1b, and V2, were expressed in the POA, though the amount of PCR products was apparently different among them. Using in situ hybridization histochemistry, we found that AVP V1a receptor mRNA was expressed in the POA, especially in the anteroventral periventricular nucleus of the POA. This was in good accord with a previous report (Ostrowski et al., 1994). In contrast, AVP V1b and V2 receptor mRNAs were not abundant in this area. Northern blot analysis revealed that estrogen significantly increased the expression of AVP V1a receptor mRNA in the POA of young ovariectomized female rats.

Can POA time the LH surge without the SCN?

As described above, the SCN is important for timing the LH surge at least in rats. However, we cannot generalize this concept for rats to other species. For example, the surge of LH secretion occurs as a function of time after estrogen priming in sheep (Goodman et al., 1981). But how does the sheep brain count the time after estrogen treatment without a clock? Is the SCN essential for inducing the LH surge in rats? To answer this question, we examined whether GnRH neurons by themselves can release GnRH as a function of time after estrogen treatment. To this end, we measured GnRH release in cultures of GT1-7 cells after applying serum shock, which has been shown to induce a circadian rhythm in another cell lines (Balsalobre et al., 1998; Akashi and Nishida, 2000). As a result, a bulk release of GnRH was observed 12 h after the serum shock (Fig. 3). This preliminary result suggests that an increase in GnRH secretion can be induced by a certain stimulation without any other synaptic input neurons. It needs to be determined, however, whether

170

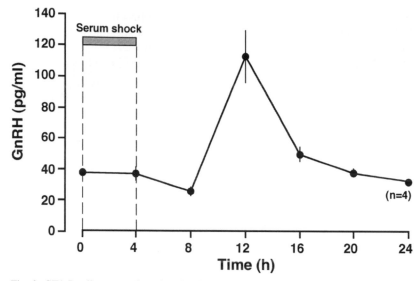

Fig. 3. GT1-7 cells were cultured and maintained in Dulbecco's modified Eagle's medium supplemented with 10% fetal bovine serum (FBS) in an atmosphere containing 5% CO_2 at 37°C. When they reached 60–80% confluency, GT1-7 cells were dissociated and the cell suspension (5×10^4) was plated in 60 mm dishes. These cells were maintained in the D-MEM/F12 medium supplemented with 10% FBS. Then the medium was replaced with one containing 50% FBS (serum shock). Thereafter, all medium was removed and replaced with fresh medium containing 2% FBS at 4-h intervals. Concentration of GnRH was measured in 50 μl of medium. Data are shown as the mean ± SEM ($n = 4$). As shown here, a bolus of GnRH was induced by serum shock.

such increase in GnRH secretion resembles the surge of GnRH secretion, or, most importantly, whether estrogen can produce the same effect as the serum shock.

Acknowledgements

We are grateful to Dr. K. Wakabayashi and the NIDDK for providing radioimmunoassay materials. We also thank Dr. T. Nett for antibody to GnRH (CR-R11B71). GT1-7 cells were generously donated by Drs R.I. Weiner and P.L. Mellon. Part of the present study was supported by a Grant-in-Aid for Scientific Research (C:11670075 to FK and C:11680790 to TF) and for Exploratory Research (13877011 to FK) from the Ministry of Education, Culture, Sports, Science and Technology, Japan.

References

Adler, B.A. and Crowley, W.R. (1986) Evidence for γ-aminobutyric acid modulation of ovarian hormonal effects on luteinizing hormone secretion and hypothalamic catecholamine activity in the female rat. *Endocrinology*, 118: 91–97.

Akashi, M. and Nishida, E. (2000) Involvement of the MAP kinase cascade in resetting of the mammalian circadian clock. *Genes Dev.*, 14: 645–649.

Balsalobre, A., Damiola, F. and Schibler, U. (1998) A serum shock induces circadian gene expression in mammalian tissue culture cells. *Cell*, 93: 929–937.

Caraty, A., Locatelli, A. and Martin, G.B. (1989) Biphasic response in the secretion of gonadotrophin-releasing hormone in ovariectomized ewes injected with oestradiol. *J. Endocrinol.*, 123: 375–382.

Clarke, I.J. and Cummins, J.T. (1982) The temporal relationship between gonadotropin releasing hormone (GnRH) and luteinizing hormone (LH) secretion in ovariectomized ewes. *Endocrinology*, 111: 1737–1739.

Demling, J., Fuchs, E., Baumert, M. and Wuttke, W. (1985) Preoptic catecholamine, GABA, and glutamate release in ovariectomized estrogen-primed rats utilizing a push–pull cannula technique. *Neuroendocrinology*, 41: 212–218.

Donoso, A.O. and Banzan, A.M. (1986) Blockade of the LH surge and ovulation by GABA-T inhibitory drugs that increase brain GABA levels in rats. *Psychoneuroendocrinology*, 11: 429–435.

Everett, J.W. (1965) Ovulation in rats from preoptic stimulation through platinum electrodes. Importance of duration and spread of stimuli. *Endocrinology*, 76: 1195–1201.

Everett, J.W. and Sawyer, C.H. (1950) A 24-hour periodicity in the 'LH-release apparatus' of female rats, disclosed by barbiturate sedation. *Endocrinology*, 47: 198–218.

Flügge, G., Oertel, W.H. and Wuttke, W. (1986) Evidence for estrogen-receptive GABAergic neurons in the preoptic/anterior hypothalamic area of the rat brain. *Neuroendocrinology*, 43: 1–5.

Freeman, M.E. (1994) The neuroendocrine control of the ovarian cycle of the rat. In: E. Knobil and J. Neill (Eds.), *The Physiology of Reproduction*. Vol. 2, Raven Press, New York, pp. 613–658.

Funabashi, T., Hashimoto, R., Jinnai, K. and Kimura, F. (1994) Microinjection of LHRH and its antagonistic analog into the medial preoptic area does not affect pulsatile secretion of LH in ovariectomized rats. *Endocr. J.*, 41: 559–563.

Funabashi, T., Jinnai, K. and Kimura, F. (1997a) Bicuculline infusion advances the timing of Fos expression in LHRH neurons in the preoptic area of proestrous rats. *NeuroReport*, 8: 771–774.

Funabashi, T., Jinnai, K. and Kimura, F. (1997b) Fos expression by naloxone in LHRH neurons of the mediobasal hypothalamus and effects of pentobarbital sodium in the proestrous rat. *J. Neuroendocrinol.*, 9: 87–92.

Funabashi, T., Aiba, S., Sano, A., Shinohara, K. and Kimura, F. (1999) Intracerebroventricular injection of arginine-vasopressin V1 receptor antagonist attenuates the surge of luteinizing hormone and prolactin secretion in proestrous rats. *Neurosci. Lett.*, 260: 37–40.

Funabashi, T., Shinohara, K., Mitsushima, D. and Kimura, F. (2000a) Gonadotropin-releasing hormone exhibits circadian rhythm in phase with arginine-vasopressin in co-cultures of the female rat preoptic area and suprachiasmatic nucleus. *J. Neuroendocrinol.*, 12: 521–528.

Funabashi, T., Shinohara, K., Mitsushima, D. and Kimura, F. (2000b) Estrogen increases arginine-vasopressin V1a receptor mRNA in the preoptic area of young but not of middle-aged female rats. *Neurosci. Lett.*, 285: 205–208.

Funabashi, T., Shinohara, K. and Kimura, F. (2002) Neuronal control circuit for the gonadotropin-releasing hormone surge in rats. In: R.J. Handa, S. Hayashi, E. Terasawa and M. Kawata (Eds.), *Neuroplasticity, Development, and Steroid Hormone Action*. CRC Press, New York, pp. 169–176.

Goodman, R.L., Legan, S.J., Ryan, K.D., Foster, D.L. and Karsch, F.J. (1981) Importance of variations in behavioural and feedback actions of oestradiol to the control of seasonal breeding in the ewe. *J. Endocrinol.*, 89: 229–240.

Gorski, R.A. (1971) Gonadal hormones and the perinatal development of neuroendcrine function. In: L. Martini and W.F. Ganong (Eds.), *Frontiers in Neuroendocrinology*. Oxford Press, New York, pp. 237–290.

Halász, B. and Gorski, R.A. (1967) Gonadotrophic hormone secretion in female rats after partial or total interruption of neural afferents to the medial basal hypothalamus. *Endocrinology*, 80: 608–622.

Han, Y.G., Kang, S.S., Seong, J.Y., Geum, D., Suh, Y.H. and Kim, K. (1999) Negative regulation of gonadotropin-releasing hormone and gonadotropin-releasing hormone receptor gene expression by a gonadotropin-releasing hormone agonist in the rat hypothalamus. *J. Neuroendocrinol.*, 11: 195–201.

Harney, J.P., Scarbrough, K., Rosewell, K.L. and Wise, P.M. (1996) In vivo antisense antagonism of vasoactive intesti-

nal peptide in the suprachiasmatic nuclei causes aging-like changes in the estradiol-induced luteinizing hormone and prolactin surges. *Endocrinology*, 137: 3696–3701.

Herbison, A.E. and Dyer, R.G. (1991) Effect on luteinizing hormone secretion of GABA receptor modulation in the medial preoptic area at the time of proestrous luteinizing hormone surge. *Neuroendocrinology*, 53: 317–320.

Hiruma, H., Funabashi, T. and Kimura, F. (1989) LHRH injected into the medial preoptic area potentiates LH secretion in ovariectomized estrogen-primed and proestrous rats. *Neuroendocrinology*, 50: 421–426.

Hyyppa, M., Motta, M. and Martini, L. (1971) 'Ultrashort' feedback control of follicle-stimulating hormone-releasing factor secretion. *Neuroendocrinology*, 7: 227–235.

Jarry, H., Hirsch, B., Leonhardt, S. and Wuttke, W. (1992) Amino acid neurotransmitter release in the preoptic area of rats during the positive feedback actions of estradiol on LH release. *Neuroendocrinology*, 56: 133–140.

Jarry, H., Leonhardt, S., Schwarze, T. and Wuttke, W. (1995) Preoptic rather than mediobasal hypothalamic amino acid neurotransmitter release regulates GnRH secretion during the estrogen-induced LH surge in the ovariectomized rat. *Neuroendocrinology*, 62: 479–486.

Jung, H., Shannon, E.M., Fritschy, J.-M. and Ojeda, S.R. (1998) Several GABA$_A$ receptor subunits are expressed in LHRH neurons of juvenile female rats. *Brain Res.*, 780: 218–229.

Kalra, S.P. (1993) Mandatory neuropeptide-steroid signaling for the preovulatory luteinizing hormone-releasing hormone discharge. *Endocrine Rev.*, 14: 507–538.

Kalsbeek, A., Buijs, R.M., Engelmann, M., Wotjak, C.T. and Landgraf, R. (1995) In vivo measurement of a diurnal variation in vasopressin release in the suprachiasmatic nucleus. *Brain Res.*, 682: 75–82.

Kawakami, M., Terasawa, E., Seto, K. and Wakabayashi, K. (1971) Effect of electrical stimulation of the medial preoptic area on hypothalamic multiple unit activity in relation to LH release. *Endocr. Japon.*, 18: 13–21.

Kawano, H. and Daikoku, S. (1981) Immunohistochemical demonstration of LHRH neurons and their pathways in the rat hypothalamus. *Neuroendocrinology*, 32: 179–186.

Kimura, F. and Funabashi, T. (1998) Two subgroups of gonadotropin-releasing hormone neurons control gonadotropin secretion in rats. *News Physiol. Sci.*, 13: 225–231.

Kimura, F. and Jinnai, K. (1994) Bicuculline infusions advance the timing of luteinizing hormone surge in proestrous rats: Comparisons with naloxone effects. *Horm. Behav.*, 28: 424–430.

King, J.C., Tobet, S.A., Snavely, F.L. and Arimura, A.A. (1982) LHRH immunopositive cells and their projections to the median eminence and organum vasculosum of the lamina terminalis. *J. Comp. Neurol.*, 209: 287–300.

Krajnak, K., Kashon, M.L., Rosewell, K.L. and Wise, P.M. (1998a) Aging alters the rhythmic expression of vasoactive intestinal polypeptide mRNA but not arginine-vasopressin mRNA in the suprachiasmatic nuclei of female rats. *J. Neurosci.*, 18: 4767–4774.

Krajnak, K., Kashon, M.L., Rosewell, K.L. and Wise, P.M.

172

(1998b) Sex differences in the daily rhythm of vasoactive intestinal polypeptide but not arginine vasopressin messenger ribonucleic acid in the suprachiasmatic nuclei. *Endocrinology*, 139: 4189–4196.

Lee, W.-S., Smith, M.S. and Hoffman, G.E. (1990a) Luteinizing hormone-releasing hormone neurons express Fos protein during the proestrous surge of luteinizing hormone. *Proc. Natl. Acad. Sci. USA*, 87: 5163–5167.

Lee, W.-S., Smith, M.S. and Hoffman, G.E. (1990b) Progesterone enhances the surge of luteinizing hormone by increasing the activation of luteinizing hormone-releasing hormone neurons. *Endocrinology*, 127: 2604–2606.

Lee, W.-S., Smith, M.S. and Hoffman, G.E. (1992) cFos activity identifies recruitment of luteinizing hormone-releasing hormone neurons during the ascending phase of the proestrous luteinizing hormone surge. *J. Neuroendocrinol.*, 4: 161–166.

Legan, S.J. and Karsch, F.J. (1975) A daily signal for the LH surge in the rat. *Endocrinology*, 96: 57–62.

Leranth, Cs., Segura, L.M.G., Palkovits, M., MacLusky, N.J., Shanabrough, M. and Naftolin, F. (1985) The LH-RH-containing neuronal network in the preoptic area of the rat: Demonstration of LH-RH-containing nerve terminals in synaptic contact with LH-RH neurons. *Brain Res.*, 345: 332–336.

Levine, J.E. and Ramirez, V.D. (1982) Luteinizing hormone-releasing hormone release during the rat estrous cycle and after ovariectomy, as estimated with push–pull cannulae. *Endocrinology*, 111: 1439–1448.

Levine, J.E., Norman, R.L., Gliessman, P.M., Oyama, T.T., Bangsberg, D.R. and Spies, H.G. (1985) In vivo gonadotropin-releasing hormone release and serum luteinizing hormone measurements in ovariectomized, estrogen-treated rhesus macaques. *Endocrinology*, 117: 711–721.

Meijer, J.H. and Rietveld, W.J. (1989) Neurophysiology of the suprachiasmatic circadian pacemaker in rodents. *Physiol. Rev.*, 69: 671–707.

Merchenthaler, I., Görcs, T., Sétáló, G., Petrusz, P. and Flerkó, B. (1984) Gonadotropin-releasing hormone (GnRH) neurons and pathways in the rat brain. *Cell Tissue Res.*, 237: 15–29.

Mitsushima, D., Jinnai, K. and Kimura, F. (1997) Possible role of the γ-aminobutyric acid-A receptor system in the timing of the proestrous luteinizing hormone surge in rats. *Endocrinology*, 138: 1944–1948.

Mitsushima, D., Tin-Tin-Win-Shwe, Funabashi, T., Shinohara, K. and Kimura, F. (2002) GABA release in the medial preoptic area of cyclic female rats. *Neuroscience*, 113: 109–114.

Moenter, S.M., Brand, R.C. and Karsch, F.J. (1992) Dynamics of gonadotropin-releasing hormone (GnRH) secretion during the GnRH surge: Insights into the mechanism of GnRH surge induction. *Endocrinology*, 130: 2978–2984.

Moss, R.L. (1979) Actions of hypothalamic–hypophysiotropic hormones on the brain. *Ann. Rev. Physiol.*, 41: 617–631.

Ostrowski, N.L., Lolait, S.J. and Young III, W.S. (1994) Cellular localization of vasopressin V1a receptor messenger ribonucleic acid in adult male rat brain, pineal, and brain vasculature. *Endocrinology*, 135: 1511–1528.

Palm, I.F., van der Beek, E.M., Wiegant, V.M., Buijs, R.M. and Kalsbeek, A. (1999) Vasopressin induces a luteinizing hor-

mone surge in oavriectomized, estradiol-treated rats with lesions of the suprachiasmatic nucleus. *Neuroscience*, 93: 659–666.

Palm, I.F., van der Beek, E.M., Wiegant, V.M., Buijs, R.M. and Kalsbeek, A. (2001) The stimulatory effect of vasopressin on the luteinizing hormone surge in ovariectomized, estradiol-treated rats is time-dependent. *Brain Res.*, 901: 109–116.

Pau, K.Y.F., Berria, M., Hess, D.L. and Spies, H.G. (1993) Pre-ovulatory gonadotropin-releasing hormone surge in ovarian-intact rhesus macaques. *Endocrinology*, 133: 1650–1656.

Pelletier, G. (1987) Demonstration of contacts between neurons staining for LHRH in the preoptic area of the rat brain. *Neuroendocrinology*, 46: 457–459.

Petersen, S.L., McCrone, S., Coy, D., Adelman, J.P. and Mahan, L.C. (1993a) GABA$_A$ receptor subunit mRNAs in cells of the preoptic area: Colocalization with LHRH mRNA using dual-label in situ hybridization histochemistry. *Endocr. J.*, 1: 29–34.

Petersen, S.L., McCrone, S. and Shores, S. (1993b) Localized changes in LHRH mRNA levels as cellular correlates of the positive feedback effects of estrogen on LHRH neurons. *Amer. Zool.*, 33: 255–265.

Sarkar, D.K., Chiappa, S.A., Fink, G. and Sherwood, N.M. (1976) Gonadotropin-releasing hormone surge in pro-oestrous rats. *Nature*, 264: 461–463.

Schwanzel-Fukuda, M., Jorgenson, K.L., Bergen, H.T., Weesner, G.D. and Pfaff, D.W. (1992) Biology of normal luteinizing hormone-releasing hormone neurons during and after their migration from olfactory placode. *Endocr. Rev.*, 13: 623–634.

Shinohara, K., Honma, S., Katsuno, Y., Abe, H. and Honma, K. (1994) Circadian rhythms in the release of vasoactive intestinal polypeptide and arginine-vasopressin in organotypic slice culture of rat suprachiasmatic nucleus. *Neurosci. Lett.*, 170: 183–186.

Shinohara, K., Honma, S., Katsuno, Y., Abe, H. and Honma, K. (1995) Two distinct oscillators in the rat suprachiasmatic nucleus in vitro. *Proc. Natl. Acad. Sci. USA*, 92: 7396–7400.

Shivers, B.D., Harlan, R.E., Morrell, J.I. and Pfaff, D.W. (1983) Immunocytochemical localization of luteinizing hormone-releasing hormone in male and female rat brains. *Neuroendocrinology*, 36: 1–12.

Silverman, A.-J., Livne, I. and Witkin, J.W. (1994) The gonadotropin-releasing hormone (GnRH), neuronal systems: Immunocytochemistry and in situ hybridization. In: E. Knobil and J.D. Neill (Eds.), *The Physiology of Reproduction*. Vol. 1, Raven Press, New York, pp. 1683–1710.

Sim, J.A., Skynner, M.J., Rape, J.-R. and Herbison, A.E. (2000) Late postnatal reorganization of GABA$_A$ receptor signalling in native GnRH neurons. *Eur. J. Neurosci.*, 12: 3497–3504.

van der Beek, E.M. (1996) Circadian control of reproduction in the female rat. In: R.M. Buijs, A. Kalsbeek, H.J. Romijn, C.M.A. Pennartz and M. Mirmiran (Eds.), *Progress in Brain Resesrch*. Vol. 111, Elsevier Science BV, pp. 295–320.

Watts, A.G. and Swanson, L.W. (1987) Efferent projections of the suprachiasmatic nucleus: II. Studies using retrograde transport of fluorescent dyes and simultaneous peptide immunohistochemistry in the rat. *J. Comp. Neurol.*, 258: 230–252.

Weesner, G.D. and Pfaff, D.W. (1994) Disruption of estrous cyclicity following administration of a luteinizing hormone-releasing hormone antagonist to the preoptic area of the rat. *Biol. Reprod.*, 50: 1178–1182.

Wiegand, S.J., Terasawa, E. and Bridson, W.E. (1978) Persistent estrus and blockade of progesterone-induced LH release follows lesions which do not damage the suprachiasmatic nucleus. *Endocrinology*, 102: 1645–1648.

Wiegand, S.J., Terasawa, E., Bridson, W.E. and Goy, R.W. (1980) Effects of discrete lesions of preoptic and suprachiasmatic structures in the female rat. Alterations in the feedback regulation of gonadotropin secretion. *Neuroendocrinology*, 31: 147–157.

Woller, M.J., McDonald, J.K., Reboussin, D.M. and Terasawa, E. (1992) Neuropeptide Y is a neuromodulator of pulsatile luteinizing hormone-releasing hormone release in the gonadectomized rhesus monkey. *Endocrinology*, 130: 2333–2342.

I.S. Parhar (Ed.)
Progress in Brain Research, Vol. 141
© 2002 Elsevier Science B.V. All rights reserved

CHAPTER 13

Mechanisms of inhibition of LHRH release by alcohol and cannabinoids

Valeria Rettori [1,*], Alejandro Lomniczi [1], Claudia Mohn [1], Camila Scorticati [1], Paula Vissio [1], Mercedes Lasaga [2], Ana Franchi [1], Samuel M. McCann [3]

[1] Centro de Estudios Farmacológicos y Botánicos CONICET, Serrano 669, 1414, Buenos Aires, Argentina
[2] Centro de Investigaciones en Reproducción, Fac. Medicina, UBA, Buenos Aires, Argentina
[3] Pennington Biomedical Research Center, LSU, Baton Rouge, LA, USA

Introduction

This paper will review our research and that of our associates on the effects of alcohol and cannabinoids on reproduction. Today, alcohol is classified as a psychotropic drug, like delta-9 tetrahydrocannabinol (THC), the active ingredient of marihuana. A large number of neurotransmitters not only the classical ones, but also a host of neuropeptides that can act as neutransmitters or neuromodulators, exist in the central nervous system (CNS) and there is abundant evidence that alcohol and cannabinoids can affect a number of them. Therefore, mechanisms of action of these drugs in the CNS are very complex.

Addiction to alcohol or marihuana produces numerous deleterious effects in the organism. Among these alterations is the suppression of reproduction in humans, monkeys and small rodents by inhibition of the release of luteinizing hormone (LH). This inhibition of LH secretion is caused mainly by hypothalamic action to inhibit the release of luteinizing hormone-releasing hormone (LHRH), in vivo and in vitro. In conscious, ovariectomized rats, intragastric administration of alcohol (3 g/kg), a dose that

* Correspondence to: V. Rettori, CEFYBO-CONICET, Serrano 669, 1414, Buenos Aires, Argentina. Fax: +5411-4963-4473; Tel: +5411-4855-7204;
E-mail: vrettori@yahoo.com

causes mild intoxication, has been shown to produce a marked decrease in plasma LH concentrations on comparison with the unaffected LH values of diluent administered animals (Dees et al., 1985). There was a highly significant decrease in the area under the secretion curve of LH, with a reduction of LH pulses. On the other hand, following a challenge with exogenous LHRH, the response of LH was the same as in controls, indicating that pituitary responsiveness was the same for alcohol and saline groups. In contrast to LH, alcohol did not significantly alter pulsatile FSH secretion, indicating that alcohol selectively inhibited pulsatile release of LHRH but not the putative FSHRF. Furthermore, the secretion of LH by pituitaries incubated in vitro in the presence of different concentrations of alcohol (50–100 mM) was the same as without alcohol.

Similar results were obtained when THC was injected into the third cerebral ventricle. A single dose of THC (2 μl of 10^{-6} M) significantly decreased serum LH levels as compared to values in vehicle injected rats (Wenger et al., 1987). Also, there was no effect on plasma FSH levels as seen with alcohol. Furthermore, the response to a challenge dose of LHRH on LH secretion by cultured dispersed pituitary cells was the same in the presence of THC or vehicle. Therefore, we investigated the effect of both drugs on LHRH release from medial basal hypothalami (MBH) incubated in an in vitro system.

Effect of alcohol on LHRH release

To understand the inhibitory pathways involved in the inhibition of LHRH release by alcohol, it is necessary to put them into the context of our previous work on the NOergic control of LHRH release. Our previous work indicated that the release of LHRH is controlled by nitric oxide (NO) (Rettori et al., 1993; Canteros et al., 1996). NO is formed by conversion of arginine to citrulline and NO in equimolar concentrations by the action of nitric oxide synthase (NOS). There are three isoforms of NOS, neuronal NOS (nNOS) and endothelial NOS (eNOS) are constitutive, and need the presence of Ca^{2+} to form NO, the inducible NOS (iNOS) is Ca^{2+} independent and is induced mostly by endotoxins as well by cytokines (Moncada et al., 1991). In the present work we will be referring to constitutive NOS, mainly to nNOS. nNOS has been demonstrated by immunocytochemical methods in neurons in some areas of the CNS including some regions of hypothalamus, as the median eminence–arcuate region. Previous research has indicated that NO stimulates the release of LHRH both in vivo and in vitro. On the basis of in vitro experiments using incubations of MBH in a static incubation system, it has been determined that norepinephrine (NE) activates constitutive NOS in this region. The NO released from these neurons diffuses to LHRH terminals, where it induces the release of LHRH. It has been shown that NO not only activates guanylate cyclase followed by increased cyclic guanosine monophosphate (cGMP) release but also activates cyclooxygenase (COX) that increases release of eicosanoids (Rettori et al., 1992). Prostaglandin E2 (PGE2) by activating adenylate cyclase (Ojeda et al., 1979) with consequent increase in cyclic adenosine monophosphate (cAMP) evokes exocytosis of LHRH granules by activation of protein kinase A. The LHRH released diffuses into the hypophyseal portal vessels that deliver it to the anterior pituitary gland where it acts on gonadotropes to release LH. Support for this theoretical pathway stems from the ability of inhibitors of NOS, such as NG-monomethyl-L-arginine, to inhibit LHRH release, whereas releasers of NO, such as sodium nitroprusside (NP), induce LHRH release as well as that of PGE2 from MBH (Rettori et al., 1992).

The release of LHRH is not only under the control of stimulatory neurotransmitters such as NE (Rettori et al., 1993) and glutamic acid (Rettori et al., 1994) via NO, but is also under the control of inhibitory neurotransmitters such as gamma-amino butyric acid (GABA) (Masotto et al., 1989) and beta-endorphin (Lomniczi et al., 2000) both of which inhibit LHRH release. The inhibitory action of GABA on LHRH release could be prevented by hemoglobin (a scavenger of NO), indicating that NO has a stimulatory action on GABA release. It is possible that the increase in GABA release during LHRH secretion induced by NO could be a mechanism to terminate the pulses of LHRH (Seilicovich et al., 1995).

Beta-endorphin also can inhibit LHRH release, probably by stimulating μ-opiate receptors on NOergic neurons because we have shown that beta-endorphin inhibits the activity of NOS in MBH, whereas naltrexone, a μ-opiate receptor antagonist, increased the activity of NOS in this tissue. Beta-endorphin also blocked the action of sodium nitroprussiate (NP) (a NO donor) on PGE2 release and consequently LHRH secretion (Faletti et al., 1999).

Since alcohol has been shown to increase the release of GABA and of beta-endorphin we studied the interactions between GABA, beta-endorphin and alcohol. We confirmed that alcohol increases the release of GABA and beta-endorphin. Furthermore, beta-endorphin also stimulated GABA release, but GABA had no stimulatory action on beta-endorphin release and NP significantly inhibited the release of beta-endorphin from MBH in our in vitro experimental model. Furthermore, alcohol diminished significantly the N-methyl-D-aspartate (NMDA) stimulated NOS and this inhibition could be reversed by addition of naltrexone and bicuculline (a GABA-A receptor antagonist). However, bicuculline (10^{-4} M) could reverse the alcohol induced block of NMDA-stimulated LHRH release only when the concentration of alcohol used was 50 mM. At a higher concentration of alcohol such as 100 mM, the inhibitory action of alcohol could be reversed by naltrexone (10^{-6} M) and not by bicuculline. These results suggest that the primary effect of alcohol is to stimulate beta-endorphin release which in turn stimulates GABA release. Both of these inhibitory neurotransmiters then act together to suppress LHRH release. Since beta-endorphin decreases NOS activity while GABA and alcohol are without effect on NOS activity, they are acting down stream from NOS that is, they block the NOergic activation of cyclooxigenase (COX), that is necessary for LHRH release. This conclusion was confirmed, since the addition of an effective concentration of 8-bromo cGMP (a stable analogue of cGMP) to correct for a possible blockade of guanylate cyclase did not reverse the action of GABA or alcohol (Lomniczi et al., 2000). Addition of arachidonate to provide more substrate for COX also failed to reverse this inhibition, suggesting that the primary action of both, GABA and alcohol was by inhibition of COX. This conclusion is supported also by the fact that addition of PGE_2 reversed the alcohol block on LHRH release. Recent preliminary results show that alcohol not only inhibits the activity of COX as measured by radio-conversion assay of ^{14}C-arachidonic acid to eicosanoids, such as PGE_2 (Canteros et al., 1995) but also decreases COX content in MBH (unpublished observation from a collaborative work with V. Svivaslava and W.L. Dees). In Fig. 1 we present the diagrammatic representation of the postulated mechanism of action of alcohol to suppress NMDA-stimulated LHRH release that we described above.

Effect of the active cannabinoid, delta-9-tetrahydrocannabinol (THC) on LHRH release

All previous studies including ours in several species (Ayalon et al., 1977; Almirea et al., 1983; Wenger et al., 1987) indicate that the inhibitory effect of THC on the reproductive axis is exerted mainly at the hypothalamic level with the inhibition mainly of LH secretion by the pituitary and consequent alteration of reproductive function. Studies performed by our group (Rettori et al., 1990) using a static incubation system to incubate medial basal hypothalamic (MBH) explants in the presence of different concentrations of THC (10^{-11} to 10^{-8} M) showed that THC was without effect on basal release of LHRH. Since it was reported previously (Negro-Vilar et al., 1979), that catecholamines stimulate LHRH release, we used this approach to study the effect of THC on stimulated LHRH release and found as expected, that norepinephrine (NE) (5×10^{-5}) as well as dopamine (5×10^{-5} M) stimulated significantly the release of LHRH and that it was inhibited by the addition of THC (10^{-8} M). Since it is also established that PGE_2 is stimulated by NE and is also part of the

secretory pathway of LHRH release, we measured the release of PGE_2 from MBH in the presence of THC and found that THC (10^{-7} M) lowered significantly the release of PGE_2 from MBH as measured by RIA. The inhibition of PGE_2 release could also be due to an inhibition of COX activity. Therefore, we performed radio-conversion studies using ^{14}C-arachidonic acid and measured the eicosanoids that are formed by the action of COX and found that addition of THC to MBH incubated with ^{14}C-arachidonic acid had a dramatic inhibitory effect on COX activity, since all the eicosanoids measured such as 6-keto $F_{1\alpha}$, $PGF_{2\alpha}$, PGE_2 and TxB_2 were highly significantly inhibited by THC (10^{-8} and 10^{-7} M) as compared to values in controls (Rettori et al., 1990).

Our findings are in agreement with the known fact that THC and other cannabinoids inhibit adenylate cyclase in a reversible, dose-dependent and stereoselective manner (Bidau-Russell et al., 1990). These facts taken together (inhibition of PGE_2 and PGE_2-stimulated cAMP release) explain the inhibition of LHRH release with the consequent lowering of plasma LH levels.

The effects of THC as well as of other cannabinoids were believed to be due to a non-specific interaction with the membrane lipids, since cannabinoids are highly lipophilic molecules. But since the discovery of cannabinoid receptor CB1 in the brain (Devane et al., 1988), its abundance and anatomical localization, together with the behavioral effects of THC provide the molecular basis for the action of cannabinoids. Till now, two cannabinoid receptors: CB1 localized mostly in the brain (Herkenham et al., 1990) and CB2 receptor localized mostly in peripheral tissues and immune cells (Galiègue et al., 1995) have been described. Since then, complimentary DNAs have been cloned, the expression of their genes, and their functional domains have been described. The structure of the CB1 receptor exhibits the basic structure of a G-protein-coupled receptor with a molecular weight of 64 kDA (Ameri, 1999). The distribution of CB1 has been well described in rat brain by Herkenham et al. (1991) who found that these binding sites are not homogenously distributed. Although they present high density binding by radioautographic method in some areas such as hippocampus, the hypothalamus present sparse binding that is slightly elevated in the ventromedial nu-

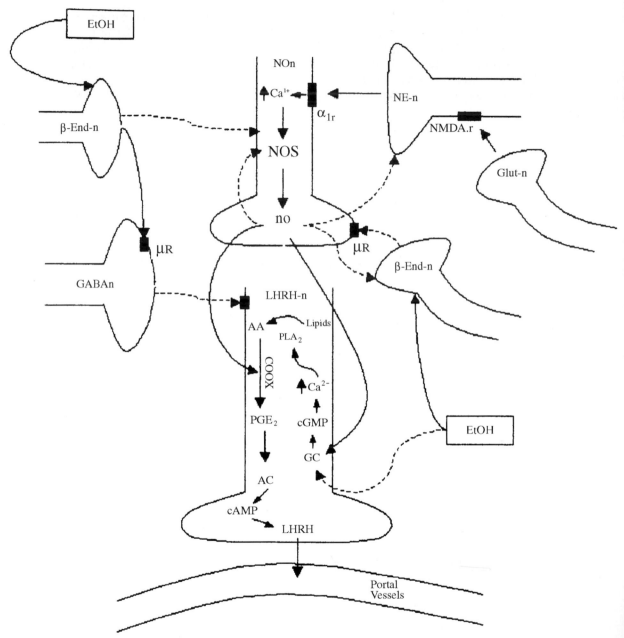

Fig. 1. Diagrammatic representation of the postulated mechanism of action of alcohol (EtOH) to suppress NMDA-stimulated LHRH release. For explanation, see text. β-End, beta-endorphin; μR, μ-opioid receptor; GABA-n, GABA neuron; NO-n, NO-ergic neuron; NE-n, noradrenergic neuron; α_{1r}, α_1 adrenergic receptor; NMDA-r, NMDA receptor; Glut-n, glutamergic neuron; LHRH-n, LHRH neuronal terminal; lipids, membrane phospholipids; PLA$_2$, phospholipase A$_2$; GC, guanylate cyclase; AC, adenylate cyclase. Solid arrow indicates stimulation, Dashed arrow indicates inhibition. (From Lomniczi et al., 2000, *Proc. Natl. Acad. Sci. USA*, 97: 2337–2342.)

cleus (Mailleux and Vanderhaeghen, 1992). In order to find out if the CB1 receptors in the hypothalamus are localized on LHRH neurons and terminals, we performed immunohistochemical studies using an antiserum raised in rabbits against CB1 receptors (anti-rat CB1, working dilution 1:200, kindly do-

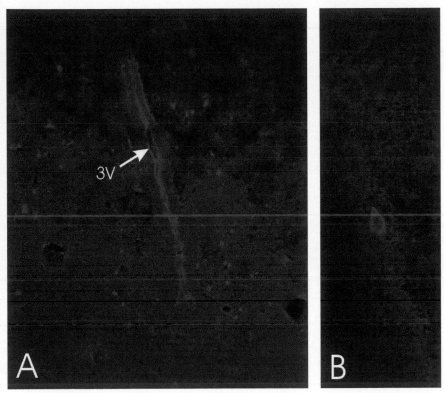

Fig. 2. Transversal sections (20 μm) of rat hypothalamus immunostained with anti-CB1 receptor (red). (A) 3v: third cerebral ventricle. (B) Detail of a neuron with CB1 receptors.

nated by Dr. K. Mackie, Dept. of Anesthesiology, Univ. of Washington, Seattle, WA, USA). We observed a scattered distribution in the hypothalamus (Fig. 2A and B). Although ir-CB1 receptors were distributed in the hypothalamic area where LHRH neurons are found, using double immunohistochemistry techniques, we did not observed colocalization of CB1 receptors with LHRH (data not shown).

These studies suggest that THC most probably is acting on LHRH release by affecting neurotransmitters that are involved in the pathway of LHRH release (Murphy et al., 1998). There is evidence that cannabinoids can inhibit pre-synaptic release of glutamate in rat hippocampus (Shen et al., 1996).

Effect of endogenous cannabinoid, anandamide, on LHRH release

The discovery of specific cannabinoid receptors mediating the effects of marihuana raised the possibility of the existence of endogenous ligands, similar to the endogenous ligands for opiate receptors, such as beta-endorphin. Endogenous substances that bind to cannabinoid receptors and mimic the action of THC isolated from nervous and peripheral tissues are amides and esters of eicosanoid-like fatty acids. The first such substance isolated from porcine brain by Devane et al., 1992 is N-arachidonylethanolamine, named 'anandamide' ('ananda' means 'inner bliss' in Sanskrit) and amide for chemical bonding. Anandamide possess all the properties of a cannabinoid agonist for CB1 and CB2 receptors (Felder et al., 1993) and as THC causes inhibition of adenylate cyclase. Another endocannabinoid is 2-arachidonylglycerol (Mechoulam et al., 1998). The pathway of anandamide formation is from the hydrolysis of N-arachidonoyl-phophatidylethanolamine catalyzed by a phospholipase D-like enzyme (Di Marzo et al., 1994). This pathway suggests that anandamide is formed 'on demand' by stimulated cells and accounts for very low levels found in the brain.

As with THC, anandamide was found to be able

to lower plasma LH levels (Wenger et al., 1995). Therefore, we studied the effect of different concentrations of anandamide on LHRH release from MBH in the in vitro system described here. As seen with THC, anandamide (10^{-9} to 10^{-6} M) did not modify significantly the basal release of LHRH from MBH incubated in vitro. When LHRH release was stimulated with NMDA (20 mM) the addition of anandamide together with NMDA inhibited the increase in LHRH release that was evoked by NMDA. This inhibition could be completely reversed by addition of GABA-A antagonist (bicuculline, 10^{-5} M) but not by the opioid antagonist naltrexone (10^{-6} M). Therefore we studied the effect of anandamide on GABA release from MBH in vitro. The addition of anandamide (10^{-9} M) stimulated highly significantly GABA release from MBH. Studies on the effect of anandamide on LH secretion from pituitaries incubated in vitro, showed that only the concentration of 10^{-9} M induced a significant decrease of LH secretion and did not modify the stimulatory response to exogenous LHRH. All these results indicate that anandamide lowers plasma LH mainly by inhibiting the pulsatility of LHRH (since basal release was not modified) and that the pathway is by increased GABA release, a well known inhibitory neurotransmitter that inhibits LHRH release.

In conclusion, all these studies indicate that alcohol and plant derived cannabinoid (THC) as well as endogenous compound (anandamide) have a deleterious effect on reproduction in adult male rats, mainly by lowering plasma LH levels exerting its action mainly at the hypothalamic level by inhibiting LHRH release. Also, they share a common pathway by stimulating inhibitory neurotransmitters such as GABA and consequently inhibiting cyclooxygenase with consequent decrease of PGE_2 release and inhibition of adenylate cyclase with inhibition of cAMP that is necessary for the extrusion of LHRH from its terminals into hypophyseal portal vessels in order to reach the pituitary gland and release LH from gonadotropes.

Acknowledgements

This work was supported by Ministery of Public Health 'Carrillo-Oñativia' grant 2001 and BID 802/OC-AR PICDCT No. 5-6117.

References

Almirea, R.G., Smith, C.G. and Asch, R.H. (1983) The effects of marihuana and delta-9-tetrahydroannabinol on luteal function in the rhesus monkey. Fert. Steril., 39: 212–217.

Ameri, A. (1999) The effects of cannabinoids on the brain. Prog. Neurobiol., 58: 315–348.

Ayalon, D., Nir, L., Cordova, T., Bauminger, S., Puder, M., Naor, Z., Kashi, B., Zor, U., Harrell, A. and Lindner, H.R. (1977) Acute effect of delta-9-tetrahydrocannabinol on hypothalamo–pituitary–ovarian axis in the rat. Neuroendocrinology, 23: 31–42.

Bidau-Russell, M., Devane, W.A. and Howlett, A.C. (1990) Cannabinoid receptors and modulation of cyclic AMP accumulation in the rat brain. J. Neurochem., 55: 21–26.

Canteros, G., Rettori, V., Franchi, A., Genaro, A., Cebral, E., Faletti, A., Gimeno, M. and McCann, S.M. (1995) Ethanol inhibits luteinizing hormone-releasing hormone (LHRH) secretion by blocking the response of LHRH neuronal terminals to nitric oxide. Proc. Natl. Acad. Sci. USA, 92: 3416–3420.

Canteros, G., Rettori, V., Genaro, A., Suburo, A., Gimeno, M. and McCann, S.M. (1996) Nitric Oxide synthase content of hypothalamic explants: Increased by norepinephrine and inactivated by NO and cGMP. Proc. Natl. Acad. Sci. USA, 93: 4246–4250.

Dees, W.L., Rettori, V., Kozzlowski, G.P. and McCann, S.M. (1985) Ethanol and the pulsatile release of luteinizing hormone, follicle stimulating hormone and prolactin in ovariectomized rats. Alcohol, 2: 641–646.

Devane, W.A., Dysarz, F.A., Johnson, M.R., Melvin, L.S. and Howlett, A.C. (1988) Determination and characterization of a cannabinoid receptor in rat brain. Mol. Pharmacol., 34: 605–613.

Devane, W.A., Hanus, L., Breuer, A., Pertwee, R.G., Stevenson, L.A., Griffin, G., Gibson, D., Mandelbaum, A., Etinger, A. and Mechoulam, R. (1992) Isolation and structure of a brain constituent that binds to the cannabinoid receptor. Science, 258: 1946–1949.

Di Marzo, V., Fontana, A., Cadas, H., Schinelli, S., Cimino, G., Schwartz, J.C. and Piomelli, D. (1994) Formation and inactivation of endogenous cannabinoid anandamide in central neurons. Nature, 372: 686–691.

Faletti, A., Mastronardi, C.A., Lomniczi, A., Seilicovich, A., Gimeno, M., McCann, S.M. and Rettori, V. (1999) Beta-endorphin blocks luteinizing hormone-releasing hormone release by inhibiting the nitric oxidergic pathway controlling its release. Proc. Natl. Acad. Sci. USA, 96: 1722–1726.

Felder, C.C., Briley, E.M., Axelrod, J., Simpson, J.T., Mackie, K. and Devane, W.A. (1993) Anandamide, an endogenous cannabinomimetic eicosanoid, binds to the cloned human cannabinoid receptor and stimulates receptor-mediated signal transduction. Proc. Natl. Acad. Sci. USA, 90: 7656–7660.

Galiègue, S., Mary, S., Marchand, J., Dussossoy, D., Carriére, D., Carayon, P., Bouaboula, M., Shire, D., Le Fur, G. and Casellas, P. (1995) Expression of central and peripheral cannabinoid receptors in human immune tissues and leukocyte subpopulations. Eur. J. Biochem., 232: 54–61.

Herkenham, M., Lynn, A.B., Little, M.D., Johnson, M.R., Melvin, L.S., de Costa, B.R. and Rice, K.C. (1990) Cannabinoid receptor localization in brain. *Proc. Natl. Acad. Sci. USA*, 87: 1932–1936.

Herkenham, M., Lynn, A.B., Johnson, M.R., Melvin, L.S., de Costa, B.R. and Rice, K.L. (1991) Characterization and localization of cannabinoid receptors in the rat brain: a quantitative in vitro autoradiographic study. *J. Neurosci.*, 11: 563–583.

Lomniczi, A., Mastronardi, C.A., Faletti, A.G., Seilicovich, A., De Laurentiis, A., McCann, S.M. and Rettori, V. (2000) Inhibitory pathways and the inhibition of luteinizing hormone-releasing hormone release by alcohol. *Proc. Natl. Acad. Sci. USA*, 97: 2337–2342.

Mailleux, P. and Vanderhaeghen, J.J. (1992) Distribution of the neuronal cannabinoid receptor in the adult rat brain: a comparative receptor binding radioautography and in situ hybridization histochemistry. *Neuroscience*, 48: 655–688.

Masotto, C., Wisnieski, G. and Negro-Vilar, A. (1989) Different gamma-aminobutyric acid receptor subtypes are involved in the regulation of opiate-dependent and independent luteinizing hormone-releasing hormone secretion. *Endocrinology*, 125: 548–553.

Mechoulam, R., Fide, E. and Di Marzo, V. (1998) Endocannabinoids. *Eur. J. Pharmacol.*, 359: 1–18.

Moncada, S., Palmer, R.M.J. and Higgs, E.A. (1991) Nitric oxide: physiology, pathophysiology and pharmacology. *Pharmacol. Rev.*, 43: 109–115.

Murphy, L.L., Muñoz, R.M., Adrian, B.A. and Villanúa, M.A. (1998) Function of cannabinoid receptors in the neuroendocrine regulation of hormone secretion. *Neurobiol. Dis.*, 5: 432–446.

Negro-Vilar, A., Ojeda, S.R. and McCann, S.M. (1979) Catecholaminergic modulation of luteinizing hormone-releasing hormone release by median eminence terminals in vitro. *Endocrinology*, 104: 1749–1757.

Ojeda, S.R., Negro-Vilar, A. and McCann, S.M. (1979) Release of prostaglandin E$_2$ by hypothalamic tissue: evidence for their involvement in catecholamine-induced luteinizing hormone-releasing hormone release. *Endocrinology*, 104: 617–624.

Rettori, V., Aguila, M.C., Gimeno, M.F., Franchi, A.M. and McCann, S.M. (1990) In vitro effect of delta-9-tetrahydrocannabinol to stimulate somatostatin release and block that of luteinizing hormone-releasing hormone by suppression of the release of prostaglandin E$_2$. *Proc. Natl. Acad. Sci. USA*, 87: 10063–10066.

Rettori, V., Gimeno, M., Lyson, K. and McCann, S.M. (1992) Nitric oxide mediates norepinephrine-induced prostaglandin E$_2$ release from the hypothalamus. *Proc. Natl. Acad. Sci. USA*, 89: 11543–11546.

Rettori, V., Belova, N., Dees, W.L., Lyson, K., Dees, W.L., Gimeno, M. and McCann, S.M. (1993) Role of nitric oxide in the control of luteinizing hormone-releasing hormone release in vivo and in vitro. *Proc. Natl. Acad. Sci. USA*, 90: 10130–10134.

Rettori, V., Kamat, A. and McCann, S.M. (1994) Nitric oxide mediates the stimulation of luteinizing-hormone releasing hormone release induced by glutamic acid in vitro. *Brain Res. Bull.*, 33: 501–503.

Seilicovich, A., Duvilanski, B.H., Pisera, D., Thea, S., Gimeno, M., Rettori, V. and McCann, S.M. (1995) Nitric oxide inhibits hypothalamic luteinizing hormone-releasing hormone release by releasing gamma-aminobutyric acid. *Proc. Natl. Acad. Sci. USA*, 92: 3421–3424.

Shen, M., Piser, T.M., Seybold, V.S. and Thayer, S.A. (1996) Cannabinoid receptor agonists inhibit glutamatergic synaptic transmission in rat hippocampal cultures. *J. Neurosci.*, 16: 4322–4334.

Wenger, T., Rettori, V., Snyder, G.D., Dalterio, S. and McCann, S.M. (1987) Effects of delta-9-tetrahydrocannabinol on the hypothalamic–pituitary control of luteinizing hormone and follicle-stimulating-hormone secretion in adult male rats. *Neuroendocrinology*, 46: 488–493.

Wenger, T., Toth, B.E. and Martin, B.R. (1995) Effects of anandamide (endogen cannabinoid) on anterior pituitary hormone secretion in adult ovariectomized rats. *Life Sci.*, 56: 2057–2063.

I.S. Parhar (Ed.)
Progress in Brain Research, Vol. 141
© 2002 Elsevier Science B.V. All rights reserved

CHAPTER 14

Glutamatergic regulation of gonadotropin-releasing hormone neurons

Lothar Jennes *, Wensheng Lin, Shruti Lakhlani

Department of Anatomy and Neurobiology, University of Kentucky, College of Medicine, 430 Health Science Research Building, Lexington, KY 40536, USA

Introduction

The control of the secretory activity of gonadotropin-releasing hormone (GnRH) producing neurons requires complex integration of sensory, seasonal and behavioral cues as well as feedback interactions with the gonadal steroids. Throughout the days of estrus, diestrus I and II of the estrous cycle of the female rat, estradiol exerts an inhibitory effect on the GnRH neurons which release the GnRH peptide at low frequency and low amplitude spikes that maintain basal levels of LH and FSH secretion from the anterior pituitary. Only during proestrus, rising circulating estradiol levels cause an increase in the frequency and amplitude of GnRH release which results in a massive secretion of the luteinizing hormone (LH) and follicle-stimulating hormone (FSH) from the anterior pituitary gonadotropes which will induce ovulation (Freeman, 1994; Kordon et al., 1994). It had been postulated for a long time that the effects of the gonadal steroids on GnRH release are indirect and that other neurons in the central nervous system convey the stimulatory or inhibitory signals of the steroids to the GnRH neurons. Recently, it was shown, however that certain GnRH neurons do express estrogen receptor-β (Hrabovszky et al., 2001)

which suggests that some effects of estradiol are exerted by directly altering gene transcription in GnRH neurons. These possible direct effects of estradiol are, however, not sufficient to drive the estrous cycle and afferent input from non-GnRH neurons is required for GnRH neurons to release adequate quantities of GnRH into the fenestrated capillaries in the median eminence to induce a preovulatory gonadotropin surge.

Over the past 50 years, numerous studies have implicated almost every neurotransmitter or neuropeptide to have some influence on GnRH release and it is beyond the scope of this article to review all these neuroactive substances. Instead, we will focus on the role of glutamate as a stimulatory neurotransmitter the release of which causes an increase in the secretory activity of GnRH neurons.

Glutamate

Glutamate is the most abundant excitatory neurotransmitter in the brain and its functions and mechanisms of action have been extensively studied in extrahypothalamic sites such as the hippocampus and the Purkinje cells in the cerebellum (for review see Curtis and Johnston, 1974; Monaghan et al., 1989). More recently, several studies have defined the roles of glutamate in the hypothalamus where glutamate is released from nerve terminals in a Ca^{2+} dependent fashion and causes a dramatic increase in Ca^{2+} influx in the postsynaptic neurons (van den Pol, 1991) as well as the generation of strong EP-

---

* Correspondence to: L. Jennes, Department of Anatomy and Neurobiology, University of Kentucky, College of Medicine, 430 Health Science Research Building, Lexington, KY 40536, USA. Tel.: +1-859-257-1093; Fax: +1-859-323-5946; E-mail: ljenn0@pop.uky.edu

SPs (van den Pol and Trombley, 1993; Belousov and van den Pol, 1997). The view that glutamate is indeed an endogenous neurotransmitter in the hypothalamus is further supported by immunohisto-chemical studies which clearly show the presence of this amino acid in presynaptic boutons (van den Pol et al., 1990; Decavel and van den Pol, 1992) as has been shown previously in other regions of the brain (Clements et al., 1990). After its release from a presynaptic bouton, glutamate can bind to 2 major classes of postsynaptic receptors, the ion channel forming ionotropic class and the G-protein-coupled metabotropic receptor class. Based upon preferential binding of agonists, the ionotropic class is further divided into NMDA-, AMPA-, and kainate-preferring receptor subunit families which need to assemble with other subunits of the same family to form functional receptor channels. Each family contains several members such as NMDA-R1, -R2A and 2B; Glu-R1, -R2, -R3 and -R4 for the AMPA family and kainate-1 and -2, Glu-R5, -R6 and -R7 for the kainate family (Seeburg, 1993; Hollmann and Heinemann, 1994). Similarly, the metabotropic class of glutamate receptors is divided into three families, depending upon signal transduction mechanisms and agonist potencies. Thus, group I receptors include mGlu-R1 and mGlu-R5 whose activation stimulates phospholipase C, group II receptors include mGlu-R2 and -R3 which inhibit adenylate cyclase and group III which includes mGlu-R4, -R6, -R7 and -R8 which also inhibit cAMP production, however, this group of receptors is activated by the agonist L-AP4 while Group II receptors are activated by trans-MCG-I (Nakanishi, 1994).

Role of glutamate in the regulation of GnRH secretion

Since the mid-seventies it has been known that administration of glutamate can enhance circulating LH levels and that this effect is exerted in the central nervous system since exposure of pituitary slices or dispersed pituitary cells to glutamate or injections of glutamate directly into the pituitary do not alter LH release (Ondo et al., 1976; Schainker and Cicero, 1980; Ondo, 1981; Tal et al., 1983). While these studies offered important first clues about an important participation of glutamate in the regulation

of the brain–pituitary–gonadal axis, no follow-up studies were conducted until renewed interest and new tools became available in the nineties. Since then, numerous studies have been conducted to define the sites and mechanisms of action of glutamate in the regulation of gonadotropin-releasing hormone secretion which have been summarized in several excellent recent reviews (Brann and Mahesh, 1994, 1997). In general, the studies have examined various aspects of an involvement of ionotropic glutamate receptor channels while data on a possible role of the metabotropic receptor class in the control of GnRH neurons are not available. As a brief summary, activation of NMDA, AMPA or kainate receptors by specific agonists or by glutamate enhances GnRH-stimulated LH release in both, male and female animals in a dose-dependent manner. These effects are dependent upon the presence of gonadal steroids since either no or an inhibitory effect of the glutamate agonist NMDA has been observed in ovariectomized animals. Conversely, administration of ionotropic glutamate receptor channel antagonists prevents the preovulatory and the steroid-induced LH surge as well as pulsatile basal LH release which suggests that endogenous glutamate is required for adequate maintenance of basal LH release and preovulatory LH surge generation. This view is supported by push–pull perfusion studies which measured greatly enhanced extracellular glutamate levels in the preoptic area of female ovariectomized and steroid treated rats that paralleled the steroid-induced LH surge which suggests that endogenous glutamate had been released from nerve terminals and activated the GnRH neurons (Jarry et al., 1992).

Localization of glutamate in the hypothalamus

In an effort to determine where in the hypothalamus the glutamate neurons are located that are relevant for the regulation of GnRH release we employed immunohistochemical staining procedures using various mono- and polyclonal antibodies to glutamate (Clements et al., 1990; Petrusz et al., 1990). The results of these studies show that glutamatergic neurons are abundant throughout the hypothalamus, especially in the medial preoptic nucleus and area, median preoptic and preoptic periventricular nucleus and anterodorsal preoptic nucleus, paraventricular and supraoptic nuclei as well as in the arcuate, dorsomedial and ventromedial nuclei. These results are in agreement with the findings of Ottersen and Storm-Mathisen (1984) who also examined the rat brain and with the data provided by Goldsmith et al. (1994) and Thind and Goldsmith (1995) who examined the brain of cynomolgus monkeys. Dual immunohistochemistry for glutamate and GnRH revealed that many GnRH perikarya and axon terminals in the median eminence are closely apposed by glutamatergic fibers which indicates that glutamatergic input to the GnRH neurons is likely to occur (Eyigor and Jennes, 1996). Interestingly, many of the immunoreactive glutamatergic neurons in the rat (Moore et al., 1999) and in the monkey (Thind and Goldsmith, 1997) also express estrogen receptor-α and, to a lesser extent estrogen receptor-β which suggests that estradiol could directly activate these neurons which in turn could convey the estradiol signal to the GnRH perikarya in the medial septum–diagonal band–preoptic area or to the axon terminals in the median eminence.

Neuron-specific glutamate transporters

Recently, two neuron-specific glutamate transporters have been identified and the cDNA and amino acid sequences have been determined. These transporters have been named VGLUT 1 (vesicular glutamate transporter 1) or BNPI (brain specific Na$^+$-dependent inorganic phosphate transporter) (Ni et al., 1994; Bellocchio et al., 2000; Takamori et al., 2000) and VGLUT 2 or DNPI (Na$^+$-dependent inorganic phosphate cotransporter) (Aihara et al., 2000; Hisano et al., 2000). These transporters are specific for mediating glutamate uptake by transfected cells while they do not interact with related amino acids such as glutamine or aspartate. Electron microscopic studies have recently confirmed that the glutamate transporters are localized to presynaptic terminals (Bellocchio et al., 1998; Fujiyama et al., 2001). We recently used specific antibodies to both transporters in order to identify more precisely the sites of interactions between the glutamatergic and GnRH-neuronal system. In general, immunoreactive VGLUT 1 and 2 are present in small varicosities or boutons throughout the septum-diagonal band-hypothalamus in extremely dense distribution patterns. It appears that most neurons in the hypothalamus are juxtaposed by several glutamate transporter containing boutons while in the septum–diagonal band the number of immunoreactive boutons is slightly lower. In general, VGLUT 2 immunoreactive boutons appear slightly larger and more numerous throughout the hypothalamus when compared to VGLUT 1 boutons. Both VGLUT 1 and VGLUT 2 immunoreactive boutons are present in all areas of the ventral diencephalon, however several region-specific differences in the number of boutons containing the transporters are noted. Thus, the medial septum–diagonal band appears to be less densely innervated by VGLUT 1 containing boutons than by VGLUT 2 containing axons as is the mediobasal hypothalamus where the arcuate nucleus contains only relatively sparse VGLUT 1 immunoreactivities and the median eminence which is almost devoid of VGLUT 1 positive structures. In contrast, both the arcuate nucleus and the median eminence are very densely innervated by VGLUT 2 containing boutons which are in many instances juxtaposed to GnRH axon terminals (Fig. 1). Important for the control of GnRH neurons are also the findings that both, immunoreactive VGLUT 1 and VGLUT 2 boutons are found at GnRH perikarya (Fig. 2) which indicates that excitatory glutamatergic input can occur at the GnRH cell bodies as well as at the sites of release of the peptide into the fenestrated capillaries. Based upon the extremely high density of immunoreactive boutons it would be expected that a large number of immunoreactive perikarya should be identified, however, this does not appear to be the case. Only a few immunoreactive cell are visible in the diagonal band, preoptic region and mediobasal hypothalamus which indicates that the transporter protein is either synthesized at a very slow rate or that the proteins are rapidly transported out of the perikaryal region. Future studies are needed to determine which possibility applies.

In order to identify the location of the perikarya that synthesize VGLUT 1 or VGLUT 2 we cloned the cDNA of both transporters and generated cRNA probes for in situ hybridization. The results of these studies confirmed that both transporter mRNAs were present in the central nervous system and that the distribution patterns were more or less complimentary in that VGLUT 1 mRNA was the predominant transporter mRNA in the cortex and hippocampus

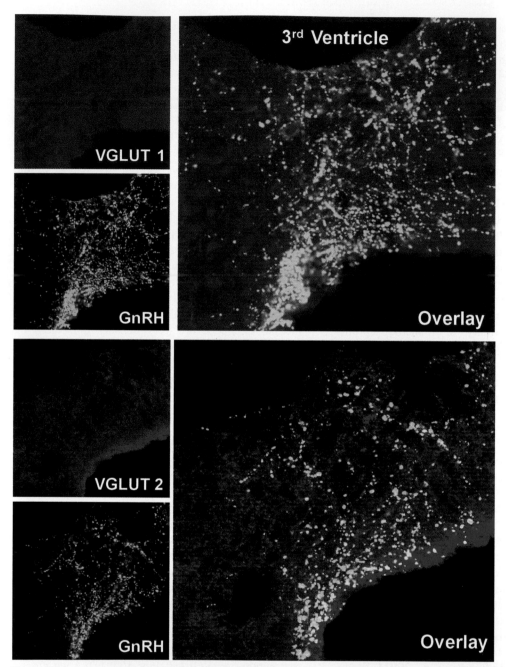

Fig. 1. Immunofluorescence for VGLUT 1 (red) and GnRH (green; top figure) and VGLUT 2 (red) and GnRH (green; bottom figure) and overlays in the rat median eminence showing absence of VGLUT 1 staining while VGLUT 2 is present in numerous boutons many of which are juxtaposed to GnRH axons.

while VGLUT 2 mRNA was much more abundant in the thalamus and hypothalamus. Within the hypothalamus, relatively small amounts of VGLUT 1 mRNA were seen throughout the preoptic region and most of the rostral and central nuclei while VGLUT 2 mRNA was found in slightly larger amounts in

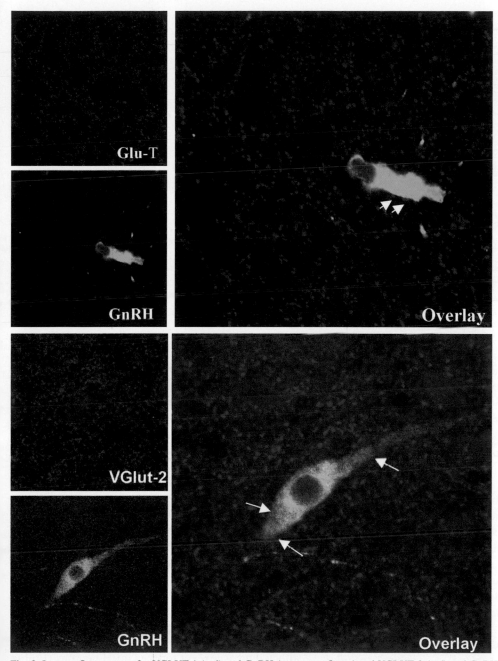

Fig. 2. Immunofluorescence for VGLUT 1 (red) and GnRH (green; top figure) and VGLUT 2 (red) and GnRH (green; bottom figure) and overlays in the ventral septum showing that many VGLUT 1 and VGLUT 2 immunoreactive boutons are juxtaposed to GnRH perikarya (arrows).

the preoptic region as well as the rostral and central nuclei, especially the ventromedial nucleus (Fig. 3). Interestingly, VGLUT 2 mRNA was abundant in the mammillary complex which did not contain measurable amounts of VGLUT 1 mRNA. The results show that glutamatergic neurons are located in the regions

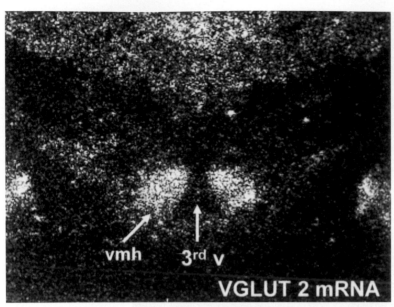

Fig. 3. Autoradiogram after in situ hybridization with ^{35}S-labeled cRNA probes encoding VGLUT 2 in the mediobasal hypothalamus showing abundant VGLUT 2 mRNA in the ventromedial nucleus (vmh).

of the hypothalamus that are relevant for the regulation of GnRH neurons which further strengthens the view that glutamate is intimately involved in the neuroendocrine control of reproduction.

Expression of glutamate receptor subunits in GnRH neurons

Before glutamate can be accepted to be a neurotransmitter that regulates the activity of GnRH neurons, it needs to be established that the GnRH neurons express the relevant postsynaptic glutamate receptors. After the cDNA sequences of the various ionotropic glutamate receptor subunits became available we used in situ hybridization procedures to identify the distribution of each subunit mRNA in the hypothalamus (Eyigor et al., 2001). The results show that all NMDA, AMPA and kainate receptor subunit mRNAs were widely distributed in a subunit-specific distribution pattern and that the NMDA-R1, -R2A and -R2B subunits were the most abundant in the hypothalamus. Interestingly, the extensive overlap in the distribution patterns of the various subunit mRNAs suggests that many neurons can express glutamate receptor channels that belong to different receptor families. Thus, it is likely that these multi-

receptor neurons can integrate the glutamate signal in many different but cell-specific manners.

In order to determine which subunits were expressed in GnRH neurons we applied dual in situ hybridization procedures. The results of these studies showed that the mandatory NMDA-R1 subunit mRNA was apparently absent in most GnRH neurons (Abbud and Smith, 1995; Eyigor and Jennes, 1996) while the NMDA-R2A subunit was detected in 17% of the GnRH neurons. This approach was unable to identify the mRNAs for NMDA-R2B and -R2C in GnRH cells (Eyigor and Jennes, 1996). These results, although unexpected because of the known stimulatory effects of NMDA on GnRH release were supported by the finding that administration of cytotoxic concentrations of the glutamate agonists NMDA or kainic acid into the septum–preoptic region destroys most of the neurons in the affected areas, except for the GnRH neurons (Ebling et al., 1998). These data were interpreted to indicate an absence of NMDA receptors on GnRH neurons. However, more recent studies which employed a more sensitive in situ hybridization or immunohistochemical procedures were able to detect the NMDA-R1 subunit mRNA and protein in the majority of GnRH neurons (Ottem and Petersen, 2000; see also

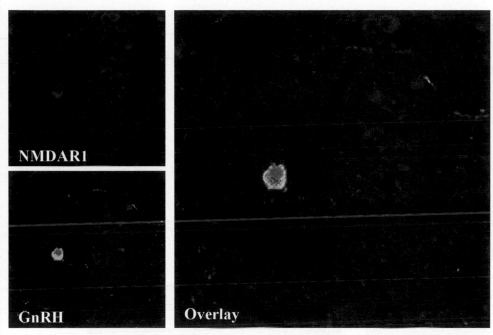

Fig. 4. Immunofluorescence for NMDAR1 receptor subunit (red), GnRH (green) and overlay showing co-localization of this mandatory subunit in a GnRH perikaryon.

Fig. 4) and future studies are needed to show that this subunit assembles with other subunits to form functional channels.

Dual in situ hybridization studies also revealed that less than 5% of the GnRH neurons express the AMPA subunits GluR1–4, however, kainate-2 subunit mRNA was detected in about 50% of the GnRH neurons (Eyigor and Jennes, 1996). Since kainate-2 subunits need to assemble with other kainate-preferring subunits to form functional receptor channels, we examined GnRH neurons for the presence of GluR5, -6 and -7 subunit mRNA and protein. The results show that kainate-2 subunit mRNA containing GnRH neurons also express GluR5 subunit mRNA while GluR6 and -7 mRNAs were not detected. These results have been confirmed with multiple immunohistochemical stainings indicating that the kainate receptor subunit mRNAs are translated into protein and it can be expected that in GnRH neurons, kainate 2 subunits assemble with GluR5 subunits to form functioning receptor channels (Eyigor and Jennes, 2000; see also Fig. 5).

These receptors may be important for the control of the preovulatory or steroid-induced LH surge since most kainate receptor expressing GnRH neurons belong to the subpopulation of GnRH neurons that synthesizes the transcription factor *fos* during the surge (Eyigor and Jennes, 2000). It is thought that this subpopulation of GnRH neurons is functionally different from the *fos*-negative GnRH neurons and that the presence of *fos* in these cells is an indicator of cellular activation (for review see Hoffman et al., 1993). Thus, GnRH neurons that express *fos* have a higher GnRH mRNA content than *fos*-negative GnRH neurons (Wang et al., 1995) and the number of *fos*-positive GnRH neurons parallels magnitude of the LH surge (Attardi et al., 1997). The induction of *fos* in GnRH neurons during the LH surge suggests that this transcription factor may control the replenishing of GnRH peptide in depleted neurons since an activation protein (AP)-1 site is present in the GnRH promoter (Bond et al., 1989) to which a *fos* heterodimer could bind.

Recently, a transgenic mouse that expresses green fluorescent protein under the control of the GnRH promotor has been generated and used to determine the electrophysiological properties of identified GnRH neurons in slice preparations (Spergel et al.,

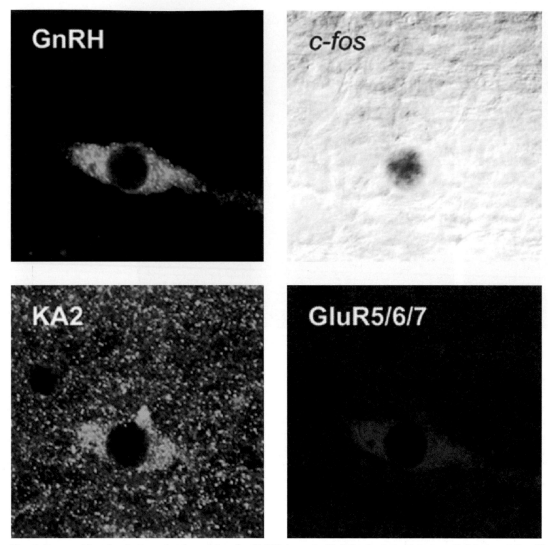

Fig. 5. Immunohistochemical quadruple staining for GnRH (green), GluR5, -6, -7 (red), kainate-2 (blue) and *fos* (black) showing an 'activated' GnRH neuron in the ventral septum that co-expresses two kainate receptor subunits.

1999). These studies clearly showed that functional glutamate receptor channels are present in the plasma membrane of the GnRH neurons and, based upon the effects of antagonists it appears that in the mouse most of the responses to glutamate were apparently mediated by AMPA receptors and less by NMDA receptors. Since these electrophysiological data obtained in the mouse are very different from the results obtained in the rat with in situ hybridization, it remains to be determined if there is a significant species difference in the expression patterns of gluta-

mate receptor subunits in GnRH neurons or if other explanations, such as differences in the sensitivity of the detection methods apply.

Conclusion

The data presented in this article support the view that glutamate is an important neurotransmitter in the control of GnRH neuronal activity. Based upon the morphological data, it is suggested that certain glutamatergic neurons express estrogen receptor-α and these neurons are present in the critical regions of the hypothalamus that are known to be involved in the neuroendocrine control of reproduction, such as the preoptic region, ventromedial nucleus and arcuate nucleus. Furthermore, glutamatergic axons appear to contact GnRH perikarya in the medial septum–diagonal band–rostral preoptic region as well as GnRH containing axons in the median eminence. Since GnRH neurons express two kainate preferring receptor subunits, the GluR5 and kainate 2 subunits, it is likely that some of the effects of glutamate on GnRH neurons are mediated by kainate receptor channels. Recent evidence also suggests that NMDA receptors may be involved since many GnRH neurons express the NMDAR1 subunit while at present the identity of the companion subunits is not clear.

Acknowledgements

This work was supported by NIH grants AG17164 and MH59890 (LJ).

References

Aihara, Y., Mashima, H., Onda, H., Hisano, S., Kasuya, H., Hori, T., Yamada, S., Tomura, H., Yamada, Y., Inoue, I., Kojima, I. and Takeda, J. (2000) Molecular cloning of a novel brain-type Na(+) dependent inorganic phosphate cotransporter. *J. Neurochem.*, 74: 2622–2625.

Abbud, R. and Smith, M.S. (1995) Do GnRH neurons express the gene for the NMDA receptor? *Brain Res.*, 690: 117–120.

Attardi, B., Klatt, B., Hoffman, G.E. and Smith, M.S. (1997) Facilitation or inhibition of the estradiol-induced gonadotropin surge in the immature rat by progesterone: regulation of GnRH and LH messenger RNAs and activation of GnRH neurons. *J. Neuroendocrinol.*, 9: 589–599.

Bellocchio, E.E., Hu, H., Pohorille, A., Chan, J., Pickel, V.M. and Edwards, R. (1998) The localization of the brain-specific inorganic phosphate transporter suggests a specific presynaptic role in glutamatergic transmission. *J. Neurosci.*, 18: 8648–8659.

Bellocchio, E.E., Reimer, R.J., Fremeau, R.T. Jr. and Edwards, R.H. (2000) Uptake of glutamate into synaptic vesicles by an inorganic phosphate transporter. *Science*, 289: 957–960.

Belousov, A.B. and van den Pol, A.N. (1997) Local synaptic release of glutamate from neurons in the rat hypothalamic arcuate nucleus. *J. Physiol. (London)*, 499: 747–761.

Bond, C.T., Hayflick, J.S., Seeburg, P.H. and Adelman, J.P. (1989) The rat gonadotropin-releasing hormone: SH locus: structure and hypothalamic expression. *Mol. Endocrinol.*, 3: 1257–1262.

Brann, D.W. and Mahesh, V.D. (1994) Excitatory amino acids: function and significance in reproduction and neuroendocrine regulation. *Front. Neuroendocrinol.*, 15: 3–49.

Brann, D.W. and Mahesh, V.B. (1997) Excitatory amino acids: Evidence for a role in the control of reproduction and anterior pituitary hormone secretion. *Endocr. Rev.*, 18: 678–700.

Clements, J.R., Magnusson, K.R. and Beitz, A.J. (1990) Ultrastructural description of glutamate-, aspartate-, taurine-, and glycine-like immunoreactive terminals from five rat brain regions. *J. Electron Microsc. Tech.*, 15: 49–66.

Curtis, D.R. and Johnston, G.A. (1974) Amino acid transmitters in the mammalian central nervous system. *Ergeb. Physiol.*, 69: 97–188.

Decavel, C. and van den Pol, A.N. (1992) Converging GABA- and glutamate-immunoreactive axons make synaptic contact with identified hypothalamic neurosecretory neurons. *J. Comp. Neurol.*, 316: 104–116.

Ebling, F.J., Cronin, A.S. and Hastings, M.H. (1998) Resistance of gonadotropin-releasing hormone neurons to glutamatergic neurotoxicity. *Brain Res. Bull.*, 47: 575–584.

Eyigor, O. and Jennes, L. (1996) Identification of glutamate receptor subtype mRNAs in gonadotropin-releasing hormone neurons in rat brain. *Endocrine*, 4: 133–139.

Eyigor, O. and Jennes, L. (2000) Kainate receptor subunit-positive gonadotropin-releasing hormone neurons express c-Fos during the steroid-induced luteinizing hormone surge in the female rat. *Endocrinology*, 141: 779–786.

Eyigor, O., Centers, A. and Jennes, L. (2001) Distribution of ionotropic glutamate receptor subunit mRNA in the rat hypothalamus. *J. Comp. Neurol.*, 434: 101–124.

Freeman, M.E. (1994) The neuroendocrine control of the ovarian cycle of the rat. In: E. Knobil and J.D. Neill (Eds.), *The Physiology of Reproduction*. Raven Press, New York, NY, pp. 613–658.

Fujiyama, F., Furuta, T. and Kaneko, T. (2001) Immunocytochemical localization of candidates for vesicular glutamate transporters in the rat cerebral cortex. *J. Comp. Neurol.*, 435: 379–387.

Goldsmith, P.C., Thind, K.K., Perera, A.D. and Plant, T. (1994) Glutamate-immunoreactive neurons and their gonadotropin releasing hormone-neuronal interactions in the monkey hypothalamus. *Endocrinology*, 134: 858–868.

Hisano, S., Hoshi, K., Ikeda, Y., Maruyama, D., Kanemoto, M., Ichijo, H., Kojima, I., Takeda, J. and Nogami, H. (2000) Regional expression of a gene encoding a neuron-specific Na(+)-dependent inorganic phosphate cotransporter (DNPI) in the rat forebrain. *Brain Res. Mol. Brain Res.*, 83: 34–43.

Hoffman, G.E., Smith, M.S. and Verbalis, J.G. (1993) c-Fos and related immediate early gene products as markers of activity in neuroendocrine systems. *Front. Neuroendocrinol.*, 14: 173–213.

Hollmann, M. and Heinemann, S. (1994) Cloned glutamate receptors. *Annu. Rev. Neurosci.*, 17: 31–108.

Hrabovszky, E., Steinhauser, A., Barabas, K., Shughrue, P.J., Petersen, S.L., Merchenthaler, I. and Liposits, Z. (2001) Estrogen receptor-beta immunoreactivity in luteinizing hormone-releasing hormone neurons of the rat brain. *Endocrinology*, 142: 3261–3264.

Jarry, H., Hirsch, B., Leonhardt, S. and Wuttke, W. (1992) Amino acid neurotransmitter release in the preoptic area of rats during the positive feedback actions of estradiol on LH release. *Neuroendocrinology*, 56: 133–140.

Kordon, C., Drouva, S.V., Martinez de la Escalera, G. and Weiner, R.I. (1994) Role of classic and peptide neurotransmitters in the neuroendocrine regulation of luteinizing hormone and prolactin. In: E. Knobil and J.D. Neill (Eds.), *The Physiology of Reproduction*. Raven Press, New York, NY, pp. 1621–1681.

Monaghan, D.T., Bridges, R.J. and Cotman, C.W. (1989) The excitatory amino acid receptors: their classes, pharmacology, and distinct properties in the function of the central nervous system. *Annu. Rev. Pharmacol. Toxicol.*, 29: 365–402.

Moore, C.T., Lee, E., Tuggle, B., Eyigor, O. and Jennes, L. (1999) Glutamategic and adrenergic innervation of gonadotropin releasing hormone neurons in rat brain. *Adv. Reprod.*, 3: 293–302.

Nakanishi, S. (1994) Metabotropic glutamate receptors: synaptic transmission, modulation, and plasticity. *Neuron*, 13: 1031–1037.

Ni, B., Rosteck, P.R. Jr., Nadi, N.S. and Paul, M.S. (1994) Cloning and expression of a cDNA encoding a brain-specific Na^+-dependent inorganic phosphate cotransporter. *Neurobiology*, 91: 5607–5611.

Ondo, J.G. (1981) Effects of the neuroexcitatory amino acids aspartate and glutamate on LH secretion. *Brain Res. Bull.*, 7: 333–335.

Ondo, J.G., Pass, K.A. and Baldwin, R. (1976) The effects of neurally active amino acids on pituitary secretion. *Neuroendocrinology*, 21: 79–87.

Ottem, E.N. and Petersen, S.L. (2000) The majority of LHRH neurons in the MEPO/OVLT express NMDAR1 mRNA. *Soc. Neuroscience 30th Annual Meeting*, 504.4 (abstract).

Ottersen, O.P. and Storm-Mathisen, J. (1984) Neurons containing or accumulating transmitter amino acids. In: A. Björklund, T. Hokfelt and M.J. Kuhar (Eds.), *Handbook of Chemical Neuroanatomy*. Elsevier, Amsterdam, pp. 141–246.

Petrusz, P., Van Eyck, S.L., Weinberg, R.J. and Rustioni, A. (1990) Antibodies to glutamate and aspartate recognize non-endogenous ligands for excitatory amino acid receptors. *Brain Res.*, 529: 339–344.

Schainker, B.A. and Cicero, T.J. (1980) Acute central stimulation of luteinizing hormone by parenterally administered N-methyl-D,L-aspartic acid in the male rat. *Brain Res.*, 184: 425–437.

Seeburg, P.H. (1993) The TINS/TIPS Lecture. The molecular biology of mammalian glutamate receptor channels. *Trends Neurosci.*, 16: 359–365.

Spergel, D.J., Kruth, U., Hanley, D.F., Sprengel, R. and Seeburg, P.H. (1999) GABA- and Glutamate-activated channels in green fluorescent protein-tagged gonadotropin-releasing hormone neurons in transgenic mice. *J. Neurosci.*, 19: 2037–2050.

Takamori, S., Rhee, J.S., Rosenmund, C. and Jahn, R. (2000) Identification of a vesicular glutamate transporter that defines a glutamatergic phenotype in neurons. *Nature*, 407: 189–194.

Tal, J., Price, M.T. and Olney, J.W. (1983) Neuroactive amino acids influence gonadotrophin output by a suprapituitary mechanism in either rodent or primate. *Brain Res.*, 273: 179–182.

Thind, K.K. and Goldsmith, P.C. (1995) Glutamate and gabaergic neurointeractions in the monkey hypothalamus: a quantitative immunomorphological study. *Neuroendocrinology*, 61: 471–485.

Thind, K.K. and Goldsmith, P.C. (1997) Expression of estrogen and progesterone receptors in glutamate and GABA neurons of the pubertal female monkey hypothalamus. *Neuroendocrinology*, 65: 314–324.

van den Pol, A.N. (1991) Glutamate and aspartate immunoreactivity in hypothalamic presynaptic axons. *J. Neurosci.*, 11: 2087–2101.

van den Pol, A.N. and Trombley, P.Q. (1993) Glutamate neurons in hypothalamus regulate excitatory transmission. *J. Neurosci.*, 13: 2829–2836.

van den Pol, A.N., Wuarin, J.-P. and Dudek, F.E. (1990) Glutamate, the dominant excitatory transmitter in neuroendocrine regulation. *Science*, 250: 1276–1278.

Wang, H., Hoffman, G.E. and Smith, M.S. (1995) Increased GnRH mRNA in the GnRH neurons expressing cFos during the proestrus LH surge. *Endocrinology*, 136: 3673–3676.

I.S. Parhar (Ed.)
Progress in Brain Research, Vol. 141

CHAPTER 15

Gonadotropin-releasing hormone (GnRH) neurons: gene expression and neuroanatomical studies

Andrea C. Gore [*]

Kastor Neurobiology of Aging Laboratories, Fishberg Research Center for Neurobiology, and Brookdale Department of Geriatrics and Development, Mount Sinai School of Medicine, New York, NY, USA

Introduction

Gonadotropin-releasing hormone (GnRH) neurons are the key cells regulating reproductive function in all vertebrate organisms. In mammals, birds, reptiles and amphibians, the GnRH-1 decapeptide is released in a pulsatile manner from neuroterminals in the median eminence into the portal capillary vasculature, where it binds to receptors located on pituitary gonadotropes to regulate the synthesis and secretion of the gonadotropins. In fish, GnRH-1 is released directly into the anterior pituitary gland where it stimulates gonadotropin release. The gonadotropins (luteinizing hormone, LH, and follicle-stimulating hormone, FSH) are secreted into the peripheral circulation where they act at the gonads to regulate steroidogenesis, spermatogenesis in males, and folliculogenesis and oogenesis in females.

Although each level of the hypothalamic–pituitary–gonadal axis is critical for normal reproductive function, the GnRH neurons are the *primary* regulator of this axis. In the absence of GnRH neurons, as in the genetic mutations of Kallmann's syndrome in humans, or the hypogonadal (hpg) mouse, reproductive development never occurs spontaneously

(Kallmann et al., 1944; Gibson et al., 1997). GnRH neurons in turn are regulated by inputs from other neurotransmitters, neuropeptides, neurotrophic factors, as well as by feedback from gonadal steroid hormones. Thus, GnRH neurons serve to integrate crucial information about the internal and external environment, thereby coordinating the timing of reproductive physiology with the proper behavioral, environmental and homeostatic cues.

The goal of this chapter is to review the work from my laboratory and others on GnRH-1 gene expression and neuroanatomy. In the first section, studies on GnRH gene expression have been undertaken in order to determine the relationship between GnRH biosynthesis and secretion. GnRH release varies depending upon a number of physiological stimuli, including the time of year in seasonal breeders, the time of day (i.e., circadian or diurnal fluctuations), the stage of the cycle in females (e.g., menstrual or estrous cycles), as well as the age of the animal (e.g., prepubertal vs. post-pubertal, and adult vs. aging animals). It seems reasonable to predict that GnRH gene expression is affected in a similar manner as GnRH release. However, as I will discuss below, in many cases, alterations in GnRH gene expression do not necessarily parallel these alterations in GnRH release, or the former may be much less robust than the latter. Moreover, the mechanisms for the regulation of GnRH mRNA levels can vary depending upon the experimental protocol or the physiological condition of the organism.

[*] Correspondence to: A.C. Gore, Neurobiology of Aging Laboratories, Box 1639, Mount Sinai School of Medicine, 1425 Madison Avenue, New York, NY 10029, USA. Tel.: +1-212-659-5909; Fax: +1-212-849-2510; E-mail: andrea.gore@mssm.edu

The second part of this chapter will focus on the neuroanatomical regulation of GnRH-1 neurons. Many laboratories have reported in diverse species that the number of GnRH-1 neurons does not vary substantially during postnatal life, even during those periods when it might be predicted that such changes may occur, including puberty or reproductive senescence (Hoffman and Finch, 1986; Witkin, 1986; Wray and Hoffman, 1986; Rubin and Bridges, 1989; Gore et al., 1996). Moreover, numbers of GnRH cells are similar among males and females of a species (Wray and Hoffman, 1986). Thus, the regulation of GnRH neurons probably occurs primarily by changes in inputs to GnRH cells from other neurons in the brain. I have been using two model systems for studying this phenomenon: the neurotransmitter glutamate, acting through the N-methyl-D-aspartate (NMDA) receptor, and the neurotrophic factor insulin-like growth factor-I (IGF-I); this chapter will focus on these two substances, with an emphasis on NMDA receptor studies. Nevertheless, many other neurotransmitters and neurotrophic factors are extremely important in the regulation of GnRH cellular functions, and it is necessary to take these other factors into consideration in understanding the 'big picture' of the GnRH neurosecretory system.

I would like to add a brief note on the nomenclature used in this review paper. The GnRH neurons under investigation in studies in my laboratory are the hypophysiotropic, or GnRH-1 population of neurons (Fernald and White, 1999; Gore, 2002). These cells are found predominantly in the preoptic area (POA), organum vasculosum of the lamina terminalis (OVLT), septum and anterior hypothalamus. GnRH-1 neurons project a neuroterminal to the median eminence, where the GnRH decapeptide is released. Two other populations of GnRH neurons have been found in the brain: the GnRH-2 cells, located predominantly in the midbrain, hindbrain and/or non-reproductive hypothalamic regions, and the GnRH-3 cells, associated with olfactory regions and the terminal nerve (Fernald and White, 1999; Parhar et al., 2000). The functions and anatomy of the GnRH-2 and GnRH-3 populations are probably quite different from those of the GnRH-1 cells, and therefore this chapter will focus only on GnRH-1 neurons.

Regulation of GnRH gene expression

General observations

GnRH mRNA levels vary during development, across reproductive cycles of females, and are regulated by neurotransmitters, neurotrophic factors and steroid hormone feedback. Several general conclusions on the regulation of GnRH gene expression can be drawn from the literature. First, changes in GnRH mRNA levels in vivo commonly occur in response to experimental manipulations (e.g. following gonadectomy or steroid hormone replacement, in response to pharmacological agents) and during reproductive development. With the exception of developmental changes, most alterations in GnRH mRNA levels are generally not very large (i.e., on the order of 20–50%) and thus they are relatively small in comparison to concomitant changes in GnRH release. Second, much of the regulation of GnRH mRNA levels appears to occur at a post-transcriptional level (Gore and Roberts, 1997). Those studies that have measured GnRH gene transcription together with GnRH mRNA levels in vivo often report that GnRH mRNA levels change independently of GnRH gene transcription Third, GnRH mRNA levels can be regulated quite rapidly in the animal, as early as fifteen minutes after treatment with a pharmacological agent such as NMDA (Petersen et al., 1991; Liaw and Barraclough, 1993; Gore and Roberts, 1994). Such a rapid response is consistent with a post-transcriptional change in GnRH mRNA stability.

Developmental changes in GnRH mRNA levels

GnRH release changes profoundly during reproductive development. During the late gestational and early postnatal periods, GnRH release is elevated, and may be disorganized (Donovan et al., 1975; Doecke et al., 1978; Ramaley, 1979; Plant, 1988; Gore, 2000). Then, a period of low GnRH release ensues, often called the 'prepubertal hiatus'. Although the mechanisms for the hiatus are not completely understood, this relative quiescence of the GnRH neurosecretory system may be attributable to inhibitory inputs from other neurons that suppress GnRH release. Following this quiescent period, GnRH release begins to increase at the onset of puberty, and during the progression of puberty, the amplitude and frequency of GnRH pulses continue to increase until adult reproductive function is attained (Urbanski and Ojeda, 1985; Plant, 1988; Watanabe and Terasawa, 1989; Gore, 2000; Sisk et al., 2001). Because the physiological demand for GnRH varies depending upon the life stage of the organism, it is of interest to determine how GnRH biosynthesis, measured by levels of GnRH gene expression, changes during reproductive maturation. Studies in rodents have consistently shown that GnRH mRNA levels increase during the pubertal process, and this may play a role in regulating the increasing GnRH peptide levels during puberty.

The relationship between GnRH release and gene expression has been studied in my laboratory by measuring changes in GnRH gene expression in developing male and female C57bl/6 mice, beginning at embryonic day (E) 16 through postnatal day (P) 60 (Gore et al., 1999). Using the highly quantitative RNase protection assay, we reported that GnRH mRNA levels increase gradually and significantly, by approximately 15-fold during this developmental period, peaking on P60 in females, and P40 in males (Fig. 1; Gore et al., 1999). Most of the changes in GnRH mRNA levels are associated with the pubertal period of development (i.e., beginning in the third to fourth week of postnatal life), whereas few changes in GnRH mRNA levels occur earlier in development, i.e., during the first three postnatal weeks of life (Adams et al., 1999; Gore et al., 1999). Most other reports support the concept that GnRH mRNA levels increase during the pubertal period in rats, mice and hamsters (Jakubowski et al., 1991; Dutlow et al., 1992; Gore et al., 1996, 1999; Parfitt et al., 1999). Our laboratory also demonstrated an association between the increase in GnRH mRNA levels during pubertal development in female rats and the timing of vaginal opening, an index of reproductive maturation (Gore et al., 1996), and this suggests a physiological relevance to the pubertal increase in GnRH mRNA levels. In addition, to my knowledge, this developmental increase in GnRH mRNA levels is the most dramatic such change identified in vivo.

Mechanisms for the increase in GnRH mRNA during development

In order to determine the mechanism for the increase in GnRH mRNA levels during maturation, we measured GnRH primary transcript levels in the same animals in which we had measured GnRH mRNA (Fig. 1; Gore et al., 1999). GnRH primary transcript levels, which are measured by RNase protection assay using a probe containing the intron B-exon 2-intron C junction of the proGnRH gene, representing the unprocessed GnRH RNA precursor (Jakubowski and Roberts, 1994), are used as an index of GnRH gene transcription in vivo (Yeo et al., 1996). Overall, developmental changes in GnRH primary transcript are quite different from those of GnRH mRNA, and moreover, are sexually dimorphic during the neonatal period. In males, GnRH primary transcript levels are low until P5; they then undergo a 4-fold increase between P5 and P7. Levels continue to increase through P15, at which time they reach a plateau that is maintained through adulthood (Gore et al., 1999). In females, GnRH primary transcript levels are high at E16, decrease to a nadir at P5, and then increase 5-fold from P5 to P7, at which time adult levels are attained (Gore et al., 1999). These results are similar to those obtained using a transgenic mouse model which undergoes large increases in GnRH gene transcription from the first to second postnatal week of life (Wolfe et al., 1996). Thus, the increase in GnRH gene transcription occurs much earlier in development than the increase in GnRH mRNA levels. Taken together, these findings support the concept that the pubertal increase in GnRH mRNA levels is regulated to a large extent by a post-transcriptional mechanism, as GnRH gene transcription has already reached adult levels at that time.

Regulation of GnRH gene expression by the NMDA receptor

The excitatory amino acid glutamate, acting through both N-methyl-D-aspartate (NMDA) and non-NMDA receptors, stimulates GnRH and LH release (Brann, 1995; Gore, 2000, 2001). Glutamate receptors have also been shown to regulate GnRH gene expression. Effects of NMDA receptor activation on GnRH mRNA levels in rats have been reported by

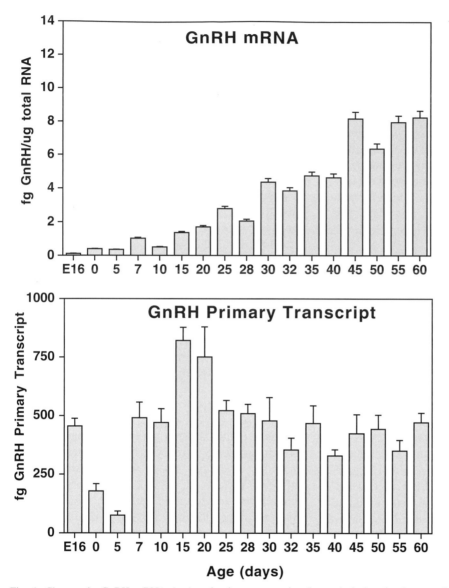

Fig. 1. Changes in GnRH mRNA (top) and primary transcript (bottom) during development in female mice. GnRH mRNA levels, normalized to total μg of RNA in the POA, increase gradually and significantly during postnatal development. GnRH primary transcript levels in the POA are high at E16, decrease to a nadir at P5, then increase again by P7, at which time levels are maintained through adulthood. Abbreviations: E: embryonic day; P: postnatal day; POA: preoptic area. Modified from Gore et al., 1999.

several laboratories, and the results have consistently demonstrated stimulatory effects of NMDA receptor activation on GnRH mRNA levels. In young adult male and female rats, GnRH mRNA levels are significantly elevated 15 minutes to one hour after treatment with NMDA agonists (Petersen et al., 1991; Liaw and Barraclough, 1993; Gore and Roberts, 1994; Gore et al., 2000a). Activation of non-NMDA (kainate) receptors also rapidly elevates GnRH mRNA levels in male rats (Gore and Roberts, 1994).

Because NMDA receptor activation causes such a rapid increase in GnRH mRNA levels, we speculated that this may be due to a post-transcriptional effect

(Gore and Roberts, 1994). The mechanism for the stimulation of GnRH mRNA levels was examined by measuring steady-state GnRH primary transcript levels as an index of GnRH gene transcription (Yeo et al., 1996). Following NMDA or kainate receptor activation, GnRH mRNA levels are elevated, while GnRH primary transcript levels remain unchanged. Thus, the increase in GnRH mRNA levels occurs by a post-transcriptional mechanism, probably mRNA stability, as GnRH gene transcription is not affected by activation of glutamate receptors.

NMDA receptor regulation of GnRH gene expression during development and puberty

Although an increase in pulsatile GnRH release is the rate-limiting step for the onset and progression of puberty, this is subject to regulation by afferent inputs. Glutamate is one of the primary excitatory neurotransmitters implicated in playing such a role in puberty as well as during adulthood, and the NMDA receptor has been most widely studied in mediating this effect. A role for excitatory glutamatergic input in the stimulation of puberty has been suggested by reports demonstrating that activation of the NMDA receptor advances the timing of puberty, and that blockade of NMDA receptors delays the timing of puberty (Gay and Plant, 1987; Urbanski and Ojeda, 1990; Smyth and Wilkinson, 1994; Gore et al., 1996). Treatment of prepubertal animals with NMDA receptor agonists stimulates GnRH and LH release (Gay and Plant, 1987; Urbanski and Ojeda, 1990; Smyth and Wilkinson, 1994; Gore et al., 1996), and the sensitivity of GnRH neurons to NMDA receptor activation peaks during the peripubertal period (Cicero et al., 1988; Smyth and Wilkinson, 1994).

Effects of NMDA receptor activation on GnRH gene expression have been studied from a developmental perspective. Neonatal male and female rats are not very sensitive to treatment with an NMDA receptor agonist or antagonist, which exert only small effects on GnRH gene expression (Adams et al., 1999). Nevertheless, those effects that are observed are consistent with a stimulatory effect of NMDA, and inhibitory effect of MK-801, on GnRH mRNA levels (Adams et al., 1999), similar to effects in mature animals.

We investigated the relevance of the increase in GnRH mRNA to the timing of puberty. To do this, we induced precocious puberty by activation of the NMDA receptor. As discussed above, we had found that GnRH mRNA levels increase during the pubertal period, and that this is associated with the timing of vaginal opening in female rats (Gore et al., 1996). In subsequent studies, we administered an NMDA receptor agonist, N-methyl-D,L-aspartate (NMA), to prepubertal female rats. This causes an advancement in the timing of the onset of puberty (vaginal opening and first diestrus) by approximately five days in our rats, similar to other studies (Gay and Plant, 1987; Urbanski and Ojeda, 1990; Smyth and Wilkinson, 1994; Gore et al., 1996). NMA treatment causes an increase in GnRH mRNA levels coincidentally with the day of vaginal opening (Gore et al., 1996). We also observed that GnRH primary transcript levels are unaltered by this NMA treatment regime, indicating that the increase in GnRH mRNA levels in rats undergoing precocious puberty is due to a post-transcriptional mechanism (Gore et al., 1996). These data support the hypothesis that an increase in GnRH mRNA levels is associated with, and may be causally related to, the process underlying pubertal maturation and the attainment of adult reproductive function. In addition, the change in GnRH mRNA levels is probably due to a post-transcriptional event such as an enhancement of GnRH mRNA stability.

Regulation of GnRH gene expression by ovarian steroid hormones

The regulation of GnRH gene expression by sex steroid hormones has been intensively studied (Table 1; reviewed in Pfaff, 1986; Sagrillo et al., 1996; Gore and Roberts, 1997). Nevertheless, even the most seemingly straightforward experimental models have yielded considerably different results, depending upon the age, hormonal treatment, and method of detection of the GnRH mRNA. Several generalities may be drawn from the literature on intact female rats. First, most reports agree that GnRH gene expression varies across the estrous cycle (Pfaff, 1986; Sagrillo et al., 1996; Gore and Roberts, 1997; Gore, 2002). Second, a peak in GnRH mRNA levels occurs shortly before or during the preovulatory GnRH/LH surge (Table 1; Zoeller and Young, 1988; Park et

TABLE 1

GnRH gene expression: regulation by estrogen in female rats [a]

A. Intact female rats (estrous cycle and GnRH/LH surge)
  – GnRH mRNA levels peak at 1200 h on proestrus (ISH) (Porkka-Heiskanen et al., 1994)
  – GnRH mRNA levels peak at 1500 h on proestrus, and change during the estrous cycle (RPA) (Gore and Roberts, 1995; Gore et al., 2000b)
  – GnRH mRNA levels peak at 1800 h on proestrus (ISH) (Park et al., 1990)
  – GnRH mRNA levels peak at 1900 h on proestrus, and vary across the estrous cycle (ISH) (Zoeller and Young, 1988)
  – Two peaks in GnRH mRNA, at 1100 h on proestrus and estrus (RT-PCR) (Suzuki et al., 1995)
  – No change in GnRH mRNA between 1200 and 1800 h on proestrus (ISH) (Marks et al., 1993)
  – No differences in GnRH mRNA across estrous cycle (ISH) (Malik et al., 1991)

B. Effects of estrogen in OVX rats
Stimulatory effects of estrogen:
  – E2 (1 d) increases in GnRH mRNA (ISH) (Rosie et al., 1990)
  – E2 (1 or 4 d) increases GnRH mRNA (RPA) (Roberts et al., 1989)
  – E2 (2 d) increases GnRH mRNA (ISH) (Park et al., 1990)
  – E2 (2 d) increases GnRH mRNA with peak at 1200 h (ISH) (Petersen et al., 1995, 1996)
  – E2 (1–2 d) increases GnRH mRNA in immature rats (RPA) (Attardi et al., 1997)
  – E2 (7 d) increases GnRH mRNA (ISH) (Pfaff, 1986; Rothfeld et al., 1989)
Inhibitory effects of estrogen:
  – E2 (2 d treatment) decreases GnRH mRNA (ISH) (Zoeller et al., 1988)
  – E2 (14 d) decreases GnRH mRNA (ISH) (Toranzo et al., 1989)
  – E2 (18 d) decreases GnRH mRNA in slice explant cultures (ISH) (Wray et al., 1989)
No effects of estrogen:
  – No change in GnRH mRNA following E2 (2 d) (RPA) (Gore and Roberts, 1995)
  – No change in GnRH mRNA following E2 (2d) (ISH) (Marks et al., 1994)

C. Effects of OVX on GnRH gene expression in female rats
  – OVX (2 d) decreases GnRH mRNA levels in immature rats (Slot-blot) (Kim et al., 1989)
  – OVX (2–24 d) decreases GnRH mRNA levels (RPA) (Roberts et al., 1989)
  – OVX (3 d) increases GnRH mRNA levels (ISH) (Li et al., 1995)
  – OVX (14 d) increases GnRH mRNA levels (ISH) (Toranzo et al., 1989)
  – OVX (14 d) has no effect on GnRH mRNA levels (ISH) (Kelly et al., 1989)

---

[a] The technique used to detect GnRH mRNA is indicated in parentheses.
Abbreviations: RPA: RNase protection assay; ISH: In situ hybridization; NB: Northern blot; E2: estrogen; OVX: ovariectomy; d: days.

al., 1990; Porkka-Heiskanen et al., 1994; Gore and Roberts, 1995; Suzuki et al., 1995). Third, these changes in GnRH mRNA levels on proestrus tend to be relatively small (e.g., approximately 30–50%); this is of much lower magnitude than changes in preovulatory GnRH release (Sarkar and Fink, 1980; Levine and Ramirez, 1982; Park and Ramirez, 1989).

I had reported that GnRH mRNA levels fluctuate during the estrous cycle, with a peak occurring on diestrus 2, the day before the preovulatory GnRH/LH surge, and a second peak occurring on proestrus at 1500 h, shortly before the GnRH/LH surge (Gore and Roberts, 1995; Gore et al., 2000b). The mechanism for these increases in GnRH mRNA levels was examined by measuring levels of GnRH

primary transcript in these same animals. GnRH primary transcript increases in parallel with GnRH mRNA at 1500 h on proestrus; this suggests that the increase in GnRH mRNA levels at this time is due at least in part to an activation of GnRH gene transcription (Gore and Roberts, 1995). I believe that this peak in GnRH gene expression (both gene transcription and mRNA levels) occurring prior to or during the preovulatory GnRH/LH surge may prepare the organism for the large preovulatory increase in GnRH release, or may help to replenish stores of GnRH peptide that may be depleted during the surge. Further support for this concept is provided by studies demonstrating that blockade of the GnRH/LH surge with pentobarbital prevents the increase in

GnRH mRNA on proestrus from occurring (Park et al., 1990; Gore and Roberts, 1995). Thus, there is a temporal relationship between GnRH biosynthesis and the preovulatory GnRH/LH surge.

GnRH gene regulation by ovariectomy (OVX) and estrogen replacement in female rats

The regulation of GnRH mRNA levels by OVX and steroid treatment has proven to be a very controversial field (Table 1). For example, stimulatory, inhibitory or no effects of OVX on GnRH gene expression have been reported. Thus, one group found no effect of OVX on GnRH mRNA levels (Kelly et al., 1989), two laboratories demonstrated OVX-induced increases (Toranzo et al., 1989; Li et al., 1995), and two groups reported OVX-induced decreases in GnRH mRNA (Kim et al., 1989; Roberts et al., 1989). Effects of estrogen on GnRH mRNA levels in OVX female rats have also been studied, and again, very different results have been obtained, although these may be more generalizable (Table 1). Some of the differences between experiments may be due to different experimental protocols for estrogen replacement (length of treatment, time after OVX at which treatment was initiated, dose, presence or absence of progesterone, age of the animal, time of day of the experiment). Nevertheless, several general trends emerge. First, short-term treatment with estrogen (e.g., 1–2 days) usually results in stimulatory effects on GnRH mRNA levels (Roberts et al., 1989; Park et al., 1990; Rosie et al., 1990; Petersen et al., 1995, 1996; Attardi et al., 1997). Second, studies using long-term treatment with estrogen (14–18 days) generally report negative effects on GnRH mRNA levels (Toranzo et al., 1989; Wray et al., 1989). Third, progesterone generally has not been shown to have consistent or strong effects on GnRH gene expression (Marks et al., 1994; Gore and Roberts, 1995; Petersen et al., 1995). Fourth, and perhaps most importantly, these effects of estrogen on GnRH gene expression are relatively small, generally in the range of 20–50%, as is also the case in intact female rats during the preovulatory GnRH/LH surge (see above). Therefore the GnRH/LH surge, either natural or steroid-induced does not require a massive induction of the GnRH gene either to accumulate or replenish stores of GnRH decapeptide (Gore, 2002).

Reproductive senescence

The role of hypothalamic GnRH neurons in the timing and progression of reproductive senescence is not well-understood. Most studies on menopause in women have focused on ovarian changes, particularly the dramatic follicular atresia that results in a precipitous decline in estrogen levels. However, studies on animals are making it increasingly clear that changes in other levels of the hypothalamic–pituitary–gonadal axis, including hypothalamic GnRH neurons, may also play a role in the process of reproductive senescence.

Female rodents, unlike primates, do not experience a phenomenon equivalent to menopause. Because rodents experience estrous cycles but not menstrual cycles, their reproductive senescence is better termed 'estropause' as opposed to the primate equivalent of menopause. As rodents age, their estrous cycles become more and more irregular, and eventually, the rodents enter a stage of acyclicity termed persistent estrus. However, this transition to acyclicity occurs independently of ovarian follicular atresia, as the ovaries of old, acyclic rodents remain capable of resuming ovulation in response to external stimuli, and contain functional follicles (Huang and Meites, 1975; Lu et al., 1977). Therefore, 'estropause' in rodents must be due to alterations in hypothalamic and/or pituitary drive to the ovaries.

Changes in GnRH gene expression during reproductive aging

Although the number of GnRH neurons does not change substantially during reproductive aging (Hoffman and Finch, 1986; Rubin and Bridges, 1989), GnRH neurons undergo several functional changes in aging rats, prior to reproductive failure. First, pulsatile GnRH and luteinizing hormone release change during aging (Rubin and Bridges, 1989; Hwang et al., 1990; Sortino et al., 1996; Zuo et al., 1996). Second, the preovulatory GnRH/LH surge is significantly attenuated in middle-aged compared to young rats (Steger et al., 1980; Wise, 1984; Nelson et al., 1992; Rubin and King, 1994). Third, the expression of the immediate early gene Fos, a marker of gene activation, is lower in GnRH neurons of middle-aged compared to young rats during the preovulatory GnRH/LH surge (Lee et al., 1990; Lloyd et al., 1994; Rubin et al., 1994). These findings are all consistent with an age-related alteration at the level of the GnRH neuron itself that may play a role in 'estropause' in female rodents.

GnRH gene expression changes during reproductive senescence in rats. Studies from our laboratory on intact female rats have demonstrated an age-related increase in GnRH mRNA levels, measured by RNase protection assay (Fig. 2; Gore et al., 2000b). This finding is consistent with that from Naomi Rance's laboratory demonstrating that GnRH mRNA levels increase in postmenopausal compared to pre-menopausal women (Rance and Uswandi, 1996). Other studies, however, have shown age-related decreases in GnRH mRNA levels in intact (but not castrate) male F344 rats (Gruenewald and Matsumoto, 1991), and in 18 month compared to 2 month old female Sprague-Dawley rats (Li et al., 1997). Moreover, GnRH gene expression has been reported to decrease during aging in ovariectomized female rats, and we have made this observation in my laboratory as well (Rubin et al., 1997; Miller and Gore, 2001). Thus, the ovarian status of the animal is extremely important for the regulation of GnRH gene expression during aging. This may be much more complicated than just effects of ovarian steroid hormones, as the ovaries contain other non-steroid peptides and proteins that could influence the neuroendocrine system during aging.

The preovulatory GnRH/LH surge is attenuated in intact middle-aged compared to young rats on proestrus, and the steroid-induced GnRH/LH surge is also diminished with aging in ovariectomized female rats (Steger et al., 1980; Wise, 1984; Nelson et al., 1992; Rubin and King, 1994). My laboratory addressed the question of the preovulatory GnRH/LH surge in young and middle-aged female Sprague-Dawley rats. We had previously reported that GnRH mRNA levels increase significantly on the afternoon of proestrus in young female rats, at approximately 1500 h, shortly before the GnRH/LH surge which occurs at approximately 1800 h (Gore and Roberts, 1995). We examined this phenomenon in middle-aged rats, and found that the increase in GnRH mRNA levels in young rats is not seen in the middle-aged animals (Gore et al., 2000b). This result is consistent with the other changes in GnRH neurons

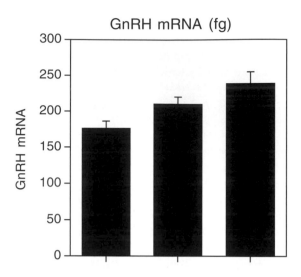

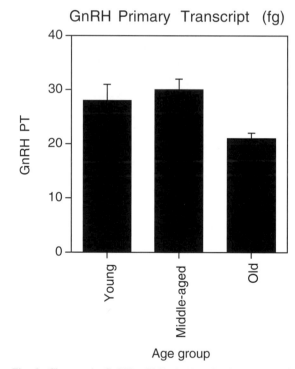

Fig. 2. Changes in GnRH mRNA (top) and primary transcript (bottom) during reproductive senescence in intact female rats. GnRH mRNA levels in the POA (expressed in fg) increase significantly during aging. GnRH primary transcript levels in the POA decrease during aging. Modified from Gore et al., 2000b.

discussed above, and may represent a substrate for the attenuation of the preovulatory GnRH/LH surge that occurs in rats, prior to reproductive failure.

NMDA receptor regulation of GnRH gene expression during aging

The regulation of the GnRH system by glutamate, acting through its NMDA receptor, changes across the reproductive life cycle. As discussed above, NMDA receptor activation stimulates GnRH release and elevates GnRH mRNA levels in young rats. A role for the NMDA receptor has also been suggested in reproductive senescence. The effects of NMDA receptor agonists on GnRH release in vivo and in vitro decrease during aging in rats (Arias et al., 1996; Zuo et al., 1996; Bonavera et al., 1998). My laboratory compared effects of NMDA receptor activation on GnRH mRNA levels between young and middle-aged rats (Gore et al., 2000a). We reported that treatment with an NMDA receptor agonist increases GnRH mRNA levels in young rats, similar to earlier studies (Petersen et al., 1991; Liaw and Barraclough, 1993; Gore and Roberts, 1994). In contrast, in middle-aged female rats, the NMDA receptor agonist actually *decreases* GnRH mRNA levels (Gore et al., 2000a). Thus, the GnRH neurosecretory system of rats varies in its responsiveness to NMDA receptor activation depending upon the age and reproductive status of the animal.

Insulin-like growth factor-I (IGF-I) regulation of GnRH neurons

IGF-I is another factor that regulates the GnRH neurosecretory system. This neurotrophic factor is synthesized in peripheral organs such as liver, and is also produced in the brain, including the hypothalamus and median eminence (Garcia-Segura et al., 1991; Miller and Gore, 2001). The IGF-I receptor is also expressed in the brain, again including hypothalamus (Werther et al., 1989; Marks et al., 1991; Pons et al., 1991), suggesting that central IGF-I may act as an autocrine or paracrine factor regulating hypothalamic functions. With respect to the hypothalamic–pituitary–gonadal axis, IGF-I can stimulate LH release and advance the timing of puberty; also, circulating IGF-I levels increase during puberty (Hiney and Dees, 1991; Hiney et al., 1996; Wilson, 1998; Suter et al., 2000). Thus, IGF-I is thought to be a metabolic signal that is associated with the timing of reproductive maturation. Moreover, because IGF-I is a target of the growth hormone axis, it may serve to coordinate the timing of the pubertal growth spurt and the onset of reproductive function.

We measured IGF-I mRNA levels in POA and medial basal hypothalamus (MBH) of rats undergoing reproductive maturation and reproductive aging. Using the RNase protection assay, we found that IGF-I mRNA levels in these two regions increase during pubertal development, and decrease during reproductive aging (Miller and Gore, 2001). In that same study, we examined effects of administration of IGF-I on GnRH mRNA levels, but did not see any effects in vivo (Miller and Gore, 2001). However, another study using hypothalamic explants from mice undergoing pubertal development showed that GnRH gene expression is stimulated by IGF-I under these in vitro conditions (Daftary et al., 2001). Differences between these results may be attributable to the experimental models. In the perifused hypothalami, IGF-I is applied to the bath, and can probably penetrate the tissue efficiently to exert its effects on the GnRH neurons (Daftary et al., 2001). In the in vivo studies, IGF-I is injected into the POA in vivo, but because this is a large region and GnRH neurons are dispersed throughout the POA, OVLT, septum and anterior hypothalamus, it is quite likely that IGF-I is not reaching all the target GnRH cells (Miller and Gore, 2001). Therefore, future studies are necessary to confirm whether IGF-I can stimulate GnRH gene expression in vivo as well as in vitro.

Post-transcriptional regulation of GnRH mRNA levels

Studies on the regulation of GnRH gene expression have led us to hypothesize that GnRH mRNA levels are predominantly regulated by a post-transcriptional mechanism in vivo. First, levels of GnRH nuclear transcripts are relatively high compared to other neuroendocrine genes, indicating that there is a large, steady-state pool of GnRH primary transcript that can result in the rapid accumulation of GnRH mRNA in the cytoplasm (reviewed in Gore and Roberts, 1997). Second, GnRH mRNA levels change extremely quickly in response to experimental manipulations, such as activation by NMDA receptor agonists, demonstrating the capacity for such a rapid accumulation (Petersen et al., 1991; Liaw and Barraclough, 1993; Gore and Roberts, 1994). Third, changes in GnRH mRNA often occur in the absence of any concomitant changes in GnRH gene transcription (Gore and Roberts, 1994, 1995, 1997; Petersen et al., 1996). Fourth, GnRH mRNA turnover can occur very rapidly (Maurer and Wray, 1997). Based on these observations, it is proposed that the major mechanism for GnRH mRNA regulation occurs via a post-transcriptional mechanism, probably via an enhancement of GnRH mRNA stability. This may involve a decrease in the degradation of the cytoplasmic GnRH mRNA pool, through changes in the poly (A) tail length of the GnRH mRNA. The poly (A) tail size is thought to be an important indicator of mRNA stability, and is clearly implicated in the regulation of GnRH mRNA levels in the GT1-7 cell line (Gore and Roberts, 1997; Gore et al., 1997). Preliminary results suggest that NMDA receptor activation increases, and blockade decreases, GnRH mRNA poly (A) tail size in rats, indicating a similar phenomenon occurs in vivo as in the GT1 cells (Sun et al., 1996).

Neuroanatomical studies on GnRH neurons

Pharmacological studies demonstrate that GnRH-1 neurons receive inputs from numerous neurotransmitters, neurotrophic factors, and sex steroid hormones. GnRH neurons express receptors for many of these substances, making them likely targets for direct effects, and in addition, they undoubtedly receive indirect inputs that are mediated by interneurons. Nevertheless, several puzzles remain about the neuroanatomical regulation of the GnRH-1 neurosecretory system. In particular, the number of synaptic inputs to GnRH perikarya in the preoptic area (POA) and rostral hypothalamus is extremely small. Relatively few synapses are found on GnRH perikarya, with only 1–3 synapses per GnRH cell detected in rostral hypothalamus and POA of the rat (Witkin, 1989; Chen et al., 1990). Only approximately 1% of the total GnRH neuronal membrane is postsynaptic, which is significantly lower than that of other non-GnRH neurons in the same brain regions (Witkin, 1989). The number of axodendritic synapses onto GnRH neurons is also low, with ~4 and 3 synapses per GnRH dendrite found in females and males, respectively (Chen et al., 1990). A study on the rhesus monkey confirmed the paucity of synapses on GnRH

perikarya (from 2 to 12 per cell; Witkin et al., 1995). Thus the question remains as to how those receptors found on GnRH neurons may be exerting their effects, and if they may be non-synaptic in nature.

Hypothalamic NMDA receptors and their co-localization on GnRH neurons

NMDA receptors (NMDARs) are pentameric in structure and contain the NMDA-R1 (NR1) subunit, which is obligatory for a functional receptor (Ishii et al., 1993). A homomeric NR1 subunit is not fully functional, so that the NR1 must combine with at least one member of the NR2 family (consisting of NR2a–d; (Monyer et al., 1992)). The stoichiometry of the NMDAR is important in determining its functional properties in terms of glutamate binding, ion flux and signal transduction (Buller et al., 1994; Tang et al., 1999). The functions of the NMDAR are further complicated by the observation that the NR1 subunit has alternative splice variants that can confer further different structural and functional properties (Kusiak and Norton, 1993; Laurie et al., 1995). In POA, at least 7 NR1 splice variants have been detected (Laurie et al., 1995).

My laboratory and others have been characterizing the expression of NMDAR and non-NMDAR subunits on hypothalamic GnRH neurons, and their regulation. In developing rats, NR1, NR2a and NR2b mRNA are detectable in the POA, and their levels increase during the first two weeks of postnatal life (Adams et al., 1999). Jennes' laboratory reported the presence of NR1, NR2a and NR2b mRNA labelled cells in septum, diagonal band and POA (Eyigor and Jennes, 1996, 1997). During reproductive aging, NR1 mRNA levels do not change much, while NR2a and NR2b mRNA levels decrease during the transition to acyclicity (Gore et al., 2000a). Therefore, it is likely that the stoichiometry of the NMDAR may change concomitantly with reproductive senescence.

The issue of whether GnRH neurons express NMDARs is extremely controversial. Initial studies looking for co-localization of GnRH with NMDARs found only very low levels (Abbud and Smith, 1995; Eyigor and Jennes, 1996, 1997). In 1996, using an improved antibody to NR1, we reported that ~20% of GnRH neurons of adult female rats express NR1 (Gore et al., 1996). More recent studies using electrophysiological recordings of identified GnRH neurons demonstrate that 20% of GnRH neurons respond to NMDAR activation with depolarization (Spergel et al., 1999). That same laboratory reported that all GnRH cells respond to glutamate, indicating that all GnRH neurons express either NMDA or non-NMDA receptors (Spergel et al., 1999). More recently, using techniques with greater sensitivity and improved antibodies, my laboratory reported that high percentages (50–80%) of GnRH neurons express NR1, NR2a and NR2b in young adult and middle-aged rats (Miller and Gore, 2002).

GnRH neuroterminals in the median eminence also coexpress NMDARs. Electron microscopy demonstrates that GnRH terminals co-localize with NR1 and non-NMDAR (KA2) receptors at the level of the median eminence (Kawakami et al., 1998a,b). Preliminary data from my laboratory, using postembed immunogold double-label electron microscopy, indicate that the GnRH and NMDA receptor molecules are co-localized in the same neuroterminals, and moreover, that they are actually co-expressed within the same dense-core vesicles (Fig. 3). This is a novel localization of the NMDAR that is under investigation in ongoing physiological and electron microscopic studies in my laboratory.

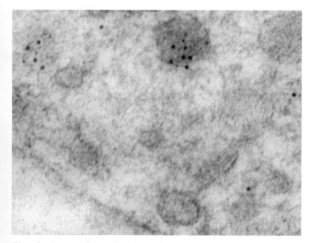

Fig. 3. Expression of the NMDA receptor in large dense-core vesicles in a GnRH neuroterminal in the median eminence. Double-label immunocytochemistry was performed using postembed immunoelectron microscopic techniques. NMDA-R1 is identified with the larger (15 nm) gold particles, and GnRH with the smaller (5 nm) gold particles.

Insulin-like growth factor-I (IGF-I) co-localization in GnRH neurons

IGF-I is synthesized in the central nervous system in glia and neurons (Rotwein et al., 1988; Baxter et al., 1989; Garcia-Segura et al., 1991; Miller and Gore, 2001). The protein and mRNA of IGF-I and the IGF-I receptor are easily detectable in hypothalamus and median eminence (Werther et al., 1989; Garcia-Segura et al., 1991; Marks et al., 1991), demonstrating synthesis in regions proximal to GnRH perikarya and neuroterminals. A physiological role for IGF-I in reproductive function has also been shown. IGF-I enhances GnRH release from rat median eminence explants in vitro (Hiney et al., 1991), and LH levels in female rhesus (Wilson, 1998). Stimulatory effects of IGF-I are also seen on GnRH-induced LH secretion (Kanematsu et al., 1991).

My laboratory has been investigating the anatomical relationship between IGF-I in the brain and GnRH neurons. Because of its stimulatory effect on GnRH release, we predicted that cells expressing IGF-I would be found in the vicinity of GnRH neurons. To our surprise, we observed that a high percentage (~75%) of GnRH neurons in adult mouse and rat brain co-express IGF-I (Daftary et al., 2001; Miller and Gore, 2001; Fig. 4). The co-expression of IGF-I in GnRH neurons is subject to developmental regulation. Thus, the percentages of GnRH neurons expressing IGF-I increase from approximately 20 to 50 to 75% in neonatal, pubertal, and adult mice, respectively (Daftary et al., 2001). We speculate that GnRH neurons may synthesize IGF-I, and that this increases during maturation. This IGF-I in GnRH cells may serve an autocrine role regulating GnRH function, or may act as a paracrine factor in the hypothalamus regulating other non-GnRH cells. Two possible candidates for this action are the somatostatin or the growth hormone-releasing hormone neurons, both of which are sensitive to IGF-I regulation.

Conclusions

The regulation of GnRH neurons involves alterations in gene expression and neuroanatomical inputs. GnRH gene expression changes in response to stimuli such as steroid hormone feedback, varies across the estrous cycle, and changes across the

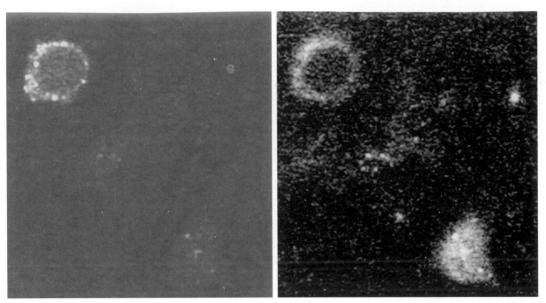

Fig. 4. Co-expression of IGF-I in a GnRH neuron of an adult female rat. A GnRH positive cell is shown in the left panel (upper left corner); the same cell, double-labeled with IGF-I, is shown in the right panel. Modified from Miller and Gore, 2001.

reproductive life cycle, particularly during pubertal maturation and reproductive aging. The most dramatic alterations in GnRH mRNA levels are seen during reproductive development in association with the pubertal process. Studies on the mechanisms for the regulation of GnRH mRNA levels suggest that a post-transcriptional process, involving an alteration in GnRH mRNA stability, possibly mediated by the GnRH poly (A) tail length, may be the primary one responsible for the control of GnRH mRNA levels. Inputs to GnRH neurons arising from neurotransmitters such as glutamate, and neurotrophic factors such as IGF-I, can affect the physiology and gene expression of the GnRH neurosecretory system. We observed that the expression of NMDA receptors on GnRH neurons is developmentally regulated, and that the responsiveness of GnRH cells to NMDA receptor activation is highest in pubertal animals and young adults, but decreases during reproductive senescence. For IGF-I, we found that GnRH cells express IGF-I, and again, this is developmentally regulated, with increases in co-localization occurring during reproductive maturation. Taken together, these studies demonstrate that changes that are intrinsic to GnRH neurons, as well as arising from extrinsic inputs to GnRH neurons, are integrated at the level of the GnRH neuron to result in the appropriate reproductive physiology of the organism.

References

Abbud, R. and Smith, M.S. (1995) Do GnRH neurons express the gene for the NMDA receptor? *Brain Res.*, 690: 117–120.

Adams, M.M., Flagg, R.A. and Gore, A.C. (1999) Perinatal changes in hypothalamic NMDA receptors and their relationship to GnRH neurons. *Endocrinology*, 140: 2288–2296.

Arias, P., Carbone, S., Szwarcfarb, B., Feleder, C., Rodriguez, M., Scacchi, P. and Moguilevsky, J.A. (1996) Effects of aging on N-methyl-D-aspartate (NMDA)-induced GnRH and LH release in female rats. *Brain Res.*, 740: 234–238.

Attardi, B., Klatt, B., Hoffman, G.E. and Smith, M.S. (1997) Facilitation or inhibition of the estradiol-induced gonadotropin surge in the immature rat by progesterone: Regulation of GnRH and LH messenger RNAs and activation of GnRH neurons. *J. Neuroendocrinology*, 9: 589–599.

Baxter, R.C., Martin, J.L. and Beniac, V.A. (1989) High molecular weight insulin-like growth factor binding protein complex: purification and properties of the acid-labile subunit from human serum. *J. Biol. Chem.*, 264: 11843–11848.

Bonavera, J.J., Swerdloff, R.S., Hikim, A.P.S., Lue, Y.H. and Wang, C. (1998) Aging results in attenuated gonadotropin releasing hormone–luteinizing hormone axis responsiveness to glutamate receptor agonist N-methyl-D-aspartate. *J. Neuroendocrinology*, 10: 93–99.

Brann, D.W. (1995) Glutamate: a major excitatory transmitter in neuroendocrine regulation. *Neuroendocrinology*, 61: 213–225.

Buller, A.L., Larson, H.C., Schneider, B.E., Beaton, J.A., Morrisett, R.A. and Monaghan, D.T. (1994) The molecular basis of NMDA receptor subtypes: native receptor diversity is predicted by subunit composition. *J. Neurosci.*, 14: 5471–5484.

Chen, W.P., Witkin, J.W. and Silverman, A.-J. (1990) Sexual dimorphism in the synaptic input to GnRH neurons. *Endocrinology*, 126: 695–702.

Cicero, T.J., Meyer, E.R. and Bell, R.D. (1988) Characterization and possible opioid modulation of N-methyl-D-aspartic acid induced increases in serum luteinizing hormone levels in the developing male rat. *Life Sci.*, 42: 1725–1732.

Daftary, S.S., Oung, T. and Gore, A.C. (2001) Insulin-like growth factor-I (IGF-I) regulation of GnRH at puberty. *Soc. for Neuroscience Meeting*, abstract.

Doecke, F., Rohde, W., Smollich, A. and Dorner, G. (1978) Hormones and brain maturation in the control of female puberty. In: G. Dorner and M. Kawakami (Eds.), *Hormones and Brain Development*. Elsevier, Amsterdam, pp. 327–340.

Donovan, B.T., ter Haar, M.B., Lockhart, A.N., MacKinnon, P.C.B., Mattock, J.M. and Peddie, M.J. (1975) Changes in the concentration of luteinizing hormone in plasma during development in the guinea pig. *J. Endocrinol.*, 64: 511–520.

Dutlow, C.M., Rachman, J., Jacobs, T.W. and Millar, R.P. (1992) Prepubertal increases in gonadotropin-releasing hormone mRNA, gonadotropin-releasing hormone precursor, and subsequent maturation of precursor processing in male rats. *J. Clin. Invest.*, 90: 2496–2501.

Eyigor, O. and Jennes, L. (1996) Identification of glutamate receptor subtype mRNAs in gonadotropin-releasing hormone neurons in rat brain. *Endocrine*, 4: 133–139.

Eyigor, O. and Jennes, L. (1997) Expression of glutamate receptor subunit mRNAs in gonadotropin-releasing hormone neurons during the sexual maturation of the female rat. *Neuroendocrinology*, 66: 122–129.

Fernald, R.D. and White, R.B. (1999) Gonadotropin-releasing hormone genes: phylogeny, structure, and functions. *Front. Neuroendocrinol.*, 20: 224–240.

Garcia-Segura, L.M., Perez, J., Pons, S., Rejas, M.T, and Torres-Aleman, I. (1991) Localization of insulin-like growth factor-1 (IGF-1) immunoreactivity in the developing and adult rat brain. *Brain Res.*, 560: 167–174.

Gay, V.L. and Plant, T.M. (1987) N-methyl-D,L-aspartate (NMA) elicits hypothalamic GnRH release in prepubertal male rhesus monkeys (*Macaca mulatta*). *Endocrinology*, 120: 2289–2296.

Gibson, M.J., Wu, T.J., Miller, G.M. and Silverman, A.-J. (1997) What nature's knockout teaches us about GnRH activity: hypogonadal mice and neuronal grafts. *Horm. Behav.* 31: 212–220.

Gore, A.C. (2000) Modulation of the GnRH gene and onset of puberty. In: J.-P. Bourguignon and T.M. Plant (Eds.), *The Onset of Puberty in Perspective*. Elsevier, Amsterdam, pp. 25–35.

Gore, A.C. (2001) Gonadotropin-releasing hormone neurons, NMDA receptors, and their regulation by steroid hormones across the reproductive life cycle. *Brain Res. Rev.*, 37: 728–736.

Gore, A.C. (2002) *GnRH: The Master Molecule of Reproduction*. Kluwer Academic Publishers, Boston, MA.

Gore, A.C. and Roberts, J.L. (1994) Regulation of gonadotropin-releasing hormone gene expression by the excitatory amino acids kainic acid and N-methyl-D,L-aspartate in the male rat. *Endocrinology*, 134: 2026–2031.

Gore, A.C. and Roberts, J.L. (1995) Regulation of gonadotropin-releasing hormone gene expression in the rat during the luteinizing hormone surge. *Endocrinology*, 136: 889–896.

Gore, A.C. and Roberts, J.L. (1997) Regulation of GnRH gene expression in vivo and in vitro. *Front. Neuroendocrinol.*, 18: 209–245.

Gore, A.C., Wu, T.J., Rosenberg, J.J. and Roberts, J.L. (1996) Gonadotropin-releasing hormone and NMDA-R1 gene expression and colocalization change during puberty in female rats. *J. Neurosci.*, 16: 5281–5289.

Gore, A.C., Yeo, T.T., Ho, A. and Roberts, J.L. (1997) Post-transcriptional regulation of the gonadotropin-releasing hormone gene in GT1-7 cells. *J. Neuroendocrinol.*, 9: 271–277.

Gore, A.C., Roberts, J.L. and Gibson, M.J. (1999) Mechanisms for the regulation of gonadotropin-releasing hormone gene expression in the developing mouse. *Endocrinology*, 140: 2280–2287.

Gore, A.C., Yeung, G., Morrison, J.H. and Oung, T. (2000a) Neuroendocrine aging in the female rat: The changing relationship of hypothalamic gonadotropin-releasing hormone neurons and N-methyl-D-aspartate receptors. *Endocrinology*, 141: 4757–4767.

Gore, A.C., Oung, T., Yung, S., Flagg, R.A. and Woller, M.J. (2000b) Neuroendocrine mechanisms for reproductive senescence in the female rat: gonadotropin-releasing hormone neurons. *Endocrine*, 13: 315–323.

Gruenewald, D.A. and Matsumoto, A.M. (1991) Age-related decreases in serum gonadotropin levels and gonadotropin-releasing hormone gene expression in the medial preoptic area of the male rat are dependent upon testicular feedback. *Endocrinology*, 129: 2442–2450.

Hiney, J.K. and Dees, W.L. (1991) Ethanol inhibits luteinizing hormone-releasing hormone release from the median eminence of prepubertal female rats in vitro: investigation of its actions on norepinephrine and prostaglandin-E2. *Endocrinology*, 128: 1404–1408.

Hiney, J.K., Ojeda, S.R. and Dees, W.L. (1991) Insulin-like growth factor I: a possible metabolic signal involved in the regulation of female puberty. *Neuroendocrinology*, 54: 420–423.

Hiney, J.K., Srivastava, V., Nyberg, C.L., Ojeda, S.R. and Dees, W.L. (1996) Insulin-like growth factor I of peripheral origin acts centrally to accelerate the onset of female puberty. *Endocrinology*, 137: 3717–3728.

Hoffman, G.E. and Finch, C.E. (1986) LHRH neurons in the female C57BL/6J mouse brain during reproductive aging: No loss up to middle age. *Neurobiol. Aging*, 7: 45–48.

Huang, H.H. and Meites, J. (1975) Reproductive capacity of aging female rats. *Neuroendocrinology*, 17: 289–295.

Hwang, C., Pu, H.-F., Hwang, C.-Y., Liu, J.-Y., Yao, H.-C., Tung, Y.-F. and Wang, P.S. (1990) Age-related differences in the release of luteinizing hormone and gonadotropin-releasing hormone in ovariectomized rats. *Neuroendocrinology*, 52: 127–132.

Ishii, T., Moriyoshi, K., Sugihara, H., Sakurada, K., Kadotani, H., Yokoi, M., Akazawa, C., Shigemoto, R., Mizuno, N., Masu, M. and Nakanishi, S. (1993) Molecular characterization of the family of the N-methyl-D-aspartate receptor subunits. *J. Biochem.*, 268: 2836–2843.

Jakubowski, M. and Roberts, J.L. (1994) Processing of gonadotropin-releasing hormone gene transcripts in the rat brain. *J. Biol. Chem.*, 269: 4078–4083.

Jakubowski, M., Blum, M. and Roberts, J.L. (1991) Postnatal development of gonadotropin-releasing hormone and cyclophilin gene expression in the female and male rat brain. *Endocrinology* 128: 2702–2708.

Kallmann, F.J., Schoenfeld, W.A. and Barrera, S.E. (1944) The genetic aspects of primary eunuchoidism. *Am. J. Ment. Deficiency*, XLVIII: 203–236.

Kanematsu, T., Irahara, M., Miyake, T., Shitsukawa, K. and Aono, T. (1991) Effect of insulin-like growth factor I on gonadotropin release from the hypothalamus–pituitary axis in vitro. *Acta Neurobiol. Exp.*, 125: 227–233.

Kawakami, S., Hirunagi, K., Ichikawa, M., Tsukamura, H. and Maeda, K.-I. (1998a) Evidence for terminal regulation of GnRH release by excitatory amino acids in the median eminence in female rats: a dual immunoelectron microscopic study. *Endocrinology*, 139: 1458–1461.

Kawakami, S., Ichikawa, M., Murahashi, K., Hirunagi, K., Tsukamura, H. and Maeda, K. (1998b) Excitatory amino acids act on the median eminence nerve terminals to induce gonadotropin-releasing hormone release in female rats. *Gen. Comp. Endocrinol.*, 112: 372–382.

Kelly, M.J., Garrett, J., Bosch, M.A., Roselli, C.E., Douglass, J., Adelman, J.P. and Ronnekleiv, O.K. (1989) Effects of ovariectomy on GnRH mRNA, proGnRH and GnRH levels in the preoptic hypothalamus of the female rat. *Neuroendocrinology*, 49: 88–97.

Kim, K., Lee, B.J., Park, Y. and Cho, W.K. (1989) Progesterone increases messenger ribonucleic acid (mRNA) encoding luteinizing hormone releasing hormone (LHRH) level in the hypothalamus of ovariectomized estradiol-primed prepubertal rats. *Mol. Brain Res.*, 6: 151–158.

Kusiak, J.W. and Norton, D.D. (1993) A splice variant of the N-methyl-D-aspartate (NMDAR1) receptor. *Mol. Brain Res.*, 20: 64–70.

Laurie, D.J., Putzke, J., Zieglgansberger, W., Seeburg, P.H. and Tolle, T.R. (1995) The distribution of splice variants of the NMDAR1 in adult rat brain. *Mol. Brain Res.*, 32: 94–108.

Lee, W.-S., Smith, M.S. and Hoffman, G.E. (1990) Luteinizing hormone-releasing hormone neurons express Fos protein during the proestrous surge of luteinizing hormone. *Proc. Natl. Acad. Sci. USA*, 87: 5163–5167.

Levine, J.E. and Ramirez, V.D. (1982) Luteinizing hormone-releasing hormone release during the rat estrous cycle and after ovariectomy, as estimated with push–pull cannulae. *Endocrinology*, 111: 1439–1448.

Li, S., Garcia de Yebenes, E. and Pelletier, G. (1995) Effects of

dehydroepiandrosterone (DHEA) on GnRH gene expression in the rat brain as studied by in situ hybridization. *Peptides*, 16: 425–430.

Li, S., Givalois, L. and Pelletier, G. (1997) Effects of aging and melatonin administration on gonadotropin-releasing hormones (GnRH) gene expression in the male and female rat. *Peptides*, 18: 1023–1028.

Liaw, J.-J. and Barraclough, C.A. (1993) N-methyl-D,L-aspartic acid differentially affects LH release and LHRH mRNA levels in estrogen-treated ovariectomized control and androgen-sterilized rats. *Mol. Brain Res.*, 17: 112–118.

Lloyd, J.M., Hoffman, G.E. and Wise, P.M. (1994) Decline in immediate early gene expression in gonadotropin-releasing hormone neurons during proestrus in regularly cycling, middle-aged rats. *Endocrinology*, 134: 1800–1805.

Lu, K.H., Huang, H.H., Chen, H.T., Kurcz, M., Mioduszewski, R. and Meites, J. (1977) Positive feedback of estrogen and progesterone on LH release in old and young rats. *Proc. Soc. Exp. Biol. Med.*, 154: 82–85.

Malik, K.F., Silverman, A.-J. and Morrell, J.I. (1991) Gonadotropin-releasing hormone mRNA in the rat: Distribution and neuronal content over the estrous cycle and after castration of males. *Anat. Rec.*, 231: 457–466.

Marks, D.L., Smith, M.S., Vrontakis, M., Clifton, D.K. and Steiner, R.A. (1993) Regulation of galanin gene expression in gonadotropin-releasing hormone neurons during the estrous cycle of the rat. *Endocrinology*, 132: 1836–1844.

Marks, D.L., Lent, K.L., Rossmanith, W.G., Clifton, D.K. and Steiner, R.A. (1994) Activation-dependent regulation of galanin gene expression in gonadotropin-releasing hormone neurons in the female rat. *Endocrinology*, 134: 1991–1998.

Marks, J.L., Porte, D.J. and Baskin, D.G. (1991) Localization of type 1 insulin-like growth factor receptor mRNA in the adult rat brain by in situ hybridization. *Mol. Endocrinol.*, 5: 1158–1168.

Maurer, J.A. and Wray, S. (1997) Luteinizing hormone-releasing hormone (LHRH) neurons maintained in hypothalamic slice explant cultures exhibit a rapid LHRH mRNA turnover rate. *J. Neurosci.*, 17: 9481–9491.

Miller, B.H. and Gore, A.C. (2001) Alterations in hypothalamic IGF-I and its associations with GnRH neurons during reproductive development and aging. *J. Neuroendocrinol.*, 13: 728–736.

Miller, B.H. and Gore, A.C. (2002) NMDA receptor subunit expression in GnRH neurons changes during reproductive senescence in the female rat. *Endocrinology*, 143: 3568–3574.

Monyer, H., Sprengel, R., Schoepfer, R., Herb, A., Higuchi, M., Lomeli, H., Burnashev, N., Sakmann, B. and Seeburg, P.H. (1992) Heteromeric NMDA receptors: Molecular and functional distinction of subtypes. *Science*, 256: 1217–1221.

Nelson, J.F., Felicio, L.S., Osterburg, H.H. and Finch, C.E. (1992) Differential contributions of ovarian and extraovarian factors to age-related reductions in plasma estradiol and progesterone during the estrous cycle of C57BL/6J mice. *Endocrinology*, 130: 805–810.

Parfitt, D.B., Thompson, R.C., Richardson, H.N., Romeo, R.D. and Sisk, C.L. (1999) GnRH mRNA increases with puberty in the male syrian hamster brain. *J. Neuroendocrinol.*, 11: 621–627.

Parhar, I.S., Soga, T. and Sakuma, Y. (2000) Thyroid hormone and estrogen regulate brain region-specific messenger ribonucleic acids encoding three gonadotropin-releasing hormone genes in sexually immature male fish, *Oreochromis niloticus*. *Endocrinology*, 141: 1618–1626.

Park, O.-K. and Ramirez, V.D. (1989) Spontaneous changes in LHRH release during the rat estrous cycle, as measured with repetitive push–pull perfusions of the pituitary gland in the same female rats. *Neuroendocrinology*, 50: 66–72.

Park, O.-K., Gugneja, S. and Mayo, K.E. (1990) Gonadotropin-releasing hormone gene expression during the rat estrous cycle: Effects of pentobarbital and ovarian steroids. *Endocrinology*, 127: 365–372.

Petersen, S.L., McCrone, S., Keller, M. and Gardner, E. (1991) Rapid increases in LHRH mRNA levels following NMDA. *Endocrinology*, 129: 1679–1681.

Petersen, S.L., McCrone, S., Keller, M. and Shores, S. (1995) Effects of estrogen and progesterone on luetinizing hormone-releasing hormone messenger ribonucleic acid levels: Consideration of temporal and neuroanatomical variables. *Endocrinology*, 136: 3604–3610.

Petersen, S.L., Gardner, E., Adelman, J. and McCrone, S. (1996) Examination of steroid-induced changes in LHRH gene transcription using 33P- and 35S-labeled probes specific for intron 2. *Endocrinology*, 137: 234–239.

Pfaff, D.W. (1986) Gene expression in hypothalamic neurons: Luteinizing hormone releasing hormone. *J. Neurosci. Res.*, 16: 109–115.

Plant, T.M. (1988) Puberty in primates. In: E. Knobil and J. Neill (Eds.), *The Physiology of Reproduction*. Raven Press, New York, NY, pp. 1763–1788.

Pons, S., Rejas, M.T. and Torres-Aleman, I. (1991) Ontogeny of insulin-like growth factor I, its receptor, and its binding proteins in the rat hypothalamus. *Devel. Brain Res.*, 62: 169–175.

Porkka-Heiskanen, T., Urban, J.H., Turek, F.W. and Levine, J.E. (1994) Gene expression in a subpopulation of luteinizing hormone-releasing hormone (LHRH) neurons prior to the preovulatory gonadotropin surge. *J. Neurosci.*, 14: 5548–5558.

Ramaley, J.A. (1979) Development of gonadotropin regulation in the prepubertal mammal. *Biol. Reprod.*, 20: 1–31.

Rance, N.E. and Uswandi, S.V. (1996) Gonadotropin-releasing hormone gene expression is increased in the medial basal hypothalamus of postmenopausal women. *J. Clin. Endocrinol. Metab.*, 81: 3540–3546.

Roberts, J.L., Dutlow, C.M., Jakubowski, M., Blum, M. and Millar, R.P. (1989) Estradiol stimulates preoptic area–anterior hypothalamic proGnRH-GAP gene expression in ovariectomized rats. *Mol. Brain Res.*, 6: 127–134.

Rosie, R., Thomson, E. and Fink, G. (1990) Oestrogen positive feedback stimulates the synthesis of LHRH mRNA in neurones of the rostral diencephalon of the rat. *J. Endocrinol.*, 124: 285–289.

Rothfeld, J.M., Hejtmancik, J.F., Conn, P.M. and Pfaff, D.W.

(1989) In situ hybridization for LHRH mRNA following estrogen treatment. *Mol. Brain Res.*, 6: 121–125.

Rotwein, P., Burgess, S.K., Milbrandt, J.D. and Krause, J.E. (1988) Differential expression of insulin-like growth factor genes in rat central nervous system. *Proc. Natl. Acad. Sci. USA*, 85: 265–269.

Rubin, B.S. and Bridges, R.S. (1989) Alterations in luteinizing hormone-releasing hormone release from the mediobasal hypothalamus of ovariectomized, steroid-primed middle-aged rats as measured by push–pull perfusion. *Neuroendocrinology*, 49: 225–232.

Rubin, B.S. and King, J.C. (1994) The number and distribution of detectable luteinizing hormone (LH)-releasing hormone cell bodies changes in association with the preovulatory LH surge in the brains of young but not middle-aged female rats. *Endocrinology*, 134: 467–474.

Rubin, B.S., Lee, C.E. and King, J.C. (1994) A reduced proportion of luteinizing hormone (LH)-releasing hormone neurons express Fos protein during the preovulatory or steroid-induced LH surge in middle-aged rats. *Biol. Reprod.*, 51: 1264–1272.

Rubin, B.S., Lee, C.E., Ohtomo, M. and King, J.C. (1997) Luteinizing hormone-releasing hormone gene expression differs in young and middle-aged females on the day of a steroid-induced LH surge. *Brain Res.*, 770: 267–276.

Sagrillo, C.A., Grattan, D.R., McCarthy, M.M. and Selmanoff, M. (1996) Hormonal and neurotransmitter regulation of GnRH gene expression and related reproductive behaviors. *Behav. Genet.*, 26: 241–277.

Sarkar, D.K. and Fink, G. (1980) Luteinizing hormone releasing factor in pituitary stalk plasma from long-term ovariectomized rats: Effects of steroids. *J. Endocrinol.*, 86: 511–524.

Sisk, C.L., Richardson, H.N., Chappell, P.E. and Levine, J.E. (2001) In vivo gonadotropin-releasing hormone secretion in female rats during peripubertal development and on proestrus. *Endocrinology*, 142: 2929–2936.

Smyth, C. and Wilkinson, M. (1994) A critical period for glutamate receptor-mediated induction of precocious puberty in female rats. *J. Neuroendocrinol.*, 6: 275–284.

Sortino, M.A., Aleppo, G., Scapagnini, U. and Canonico, P.L. (1996) Different responses of gonadotropin-releasing homrone (GnRH) release to glutamate receptor agonists during aging. *Brain Res. Bull.*, 41: 359–362.

Spergel, D.J., Krëuth, U., Hanley, D.F., Sprengel, R. and Seeburg, P.H. (1999) GABA- and glutamate-activated channels in green fluorescent protein-tagged gonadotropin-releasing hormone neurons in transgenic mice. *J. Neurosci.*, 19: 2037–2050.

Steger, R.W., Huang, H.H., Chamberlain, D.S. and Meites, J. (1980) Changes in control of gonadotropin secretion in the transition period between regular cycles and constant estrus in aging female rats. *Biol. Reprod.*, 22: 595–603.

Sun, Y., Gore, A.C. and Roberts, J.L. (1996) Regulation of gonadotropin-releasing hormone (GnRH) gene expression by glutamate in male and female rats. *Endocrine Society*, abstract P1-390, p. 232.

Suter, K.J., Pohl, C.R. and Wilson, M.E. (2000) Circulating concentrations of nocturnal leptin, growth hormone, and insulin-like growth factor-I increase before the onset of puberty in agonadal male monkeys: potential signals for the initiation of puberty. *J. Clin. Endocrinol. Metab.*, 85: 808–814.

Suzuki, M., Nishihara, M. and Takahashi, M. (1995) Hypothalamic gonadotropin-releasing hormone gene expression during rat estrous cycle. *Endocrine J.*, 42: 789–796.

Tang, Y.-P., Shimizu, E., Dube, G.R., Rampon, C., Kerchner, G.A., Zhuo, M., Liu, G. and Tsien, J.Z. (1999) Genetic enhancement of learning and memory in mice. *Nature*, 40: 63–69.

Toranzo, D., Dupont, E., Simard, J., Labrie, C., Couet, J., Labrie, F. and Pelletier, G. (1989) Regulation of pro-gonadotropin-releasing hormone gene expression by sex steroids in the brain of male and female rats. *Mol. Endocrinol.*, 3: 1748–1756.

Urbanski, H.F. and Ojeda, S.R. (1985) The juvenile–peripubertal transition period in the female rat: Establishment of a diurnal pattern of pulsatile luteinizing hormone secretion. *Endocrinology*, 117: 644–649.

Urbanski, H.F. and Ojeda, S.R. (1990) A role of N-methyl-D-aspartate (NMDA) receptors in the control of LH secretion and initiation of female puberty. *Endocrinology*, 126: 1774–1776.

Watanabe, G. and Terasawa, E. (1989) In vivo release of luteinizing hormone releasing hormone increases with puberty in the female rhesus monkey. *Endocrinology*, 125: 92–99.

Werther, A., Hogg, A., Oldfield, B.J., McKinley, M.J., Figdor, R. and Mendelsohn, F.A.O. (1989) Localization and characterization of IGF-1 receptors in rat brain and pituitary gland using in vitro autoradiography and computerized densitometry. A distinct distribution from insulin receptors. *J. Neuroendocrinol.*, 1: 369–377.

Wilson, M.E. (1998) Premature elevation in serum insulin-like growth factor-I advances first ovulation in rhesus monkeys. *J. Endocrinol.*, 158: 247–257.

Wise, P.M. (1984) Estradiol-induced daily luteinizing hormone and prolactin surges in young and middle-aged rats: Correlations with age-related changes in pituitary responsiveness and catecholamine turnover rates in microdissected brain areas. *Endocrinology*, 115: 801–809.

Witkin, J.W. (1986) Luteinizing hormone releasing hormone (LHRH) neurons in aging female rhesus macaques. *Neurobiol. Aging*, 7: 259–263.

Witkin, J.W. (1989) Synaptology of luteinizing hormone-releasing hormone neurons in the preoptic area of the male rat: effects of gonadectomy. *Neuroscience*, 29: 385–390.

Witkin, J.W., O'Sullivan, H. and Silverman, A.-J. (1995) Novel associations among gonadotropin-releasing hormone neurons. *Endocrinology*, 136: 4323–4330.

Wolfe, A.M., Wray, S., Westphal, H. and Radovick, S. (1996) Cell-specific expression of the human gonadotropin-releasing hormone gene in transgenic animals. *J. Biol. Chem.*, 271: 20018–20023.

Wray, S. and Hoffman, G. (1986) A developmental study of the quantitative distribution of LHRH neurons within the central nervous system of postnatal male and female rats. *J. Comp. Neurol.*, 252: 522–531.

Wray, S., Zoeller, R.T. and Gainer, H. (1989) Differential effects of estrogen on luteinizing hormone-releasing hormone gene expression in slice explant cultures prepared from specific rat forebrain regions. *Mol. Endocrinol.*, 3: 1197–1206.

Yeo, T.T.S., Gore, A.C., Jakubowski, M., Dong, K., Blum, M. and Roberts, J.L. (1996) Characterization of gonadotropin-releasing hormone gene transcripts in a mouse hypothalamic neuronal GT1 cell line. *Mol. Brain Res.*, 42: 255–262.

Zoeller, R.T. and Young, W.S. III (1988) Changes in cellular levels of messenger ribonucleic acid encoding gonadotropin-releasing hormone in the anterior hypothalamus of female rats during the estrous cycle. *Endocrinology*, 123: 1688–1689.

Zoeller, R.T., Seeburg, P.H. and Young, W.S. III (1988) In situ hybridization histochemistry for messenger ribonucleic acid (mRNA) encoding gonadotropin-releasing hormone (GnRH): Effect of estrogen on cellular levels of GnRH mRNA in female rat brain. *Endocrinology*, 122: 2570–2577.

Zuo, Z., Mahesh, V.B., Zamorano, P.L. and Brann, D.W. (1996) Decreased gonadotropin-releasing hormone neurosecretory response to glutamate agonists in middle-aged female rats on proestrus afternoon: a possible role in reproductive aging? *Endocrinology*, 137: 2334–2338.

I.S. Parhar (Ed.)
Progress in Brain Research, Vol. 141
© 2002 Published by Elsevier Science B.V.

CHAPTER 16

GnRH pre-mRNA splicing: role of exonic splicing enhancer

Jin Han [1], Gi Hoon Son [1], Jae Young Seong [2], Kyungjin Kim [1,*]

[1] *School of Biological Sciences, Seoul National University, Seoul 151-742, Korea*
[2] *Hormone Research Center, Chonnam National University, Kwangju 500-757, Korea*

Introduction

Gonadotropin-releasing hormone (GnRH) plays a key role in the regulation of reproduction (Seeburg et al., 1987). GnRH is synthesized in the preoptic area (POA) of the vertebrate brain and released from nerve terminals in the median eminence into the hypothalamic–pituitary portal vessel in a pulsatile fashion. GnRH then acts on gonadotrope in the anterior pituitary to stimulate the biosynthesis and secretion of pituitary gonadotropins, luteinizing hormone (LH) and follicle stimulating hormone (FSH).

The biosynthesis of GnRH can be regulated at multiple levels such as transcription rate, mRNA stability, translation rate of the mRNA into proGnRH peptide, and processing of the precursor to the mature GnRH decapeptide (Gore and Roberts, 1997; Gore et al., 1997). Recently, processing of the GnRH pre-mRNA to the mature mRNA has been reported as an important step for the regulation of GnRH biosynthesis. Various transcripts can be produced from a single GnRH gene by alternative RNA splicing (Seeburg and Adelman, 1984; Radovick et al., 1990; Zhen et al., 1997). Alternative splicing is often involved in functional diversification of proteins and quantitative control of gene expression (Lopez, 1998;

Lou and Gagel, 2001). Multiple transcripts generated by alternative splicing can serve as templates for producing various proteins that show different enzymatic function, subcellular localization, binding affinity to ligands, and electrophysiological properties (Hollmann et al., 1993; Kilpatrick et al., 1999). Generation of different 5′ or 3′ untranslated regions by alternative splicing can affect RNA stability and translational efficacy (Wang et al., 1999). Recently, we and others found that, in contrast to the mature mRNA, GnRH splice variant transcripts exhibit a severe defect in producing the GnRH peptide (Zhen et al., 1997; Seong et al., 2001). Thus, elucidation of the detailed mechanism for alternative GnRH pre-mRNA splicing is quite important for understanding the regulation of GnRH biosynthesis. In this article, we will discuss the recent progress in research regarding the enhancer-dependent GnRH pre-mRNA splicing mechanism, the possible *trans*-acting factors that may act on the splicing enhancer element, and the physiological relevance of enhanced and/or attenuated GnRH pre-mRNA splicing.

Alternative splicing of the GnRH pre-mRNA

The GnRH gene consists of 4 short exons (denoted 1, 2, 3, and 4) and 3 long introns (A, B, and C) (Mason et al., 1986; Bond et al., 1989). In the hypothalamus and GnRH-producing cell lines, all three introns are excised from the primary transcript (~4,300 bases), producing a mature mRNA of about 560 bases (Adelman et al., 1986). Despite low

---

* Correspondence to: K. Kim, School of Biological Sciences, Seoul National University, Seoul 151-742, Korea. Tel.: +82-2-880-6694; Fax: +82-2-872-1993; E-mail: kyungjin@snu.ac.kr

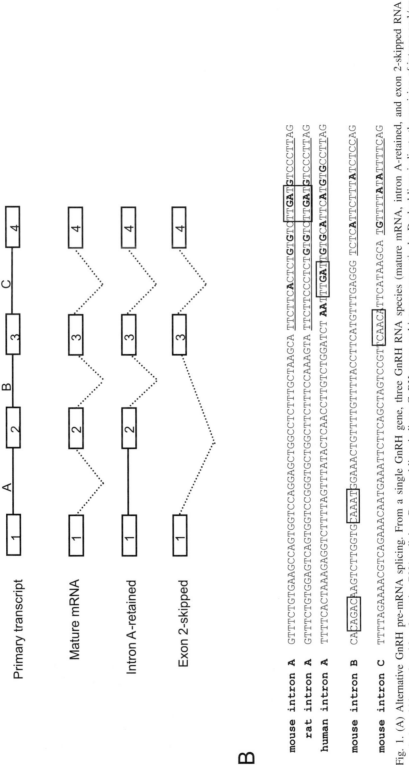

```
mouse intron A   GTTTCTGTGAAGCCAGTGGTCCAGGAGCTGGCCTCCTTGCTAAGCA TTCTTCACTCTGTGTCTTGATGTCCCTTAG

rat intron A     GTTTCTGTGGAGTCAGTGGTCCGGGTGCTGGCTTCTTCCAAAGTA TTCTTCCCTCTGTGTCTTGATGTCCCTTAG

human intron A   TTTTCACTAAAGAGGTCTTTAGTTTATACTCAACCTTGTCTGGATCT AATTGATTGTGCATTCATGTGCCTTAG

mouse intron B   CACAGACAAGTCTTGGTGTAAAATGGAAACTGTTTTGTTTTACCTTCATGTTTGAGGG TCTCATTCTTTATCTCCAG

mouse intron C   TTTTAGAAAACGTCAGAAACAATGAAATTCTTCAGCTAGTCCGTTCAACATTCATAAGCA TGTTTTATATTTTCAG
```

Fig. 1. (A) Alternative GnRH pre-mRNA splicing. From a single GnRH gene, three GnRH RNA species (mature mRNA, intron A-retained, and exon 2-skipped RNA species) could be produced by alternative RNA splicing. Boxes and lines indicate GnRH exons and introns, respectively. Dashed lines indicate the excision of introns and/or exons during alternative splicing. (B) The 3' splice sites of introns A, B, and C of the GnRH gene. Underlines indicate the pyrimidine tract. Purine bases in the pyrimidine tract were indicated in bold. Boxes represent the putative branch point site (YNRAY). Note that the putative branch point sites are within the pyrimidine tract of intron A of the mouse, rat and human GnRH gene.

level expression, GnRH transcripts are also found in the peripheral tissues including reproductive and immune-related tissues, and in some discrete brain regions such as olfactory bulb and piriform cortex (Radovick et al., 1990; Azad et al., 1991; Pagesy et al., 1992; Choi et al., 1994). Interestingly, RNA processing in extra-hypothalamic tissues appears to differ from that observed in the hypothalamus. It is noteworthy that the GnRH cDNA from the human placenta contains the first intron (intron A) (Seeburg and Adelman, 1984). Subsequent studies demonstrated that non-hypothalamic tissues, such as the placenta and reproductive organs prevalently express the intron A-containing RNA species than the mature mRNA (Radovick et al., 1990; Dong et al., 1996; Seong et al., 1999). Moreover, a GnRH transcript lacking exon 2 has been found in Gn11 cells and immune cells (Wilson et al., 1995; Zhen et al., 1997). Thus far, all of the three RNA species (intron A-retained, exon 2-skipped, and mature mRNA) have been shown to be produced from a single GnRH gene (Fig. 1A). However, the mechanism for producing three transcripts has been poorly understood. In order to better understand the molecular events, Roberts and his colleagues initiated an investigation of the GnRH pre-mRNA splicing mechanism using a sensitive RNase protection assay and RT-PCR (Jakubowski and Roberts, 1994). They found relatively high level of the intron A-containing RNA species, but low level of the intron B- and intron C-containing RNA species in the nuclear fraction, indicating the notion that introns B and C are rapidly removed from the primary transcript, while intron A is slowly excised. Consistent with this result, we recently demonstrated that, using an in vitro RNA splicing system, intron A could not be removed, while introns B and C were easily excised (Fig. 2) (Seong et al., 1999). These findings indicate that intron A excision is a rate limiting step for GnRH pre-mRNA splicing.

The cis-elements essential for intron excision are located in the exon–intron joint. At the downstream side of an exon, the 5′ splice site in higher eukaryotes conforms to the consensus sequence, AG/GURAGU (the splice site is denoted by a slash and invariant nucleotides are underlined; R = purine, Y = pyrimidine, N = any nucleotide). At the upstream side of the exon, the 3′ splice site conforms to consensus,

$Y_{(n)}AG/G$ in most vertebrate introns (Shapiro and Senapathy, 1987). The branch point site (BPS) is usually located at a distance of 18 to 40 nucleotides upstream the 3′ splice site with very loose consensus, YNYRAY (the site of branch formation is underlined) (Berglund et al., 1997). The splice sites are catalyzed by numerous proteins in a macromolecular complex, the spliceosome complex that consists of small nuclear ribonucleoproteins (snRNPs), heteronuclear RNPs (hnRNPs), and a large family of serine/arginine-rich proteins (SR proteins) (Zuo and Maniatis, 1996). The spliceosome complex would barely interact with the suboptimal splice site, while strongly interacting with well-conserved splice sites. Thus, it is proposed that the attenuation of intron A excision from the GnRH pre-mRNA is most likely due to suboptimal splice sites of intron A. Sequence analysis of the 3′ splice sites of introns A, B, and C of the GnRH gene revealed that the 3′ splice site of intron A contains many purines in the pyrimidine tract, whereas 3′ splice sites of introns B and C are highly conserved (Seong et al., 1999) as other normal 3′ splice sites (Shapiro and Senapathy, 1987). Moreover, a putative BPS is located within the pyrimidine tract but not in the upstream side of the pyrimidine tract (Fig. 1B). The 3′ spliceosome complex formation begins with U2 auxiliary factor (U2AF) recognizing the pyrimidine bases in the pyrimidine tract, thereby U2 snRNP binds to the 3′ splice site. However, many purines in the pyrimidine tract could interfere with binding to U2AF (Lavigueur et al., 1993; Tian and Maniatis, 1994). The BPS is usually located upstream the pyrimidine tract and recognized by branchpoint bridging protein (BBP) (Berglund et al., 1997). BPS within the pyrimidine tract may cause a steric hindrance of BBP and U2AF binding to the pyrimidine tract. Thus, the presence of many purine bases in the pyrimidine tract and the abnormal location of BPS appears to be responsible for the attenuation of intron A excision. This possibility is strongly supported by the observations that either mutations of purines to pyrimidines or translocation of the putative BPS to upstream of the pyrimidine tract significantly increased the splicing rate (Seong et al., 1999). The increased splicing activities caused by the mutation of the suboptimal site were also observed in growth hormone (GH) and doublesex (dsx) pre-mRNA splicing (Tian and Maniatis, 1994;

212

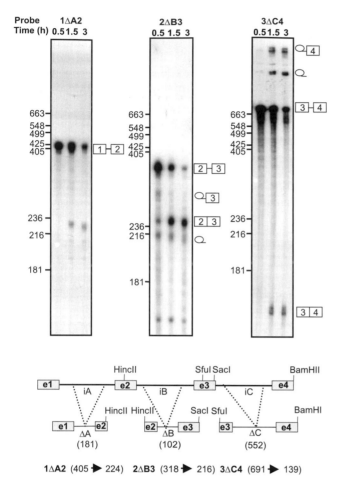

Fig. 2. Excision rate of rat GnRH introns in vitro. The ^{32}P-labeled RNA substrates consisting of each GnRH intron and its neighboring exons were synthesized by in vitro transcription. The RNAs purified from the 6% polyacrylamide gels containing 8 M urea were incubated with HeLa NE. The products were electrophoresed on the 6% polyacrylamide gels containing 8 M urea and then subjected to autoradiography. RNA substrates are shown at the *top* of the gels and the time of incubation is specified at the *top* of each lane. Illustrated RNA structures on the *right sides* of gels were identified by RNA size markers and by running on higher percentage gels. Spliced RNA (exon–exon) was confirmed by RT-PCR using gel-purified RNAs. Other bands in this and following figures are cryptic splicing intermediate(s) and denoted with asterisks. The *schematic diagram on the bottom panel* shows the restriction map of the GnRH gene and RNA substrates. The sizes of substrates containing each intron and their spliced products are shown in *parentheses*.

Dirksen et al., 1995). Although the 5′ splice site of intron A (GUAAAA) is not well conserved as other normal 5′ splice sites (GURAGU), it seems not to affect intron A excision since the mutation of the 5′ splice site of intron A toward consensus did not significantly increase intron A excision (Seong et al., 1999). Exon 2 skipping from the GnRH pre-mRNA could be explained by the same context as intron A retention. Since the 3′ spliceosome complex could not be formed on intron A, a 3′ spliceosome complex that is formed on the 3′ splice site of intron B can

directly interact with the 5′ spliceosome complex on the 5′ splice site of intron A, which allows exon 2 to be removed.

Exonic splicing enhancers (ESEs) in GnRH exons 3 and 4

The presence of a 3′ suboptimal splice site in intron A raises a fundamental question of how GnRH-producing cells generate the mature mRNA. It is well known that the suboptimal splice site could be actively recognized by the 3′ spliceosome complex in the presence of additional positive exon element, such as purine-rich enhancer (Tian and Maniatis, 1992; Sun et al., 1993; Tanaka et al., 1994). Recently, list of enhancers has been extended to include the AC-rich sequences (Coulter et al., 1997) and certain intronic sequences (Modafferi and Black, 1997). In addition to their role in constitutive pre-mRNA splicing, SR proteins are known to be involved in enhancer-dependent splicing by binding to their specific enhancer element (Tian and Maniatis, 1994; Liu et al., 1998; Schaal and Maniatis, 1999). Thus, it is postulated that the presence of additional enhancer elements in the GnRH pre-mRNA and a specific splicing factor(s) acting on the enhancer elements is a prerequisite for achieving the mature GnRH mRNA in GnRH-producing cells.

Sequence analysis of GnRH exons reveals that two purine-rich regions are located in exon 3 and exon 4 (Seong et al., 1999). The sequence of purine rich regions in GnRH exons exhibits an extremely high degree of sequence similarity among mouse, rat, and human (Fig. 3A). Purine-rich regions in exons 3 and 4 exhibit alternative repeats of guanosine (G) and adenosine (A) bases, like GAR (R = purine) or AGR. Concerning rapid and efficient excisions of introns B and C, it is postulated that exons 3 and 4 can join to exon 2, producing the 1A234 RNA species. Then, the majority of the 1A234 RNA species undergo further processing of intron A excision with the help of exonic splicing enhancers (ESEs) located in exons 3 and 4. This possibility is strongly supported by the finding that, in an in vitro RNA splicing system, intron A is partially excised when exon 3 and exon 4 are linked up with exon 2, while intron A was not excised at all in the presence of exon 2 alone. The partial excision of intron A from the 1A234 construct is attributable to the distance of the purine-rich sequences in exons 3 and 4 from the 3′ splice site of intron A. The purine-rich sequences show a strong enhancing activity when they are located close to the 3′ splice site of intron A (Fig. 3B). Thus, the closer the purine-rich sequence to the 3′ splice site of intron A, the better the splicing activity (Seong et al., 1999), which is a common feature of ESE found in other pre-mRNA splicing (Tanaka et al., 1994). Mutations in the purine-rich sequences in exons 3 and 4 decreased pre-mRNA splicing activity, while

mutation in the purine-rich sequence in exon 2 did not affect the splicing activity. These results suggest that the purine-rich sequences in exons 3 and 4 are functionally active ESEs, which are denoted ESE3 and ESE4, respectively (Seong et al., 1999). ESE4 consists of three purine repeats separated by two spacers and a putative hairpin constructing sequence. Analysis of twelve ESE4 mutants suggest that there are at least three parameters affecting intron A excision (Han et al., 2001). First, the amount of the purine residues in the purine-rich region is critical for the enhanced splicing activity. The purine-rich sequence in the middle of exon 2 contains 14 purines while ESE3 and ESE4 contains 22 and 30 purines, respectively. Enhancing activity of ESE4 was greater than ESE3, and the purine-rich sequence in exon 2 did not serve as an enhancer. The decrease in purine number in ESE4 significantly reduced enhancing activity. Thus, a certain number of purine bases in ESE is required for maintaining the enhancing activity. Second, pyrimidine residues between two long continuous purine repeats seem to be stimulatory. An increase in splicing activity was observed when a longer pyrimidine spacer region between two long purine repeats was created by mutations of three purines in the spacer to pyrimidines. This result indicates that an optimal alignment of the purine repeats and pyrimidine spacer can improve the binding affinity to a certain splicing factor(s). Third, pyrimidine residues within the continuous purine repeats are deleterious, especially when they are located very close to 3′ splice site. Mutations of purines in polypurine repeats to pyrimidines greatly reduced splicing activity, indicating that a certain number of purines (at least 6–8 purines) should be continuously repeated without interruptions by pyrimidine for enhanced splicing activity.

Presence of GnRH neuron-specific splicing factors acting on ESE4

Alternative pre-mRNA splicing relies on a large family of SR proteins. Up-to-date, more than ten SR proteins, SRp20, ASF/SF, SC35, SRp30, SRp40, SRp55, SRp75, 9G8, Tra, and Tra2 have been identified (Manley and Tacke, 1996). SR proteins are characterized by a modular composition, comprised of one or more RNA recognition motifs (RRM) and

A

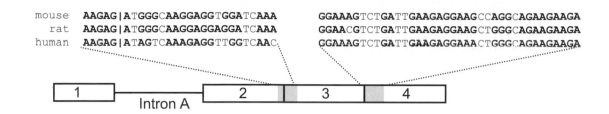

mouse   **AAGAG|ATGGGCAAGGAGG**TGGATCAAA   **GGAAAG**TCTGATT**GAAGAGGAAG**CCAGGC**AGAAGAAGA**
rat   **AAGAG|ATGGGCAAGGAGGAGGA**TCAAA   **GGAAC**GTCTGATT**GAAGAGGAAG**CTGGGC**AGAAGAAGA**
human   **AAGAG|ATAGTCAAAGAGG**TTGGTCAAC.   **GGAAAG**TCTGATT**GAAGAGGAAA**CTGGGC**AGAAGAAGA**

B

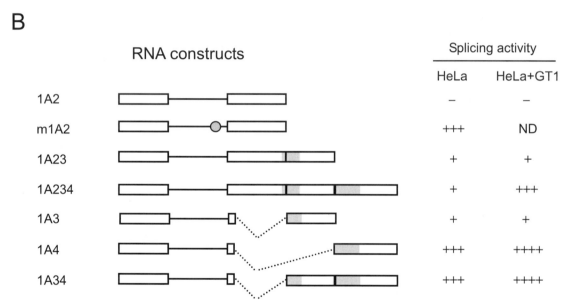

Fig. 3. Exonic splicing enhancers (ESEs) and their role in GnRH pre-mRNA splicing. (A) Schematic representation of the GnRH pre-mRNA and ESEs. Gray boxes indicate exonic splicing enhancers (ESEs) in GnRH exons 3 and 4, and their sequences are shown above the RNA structure. Purine bases are shown in bold. The vertical bars indicate the border of exons 2 and 3. The purine-rich sequences of the mouse, rat, and human GnRH gene are well conserved. (B) Splicing activities of the GnRH pre-mRNA in the presence of ESE. RNA constructs are shown in the left and their splicing activities in HeLa or HeLa + GT1 nuclear extracts are shown in the right. Splicing activity indicates % splicing; splice product/(spliced product + precursor RNA). Each + indicates the about 10% splicing activity. Splicing activity was observed in the presence of ESEs or when 3′ spliced site was mutated to the consensus (m1A2 construct). It should be noted that GT1 nuclear extract further increased splicing activity only in the presence of ESE in exon 4.

SR domain in which arginine and serine residues are repeated. It has been proposed that SR proteins can bind to each other using their SR domains, allowing protein–protein interaction to form networks of SR proteins (Lynch and Maniatis, 1996). SR proteins are expressed in all tissues (Zahler et al., 1993) and can complement splicing-deficient nuclear extracts, indicating that they are essential splicing factors for general or constitutive splicing (Zahler et al., 1992).

However, they appear to be equally important in alternative splicing being able to modulate selection of alternative splice sites in a concentration-dependent manner (Mayeda and Krainer, 1992; Caceres et al., 1994; Hanamura et al., 1998). Although earlier studies suggest that SR proteins might be functionally redundant, recent studies indicate that each protein probably performs at least some non-redundant functions (Tacke and Manley, 1999). Evidence for divergent RNA binding specificity among SR proteins has been achieved by extensive application of SE-LEX (systemic evolution of ligands by exponential enrichment), which allows the identification of high-affinity binding sites from pools of random RNA sequence (Liu et al., 1998; Schaal and Maniatis, 1999).

In a HeLa nuclear extract (NE), the 1A234 pre-mRNA exhibited a partial splicing activity since ESEs are located at a far distance from the 3′ splice site of intron A. Interestingly, addition of GT1 (GnRH-producing cell) NE further increased excision rate of intron A from the 1A234 pre-mRNA in a dose-dependent manner, while addition of KK1 (non-GnRH-producing cell) NE rather decreased it (Seong et al., 2002). These results are consistent with the in vivo experiments in which the most efficient excision of intron A was found in GT1 cells and the POA, while intron A-containing transcript was abundant in extrahypothalamic regions of the brain and peripheral tissues (Seong et al., 1999). There appears to be a GnRH neuron-specific *trans*-acting factor(s) interacting with ESE4. It is of importance to note that most splicing enhancers are active when located within 100 bases from the 3′ splice site. However, ESE4 is located at 238 bases downstream the 3′ splice site of intron A when exons 3 and 4 are linked up with the exon 2. Indeed, full enhancing activity of ESE4 was observed when ESE4 is close to the 3′ splice site of intron A. This finding raises a question of how such an efficient splicing activity of the 1A234 pre-mRNA is achieved in GnRH-producing cells. One good example accounting for the unusually increased splicing activity of the 1A234 pre-mRNA is the *Drosphila melanogaster* dsx pre-mRNA splicing mechanism. The dsx pre-mRNA has the dsx repeat element(dsxRE) that normally functions when located at 300 bases downstream of the 3′ splice site in the presence of some specific splicing factors, Tra and Tra2 (Lynch and Maniatis, 1995). In the absence of Tra/Tra2, these elements are not active when located at a long distance from the 3′ splice site. Tra/Tra2 may form a stable complex with dsxRE, recruiting a certain splicing factor to allow the enhancer to function at this distance (Lynch and Maniatis, 1995, 1996). Recently, two human homologs of Tra2, Tra2α (Dauwalder et al., 1996) and Tra2β (Beil et al., 1997) have been

identified. Interestingly, the purified Tra2 proteins can bind preferentially to RNA sequences containing GAA repeats (Tacke et al., 1998) that are also seen in ESE4 of the GnRH pre-mRNA. Northern blot analysis for mouse Tra2α indicates that the POA and GT1 cells express Tra2α mRNA, suggesting that Tra2α may participate in enhanced splicing activity of the 1A234 pre-mRNA. Indeed, we observed that purified GST-tagged Tra2α protein actively bound to the ESE4 sequence, while several other splicing factors did not (Seong et al., 2002). Moreover, Tra2α did not bind to mutant ESEs in which purines were changed to pyrimidines. This result indicates that Tra2α specifically binds to purine stretches in the ESE4 sequence, and then probably forms a stable complex with other SR protein(s) or GnRH neuron-specific splicing factor(s), allowing ESE4 to help intron A excision at a long distance. It is notable that enhanced splicing of the GnRH pre-mRNA was observed when a nuclear fraction of GnRH neurons that was precipitated at the saturation of 40–50% ammonium sulfate (ASP40–50) was added to splicing reaction (Seong et al., 2002). ASP40–50 fraction significantly increased the excision rate of intron A in the presence of HeLa NE or SR protein-rich fraction. However, ASP40–50 fraction alone could not remove intron A. Activity of ASP40–50 fraction was much greater in RNA constructs in which ESE4 is far from the 3′ splice site of intron A than in RNA constructs in which ESE4 is close to the 3′ splice site of intron A or RNA construct which is constitutively active in HeLa NE. This result suggests that a splicing factor(s) in ASP40–50 fraction is not necessary for constitutive RNA splicing but required for splicing of the pre-mRNA in which ESE is far from the 3′ splice site. A splicing factor(s) in ASP40–50 fraction seems not to bind ESE4 directly as revealed by UV cross-linking and affinity chromatography studies. This result indicates the splicing factor(s) in ASP40–50 fraction is a cofactor protein(s) that is likely to interact with the SR protein, probably Tra2α that is already bound to ESE4. We recently performed a yeast two hybrid assay using Tra2α as a bait from the GT1-1 cDNA library. One of the SR proteins, SRp30c was isolated and confirmed to bind to Tra2α. Therefore, such an interaction appears to be important for the GnRH neuron-specific enhancement of GnRH pre-mRNA splicing as schematically presented in Fig. 4.

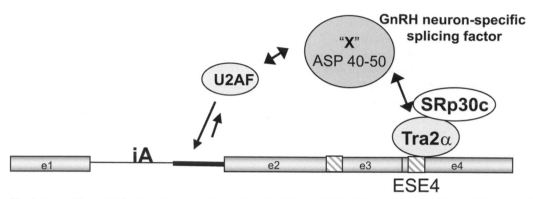

Fig. 4. A possible model for the enhanced splicing of the GnRH pre-mRNA. Efficient splicing of the GnRH pre-mRNA may be mediated by an interaction of ESE4 and ESE4-recognizing splicing factors in GnRH neurons. A 40 kDa protein, Tra2α specifically binds to ESE4. A cofactor protein (X) in ASP40–50 fraction seems to interact with Tra2α that is already bound to ESE4. Interaction of a cofactor protein(s) in ASP40–50 fraction with ESE4–Tra2α complex may recruit U2AF to recognize suboptimal 3′ splice site of intron A, leading to an enhancement of GnRH pre-mRNA splicing. In addition SRp30c interacts with Tra2α that is already bound to ESE4. This interaction of SRp30c with ESE4–Tra2α complex also seems to recruit U2AF to recognize suboptimal 3′ splice site of intron A, leading to an enhancement of GnRH pre-mRNA splicing.

Developmental changes in GnRH pre-mRNA splicing

Recent studies have demonstrated that hormonal inputs alter the splicing pattern of a variety of genes such as potassium channel, activin receptor II, myosin II heavy chain-B, dopamine D2 receptor, and GnRH receptor (Kawamoto, 1996; Cowley et al., 1998; Guivarc'h et al., 1998; Shoji et al., 1998; Xie and McCobb, 1998). Alterations in GnRH primary transcript level during mouse embryonic and postnatal development were not accompanied by similar changes in GnRH mRNA level (Gore et al., 1999). Similarly, uncoupling of transcription and posttranscription process of GnRH transcripts by steroids or neural inputs was also observed (Gore and Roberts, 1995). A potential explanation for the uncoupling of transcription and posttranscription, it has been suggested that the changes in splicing rate of the GnRH pre-mRNA may be, at least in part, involved in this process. Recently, we demonstrated that the excision rate of intron A in the POA was increased during postnatal development of normal mice, which was accompanied by an increase in mature GnRH mRNA levels (Seong et al., 2001). The intron A excision rate in the POA was significantly increased in 3-week-old mice and further increased until adulthood. In contrast, in the cortex (CTX), intron A excision rate was extremely low, further decreased

in 3-week-old mice, and remained at very low levels until adulthood. Excision rate of introns B and C in the POA was not significantly changed during development. GnRH mRNA levels in the POA gradually increased during development, similarly to the intron A excision rate. The developmental changes in intron A excision rate indicates developmental maturation of the splicing machinery to increase splicing rate of the GnRH pre-mRNA in GnRH neurons. Interestingly, mRNA levels for Tra2 can be altered by neural activation or hormonal input (Daoud et al., 1999). This finding suggests that changes in splicing regulator protein levels may affect the splicing patterns of the downstream pre-mRNAs (Stamm et al., 1999), raising the possibility that a variety of signaling may affect the expression of splicing factors in GnRH neurons, thereby allowing splicing pattern of the GnRH pre-mRNA to be altered.

Biological relevance of efficient intron A excision: analysis of hpg mouse

Splice variant transcripts produced by alternative splicing often have a distinct role compared with wild type transcripts. An alternative splice variant of growth hormone-releasing hormone (GHRH) transcripts generates a new C-terminal peptide (Perez-Riba et al., 1997). Fourth intron retention of growth hormone (GH) transcripts produces an isoform of

GH (Yamamoto et al., 1998). Alternative transcripts of nitric oxide synthase (NOS) having different 5′ untranslated regions (5′ UTR) exhibit differential translation efficiency (Wang et al., 1999). So, what is the physiological relevance of alternative splicing of the GnRH pre-mRNA? Hypogonadal (hpg) mice may provide a clue for such a question. Hpg mice are infertile due to lack of the GnRH biosynthesis in the hypothalamus (Mason et al., 1986; Seeburg et al., 1987). Hpg mice have a natural genetic defect of the GnRH gene of which exons 3 and 4 are truncated (Mason et al., 1986). Since no truncation or mutations in the GnRH promoter sequence has been observed in hpg mice and GnRH exon 2 that encodes a signal peptide, GnRH decapeptide, and a part of GnRH-associated peptide (GAP) is intact in hpg mice, despite the lack of exons 3 and 4, it could be postulated that hpg mice could produce the GnRH peptide. However, little GnRH is detectable in hpg mice (Mason et al., 1986). Thus, the reason for the low production of GnRH in the hpg mice is poorly understood. Recently, we found that intron A excision rate in the POA of hpg mice was severely lower than that of normal mice (Seong et al., 2001). Such an attenuation of intron A excision in hpg mice is most likely due to the lack of GnRH exons 3 and 4 in which ESEs are located. Our in vitro splicing assay using either HeLa or GT1 NE demonstrated that excision of intron A took place in the presence of exons 3 and 4, while exon 2 alone marginally affected the intron A excision (Seong et al., 1999). Thus, it may be proposed that low excision rate of intron A results in a larger amount of the intron A-retained transcript than the mature mRNA, and that the intron A-retained transcript has a severe defect in translation capacity, eventually leading to little biosynthesis of the GnRH peptide. In fact, sequence analysis reveals that intron A provides many AUGs and at least five conserved Kozak sequences (Kozak, 1987), suggesting the existence of multiple short open reading frames. The presence of multiple AUGs or Kozak sequences in intron A may, in part, interfere with the translation efficiency. In support of this idea, an insertion of the intron A sequence into the upstream region of the luciferase gene markedly decreased translation rate without affecting transcription rate (Seong et al., 2001). In addition to the intron A-retained RNA species, an alternative splicing variant form of the GnRH transcript lacking exon 2 was found in the mouse olfactory, hypothalamus and immortalized GnRH cell lines (Zhen et al., 1997). Since exon 2 encodes the GnRH peptide, this RNA species also cannot produce the GnRH peptide. Thus, either intron A-retained or exon 2-skipped RNA species has a defect in producing the GnRH peptide.

In conclusion, in addition to transcription, translation, and posttranslational regulation, the efficient and accurate processing of the GnRH pre-mRNA is critical for tissue-specific and/or developmental stage-specific GnRH expression, thereby maintaining normal function of GnRH neurons. Identification of GnRH neuron-specific splicing factors that may interact with ESE provides important insights into the GnRH gene regulation mechanism.

Acknowledgements

The present study was supported by the Ministry of Science and Technology through Korea Brain Science Program and the National Research Laboratory (2000-N-NL-01-C-149). J. Han, and G.H. Son are recipient for the post-doctoral and pre-doctoral fellowship from Brain Korea 21 of the Ministry of Education of Korea, respectively.

References

Adelman, J.P., Mason, A.J., Hayflick, J.S. and Seeburg, P.H. (1986) Isolation of the gene and hypothalamic cDNA for the common precursor of gonadotropin-releasing hormone and prolactin release-inhibiting factor in human and rat. *Proc. Natl. Acad. Sci. USA*, 83: 179–183.

Azad, N., Emanuele, N.V., Halloran, M.M., Tentler, J. and Kelley, M.R. (1991) Presence of luteinizing hormone-releasing hormone (LHRH) mRNA in rat spleen lymphocytes. *Endocrinology*, 128: 1679–1681.

Beil, B., Screaton, G. and Stamm, S. (1997) Molecular cloning of htra2-beta-1 and htra2-beta-2, two human homologs of tra-2 generated by alternative splicing. *DNA Cell Biol.*, 16: 679–690.

Bond, C.T., Hayflick, J.S., Seeburg, P.H. and Adelman, J.P. (1989) The rat gonadotropin-releasing hormone: SH locus: structure and hypothalamic expression. *Mol. Endocrinol.*, 3: 1257–1262.

Berglund, J.A., Chua, K., Abovich, N., Reed, R. and Rosbash, M. (1997) The splicing factor BBP interacts specifically with the pre-mRNA branchpoint sequence UACUAAC. *Cell*, 89: 781–787.

Caceres, J.F., Stamm, S., Helfman, D.M. and Krainer, A.R.

(1994) Regulation of alternative splicing in vivo by overexpression of antagonistic splicing factors. *Science*, 265: 1706–1709.

Choi, W.S., Kim, M.O., Lee, B.J., Kim, J.H., Sun, W., Seong, J.Y. and Kim, K. (1994) Presence of gonadotropin-releasing hormone mRNA in the rat olfactory piriform cortex. *Brain Res.*, 648: 148–151.

Coulter, L.R., Landree, M.A. and Cooper, T.A. (1997) Identification of a new class of exonic splicing enhancers by in vivo selection. *Mol. Cell. Biol.*, 17: 2143–2150.

Cowley, M.A., Rao, A., Wright, P.J., Illing, N., Millar, R.P. and Clarke, I.J. (1998) Evidence for differential regulation of multiple transcripts of the gonadotropin releasing hormone receptor in the ovine pituitary gland; effect of estrogen. *Mol. Cell. Endocrinol.*, 146: 141–149.

Daoud, R., Da Penha Berzaghi, M., Siedler, F., Hubener, M. and Stamm, S. (1999) Activity-dependent regulation of alternative splicing patterns in the rat brain. *Eur. J. Neurosci.*, 11: 788–802.

Dauwalder, B., Amaya-Manzanares, F. and Mattox, W. (1996) A human homologue of the Drosophila sex determination factor transformer-2 has conserved splicing regulatory functions. *Proc. Natl. Acad. Sci. USA*, 93: 9004–9009.

Dirksen, W.P., Sun, Q. and Rottman, F.M. (1995) Multiple splicing signals control alternative intron retention of bovine growth hormone pre-mRNA. *J. Biol. Chem.*, 270: 5346–5352.

Dong, K.W., Duval, P., Zeng, Z., Gordon, K., Williams, R.F., Hodgen, G.D., Jones, G., Kerdelhue, B. and Roberts, J.L. (1996) Multiple transcription start sites for the GnRH gene in rhesus and cynomolgus monkeys: a non-human primate model for studying GnRH gene regulation. *Mol. Cell. Endocrinol.*, 117: 121–130.

Gore, A.C. and Roberts, J.L. (1995) Regulation of gonadotropin-releasing hormone gene expression in the rat during the luteinizing hormone surge. *Endocrinology*, 136: 889–896.

Gore, A.C. and Roberts, J.L. (1997) Regulation of gonadotropin-releasing hormone gene expression in vivo and in vitro. *Front. Neuroendocrinol.*, 18: 209–245.

Gore, A.C., Yeo, T.T., Ho, A. and Roberts, J.L. (1997) Post-transcriptional regulation of the gonadotropin-releasing hormone gene in GT1-7 cells. *J. Neuroendocrinol.*, 9: 271–277.

Gore, A.C., Roberts, J.L. and Gibson, M.J. (1999) Mechanisms for the regulation of gonadotropin-releasing hormone gene expression in the developing mouse. *Endocrinology*, 140: 2280–2287.

Guivarc'h, D., Vincent, J.D. and Vernier, P. (1998) Alternative splicing of the D2 dopamine receptor messenger ribonucleic acid is modulated by activated sex steroid receptors in the MMQ prolactin cell line. *Endocrinology*, 139: 4213–4221.

Han, J., Seong, J.Y., Kim, K., Wuttke, W. and Jarry, H. (2001) Analysis of exonic splicing enhancers in the mouse gonadotropin-releasing hormone (GnRH) gene. *Mol. Cell. Endocrinol.*, 173: 157–166.

Hanamura, A., Caceres, J.F., Mayeda, A., Franza, B.R. Jr. and Krainer, A.R. (1998) Regulated tissue-specific expression of antagonistic pre-mRNA splicing factors. *RNA*, 4: 430–444.

Hollmann, M., Boulter, J., Maron, C., Beasley, L., Sullivan, J.,

Pecht, G. and Heinemann, S. (1993) Zinc potentiates agonist-induced currents at certain splice variants of the NMDA receptor. *Neuron*, 10: 943–954.

Jakubowski, M. and Roberts, J.L. (1994) Processing of gonadotropin-releasing hormone gene transcripts in the rat brain. *J. Biol. Chem.*, 269: 4078–4083.

Kawamoto, S. (1996) Neuron-specific alternative splicing of nonmuscle myosin II heavy chain-B pre-mRNA requires a cis-acting intron sequence. *J. Biol. Chem.*, 271: 17613–17616.

Kilpatrick, G.J., Dautzenberg, F.M., Martin, G.R. and Eglen, R.M. (1999) 7TM receptors: the splicing on the cake. *Trends Pharmacol. Sci.*, 20: 294–301.

Kozak, M. (1987) At least six nucleotides preceding the AUG initiator codon enhance translation in mammalian cells. *J. Mol. Biol.*, 196: 947–950.

Lavigueur, A., La Branche, H., Kornblihtt, A.R. and Chabot, B. (1993) A splicing enhancer in the human fibronectin alternate ED1 exon interacts with SR proteins and stimulates U2 snRNP binding. *Genes Dev.*, 7: 2405–2417.

Liu, H.X., Zhang, M. and Krainer, A.R. (1998) Identification of functional exonic splicing enhancer motifs recognized by individual SR proteins. *Genes Dev.*, 12: 1998–2012.

Lopez, A.J. (1998) Alternative splicing of pre-mRNA: developmental consequences and mechanisms of regulation. *Annu. Rev. Genet.*, 32: 279–305.

Lou, H. and Gagel, R.F. (2001) Alternative ribonucleic acid processing in endocrine systems. *Endocr. Rev.*, 22: 205–225.

Lynch, K.W. and Maniatis, T. (1995) Synergistic interactions between two distinct elements of a regulated splicing enhancer. *Genes Dev.*, 9: 284–293.

Lynch, K.W. and Maniatis, T. (1996) Assembly of specific SR protein complexes on distinct regulatory elements of the Drosophila doublesex splicing enhancer. *Genes Dev.*, 10: 2089–2101.

Manley, J.L. and Tacke, R. (1996) SR proteins and splicing control. *Genes Dev.*, 10: 1569–1579.

Mason, A.J., Hayflick, J.S., Zoeller, R.T., Young, W.S., Phillips, H.S., Nikolics, K. and Seeburg, P.H. (1986) A deletion truncating the gonadotropin-releasing hormone gene is responsible for hypogonadism in the hpg mouse. *Science*, 234: 1366–1371.

Mayeda, A. and Krainer, A.R. (1992) Regulation of alternative pre-mRNA splicing by hnRNP A1 and splicing factor SF2. *Cell*, 68: 365–375.

Modafferi, E.F. and Black, D.L. (1997) A complex intronic splicing enhancer from the c-src pre-mRNA activates inclusion of a heterologous exon. *Mol. Cell. Biol.*, 17: 6537–6545.

Pagesy, P., Li, J.Y., Berthet, M. and Peillon, F. (1992) Evidence of gonadotropin-releasing hormone mRNA in the rat anterior pituitary. *Mol. Endocrinol.*, 6: 523–528.

Perez-Riba, M., Gonzalez-Crespo, S. and Boronat, A. (1997) Differential splicing of the growth hormone-releasing hormone gene in rat placenta generates a novel pre-proGHRH mRNA that encodes a different C-terminal flanking peptide. *FEBS Lett.*, 402: 273–276.

Radovick, S., Wondisford, F.E., Nakayama, Y., Yamada, M., Cutler, G.B. and Weintraub, B.D. (1990) Isolation and characterization of the human gonadotropin-releasing hormone gene in the hypothalamus and placenta. *Mol. Endocrinol.*, 4: 476–480.

Schaal, T.D. and Maniatis, T. (1999) Selection and characterization of pre-mRNA splicing enhancers: identification of novel SR protein-specific enhancer sequences. *Mol. Cell. Biol.*, 19: 1705–1719.

Seeburg, P.H. and Adelman, J.P. (1984) Characterization of cDNA for precursor of human luteinizing hormone releasing hormone. *Nature*, 311: 666–668.

Seeburg, P.H., Mason, A.J., Stewart, T.A. and Nikolics, K. (1987) The mammalian GnRH gene and its pivotal role in reproduction. *Recent Prog. Horm. Res.*, 43: 69–98

Seong, J.Y., Park, S. and Kim, K. (1999) Enhanced splicing of the first intron from the gonadotropin-releasing hormone (GnRH) primary transcript is a prerequisite for mature GnRH messenger RNA: presence of GnRH neuron-specific splicing factors. *Mol. Endocrinol.*, 13: 1882–1895.

Seong, J.Y., Kim, B.W., Park, S., Son, G.H. and Kim, K. (2001) First intron excision of GnRH pre-mRNA during postnatal development of normal mice and adult hypogonadal mice. *Endocrinology*, 142: 4454–4461.

Seong, J.Y., Han, J., Park, S., Wuttke, W., Jarry, H. and Kim, K. (2002) Exonic splicing enhancer-dependent splicing of the GnRH pre-mRNA is mediated by Tra2α, a 40 kDa serine/arginine-rich (SR) protein. *Mol. Endocrinol.*, in press.

Shapiro, M.B. and Senapathy, P. (1987) RNA splice junctions of different classes of eukaryotes: sequence statistics and functional implications in gene expression. *Nucleic Acids Res.*, 15: 7155–7174.

Shoji, H., Nakamura, T., van den Eijnden-van Raaij, A.J. and Sugino, H. (1998) Identification of a novel type II activin receptor, type IIA-N, induced during the neural differentiation of murine P19 embryonal carcinoma cells. *Biochem. Biophys. Res. Commun.*, 246: 320–324.

Stamm, S., Casper, D., Hanson, V. and Helfman, D.M. (1999) Regulation of the neuron-specific exon of clathrin light chain B. *Brain Res. Mol. Brain. Res.*, 64: 108–118.

Sun, Q., Hampson, R.K. and Rottman, F.M. (1993) In vitro analysis of bovine growth hormone pre-mRNA alternative splicing. Involvement of exon sequences and trans-acting factor(s). *J. Biol. Chem.*, 268: 15659–15666.

Tacke, R. and Manley, J.L. (1999) Determinants of SR protein specificity. *Curr. Opin. Cell. Biol.*, 11: 358–362.

Tacke, R., Tohyama, M., Ogawa, S. and Manley, J.L. (1998) Human Tra2 proteins are sequence-specific activators of pre-mRNA splicing. *Cell*, 93: 139–148.

Tanaka, K., Watakabe, A. and Shimura, Y. (1994) Polypurine sequences within a downstream exon function as a splicing enhancer. *Mol. Cell. Biol.*, 14: 1347–1354.

Tian, M. and Maniatis, T. (1992) Positive control of pre-mRNA splicing in vitro. *Science*, 256: 237–240.

Tian, M. and Maniatis, T. (1994) A splicing enhancer exhibits both constitutive and regulated activities. *Genes Dev.*, 8: 1703–1712.

Wang, Y., Newton, D.C., Robb, G.B., Kau, C.L., Miller, T.L., Cheung, A.H., Hall, A.V., VanDamme, S., Wilcox, J.N. and Marsden, P.A. (1999) RNA diversity has profound effects on the translation of neuronal nitric oxide synthase. *Proc. Natl. Acad. Sci. USA*, 96: 12150–12155.

Wilson, T.M., Yu-Lee, L.Y. and Kelley, M.R. (1995) Coordinate gene expression of luteinizing hormone-releasing hormone (LHRH) and the LHRH-receptor after prolactin stimulation in the rat Nb2 T-cell line: implications for a role in immunomodulation and cell cycle gene expression. *Mol. Endocrinol.*, 9: 44–53.

Xie, J. and McCobb, D.P. (1998) Control of alternative splicing of potassium channels by stress hormones. *Science*, 280: 443–446.

Yamamoto, K., Hashimoto, H., Hagihara, N., Nishino, A., Fujita, T., Matsuda, T. and Baba, A. (1998) Cloning and characterization of the mouse pituitary adenylate cyclase-activating polypeptide (PACAP) gene. *Gene*, 211: 63–69.

Zahler, A.M., Lane, W.S., Stolk, J.A. and Roth, M.B. (1992) SR proteins: a conserved family of pre-mRNA splicing factors. *Genes Dev.*, 6: 837–847.

Zahler, A.M., Neugebauer, K.M., Lane, W.S. and Roth, M.B. (1993) Distinct functions of SR proteins in alternative pre-mRNA splicing. *Science*, 260: 219–222.

Zhen, S., Dunn, I.C., Wray, S., Liu, Y., Chappell, P.E., Levine, J.E. and Radovick, S. (1997) An alternative gonadotropin-releasing hormone (GnRH) RNA splicing product found in cultured GnRH neurons and mouse hypothalamus. *J. Biol. Chem.*, 272: 12620–12625.

Zuo, P. and Maniatis, T. (1996) The splicing factor U2AF35 mediates critical protein–protein interactions in constitutive and enhancer-dependent splicing. *Genes Dev.*, 10: 1356–1368.

I.S. Parhar (Ed.)
Progress in Brain Research, Vol. 141
© 2002 Published by Elsevier Science B.V.

CHAPTER 17

Pro-GnRH processing

William C. Wetsel *, Sudha Srinivasan

Departments of Psychiatry and Behavioral Sciences, Cell Biology, and Medicine (Endocrinology), Mouse Behavioral and Neuroendocrine Analysis Core Facility, Duke University Medical Center, Durham, NC 27710, USA

Introduction

The anterior pituitary has long been known to regulate stress, reproduction, lactation, growth, and metabolism, while the posterior pituitary controls osmolarity and fluid volume, milk ejection from the breast, and uterine contractions in labor (Reeves et al., 1998; Thorner et al., 1998). Due to its multiplicity of functions, the pituitary was thought to be the 'master gland'. In the early part of the twentieth century, Erdheim and Stumme (1909) used pathological analyses to show that pituitary insufficiency in humans could be produced by hypothalamic damage. Later, Aschner (1912) reported that hypothalamic lesions that spared the pituitary could disrupt reproductive functions in dogs. Despite these early studies, it was not until the classic experiments of Harris and Jacobsohn (1952) that an endocrine role for hypothalamic control of anterior pituitary function was shown. In this case, pituitary stalk section in the rat produced a loss of reproductive function and the condition was reversed when the hypophyseal blood vessels regenerated. Transplantation experiments further emphasized the importance of the hypothalamus in the control of reproduction

* Correspondence to: W.C. Wetsel, Departments of Psychiatry and Behavioral Sciences, Medicine (Endocrinology), and Cell Biology, Mouse Behavioral and Neuroendocrine Analysis Core Facility, Duke University Medical Center, Box 3497, 028 CARL Building, Durham, NC 27710, USA. Tel.: +1-919-684-4574; Fax: +1-919-684-3071; E-mail: wetse001@mc.duke.edu.

(Nikitovitch-Winer and Everett, 1958). Removal of the pituitary to the renal capsule or temporal lobe was unable to correct the deficit. However, when the pituitary was placed beneath the median eminence of the hypothalamus and revascularization occurred, reproductive function was restored. Additionally, transplantation of pituitary fragments into the mediobasal hypothalamus further demonstrated a trophic role for this region of the brain (Halasz et al., 1962; Knigge, 1962).

Since an intact hypophyseal–portal circulation was necessary to re-establish anterior pituitary function, many investigators began to try to identify these hypothalamic hormones or releasing factors. Using extracts from hypothalamus, McCann (McCann et al., 1960) and Harris (Campbell et al., 1964) identified a biologically active material that was able to stimulate the release of gonadotropin hormones from the pituitary. In 1971, Schally and coworkers (Matsuo et al., 1971; Schally et al., 1973) isolated and sequenced a decapeptide from thousands of porcine hypothalami and showed that this and the synthetic peptide could release luteinizing hormone (LH) and follicle-stimulating hormone (FSH) from the pituitary. These findings were replicated using extracts from ovine hypothalamus by Guillemin and collaborators (Burgus et al., 1972). The peptide with the LH- and FSH-releasing activities was termed gonadotropin-releasing hormone (GnRH) or luteinizing hormone-releasing hormone. This hormone contains 10 amino acids where the N-terminal residue consists of a pyroglutamate and the C-terminal has a glycine-amide (see Table 1). These two modifications serve to retard degradation of the peptide in

TABLE 1

Molecular forms of GnRH

	Amino acid position [a,b]									
	1	2	3	4	5	6	7	8	9	10
mGnRH [c]	pGlu-His-Trp-Ser-Tyr-Gly-Leu-Arg-Pro-Gly-NH$_2$									
cGnRH-I	pGlu-His-Trp-Ser-Tyr-Gly-Leu-**Gln**-Pro-Gly-NH$_2$									
sbGnRH	pGlu-His-Trp-Ser-Tyr-Gly-Leu-**Ser**-Pro-Gly-NH$_2$									
fGnRH	pGlu-His-Trp-Ser-Tyr-Gly-Leu-**Trp**-Pro-Gly-NH$_2$									
pjGnRH	pGlu-His-Trp-Ser-**Phe**-Gly-Leu-**Ser**-Pro-Gly-NH$_2$									
hrGnRH	pGlu-His-Trp-Ser-**His**-Gly-Leu-**Ser**-Pro-Gly-NH$_2$									
cfGnRH	pGlu-His-Trp-Ser-**His**-Gly-Leu-**Asn**-Pro-Gly-NH$_2$									
sGnRH	pGlu-His-Trp-Ser-Tyr-Gly-**Trp**-**Leu**-Pro-Gly-NH$_2$									
gpGnRH	pGlu-**Tyr**-Trp-Ser-Tyr-Gly-**Val**-Arg-Pro-Gly-NH$_2$									
cGnRH-II	pGlu-His-Trp-Ser-**His**-Gly-**Trp**-**Tyr**-Pro-Gly-NH$_2$									
dfGnRH	pGlu-His-Trp-Ser-**His**-Gly-**Trp**-**Leu**-Pro-Gly-NH$_2$									
lGnRH-III	pGlu-His-Trp-Ser-**His**-**Asp**-**Trp**-**Lys**-Pro-Gly-NH$_2$									
tGnRH-I	pGlu-His-Trp-Ser-**Asp**-**Tyr**-**Phe**-**Lys**-Pro-Gly-NH$_2$									
tGnRH-II	pGlu-His-Trp-Ser-**Leu**-**Cys**-**His**-**Ala**-Pro-Gly-NH$_2$									
lGnRH-I	pGlu-His-**Tyr**-Ser-**Leu**-**Glu**-**Trp**-**Lys**-Pro-Gly-NH$_2$									

[a] The amino acids are designated by three-letter abbreviations; pGlu refers to pyroglutamate, Gly–NH$_2$ is glycine amide.
[b] Amino acid differences from the mammalian GnRH are highlighted.
[c] *Mammalian (m):* Matsuo et al., 1971; Burgus et al., 1972; *chicken (c) I:* King and Millar, 1982; *seabream (sb):* Powell et al., 1994; *frog (f):* Yoo et al., 2000; *pejerry (pj):* Montaner et al., 2001; *herring (hr):* Carolsfeld et al., 2000; *catfish (cf):* Ngamvongchon et al., 1992; *salmon (s):* Sherwood et al., 1983; *guinea pig (gp):* Jimenez-Liñan et al., 1997; *chicken (c) II:* Miyamoto et al., 1984; *dogfish (df):* Lovejoy et al., 1992; *lamprey (l) III:* Sower et al., 1993; *tunicate (t) I* and *II:* Powell et al., 1996; *lamprey (l) I:* Sherwood et al., 1986.

blood (Chertow, 1981).

Eleven years following the identification of the sequence for mammalian GnRH, a second form was isolated from chickens (King and Millar, 1982) and this was soon followed by the purification of another form from the same species (Miyamoto et al., 1984). Presently, fifteen different molecular forms of the decapeptide have been isolated and sequenced from a variety of different species of animals (Matsuo et al., 1971; Burgus et al., 1972; King and Millar, 1982; Sherwood et al., 1983, 1986; Miyamoto et al., 1984; Lovejoy et al., 1992; Ngamvongchon et al., 1992; Sower et al., 1993; Powell et al., 1994, 1996; Jimenez-Liñan et al., 1997; Carolsfeld et al., 2000; Yoo et al., 2000; Montaner et al., 2001). A listing of these molecular forms can be found in Table 1, where the amino acid sequences are compared to that of the mammalian form. It should be noted that substitutions have occurred evolutionarily in all locations except those occupied by the pGlu[1] at the N-terminal, the Ser[4], the Pro[9], and the glycine-amide at the C-terminal. These data suggest these 4 residues are important for ligand binding and/or biological activity. Interestingly, mammalian GnRH is active at low concentrations, whereas the remaining vertebrate forms — with the exception of chicken II — show reduced binding affinity and gonadotropin release (Sealfon et al., 1997). The N- and C-terminal domains of GnRH are both involved in receptor binding, while the former also participates in receptor activation and contributes to the biological activity of the decapeptide (Okada et al., 1973). The reason for the conservation of the Ser[4] residue is unclear since substitutions by a number of different amino acids will retain activity (Sealfon et al., 1997). In contrast, both the Pro[9] and Gly-amide residues are critical for biological activity and alterations in the C-terminal end of the peptide are known to confer potent agonist or antagonist actions (Karten and Rivier, 1986). In this regard, it is noteworthy that the Pro[9] residue can be hydroxylated and this peptide can stimulate LH and FSH secretion, albeit less potently than GnRH (Gautron et al., 1995).

With the identification of the various molecular forms of GnRH, it became clear that multiple species of the peptide co-existed in the brains of reptiles, amphibians, fish, and birds (Fernald and White, 1999). In many cases at least three different isoforms are present and they are typically found in unique locations in brain that include the midbrain, hypothalamus, and terminal nerve/olfactory system in the telencephalon. Despite this fact, for many years it was thought that only one form of GnRH was present in mammalian brain. However, the musk shrew brain was found to contain a second form (Dellovade et al., 1993; Kasten et al., 1996). Since that time, multiple LHRH species have been identified in rodent (Chen et al., 1998), non-human primate (Lescheid et al., 1997), and human brain (Chen et al., 1998; White et al., 1998a; Yahalom et al., 1999). The hypothalamic form of GnRH is clearly involved in the endocrine regulation of reproduction since a mutation in this gene in mice renders the animal infertile (Cattanach et al., 1977; Mason et al., 1986a) and gene therapy restores reproductive function (Mason et al., 1986b). By contrast, the midbrain form is thought to play some role in sexual behavior because injection of high concentrations of mammalian GnRH into this brain region of rats induces lordosis (Sakuma and Pfaff, 1983). Moreover, exposure of female musk shrews to males induces rapid changes in specific populations of GnRH neurons in the midbrain (Dellovade et al., 1993). Currently, both chicken II and the salmon forms of the decapeptide have been found in the midbrains of mammals (Lescheid et al., 1997; Chen et al., 1998; and Yahalom et al., 1999).

The cDNA for GnRH was first cloned in humans by Seeburg and Adelman (1984). Deduction of the primary sequence from the cDNA revealed that the prepro-hormone consisted of 92 amino acids (Fig. 1). At the N-terminal, the signal sequence is composed of 23 amino acids and it is separated from a pro-peptide of 69 amino acids. The GnRH is located at the N-terminal end of the prohormone and is separated by a putative processing site from the remainder of the molecule. The C-terminal portion of the pro-GnRH protein is termed GnRH-associated peptide (GAP). Since some peptide precursors can

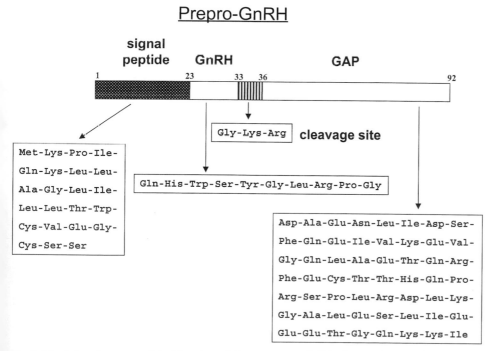

Fig. 1. The deduced structure of the 92 amino acid human prepro-GnRH. The organization of the precursor is shown. The positions of the amino acids that comprise the signal peptide, GnRH, putative cleavage site, and GAP-(1–56) are displayed. A three-letter abbreviation is given for the amino acids.

TABLE 2

Comparison of the deduced amino acid sequences of the pro-GnRH from various species [a,b]

Mammalian Pro-GnRH [c]

Positions: 1 · 10 · 14 · 20 · 30 · 40 · 50 · 60 · 69

Human QHWSYGLRPG GKR DAENLIDSFQEIVKEVGQLAETQRFECTTHQPRSPLRDLKGALESLIEETGQKKI

Rat QHWSYGLRPG GKR NTEHLVDSFQEMGKEEDQMAEPQNFECTVHWPRSPLRDLRGALERLIEEEAGQKKM

Mouse QHWSYGLRPG GKR NTEHLVESFQEMGKEVDQMAEPQHFECTVHWPRSPLRDLRGALESLIEEEARQKKM

Chicken I Pro-GnRH [d]

Positions: 1 · 10 · 14 · 20 · 30 · 40 · 50 · 60 · 69

Chicken QHWSYGLQPG GKR NAENLVESFQEIANEMESLGEGQKAECPGSYQHPRLSDLKETMASLIEGEARRKEI

Chicken II Pro-GnRH [e]

Positions: 1 · 10 · 14 · 20 · 30 · 40 · 50 · 60 · 70 · 80 · 90 · 97

Human QHWSHGWYPG GKR ALSSAQDPQNALRPPGRALDTAAGSPVQTAHGLPSDALAPLDDSMPWEGRTTAQWSLHRKRHLARTLLTAAREPRPAPPSSNKV

Rhesus QHWSHGWYPG GKR ALSSAQDPQNALRPPAGSPAQATYGLPSDALAHLEDSMPWEGRTMAVWSLRRKRYLAQTLLTAAREPRPVPPSSNKV

Salmon Pro-GnRH [f]

Positions: 1 · 10 · 14 · 20 · 30 · 40 · 50 · 59

Salmon QHWSYGWLPG GKR SVGELEATIKMMDTGGVVALPEETSAHVSERLRPYDVILKKWMPHK

Seabream Pro-GnRH [g]

Positions: 1 · 10 · 14 · 20 · 30 · 40 · 50 · 60 · 70

Seabream QHWSYGLSPG GKR DLDSLSDTLGNIIERFPHVDSPCSVLGCVEEPHVPRMYRMKGFIGSERDIGHRMYKK

Frog Pro-GnRH [h]

Positions: 1 · 10 · 14 · 20 · 30 · 40 · 50 · 60 · 66

Frog QHWSYGLWPG GKR EVEGLQESYSEVPNEVSFTDPQHFERSIPQNRISLVREALMNWLEGENTRKKI

Pejerry Pro-GnRH [i]

Positions: 1 · 10 · 14 · 20 · 30 · 40 · 50 · 60 · 70

Pejerry QHWSFGLSPG GKR ELKYFPNTLENQIRLLNSNTPCSDLSHLEESSLAKIYRIKGLLGSVTEAKNGYRTYK

Lamprey I Pro-GnRH [j]

Positions: 1 · 10 · 14 · 20 · 30 · 40 · 50 · 60 · 63

Lamprey QHYSLEWKPG GKR DLEQELEPPSNAFECDGPECAFSRVPNTKLIRELASYLSQRNYDRKGALK

Catfish Pro-GnRH [k]

Positions: 1 · 10 · 14 · 20 · 30 · 40 · 50 · 59

Catfish QHWSHGLNPG GKR AVMQESAEEIPRRSGYLCDYVAVSPRNKPFRLKDLLTPVAGREIEE

[a] The amino acid sequences are provided as single letter abbreviations; differences between subspecies are only highlighted for mammalian pro-GnRH and chicken II pro-GnRH.

[b] The N-terminal containing the GnRH moiety is separated by a space from a putative processing site and the GnRH-associated peptide; the numbers over the peptide sequence represent the positions of the amino acids.

[c] Mammalian pro-GnRH: human – Seeburg and Adelman, 1984; rat – Adelman et al., 1986; mouse – Mason et al., 1986a.

[d] Chicken I pro-GnRH: chicken – Dunn et al., 1993.

[e] Chicken II pro-GnRH: human – White et al., 1998a; rhesus monkey – White et al., 1998b.

[f] Salmon pro-GnRH: salmon – Klungland et al., 1992.

[g] Seabream pro-GnRH: seabream – Gothilf et al., 1995.

[h] Frog pro-GnRH: frog – Yoo et al., 2000.

[i] Pejerry pro-GnRH: pejerry – Okubo et al., 2000.

[j] Lamprey I pro-GnRH: lamprey – Suzuki et al., 2000.

[k] Catfish pro-GnRH: catfish.

be processed into a variety of different bioactive peptides, several investigators have studied the activity of GAP in pituitary cultures. In the initial report, a recombinant GAP protein was biosynthesized in bacteria, purified, and added to primary rat pituitary cell cultures (Nikolics et al., 1985). At 10 µM the recombinant GAP influenced secretion of LH, FSH, and prolactin, but it exerted no effects on adrenocorticotropin or growth hormone release. GAP stimulated LH release, albeit less effectively than GnRH while it was equipotent with the decapeptide for FSH secretion. By comparison, GAP potently inhibited prolactin secretion and this inhibition was not synergistic with dopamine. It was noted that during the process of raising GAP antisera in rabbits, serum prolactin levels became depressed. Due to these collective effects, GAP was described as a prolactin-inhibiting factor. Since that time, GAP has been shown to be co-released with GnRH from rat median eminence fragments in vitro (Valença et al., 1988; Wetsel and Negro-Vilar, 1989) and into the ovine hypophyseal portal circulation in vivo (Clarke et al., 1987). Additionally, there is some evidence that some GAP-(1–56) may be processed into smaller fragments in rat hypothalamus (Ackland et al., 1988; Wetsel et al., 1988). GAP has been reported to decrease the intracellular levels of [Ca^{2+}] from human prolactin-secreting adenoma cells, rat pituitary GH3 cells, and primary rat pituitary cells (Vacher et al., 1991). It has also been observed to reduce cAMP levels in GH3 cells (van Chuoi et al., 1993).

In comparison to secretion and peptide processing studies, the physiological role of GAP is controversial. Human recombinant GAP-(1–56) was initially reported to inhibit prolactin secretion from rat pituitary cells (Nikolics et al., 1985). On the other hand, the synthetic peptide can stimulate prolactin secretion from tilapia pituitary (Planas et al., 1990). Interestingly, different domains of the GAP molecule appear to possess differing biological activities. In tilapia, the stimulatory effect seemed to reside within the GAP-(28–36) and GAP-(51–66) fragments, whereas GAP-(38–49) was inhibitory. In rat pituitary cell cultures, various synthetic GAP fragments have been observed to influence gonadotropin (Milton et al., 1986) and prolactin release (Wormald et al., 1989). Additional studies conducted with rats in vivo have shown GAP-(1–56) can inhibit prolactin release, while the effects on LH and FSH secretion are very limited (Yu et al., 1988). On the other hand, GAP-(1–13) and GAP-(1–23) may have some gonadotropin-releasing activities (Yu et al., 1989). By contrast, in ovariectomized ewes, GAP-(1–56) exerts no effects on secretion of LH, FSH, or prolactin (Thomas et al., 1988a). A similar failure of GAP to inhibit prolactin release in rats (Chandrashekar et al., 1988) and cells from human pituitary adenomas has also been noted (Ishibashi et al., 1987). In summary, our knowledge on the physiology of GAP remains obscure.

Irrespective of a possible physiological role for GAP, it should be noted that during protein synthesis the ribosome encloses approximately 30–35 amino acids of the nascent protein (Lewin, 1996). Since the mammalian prepro-GnRH contains 92 amino acids and the first 23 amino acids contain the signal sequence, one purpose of GAP-(1–56) may be to serve as a linker to the decapeptide so that the pro-GnRH can be routed out of the endoplasmic reticulum and into the secretory pathway.

Although 15 different molecular forms for GnRH have been identified at the peptide level, only nine of the complementary DNAs for these forms have been cloned (Table 2; Seeburg and Adelman, 1984; Adelman et al., 1986; Mason et al., 1986a; Klungland et al., 1992; Dunn et al., 1993; Bogerd et al., 1994; Gothilf et al., 1995; White et al., 1998a,b; Okubo et al., 2000; Suzuki et al., 2000; Yoo et al., 2000). Examination of the deduced amino acid sequences reveals that in all cases, GnRH is located at the N-terminal end of the precursor and it is separated by a putative processing site from GAP. In all of the deduced sequences for the pro-GnRH, the least conservation of amino acids occurs within the GAP portion of the molecule. The rat and mouse forms of GAP in the mammalian GnRH are only 70% homologous to the human GAP-(1–56) sequence. For the chicken II GAP proteins, divergence is even greater among the deduced sequences for the human and rhesus monkey. The divergence of the GAP sequence between even closely related species suggests that, if this peptide has bioactivity, it may be different among these species.

Scheme for Pro-GnRH processing

Inspection of the deduced amino acid sequences from the nine complementary pro-GnRH forms demonstrates several points. First, all pro-GnRH proteins contain GnRH at their N-termini and the peptide is separated by a putative processing site from GAP. The organization and spatial arrangement of these sequences suggests that the enzymes that process the pro-GnRH in a given cell may be similar across all species. Second, excision of the GnRH peptide from the precursor must be performed by an endopeptidase. Since most peptide precursors appear to be processed at monobasic or dibasic residues (Douglass et al., 1984), the endopeptidase could cleave the pro-GnRH after the Lys^{12} or Arg^{13} residues. Third, because the decapeptide does not contain these basic residues at its C-terminal, there must be some exopeptidase that can remove them. Fourth, an examination of the decapeptide shows that the first amino acid of the GnRH is always Gln^1. Since the fully processed GnRH contains a $pGlu^1$ at its N-terminal, some enzyme must be present to process the Gln^1 to $pGlu^1$. Fifth, because the C-terminal of GnRH is amidated, some enzyme must be responsible for converting the Gly^{11} to an amide. Finally, there is some evidence that the Pro^9 residue can be hydroxylated to produce $[Hyp^9]GnRH$. A depiction of this processing scheme can be found in Fig. 2.

Prohormone convertases

With the cloning of many peptide hormone and neuropeptide cDNAs and following the deduction of their amino acids sequences, it became clear that bioactive peptides are first biosynthesized as precursor proteins. Typically, the N- and/or C-terminals of the peptides are bounded or separated from the remainder of the precursor by monobasic or dibasic residues (Douglass et al., 1984). Since trypsin can cleave proteins at these residues, investigators searched for many years for native trypsin-like enzymes that could process pro-peptides. The first breakthrough in the discovery of prohormone processing enzymes came with the identification of kexin (an enzyme encoded by the *Kex2* gene); this enzyme was responsible processing α-mating factor in *Saccharomyces cerevisiae* (Julius et al., 1984).

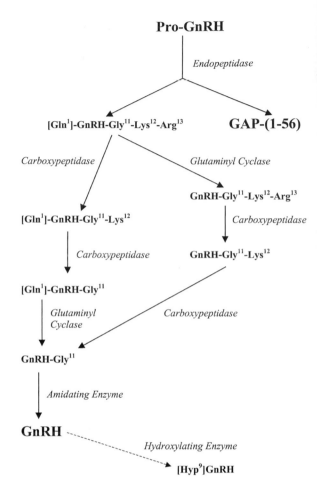

Fig. 2. A processing scheme for the pro-GnRH to bioactive GnRH. The prohormone is processed by an endopeptidase to yield a GnRH intermediate and GAP-(1–56). The intermediate has several fates. The C-terminal basic amino acids can be removed by a carboxypeptidase, the N-terminal glutamine can be converted to a pyroglutamate by a glutaminyl cyclase, and the C-terminal glycine can donate its amide group through the action of an amidating enzyme to produce GnRH. Alternatively, the N-terminal of the intermediate can be processed to the pyroglutamate, the C-terminal basic amino acids removed, and the GnRH can be amidated. There is also some evidence that the proline residue of GnRH can be hydroxylated by some hydroxylating enzyme and this step is placed at the end of the processing cascade only for convenience. At this time, it is unclear at which step(s) the hydroxylation reaction occurs. The N- and C-terminal amino acids of the intermediate are designated by three-letter abbreviations; the numbers signify the amino acid positions in the intermediate.

Subsequent studies demonstrated the ability of kexin to correctly process exogenously expressed mammalian pro-hormones in yeast (Thim et al., 1986)

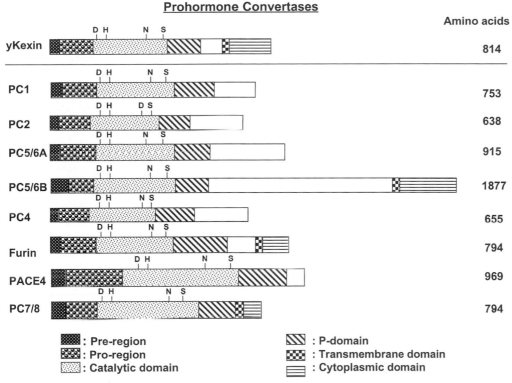

Fig. 3. A depiction of the yeast Kex2 enzyme and the mammalian prohormone convertases. The organization of the enzymes is depicted in the diagram. The catalytic sites on the endopeptidases are identified by the single amino acid designations. The significance of the shading for each enzyme can be found at the bottom of the figure. The pre-region contains the signal peptide. The P-domain is important for the correct folding of the convertases in the endoplasmic reticulum. The open bars refer to 'spacer' regions. The numbers of amino acids that comprise the proenzymes are designated to the right of each enzyme.

and in various mammalian cell lines (Thomas et al., 1988b). Later, in the course of cloning the *c-fes/fps* oncogene, it was noted that a potential open reading frame for another gene was present upstream of the oncogene and that its sequence bore some homology to that of Kex2 (Roebroek et al., 1986). This putative protein was designated *fes/fps* upstream region or furin. To date, eight members of the family of the prohormone convertases (PCs) have been identified (Fig. 3) and these include furin (also termed Paired Amino Acid Converting Enzyme or PACE), PC1/3 (Seidah et al., 1990, 1991), PC2 (Seidah et al., 1990; Smeekens and Steiner, 1990), PC4 (Nakayama et al., 1992; Seidah et al., 1992), PACE4 (Kiefer et al., 1991), PC5/6 (Lusson et al., 1993; Nakagawa et al., 1993a,b), PC7/8 (Bruzzaniti et al., 1996; Seidah et al., 1996), and Ski-1 (Seidah et al., 1999a). Since the latter enzyme has been shown to process growth

factors, receptors, and viral glycoproteins, it will not be discussed further.

The PCs differ in their tissue distributions and intracellular locations (Steiner, 1998; Seidah and Chrétien, 1999). Furin and PC7/8 are ubiquitous in their expression (Schalken et al., 1987; Bruzzaniti et al., 1996; Seidah et al., 1996), while both PC1 and PC2 are localized exclusively in neural and endocrine tissues (Seidah et al., 1990, 1991; Smeekens and Steiner, 1990). Within the central nervous system, PC2 is more widely expressed than PC1 with both convertases being expressed in the hypothalamus (Seidah et al., 1991; Schäfer et al., 1993). PC4 expression has been detected exclusively in reproductive tissues. More specifically, it is located in round spermatids and spermatozoa (Seidah et al., 1992) and in macrophage-like cells of the ovary (Tadros et al., 2001). PACE4 and PC5/6 reside in both endocrine and non-endocrine cells (Kiefer et al., 1991; Lusson et al., 1993). PACE4 is highly expressed in the pituitary and chondrocytes. PC5/6 exists in two splice forms; PC5/6A is distributed more widely than PC5/6B and it is expressed in both endocrine and non-endocrine cells (Nakagawa et al., 1993a,b; Dong et al., 1995). In rat brain, PC5/6A immunoreactivity is more restricted in comparison to that for PC1/3 and PC2. It is present throughout the forebrain, as well as in the diagonal band of Broca and hypothalamus. As is apparent from the pattern of expression, each convertase has its characteristic tissue distribution. However, there is an overlap in the expression of two or more convertases in several tissues. This finding implies that there may be considerable redundancy of function among certain of the enzymes.

The PCs are an evolutionarily conserved family of calcium-dependent proteases that are related to bacterial serine proteases of the subtilisin family of convertases (Steiner, 1998; Seidah and Chrétien, 1999). Due to this relationship and because the PCs are also associated with Kex2, the PC family of enzymes has also been called subtilisin-like pro-protein convertases or subtilisin/Kex2 family of enzymes. All of these enzymes are synthesized as precursors that contain a pre-domain, a pro-domain, a catalytic domain, a P-domain, and an enzyme-specific C-terminal domain (see Fig. 3). The pre-region contains the signal peptide. The pro-region serves as an intra-molecular chaperone, as well as a transient intra-molecular inhibitor of enzymatic activity. The conserved catalytic domain contains a catalytic triad consisting of Asp, His and Ser residues that are involved in oxyanion stabilization and are important in catalysis. The catalytic domain for PC2 contains an Asn instead of the Asp residue. The P-domain is important for the correct folding of the convertases in the endoplasmic reticulum. The C-terminal regions of these enzymes are relatively variable (Bergeron et al., 2000). While furin, PACE4, and PC5/6B contain cysteine-rich regions, PC1 and PC2 have amphipathic helices, and PC7/8 contains a serine–threonine-rich region in the C-terminal. Furin, PACE4E, PC5/6B, and PC7/8 also have transmembrane domains in this region.

As noted above, the PCs are first biosynthesized as inactive precursors and they must undergo proteolytic processing to achieve full enzymatic activity (Bergeron et al., 2000). The pro-segment of the enzyme can be removed by an autocatalytic reaction. This removal plays a role in enzymatic activation and it is necessary for sorting the convertase into the appropriate secretory pathway. Besides pro-region retardation of enzymatic activity, there are also endogenous proteins that can inhibit PC1/3 and PC2. A specific endogenous inhibitor for PC1/3 has been recently identified as a polypeptide termed pro-SAAS (Fricker et al., 2000). Pro-SAAS only inhibits PC1/3 and none of the other PC members (Fricker et al., 2000; Qian et al., 2000). Pro-SAAS is expressed in all regions where PC/3 resides, as well as in additional regions, suggesting that it may have some undisclosed functions (Feng et al., 2001; Lanoue and Day, 2001; Sayah et al., 2001). Similar to PC1/3, PC2 also interacts with an additional protein, 7B2 (Braks and Martens, 1994; Lindberg et al., 1995; Zhu and Lindberg, 1995). The 7B2 is a neural and endocrine polypeptide that binds to inactive pro-PC2. It facilitates the transport of the convertase from the endoplasmic reticulum to later compartments in the secretory pathway where PC2 can undergo proteolytic maturation. The C-terminal peptide of 7B2 can inhibit PC2. All cells that express PC2 also express 7B2. However, the converse is not true (Seidah et al., 1999b), suggesting that 7B2 might have some additional roles in the neuronal and endocrine cells. Mice lacking the 7B2 gene have impaired processing of pro-glucagon, pro-insulin, pro-enkephalin, and pro-opiomelanocortin and do not have any detectable PC2 activity (Westphal et al., 1999). These mutants display Cushing's syndrome and die within 9 weeks of age.

Research with a number of different systems has clearly established a role for the PCs in processing precursor hormones and neuropeptides. Since endoproteolysis represents the first step in pro-GnRH processing (see Fig. 2), we wanted to determine whether any of these enzymes could process the prohormone. At that time, only furin, PC1/3 and PC2 had been cloned. Since the immortalized hypothalamic LHRH neuronal cell lines can process the precursor (Wetsel et al., 1991), we selected these neurons for study. Results from Northern blots revealed that the GT1 cells contained transcripts for furin and PC2 (Wetsel et al., 1995). No transcripts were evident for PC1/3 even when poly(A)$^+$ RNA was used. These results are consistent with an in-situ hybridization study in which PC2 transcripts were found in LHRH neurons in rat hypothalamus (Voigt et al., 1996). To determine whether PC2 could process the pro-GnRH, vaccinia virus recombinants of the prohormone, furin, PC1/3, and PC2 were used in studies conducted in collaboration with Dr. Gary Thomas at the Oregon Health Sciences University in Portland Oregon. Preliminary results conducted in BSC-40 (kidney) cells demonstrate that furin, PC1/3 and PC2 can all process the pro-GnRH. Whereas furin and PC1/3 are very inefficient in converting the pro-hormone to GnRH and GAP-(1–56), PC2 is efficient in this process.

Another way to evaluate the role of the PCs in processing the pro-GnRH is through the use of mutants where the genes for these enzymes have been selectively deleted. Since furin (Roebroek et al., 1998) and PACE4 (Constam and Robertson, 2000) null mice exhibit embryonic lethality during midgestation, the effects of these enzymes on pro-GnRH processing and reproduction cannot be easily evaluated in vivo. By comparison, PC4 expression is exclusively gonadal and is not expressed in brain (Seidah et al., 1992). PC4 null males are infertile and there is reduced fertility in females (Mbikay et al., 1997). There are no published reports on PC5/6 or PC7/8 null mice.

Preliminary analyses of PC1/3 null mutants indicate that these animals die in utero (see Seidah and Chrétien, 1999). Interestingly, an adult human female with compound heterozygous mutations for PC1/3 (a missense mutation in one allele and a splicing mutation in the other allele) has been identified (O'Rahilly et al., 1995; Jackson et al., 1997). The pro-PC1/3 derived from the missense mutation would be expected to be retained in the endoplasmic reticulum, while the product derived from aberrant splicing at exon 5 would result in a frameshift with premature termination of translation in the catalytic domain. This patient had childhood obesity, hypogonadotropic hypogonadism, hypocortisolism, and elevated levels of circulating pro-insulin. The discrepancy in viability of the human patient and the PC1/3 knockout mice may be attributed to the possibility that there may be residual enzyme activity in the human patient despite mutations in both alle-

les. Alternatively, PC1/3 may have acquired slightly different functions in humans.

PC2 null mice have been described (Furuta et al., 1997, 2001). These mutants exhibit β-cell defects, mild hypoglycemia, and a small decrease in the rate of postnatal growth. The animals display a deficiency, but not complete blockade, in pro-insulin, pro-glucagon, and pro-somatostain processing. The PC2 null mice also demonstrate an impairment in processing pro-enkephalin (Johanning et al., 1998), pro-cholecystokinin (Vishnuvardhan et al., 2000), pro-dynorphin (Berman et al., 2000), pro-opiomelanocortin and pro-orphanin FQ/nociceptin (Allen et al., 2001). Preliminary results from our laboratory have demonstrated that PC2 null mice exhibit mild reproductive dysfunction as exemplified by delayed onset of puberty, irregular estrous cycles, and impaired ability of the dams to nurture their offspring. Additional results suggest that there may be some deficiency in pro-GnRH processing, but this deficit is not complete. Hence, these data indicate that some additional convertase may be compensating for the loss of PC2. Recent preliminary findings from our lab suggest that PC5/6A may be a compensating enzyme.

Carboxypeptidase

Following endoproteolytic cleavage of a peptide precursor by prohormone convertase(s), the C-terminal basic amino acids must be removed. Carboxypeptidases hydrolyze amino acids from the C-terminal of peptides and proteins. A family of 25 carboxypeptidases have been identified and their mechanisms of cleavage use an active site at serine or cysteine residues, or at zinc (Reznik and Fricker, 2001). From the enzymes comprising the latter group, approximately 13 different metallocarboxypeptidase genes have been cloned (Fig. 4). A carboxypeptidase that is expressed almost exclusively in neural and endocrine tissues is carboxypeptidase E (CPE; Fricker et al., 1986; Rodriguez et al., 1989; Fricker, 1991; Schäfer et al., 1993; Zheng et al., 1994). This carboxypeptidase was first identified as enkephalin convertase (Fricker and Snyder, 1972) and it has also been termed carboxypeptidase H (or EC3.4.17.10; Parkinson, 1990; Fricker, 1998). While the enzyme shows broad peptide specificity, it is highly specific in its

Carboxypeptidases

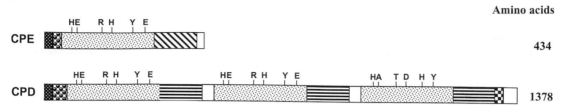

Fig. 4. A diagram of carboxypeptidases E and D. The organization of the enzymes is depicted in the diagram (see Fig. 3 for details). The sites on the carboxypeptidases (e.g., H, E, and H bind Zn^{2+}, R and Y bind the substrate, and with the E all sites participate in catalysis) are identified by the single amino acid designations. The significance of the shading for each enzyme can be found at the bottom of Figure 3. The open bars refer to 'spacer' regions, whereas the region with the dark horizontal bars represents transthyretin-like domains. The numbers of amino acids that comprise the proenzymes are designated to the right of each enzyme.

removal of C-terminal basic amino acids. Excision of arginine and lysine residues occurs more readily than histidine with this enzyme (Fricker and Snyder, 1972; Smyth et al., 1989). CPE is biosynthesized as a pro-enzyme and this precursor possesses full activity (Parkinson, 1990). The N-terminal extension of the enzyme is removed in a post-Golgi compartment or secretory vesicle (Song and Fricker, 1995a) to yield a glycosylated metalloenzyme of approximately 55,000 kDa in size. The enzyme displays optimal activity between pH 5–5.5 and is inactive at neutral pH. Since the secretory vesicle has an acidic pH and because endoproteolysis of many peptides occurs within the *trans*-Golgi network and secretory vesicle (Douglass et al., 1984), juxtaposition of CPE with peptide intermediates in the late secretory pathway is optimal for its activity (Fricker, 1998). Within the regulated secretory pathway, CPE is present both as soluble and membrane-bound forms that arise through differential posttranslational processing (Fricker et al., 1990). Despite this fact, the catalytic activities of these forms are similar.

In early studies of pro-GnRH processing in rat hypothalamus, we proposed that a carboxypeptidase may be involved in converting pro-GnRH intermediates to GnRH (see Fig. 2; Wetsel et al., 1988). Later, we showed in the immortalized GnRH cell lines that the prohormone was cleaved between the Arg^{13} and Asn^{14} residues to produce Gln^1–GnRH–$Gly^{11}Lys^{12}Arg^{13}$ and GAP-(1–56) (Wetsel et al., 1991). Since we also found evidence that the Arg^{13} and Lys^{12} residues could be removed from the GnRH intermediate and because CPE was expressed in brain, we postulated that this enzyme

was responsible for removing these C-terminal basic amino acids from the precursor. In subsequent work, we showed that the immortalized GnRH cells and the brain regions (e.g., preoptic area and hypothalamus) that contains these neurons possessed transcripts for CPE (Wetsel et al., 1995). To demonstrate that CPE could actually process these C-terminally extended peptides, we conducted two different experiments. First, we incubated in vitro synthetic GnRH–$Gly^{11}Lys^{12}Arg^{13}$ or GnRH–$Gly^{11}Lys^{12}$ with purified recombinant baculovirus-derived CPE. Both peptides were processed to GnRH–Gly^{11} by the enzyme. Second, we used vaccinia viral recombinants of CPE to examine processing of the pro-GnRH in mammalian cells. In this case, recombinant cDNAs for pro-GnRH, PC2, and CPE were expressed in kidney cells that do not normally express any of these genes. The PC2 was expressed so that the prohormone could be endoproteolytically cleaved to an intermediate with C-terminally extended basic amino acids. CPE was found to remove the Arg^{13} and Lys^{12} residues from this intermediate. Collectively, these findings clearly established a role for CPE in pro-GnRH processing, however, an in vivo role in mammals remained to be determined.

More than 10 years ago, a spontaneous autosomal recessive mutation was identified in a colony of mice at Jackson Laboratories (Coleman and Eichler, 1990). These animals were found to be obese, diabetic, and infertile and they were called *fat/fat* mice. Since the animals begin to develop obesity after puberty, they have been used as an animal model for adult-onset obesity for many years (Leiter and Herberg, 1997). Subsequent mapping studies have revealed that the locus of the *fat* mutation is on chromosome 8 and that it is close to the murine *Cpe* gene (Prochazka et al., 1991). Additional experiments have demonstrated that the CPE gene in *fat/fat* mice is mutated at a single nucleotide that results in a single amino acid change, $Ser^{202}Pro$. Due to identification of the mutation in these mice, the nomenclature for these animals was changed to Cpe^{fat}. Interestingly, CPE mRNA levels are similar in wild type (WT), heterozygous, and homozygous mice. Despite this fact, the CPE protein is undetectable in the mutants because the mutated enzyme is unstable and is rapidly degraded in the endoplasmic reticulum (Naggert et al., 1995; Varlamov et al., 1996). As a result, CPE activity in the pancreas and pituitaries of the homozygotes is very low and is reduced by at least 20-fold from that of the WT animals.

Since we had found that CPE transcripts were present in the immortalized GnRH neurons and in brain regions where GnRH neurons reside in situ (Wetsel et al., 1995), we postulated that the infertility of the Cpe^{fat} mice might be due to a processing deficit for pro-GnRH. To this purpose, we dissected hypothalami from WT and $Cpe^{fat/fat}$ animals, separated the materials by high pressure liquid chromatography, and screened the fractions with an antiserum that recognized all GnRH intermediate peptides. Our analysis revealed that the homozygotes were deficient in processing the pro-GnRH such that there was a substantial increase in the quantities of $GnRH–Gly^{11}Lys^{12}Arg^{13}$ and $GnRH–Gly^{11}Lys^{12}$. Some fully processed GnRH was detected in mutant hypothalamus, however, it was reduced by at least 75%. These data clearly show that CPE plays a major role in vivo in processing the pro-GnRH.

Since the Cpe^{fat} mouse is deficient in pro-insulin processing (Naggert et al., 1995), various investigators have tried to determine whether human diabetic patients may also have mutations in *Cpe*. While a number of different polymorphisms in the *Cpe* gene have been described in these patients (Utsunomiya et al., 1998), only one has been associated with diabetes (Chen et al., 2001). In this case, a single nucleotide alteration in $C^{847}T$ renders a codon change of $Arg^{283}Trp$. Since this alteration reduces the efficiency of CPE to cleave its substrates and because the enzyme is unstable at elevated temperatures, this polymorphism may predispose homozygous patients to hyperproinsulinemia and diabetes. At present, the reproductive status of the patients is unknown.

In the genetic characterization of the Cpe^{fat} mouse and in our own analysis, it was observed that while conversion of pro-insulin in pancreas or pro-GnRH in hypothalamus was perturbed, a small amount of biologically active peptide was still present. In addition, a very small amount of CPE-like activity was detected. These findings suggest that there might be some other carboxypeptidase that could convert the intermediates to bioactive peptides.

Although the family of metallocarboxypeptidases has significantly expanded over the past five years (Reznik and Fricker, 2001), there is one member of this family that may be able to process peptide precursors and it may account for the partial rescue of the Cpe^{fat} mouse. Some years ago, a surface protein in ducks that binds hepatitis B viral particles was cloned and it was termed gp180 (Kuroki et al., 1994). Independently, a carboxypeptidase activity different from that for CPE was purified from bovine pituitary and this enzyme was found to remove basic amino acids from the C-terminal of peptides (Song and Fricker, 1995b). This carboxypeptidase was called CPD (EC 3.4.17.22). Sequencing and subsequent cloning results indicated that gp180 and CPD were the same protein (Kuroki et al., 1995; Song and Fricker, 1995b; Tan et al., 1997; Xin et al., 1997). CPD is approximately 180 kDa in size and it contains three tandem-repeat carboxypeptidase-like domains (see Fig. 4); however, only the first two domains are catalytically active (Song and Fricker, 1996). The first domain is maximally active at pH 6.3–7.5 and it prefers C-terminal arginine substrates, whereas the second domain is active at pH 5.0–6.5 and it prefers C-terminal lysine (Novikova et al., 1999). Recently, the second domain of CPD has been crystallized (Gomis-Rüth et al., 1999). Since CPD and CPE have high homologies in the catalytic domains, it has been proposed that similar functional groups are responsible for substrate binding and catalysis (Aloy et al., 2001). CPD is a glycosylated enzyme, it exists in both soluble and particulate forms, and it is widely distributed in many different tissues (Song and Fricker, 1996). While CPD is highly expressed in brain, pituitary and adrenal gland, expression is even higher than CPE in certain brain regions (Song

232

and Fricker, 1996; Dong et al., 1999). Within cells the enzyme is found in the *trans*-Golgi network and immature secretory granules, however, it is absent from mature granules (Varlamov et al., 1999). At the present time it is not known whether GnRH neurons contain CPD. Nonetheless, given its location within other cells, CPD would be expected to be greatly restricted in its ability to process pro-GnRH intermediate peptides.

Amidating enzyme

An initial problem in the identification of GnRH involved its sequencing. Traditional sequencing methods were unable to resolve the C-terminal residue of the peptide. Tritiation (Matsuo et al., 1971) or hydrazinolysis (Burgus et al., 1972) were finally able to show that the C-terminal of GnRH contained a glycine amide. Since this residue is not a naturally occurring amino acid, it must be produced through the action of some proteolytic enzyme. An enzyme activity that could produce C-terminal amides from glycine was later identified (Bradbury et al., 1982) and it was shown to require molecular oxygen, copper, and ascorbic acid for full activity (Eipper et

al., 1983). This enzyme was termed peptidylglycine α-amidating monooxygenase (PAM). The cDNA for an amidating enzyme was first cloned from frog skin (Mizuno et al., 1987) and this was rapidly followed by isolation of a PAM cDNA from bovine neurointermediate pituitary (Eipper et al., 1987). PAM was also cloned from rats and it was found to consist of a single gene, to contain at least 27 exons, to span a region of more than 160 kb, and to be located on human chromosome 5 (Ouafik et al., 1992). Alternative splicing produces at least 7 different rat PAM proteins that range in size from approximately 35,000–108,000 molecular weight (see Fig. 5; Eipper et al., 1992a,b). These and other data revealed that PAM is a bifunctional enzyme (Kato et al., 1990; Katopodis et al., 1990, 1991; Perkins et al., 1990; Takahashi et al., 1990). The large PAM precursor protein or PAM-1 contains an N-terminal signal sequence, a short pro sequence, a catalytic region consisting of peptidylglycine α-hydroxylating monoxygenase (PHM or EC 1.14.17.3), a connecting domain, a second catalytic region consisting of peptidyl-α-hydroxyglycine α-amidating lyase (PAL or EC 4.3.2.5), a transmembrane domain, and a short C-terminal (see Fig. 5; Eipper et al., 1992a). PAM-

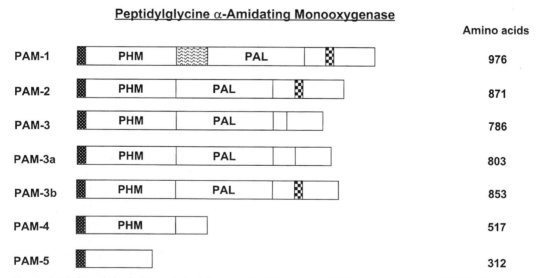

Peptidylglycine α-Amidating Monooxygenase

Amino acids

PAM-1	976
PAM-2	871
PAM-3	786
PAM-3a	803
PAM-3b	853
PAM-4	517
PAM-5	312

Fig. 5. Depiction of the 7 forms derived from peptidylglycine α-amidating monooxygenase (PAM). These enzymes are synthesized as proenzymes. The catalytic peptidylglycine α-hydroxylating monoxygenase (PHM) and peptidyl-α-hydroxyglycine α-amidating lyase (PAL) domains are shown. The PAM-1, -2, and -3b contain transmembrane domains, while the PAM-3, -3a, -4, and -5 are soluble proteins. The significance of the shading for each enzyme can be found at the bottom of Figure 3. The shading in PAM-1 between the PHM and PAL domains in encoded by an exon whose product is absent in all other PAM forms. The numbers of amino acids that comprise these enzymes are listed to the right.

1, -2, -3, -3a, and -3b all contain both PHM and PAL catalytic regions, while only PAM-1, -2, and -3b contain the transmembrane domains (Stoffers et al., 1989, 1991; Eipper et al., 1992a). PAM-4 has the PHM domain. PAM-5 is shorter than PAM-4 and is not thought to possess enzymatic activity. Its function is currently unknown.

PHM catalyzes the first step of the amidation reaction where the C-terminal glycine is hydroxylated, while PAL is responsible for the second reaction where the glyoxylate moiety is removed from the C-terminal of the intermediate and the amide group from the original hydroxyglycine is donated to the peptide (Bradbury et al., 1982; Eipper et al., 1983; Kato et al., 1990; Katopodis et al., 1990, 1991; Perkins et al., 1990). In mammals the PHM and PAL may be expressed within the same precursor, while in other organisms they may be expressed independently of each other (Prigge et al., 2000). Interestingly, in *Drosophila* deletion of the PHM gene is embryonically lethal (Kolhekar et al., 1997).

In a survey of the various bioactive peptides in tissues, all 20 amino acid residues at the C-terminal have the ability to be amidated. Besides peptides, purified PHM can catalyze O-dealkylation, N-dealkylation, and sulfoxidation reactions of non-peptide substrates (Katopodis and May, 1990). Recombinant PAM can convert fatty acyl glycines to fatty acyl amides such as oleamide that has been implicated in sleep (Cravatt et al., 1995). It can also convert nicotinuric acid to nicotinamide (Merkler et al., 1999). Since PAM is the only amidating enzyme that has been identified so far, it is presumed to be responsible for these various reactions (Eipper et al., 1992a; Prigge et al., 2000). Despite this fact, glycine-extended peptides for gastrin, adrenomedullin, and oxytocin have been found in blood (Amico and Hempel, 1990; Kaise et al., 1995; Kitamura et al., 1998). Additionally, it is recognized that gastrin-Gly[18] can bind to a receptor that is distinct from that for gastrin (Kaise et al., 1995; Dockray et al., 1996). With respect to GnRH, glycine-extended peptides have been isolated from hypothalamus and the immortalized hypothalamic GnRH neurons, however, the levels are very low (Wetsel et al., 1991). Despite this fact, PAM is expressed in the GT1 cells and in the brain region where GnRH perikarya are located (Wetsel, 1995; Wetsel et al., 1995). In addition, since

chronic treatment with diethyldithiocarbamate (an inhibitor of PAM) leads to an increase in LHRH-Gly[11]-like immunoreactivity in the hypothalamus (Kumar et al., 1994), PAM is probably intimately involved in the in vivo processing of the pro-GnRH protein to GnRH.

One substance that has been shown to act directly on GnRH neurons (GT1 cells) and to regulate pro-GnRH processing is basic fibroblast growth factor (Wetsel et al., 1996). Although the immortalized cells contain receptors for this growth factor, no effects on GnRH secretion were discerned (Voigt et al., 1996). This was a surprising result because in most systems, basic fibroblast growth factor stimulates protein kinase C activation. This kinase is known to be a strong stimulator of GnRH secretion from median eminence tissue fragments in vitro (Valença et al., 1988) and from the immortalized GnRH neurons (Bruder et al., 1992; Wetsel et al., 1993). Analysis of the media from stimulated cells reveals, however, that the conversion of GnRH–Gly[11] to GnRH was partially blocked so that the glycine-extended peptide was one of the predominant forms that was secreted (Wetsel et al., 1996). Presently, the physiological significance of the release of this peptide is unknown because it does not stimulate the secretion of LH or FSH from primary rat anterior pituitary cell cultures.

Glutaminyl cyclase

In the initial isolation and sequencing of GnRH, there was difficulty in determining the N-terminal amino acid. Mass spectrometry was required to identify the pyroglutamate in this position (Matsuo et al., 1971; Burgus et al., 1972). With the cloning of the human pro-GnRH, the deduced sequence revealed that the N-terminal amino acid for GnRH was a glutamine (Seeburg and Adelman, 1984). Since a pyroglutamyl residue is not a commonly occurring amino acid and because N-terminal glutamine is unstable, it was thought that this residue arose spontaneously under physiological conditions. Later, two groups independently isolated an enzyme that could convert glutaminyl into pyroglutaminyl peptides (Busby et al., 1987; Fischer and Spiess, 1987). The subcellular distribution of glutaminyl cyclase (QC) and PAM are virtually identical. QC activity

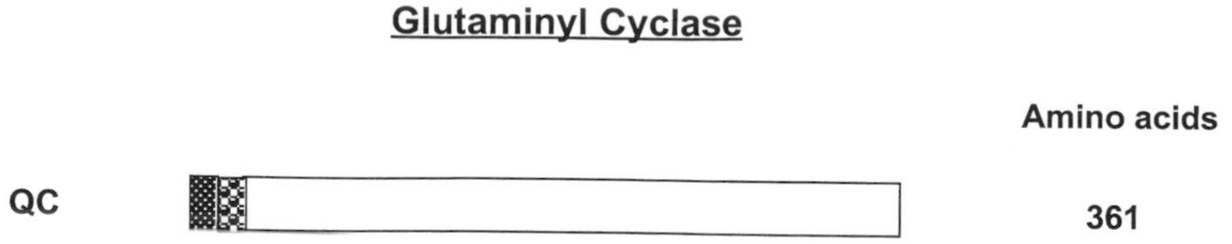

Glutaminyl Cyclase

Amino acids

QC   361

Fig. 6. A diagram showing the structure of the bovine glutaminyl cyclase. The significance of the shading for this enzyme can be found at the bottom of Figure 3. The numbers of amino acids that comprise this enzyme is listed to the right.

is especially high in secretory vesicles (Fischer and Spiess, 1987). Importantly, Gln^1–GnRH and Gln^1–TRH–Gly^4 are converted to GnRH and TRH–Gly^4, respectively, with extracts from bovine hypophysis and rat hypothalamus. The rate of this conversion is much more rapid than that which occurs spontaneously. QC (EC 2.3.2.5) was later isolated, purified, sequenced, and cloned from bovine anterior pituitary (see Fig. 6; Pohl et al., 1991). The soluble protein has a molecular weight of approximately 38,000 and it is expressed in anterior and neurointermediate pituitary, striatum, hypothalamus, and most other brain regions. More recent evidence suggests that QC may exist in multiple molecular forms and these forms appear in a tissue-specific manner (Sykes et al., 1999).

In an experiment, we showed that QC was expressed in the murine immortalized GnRH neurons and in the rat preoptic area/hypothalamus (Wetsel et al., 1995). More recently, in our investigations with the GT1 neurons, we reported that basic fibroblast growth factor stimulated a preferential release of GnRH–Gly^{11} (Wetsel et al., 1996). An examination of the lysates from the neurons revealed that conversion of the N-terminal Gln^1–GnRH to GnRH was also retarded by this factor such that release of Gln^1 peptides was increased more than 5-fold over baseline. To examine the biological activity of these GnRH intermediates, they were incubated in dispersed primary anterior pituitary cell cultures. Only Gln^1–GnRH showed some activity. Despite this fact, it is not clear whether the stimulation of LH and FSH to this peptide were due to its own actions or to some possible conversion to GnRH. Nevertheless, these data show that processing of the pro-GnRH can be regulated and that some of the intermediate peptides may possess biological activity at the pituitary. Since the GnRH neurons in the hypothalamus are also subject to negative ultrashort feedback (Valença

et al., 1987), it may be the case that some of the pro-GnRH intermediates can also control secretion of the decapeptide.

Perspective

Regulation of peptide bioactivity is a complex process. Cells have devised ingenious mechanisms to control the amounts of peptides available for release. Some of these processes include gene transcription, pre-RNA splicing, protein biosynthesis, trafficking of the precursor to the regulated secretory pathway, conversion of the pro-peptide to fully processed product, secretion, and degradation. There is considerable controversy regarding the regulation of GnRH transcription in vivo and in many cases, control appears to occur posttranscriptionally (Gore and Roberts, 1997; Seong et al., 1999, 2001; Pitts et al., 2001). The topology of the hypophysiotrophic GnRH neurons reveals that a considerable distance separates the perikarya of these neurons in the preoptic area/diagonal band of Broca from the nerve terminals in the median eminence. Hence, one might anticipate that it would take a substantial amount of time for newly synthesized and processed GnRH to appear in the nerve terminal following gene expression in the perikarya. Since relatively rapid and pronounced changes can occur in the concentrations of GnRH in the arcuate nucleus and median eminence during proestrus or other steroid-induced changes (Kalra and Kalra, 1985; Culler et al., 1987) and because the median eminence contains high levels of pro-GnRH relative to that for other neuropeptides (see Wetsel et al., 1988), it is very likely that some of these rapid alterations may be due to processing of the prohormone or its intermediates to GnRH in the nerve terminals. In this way, bioactive GnRH could be supplied relatively quickly in preparation for secretion. Since many different hormones, transmitters, peptides and lipids can affect GnRH secretion (Wetsel, 1995), it will be important to determine whether these agents can also influence the conversion of the pro-GnRH to the bioactive decapeptide. In this context, recent data on the possible roles of the PCs, CPE, PAM, and QC in pro-GnRH processing afford new perspectives on studies of GnRH function.

It has long been known that the activity of GnRH is primarily curtailed through degradation (Koch et al., 1974; O'Cuinn et al., 1990; Lew et al., 1994). Recently we have found that the immortalized GnRH neurons can degrade the decapeptide into a variety of different fragments. The appearance and patterns of fragments are consistent with our mRNA results where we find that angiotensin converting enzyme, neutral endopeptidase, metalloendopeptidase, and prolyl endopeptidase are all expressed in these neurons. Collectively, these data show not only that biosynthesis and processing of the prohormone occur within GnRH neurons, but that degradation can also occur. Inasmuch as some the GnRH fragments have biological activities on their own (Chen et al., 1993; Bourguignon et al., 1994), these findings suggest that GnRH, itself, may serve as a pro-peptide and that the pro-GnRH-derived peptides may elicit pleiotropic responses.

Abbreviations

CPD	carboxypeptidase D
CPE	carboxypeptidase E
FSH	follicle-stimulating hormone
GAP	gonadotropin-releasing hormone associated peptide
GnRH	gonadotropin-releasing hormone
[Hyp^9]GnRH	[hydroxyproline^9]GnRH
LH	luteinizing hormone
PACE	paired amino acid converting enzyme
PAL	peptidyl-α-hydroxyglycine α-amidating lyase
PAM	peptidylglycine α-amidating monooxygenase
PC	prohormone convertase
pGlu^1	pyroglutamate
PHM	peptidylglycine α-hydroxylating monooxygenase
QC	glutaminyl cyclase
WT	wild type

Acknowledgments

We would like to thank all of the past students, postdoctoral fellows, and collaborators who contributed to these studies. Some of the work described in this paper was supported by NIH grant HD36015.

References

Ackland, J.F., Nikolics, K., Seeburg, P.H. et al. (1988) Molecular forms of gonadotropin-releasing hormone associated peptide (GAP): changes within the rat hypothalamus and release from hypothalamic cells in vitro Neuroendocrinology, 48. 376–386.

Adelman, J.P., Mason, A.J., Hayflick, J.S. et al. (1986) Isolation of the gene and hypothalamic cDNA for the common precursor of gonadotropin-releasing hormone and prolactin release-inhibiting factor in human and rat. Proc. Natl. Acad. Sci. USA, 83: 179–183.

Allen, R.G., Peng, B., Pellegrino, M.J. et al. (2001) Altered processing of pro-orphanin FQ/nociceptin and pro-opiomelanocortin-derived peptides in the brains of mice expressing defective prohormone convertase 2. J. Neurosci., 21: 5864–5870.

Aloy, P., Companys, V., Vendrell, J. et al. (2001) The crystal structure of the inhibitor-complexed carboxypeptidase D domain II and the modeling of regulatory carboxypeptidases. J. Biol. Chem., 276: 16177–16184.

Amico, J.A. and Hempel, J. (1990) An oxytocin precursor intermediate circulates in the plasma of humans and rhesus monkeys administered estrogen. Neuroendocrinology, 51: 437–443.

Aschner, B. (1912) Ueber die fuktion der hypophyse. Pfluegers Arch. Ges. Physiol., 146: 1–7.

Bergeron, F., Leduc, R. and Day, R. (2000) Subtilase-like proprotein convertases: from molecular specificity to therapeutic applications. J. Mol. Endocrinol., 24: 1–22.

Berman, Y., Mzhavia, N., Polonskaia, A. et al. (2000) Defective prodynorphin processing in mice lacking prohormone convertase PC2. J. Neurochem., 75: 1763–1770.

Bogerd, J., Zandbergen, T., Andersson, E. et al. (1994) Isolation, characterization and expression of cDNAs encoding the catfish-type and chicken-II-type gonadotropin-releasing hormone precursors in the African catfish. Eur. J. Biochem., 222: 541–549.

Bourguignon, J.P., Alvarez Gonzalez, M.L. et al. (1994) Gonadotropin releasing hormone inhibitory autofeedback by subproducts antagonist at N-methyl-D-aspartate receptors: a model of autocrine regulation of peptide secretion. Endocrinology, 134: 1589–1592.

Bradbury, A.F., Finnie, D.A. and Smyth, D.G. (1982) Mechanism of C-terminal amide formation by pituitary enzymes. Nature, 298: 686–688.

Braks, J.A. and Martens, G.J. (1994) 7B2 is a neuroendocrine chaperone that transiently interacts with prohormone convertase PC2 in the secretory pathway. Cell, 78: 263–273.

Bruder, J.M., Krebs, W.D., Nett, T.M. et al. (1992) Phorbol ester activation of the protein kinase C pathways inhibits gonadotropin-releasing hormone gene expression. *Endocrinology*, 131: 2552–2558.

Bruzzaniti, A., Goodge, K., Jay, P. et al. (1996) PC8, a new member of the convertase family. *Biochem. J.*, 314: 727–731.

Burgus, R., Butcher, M., Amoss, M. et al. (1972) Primary structure of ovine luteinizing hormone-releasing factor (LRF). *Proc. Natl. Acad. Sci. USA*, 69: 278–282.

Busby, W.H. Jr., Quackenbush, G.E., Humm, J. et al. (1987) An enzyme(s) that converts glutaminyl-peptides into pyroglutamyl-peptides. Presence in pituitary, brain, adrenal medulla, and lymphocytes. *J. Biol. Chem.*, 262: 8532–8536.

Campbell, H.J., Feuer, G. and Harris, G.W. (1964) The effect of intrapituitary infusion of median eminence and other brain extracts on anterior pituitary gonadotropic secretion. *J. Physiol. (Lond.)*, 170: 474–486.

Carolsfeld, J., Powell, J.F.F., Park, M. et al. (2000) A novel form of gonadotropin-releasing hormone (GnRH) in herring sheds light on evolutionary pressures. *Endocrinology*, 141: 505–512.

Cattanach, B.M., Iddon, C.A., Charlton, H.M. et al. (1977) Gonadotropin-releasing hormone deficiency in a mutant mouse with hypogonadism. *Nature*, 269: 338–340.

Chandrashekar, V., Bartke, A. and Browning, R.A. (1988) Assessment of the effects of a synthetic gonadotropin-releasing hormone associated peptide on hormone release from the in situ and ectopic pituitaries in adult male rats. *Brain Res. Bull.*, 21: 95–99.

Chen, A., Yahalom, D., Ben-Aroya, N. et al. (1998) A second isoform of gonadotropin-releasing hormone is present in the brain of human and rodents. *FEBS Lett.*, 435: 199–203.

Chen, H., Jawahar, S., Qian, Y. et al. (2001) Missense polymorphism in the human carboxypeptidase E gene alters enzymatic activity. *Human Mutat.*, 18: 120–131.

Chen, Y., Wong, M. and Moss, R.L. (1993) Effects of a biologically active LHRH fragment on CA1 hippocampal neurons. *Peptides*, 14: 1079–1081.

Chertow, B.S. (1981) The role of lysosomes and proteases in hormone secretion and degradation. *Endocr. Rev.*, 2:137–173.

Clarke, I.J., Cummins, J.T., Karsch, F.J. et al. (1987) GnRH-associated peptide (GAP) is cosecreted with GnRH into the hypophyseal portal blood of ovariectomized sheep. *Biochem. Biophys. Res. Commun.*, 143: 665–671.

Coleman, D.L. and Eichler, E.M. (1990) Fat (*fat*) and tubby (*tub*): two autosomal recessive mutations causing obesity syndromes in mice. *J. Hered.*, 81: 424–427.

Constam, D.B. and Robertson, E.J. (2000) SPC4/PACE4 regulates a TGF-beta signaling network during axis formation. *Genes Dev.*, 14: 1146–1155.

Cravatt, B.F., Prospero-Garcia, O., Siuzdak, G. et al. (1995) Chemical characterization of a family of brain lipids that induce sleep. *Science*, 268: 1506–1509.

Culler, M.D., Culler, M.D., Wetsel, W.C. et al. (1987) Orchidectomy induces temporal and regional changes in the synthesis and processing of the LHRH prohormone in the rat brain. In: V.B. Mahesh, D.S. Dhindsa, E. Anderson and S.P. Kalra

(Eds.), *Regulation of Ovarian and Testicular Function*. Plenum Press, New York, pp. 623–628.

Dellovade, T.L., King, J.A., Millar, R.P. et al. (1993) Presence and differential distribution of distinct forms of immunoreactive gonadotropin-releasing hormone in the musk shrew brain. *Neuroendocrinology*, 58: 166–177.

Dockray, G.J., Varro, A. and Dimaline, R. (1996) Gastric endocrine cells: gene expression, processing, and targeting of active products. *Physiol. Rev.*, 76: 767–798.

Dong, W., Marcinkiewicz, M., Vieau, D. et al. (1995) Distinct mRNA expression of the highly homologous convertases PC5 and PACE4 in the rat brain and pituitary. *J. Neurosci.*, 15: 1778–1796.

Dong, W., Fricker, L.D. and Day, R. (1999) Carboxypeptidase D is a potential candidate to carry out redundant processing functions of carboxypeptidase E is based on comparative distribution studies in the rat central nervous system. *Neuroscience*, 89: 1301–1317.

Douglass, J., Civelli, O. and Herbert, E. (1984) Polyprotein gene expression: generation of diversity of neuroendocrine peptides. *Ann. Rev. Biochem.*, 53: 665–704.

Dunn, I.C., Chen, Y., Hook, C. et al. (1993) Characterization of the chicken preprogonadotropin-releasing hormone-I gene. *J. Mol. Endocrinol.*, 11: 19–29.

Eipper, B.A., Mains, R.E. and Glembotski, C.C. (1983) Identification in pituitary tissue of a peptide α-amidation activity that acts on glycine-extended peptides and requires molecular oxygen, copper, and ascorbic acid. *Proc. Natl. Acad. Sci. USA*, 80: 5144–5148.

Eipper, B.A., Park, L.P., Dickerson, I.M. et al. (1987) Structure of the precursor to an enzyme mediating COOH-terminal amidation in peptide biosynthesis. *Mol. Endocrinol.*, 1: 777–790.

Eipper, B.A., Stoffers, D.A. and Mains, R.E. (1992a) The biosynthesis of neuropeptides: peptide α-amidation. *Ann. Rev. Neurosci.*, 15: 57–85.

Eipper, B.A., Green, C.B.-R., Campbell, T.A. et al. (1992b) Alternative splicing and endoproteolytic processing generate tissue-specific forms of pituitary peptidylglycine α-amidating monooxygenase (PAM). *J. Biol. Chem.*, 267: 4008–4015.

Erdheim, J. and Stumme, E. (1909) Ueber die schwangerschaftsveranderung der hypophyse. *Beitr. Pathol. Anat. Allgem. Pathol.*, 46: 1–4.

Feng, Y., Reznik, S.E. and Fricker, L.D. (2001) Distribution of proSAAS-derived peptides in rat neuroendocrine tissues. *Neuroscience*, 105: 469–478

Fernald, R.D. and White, R.B. (1999) Gonadotropin-releasing hormone genes: phylogeny, structure, and function. *Front. Neuroendocrinology*, 20: 224–240.

Fischer, W.H. and Spiess, J. (1987) Identification of a mammalian glutaminyl cyclase converting glutaminyl into pyroglutamyl peptides. *Proc. Natl. Acad. Sci. USA*, 84: 3628–3632.

Fricker, L.D. (1991) Peptide processing exopeptidases: amino and carboxypeptidases involved with peptide biosynthesis. In: L.D. Fricker (Ed.), *Peptide Biosynthesis and Processing*. CRC Press, Boca Raton, pp. 199–229.

Fricker, L.D. (1998) Carboxypeptidase E/H. In: A.J. Barrett,

N.D. Rawlings, J.F. Woessner (Eds.), *Handbook of Proteolytic Enzymes*. Academic Press, London, pp. 1341–1344.

Fricker, L.D. and Snyder, S.H. (1972) Enkephalin convertase: purification and characterization of a specific enkephalin-synthesizing carboxypeptidase localized to adrenal chromaffin granules. *Proc. Natl. Acad. Sci. USA*, 79: 3886–3890.

Fricker, L.D., Evans, C.J., Esch, F.S. et al. (1986) Cloning and sequence analysis of cDNA for bovine carboxypeptidase E. *Nature*, 323: 461–464.

Fricker, L.D., Das, B. and Angeletti, R.H. (1990) Identification of the pH-dependent membrane anchor of carboxypeptidase E (EC 3.4.17.10). *J. Biol. Chem.*, 265: 2476–2482.

Fricker, L.D., McKinzie, A.A., Sun, J. et al. (2000) Identification and characterization of proSAAS, a granin-like neuroendocrine peptide precursor that inhibits prohormone processing. *J. Neurosci.*, 20: 639–648.

Furuta, M., Yano, H., Zhou, A. et al. (1997) Defective prohormone processing and altered pancreatic islet morphology in mice lacking active SPC2. *Proc. Natl. Acad. Sci. USA*, 94: 6646–6651.

Furuta, M., Zhou, A., Webb, G. et al. (2001) Severe defect in proglucagon processing in islet A-cells of prohormone convertase 2 null mice. *J. Biol. Chem.*, 276: 27197–27202.

Gautron, J.-P., Poulin, B., Kordon, C. et al. (1995) Characterization of [hydroxyproline[9]]luteinizing hormone-releasing hormone and its smallest precursor forms in immortalized luteinizing hormone-releasing hormone-secreting neurons (GT1–7), and evaluation of their mode of action on pituitary cells. *Mol. Cell. Endocrinol.*, 110: 161–173.

Gomis-Rüth, F.X., Companys, V., Qian, Y. et al. (1999) Crystal structure of avian carboxypeptidase D domain II: a prototype for the regulatory metallocarboxypeptidase subfamily. *EMBO J.*, 18: 5817–5826.

Gore, A.C. and Roberts, J.L. (1997) Regulation of gonadotropin-releasing hormone gene expression in vivo and in vitro. *Front. Neuroendocrinol.*, 18: 209–245.

Gothilf, Y., Elizur, A., Chow, M. et al. (1995) Molecular cloning and characterization of a novel gonadotropin-releasing hormone from the gilthead seabream (*Sparus aurata*). *Mol. Marine Biol. Biotechnol.*, 4: 27–35.

Halasz, B., Pupp, L. and Uhlarik, S. (1962) Hypophysiotrophic area in the hypothalamus. *J. Endocrinol.*, 25: 147–154.

Harris, G.W. and Jacobsohn, D. (1952) Functional grafts of anterior pituitary gland. *Proc. Roy. Soc. Ser. B*, 139: 263–276.

Ishibashi, M., Yamaji, T., Takaku, F. et al. (1987) Effect of GnRH-associated peptide on prolactin secretion from human lactotrope adenoma cells in culture. *Acta Endocrinol. (Cophen.)*, 116: 81–84.

Jackson, R.S., Creemers, J.W., Ohagi, S. et al. (1997) Obesity and impaired prohormone processing associated with mutations in the human prohormone convertase 1 gene. *Nat. Genet.*, 16: 303–306.

Jimenez-Liñan, M., Rubin, B. and King, J.C. (1997) Examination of guinea pig luteinizing hormone-releasing hormone gene reveals a unique decapeptide and existence of two transcripts in the brain. *Endocrinology*, 138: 4123–4137.

Johanning, K., Juliano, M.A., Juliano, L. et al. (1998) Specificity of prohormone convertase 2 on proenkephalin and proenkephalin-related substrates. *J. Biol. Chem.*, 273: 22672–22680.

Julius, D., Brake, A., Blair, L. et al. (1984) Isolation of the putative structural gene for the lysine-arginine-cleaving endopeptidase required for processing of yeast prepro-alpha-factor. *Cell*, 37: 1075–1089.

Kaise, M., Muraoka, A., Seva, C. et al. (1995) Glycine-extended progastrin processing intermediates induce $H^+,K(+)$-ATPase alpha-subunit gene expression through a novel receptor. *J. Biol. Chem.*, 270: 11155–11160.

Kalra, P.S. and Kalra, S.P. (1985) Control of gonadotropin secretion. In: H. Imura (Ed.), *The Pituitary Gland*. Raven Press: New York, pp. 189–220.

Karten, M.J. and Rivier, J.E. (1986) Gonadotropin-releasing hormone analog design. Structure-function studies toward the development of agonists and antagonists: rationale and perspectives. *Endocrine Rev.*, 7: 44–66.

Kasten, T.L., White, S.A., Norton, T.T. et al. (1996) Characterization of two new preproGnRH mRNAs in the tree shrew: first direct evidence for mesencephalic GnRH gene expression in a placental mammal. *Gen. Comp. Endocrinol.*, 104: 7–19.

Kato, I., Yonekura, H., Tajima, M., et al. (1990) Two enzymes concerned in peptide hormone alpha-amidation are synthesized from a single mRNA. *Biochem. Biophys. Res. Commun.*, 172: 197–203.

Katopodis, A.G. and May, S.W. (1990) Novel substrates and inhibitors of peptidylglycine alpha-amidating monooxygenase. *Biochemistry*, 29: 4541–4548.

Katopodis, A.G., Ping, D. and May, S.W. (1990) A novel enzyme from bovine neurointermediate pituitary catalyzes dealkylation of alpha-hydroxyglycine derivatives, thereby functioning sequentially with peptidylglycine alpha-amidating monooxygenase in peptide amidation. *Biochemistry*, 29: 6115–6120.

Katopodis, A.G., Ping, D.S., Smith, C.E. et al. (1991) Functional and structural characterization of peptidylamidoglycolate lyase, the enzyme catalyzing the second step in peptide amidation. *Biochemistry*, 30: 6189–6194.

Kiefer, M.C., Tucker, J.E., Joh, R. et al. (1991) Identification of a second human subtilisin-like protease gene in the fes/fps region of chromosome 15. *DNA Cell Biol.*, 10: 757–769.

King, J.A. and Millar, R.P. (1982) Structure of chicken hypothalamic luteinizing hormone-releasing hormone I. Structural determination on partially purified material. *J. Biol. Chem.*, 257: 10772–10732.

Kitamura, K., Kato, J., Kawamoto, M. et al. (1998) The intermediate form of glycine-extended adrenomedullin is the major circulating molecular form in human plasma. *Biochem. Biophys. Res. Commun.*, 244: 551–555.

Klungland, H., Lorens, J.B., Andersen, O. et al. (1992) The Atlantic salmon prepro-gonadotropin releasing hormone gene and mRNA. *Mol. Cell. Endocrinol.*, 84: 167–174.

Knigge, K.M. (1962) Gonadotropic activity of neonatal pituitary glands implanted in the rat brain. *Am. J. Physiol.*, 202: 387–391.

Koch, Y., Baram, T., Chobsieng, P. et al. (1974) Enzymatic degradation of luteinizing hormone-releasing hormone (LH-

RH) by hypothalamic tissue. *Biochem. Biophys. Res. Commun.*, 61: 95–103.

Kolhekar, A.S., Roberts, M.S., Jiang, N. et al. (1997) Neuropeptide amidation in *Drosophila*: separate genes encode the two enzymes catalyzing amidation. *J. Neurosci.*, 17: 1363–1376.

Kumar, A.M., Agarwal, R.K., Thompson, M.L. et al. (1994) Effect of chronic DDC treatment on LHRH and substance P amidation processes in the rat. *Brain Res. Bull.*, 33: 337–344.

Kuroki, K., Cheung, R., Marion, P. et al. (1994) A cell surface protein that binds avian hepatitis B virus particles. *J. Virol.*, 68: 2091–2096.

Kuroki, K., Eng, F., Ishikawa, T. et al. (1995) gp180, a host cell glycoprotein that binds duck hepatitis B virus particles, is encoded by a member of the carboxypeptidase gene family. *J. Biol. Chem.*, 270: 15022–15028.

Lanoue, E. and Day, R. (2001) Coexpression of proprotein convertase SPC3 and the neuroendocrine precursor proSAAS. *Endocrinology*, 142: 4141–4149.

Leiter, E.H. and Herberg, L. (1997) The polygenetics of diabetes in mice. *Diabetes Rev.*, 5: 131–148.

Lescheid, D.W., Teresawa, E., Abler, L.A. et al. (1997) A second form of gonadotropin-releasing hormone (GnRH) with characteristics if chicken GnRH-II is present in primate brain. *Endocrinology*, 138: 5618–5629.

Lew, R.A., Tetaz, T.J., Glucksman, M.J. et al. (1994) Evidence for a two-step mechanism of gonadotropin-releasing hormone metabolism by proylyl endopeptidase and metalloendopeptidase EC 3.4.24.15 in ovine hypothalamic extracts. *J. Biol. Chem.*, 269: 12626–12632.

Lewin, B. (1996) *Genes V.* Oxford University Press, Oxford, pp. 168.

Lindberg, I., Van den Hurk, W.H., Bui, C. et al. (1995) Enzymatic characterization of immunopurified convertase 2. Potent inhibition by a 7B2 peptide fragment. *Biochemistry*, 34: 5486–5493.

Lovejoy, D.A., Fischer, W.H., Ngamvongchon, S. et al. (1992) Distinct sequence of gonadotropin-releasing hormone (GnRH) in dogfish brain provides insight into GnRH evolution. *Proc. Natl. Acad. Sci. USA*, 89: 6373–6377.

Lusson, J., Vieau, D., Hamelin, J. et al. (1993) cDNA structure of the mouse and rat subtilisin/kexin-like PC5: a candidate proprotein convertase expressed in endocrine and neuroendocrine cells. *Proc. Natl. Acad. Sci. USA*, 90: 6691–6695.

Mason, A.J., Hayflick, J.S., Zoeller, R.T. et al. (1986a) A deletion truncating the gonadotropin-releasing hormone gene is responsible for hypogonadism in the *hpg* mouse. *Science*, 234: 1366–1371.

Mason, A.J., Pitts, S.L., Nikolics, K. et al. (1986b) The hypogonadal mouse: reproductive functions restored by gene therapy. *Science*, 234: 1372–1378.

Matsuo, H., Baba, Y., Nair, R.M.G. et al. (1971) Structure of the porcine LH- and FSH-releasing hormone. I. The proposed amino acid sequence. *Biochem. Biophys. Res. Commun.*, 43: 1334–1339.

Mbikay, M., Tadros, H., Ishida, N. et al. (1997) Impaired fertility in mice deficient for the testicular germ-cell protease PC4. *Proc. Natl. Acad. Sci. USA*, 94: 6842–6846.

McCann, S.M., Taleisnik, S. and Friedman, H.M. (1960) LH-releasing activity in hypothalamic extracts. *Proc. Soc. Exp. Biol. Med.*, 104: 432–434.

Merkler, D.J., Glufke, U., Ritenour-Rodgers, K.J. et al. (1999) Formation of nicotinamide from nicotinuric acid by peptidylglycine alpha-amidating monooxygenase (PAM): A possible alternative route from nicotinic acid (niacin) to NADP in mammals. *J. Amer. Chem. Soc.*, 121: 4904–4905.

Milton, R.C.deL., Wormald, P.J., Brandt, W. et al. (1986) The delineation of a decapeptide gonadotropin-releasing sequence in the carboxy-terminal extension of the human gonadotropin-releasing hormone precursor. *J. Biol. Chem.*, 261: 16990–16997.

Miyamoto, K., Hasegawa, Y., Nomura, M. et al. (1984) Identification of the second gonadotropin-releasing hormone in chicken hypothalamus: evidence that gonadotropin secretion is probably controlled by two distinct gonadotropin-releasing hormones in avian species. *Proc. Natl. Acad. Sci. USA*, 81: 3874–3878.

Mizuno, K., Ohsuye, K., Wada, Y. et al. (1987) Cloning and sequence of cDNA encoding a peptide C-terminal α-amidating enzyme from *Xenopus laevis*. *Biochem. Biophys. Res. Commun.*, 148: 546–552.

Montaner, A.D., Park, M.K., Fischer, W.H. et al. (2001) Primary structure of a novel gonadotropin-releasing hormone in the brain of a teleost, pejerry. *Endocrinology*, 142: 1453–1460.

Naggert, J.K., Fricker, L.D., Varlamov, O. et al. (1995) Hyperproinsulinaemia in obese *fat/fat* mice associated with a carboxypeptidase E mutation which reduces enzyme activity. *Nat. Genet.*, 10: 135–142.

Nakagawa, T., Murakami, K. and Nakayama, K. (1993a) Identification of an isoform with an extremely large Cys-rich region of PC6, a Kex2-like processing endoprotease. *FEBS Lett.*, 327: 165–171.

Nakagawa, T., Hosaka, M., Torii, S. et al. (1993b) Identification and functional expression of a new member of the mammalian Kex2-like processing endoprotease family: its striking structural similarity to PACE4. *J. Biochem. (Tokyo)*, 113: 132–135.

Nakayama, K., Kim, W.S., Torii, S. et al. (1992) Identification of the fourth member of the mammalian endoprotease family homologous to the yeast Kex2 protease. Its testis-specific expression. *J. Biol. Chem.*, 267: 5897–5900.

Ngamvongchon, S., Lovejoy, D.A., Fischer, W.H. et al. (1992) Primary structures of two forms of gonadotropin-releasing hormone, one distinct and one conserved, from catfish brain. *Mol. Cell. Neurosci.*, 3: 17–22.

Nikitovitch-Winer, M. and Everett, J.W. (1958) Functional reconstitution of pituitary grafts re-transplanted from kidney to median eminence. *Endocrinology*, 63: 916–930.

Nikolics, K., Mason, A.J., Szõnyi, E. et al. (1985) A prolactin-inhibiting factor within the precursor form human gonadotropin-releasing hormone. *Nature*, 316: 511–517.

Novikova, E.G., Eng, F.J., Yan, L. et al. (1999) Characterization of the enzymatic properties of the first and second domains of metallocarboxypeptidase D. *J. Biol. Chem.*, 274: 28887–28892.

O'Cuinn, G., O'Connor, B. and Elmore, M. (1990) Degradation of thyrotropin-releasing hormone and luteinizing hormone-releasing hormone by enzymes in brain tissue. *J. Neurochem.*, 4: 1–13.

Okada, T., Kitamura, K., Baba, Y. et al. (1973) Luteinizing hormone-releasing hormone analogs lacking N-terminal pGLU ring structure. *Biochem. Biophys. Res. Commun.*, 53: 1180–1186.

Okubo, K., Amano, M., Yoshiura, Y. et al. (2000) A novel form of gonadotropin-releasing hormone in the madaka, *oryzias latipes*. *Biochem. Biophys. Res. Commun.*, 276: 298–303.

O'Rahilly, S., Gray, H., Humphreys, P.J. et al. (1995) Brief report: impaired processing of prohormones associated with abnormalities of glucose homeostasis and adrenal function. *New Engl. J. Med.*, 333: 1386–1390.

Ouafik, L.H., Stoffers, D.A., Campbell, T.A. et al. (1992) The multifunctional peptidylglycine α-amidating monooxygenase gene: exon/intron organization of catalytic, processing, and routing domains. *Mol. Endocrinol.*, 6: 1571–1584.

Parkinson, D. (1990) Two soluble forms of bovine carboxypeptidase H have different NH_2-terminal sequences. *J. Biol. Chem.*, 265: 17101–17105.

Perkins, S.N., Husten, E.J. and Eipper, B.A. (1990) The 108-kDA peptidylglycine alpha-amidating monooxygenase precursor contains two separable enzymatic activities involved in peptide amidation. *Biochem. Biophys. Res. Commun.*, 171: 926–932.

Pitts, G.R., Nunemaker, C.S. and Moenter, S.M. (2001) Cycles of transcription and translation do not comprise the gonadotropin-releasing hormone pulse generator in GT1 cells. *Endocrinology*, 142: 1858–1864.

Planas, J., Bern, H.A. and Millar, R.P. (1990) Effects of GnRH-associated peptide and its component peptides on prolactin secretion from the tilapia pituitary in vitro. *Gen. Comp. Endocrinol.*, 77: 386–396.

Pohl, T., Zimmer, M., Mugele, K. et al. (1991) Primary structure and functional expression of a glutaminyl cyclase. *Proc. Natl. Acad. Sci. USA*, 88: 10059–10063.

Powell, J.F.F., Zohar, Y., Elizur, A. et al. (1994) Three forms of gonadotropin-releasing hormone characterized from brains of one species. *Proc. Natl. Acad. Sci. USA*, 91: 12081–12085.

Powell, J.F.F., Reska-Skinner, S.M., Om Prakash, M. et al. (1996) Two new forms of gonadotropin-releasing hormone in a protochordate and the evolutionary implications. *Proc. Natl. Acad. Sci. USA*, 93: 10461–10464.

Prigge, S.T., Mains, R.E., Eipper, B.A. et al. (2000) New insights into copper monooxygenases and peptide amidation: structure, mechanism and function. *Cell. Mol. Life Sci.*, 57: 1236–1259.

Prochazka, M., Gold, D.P., Castano, L. et al. (1991) Mapping of secretogranin 1 (*Scg-1*) to Chr 2, and carboxypeptidase H (*Cph-1*) to Chr 8. *Mouse Genome*, 89: 280.

Qian, Y., Devi, L.A., Mzhavia, N. et al. (2000) The C-terminal region of proSAAS is a potent inhibitor of prohormone convertase 1. *J. Biol. Chem.*, 275: 23596–23601.

Reeves, W.B., Bichet, D.G. and Andreoli, T.E. (1998) Posterior pituitary and water metabolism. In: J.D. Wilson, D.W. Foster, H.M. Kronenberg and P.R. Larsen (Eds.), *Williams Textbook of Endocrinology*, 9th Edition. Saunders, Philadelphia, pp. 341–388.

Reznik, S.E. and Fricker, L.D. (2001) Carboxypeptidases from A to Z: implications in embryonic development and Wnt binding. *Cell. Mol. Life Sci.*, 58: 1790–1804.

Rodriguez, C., Brayton, K.A., Brownstein, M. et al. (1989) Rat preprocarboxypeptidase H: cloning, characterization, and sequence of the cDNA and regulation of the mRNA by corticotropin-releasing factor. *J. Biol. Chem.*, 264: 5988–5995.

Roebroek, A.J., Schalken, J.A., Leunissen, J.A. et al. (1986) Evolutionary conserved close linkage of the c-fes/fps proto-oncogene and genetic sequences encoding a receptor-like protein. *EMBO J.*, 5: 2197–2202.

Roebroek, A.J., Umans, L., Pauli, I.G. et al. (1998) Failure of ventral closure and axial rotation in embryos lacking the proprotein convertase Furin. *Development*, 125: 4863–4876.

Sakuma, Y. and Pfaff, D.W. (1983) Modulation of the lordosis reflex of female rats by LHRH, its antiserum and analogs in the mesencephalic central gray. *Neuroendocrinology*, 36: 218–224.

Sayah, M., Fortenberry, Y., Cameron, A. et al. (2001) Tissue distribution and processing of proSAAS by proprotein convertases. *J. Neurochem.*, 76: 1833–1841.

Schäfer, M.K.-H., Day, R., Cullinan, W.E. et al. (1993) Gene expression of pro-hormone and proprotein convertases in the rat CNS: a comparative in situ hybridization analysis. *J. Neurosci.*, 13: 1258–1279.

Schalken, J.A., Roebroek, A.J.M., Oomen, P.P.C.A. et al. (1987) *fur* Gene expression as a discriminating marker for small cell and nonsmall cell lung carcinoma. *J. Clin. Invest.*, 80: 1545–1549.

Schally, A.V., Arimura, A. and Kastin, A.J. (1973) Hypothalamic regulatory hormones. *Science*, 179: 341–350.

Sealfon, S.C., Weinstein, H. and Millar, R.P. (1997) Molecular mechanisms of ligand interaction with the gonadotropin-releasing hormone receptor. *Endocrine Rev.*, 18: 180–205.

Seeburg, P.H. and Adelman, J.P. (1984) Characterization of the cDNA for precursor of human luteinizing hormone releasing hormone. *Nature*, 311: 666–668.

Seidah, N.G. and Chrétien, M. (1999) Proprotein and prohormone convertases: a family of subtilases generating diverse bioactive polypeptides. *Brain Res.*, 848: 45–62.

Seidah, N.G., Gaspar, L., Mion, P. et al. (1990) cDNA sequence of two distinct pituitary proteins homologous to Kex2 and furin gene products: tissue-specific mRNAs encoding candidates for pro-hormone processing proteinases. *DNA Cell Biol.*, 9: 415–424.

Seidah, N.G., Marcinkiewicz, M., Benjannet, S. et al. (1991) Cloning and primary sequence of a mouse candidate prohormone convertase PC1 homologous to PC2, Furin, and Kex2: distinct chromosomal localization and messenger RNA distribution in brain and pituitary compared to PC2. *Mol. Endocrinol.*, 5: 111–122.

Seidah, N.G., Day, R., Hamelin, J. et al. (1992) Testicular expression of PC4 in the rat: molecular diversity of a novel germ cell-specific Kex2/subtilisin-like proprotein convertase. *Mol. Endocrinol.*, 6: 1559–1570.

Seidah, N.G., Hamelin, J., Mamarbachi, M. et al. (1996) cDNA structure, tissue distribution, and chromosomal localization of rat PC7, a novel mammalian proprotein convertase closest to yeast kexin-like proteinases. *Proc. Natl. Acad. Sci. USA*, 93: 3388–3393.

Seidah, N.G., Mowla, S.J., Hamelin, J. et al. (1999a) Mammalian subtilisin/kexin isozyme SKI-1: A widely expressed proprotein convertase with a unique cleavage specificity and cellular localization. *Proc. Natl. Acad. Sci. USA*, 96: 1321–1326.

Seidah, N.G., Benjannet, S., Hamelin, J. et al. (1999b) The subtilisin/kexin family of precursor convertases. Emphasis on PC1, PC2/7B2, POMC and the novel enzyme SKI-1. *Ann. N.Y. Acad. Sci.*, 885: 57–74.

Seong, J.Y., Park, S. and Kim, K. (1999) Enhanced splicing of the first intron from the gonadotropin-releasing hormone (GnRH) primary transcript is a prerequisite for mature GnRH messenger RNA: presence of GnRH neuron-specific splicing factors. *Mol. Endocrinol.*, 13: 1882–1895.

Seong, J.Y., Kim, B.W., Park, S. et al. (2001) First intron excision of GnRH pre-mRNA during postnatal development of normal mice and adult hypogonadal mice. *Endocrinology*, 142: 4454–4461.

Sherwood, N., Eiden, L., Brownstein, M. et al. (1983) Characterization of a teleost gonadotropin-releasing hormone. *Proc. Natl. Acad. Sci. USA*, 80: 2794–2798.

Sherwood, N.M., Sower, S.A., Marshak, D.R. et al. (1986) Primary structure of gonadotropin-releasing hormone from lamprey brain. *J. Biol. Chem.*, 261: 4812–4819.

Smeekens, S.P. and Steiner, D.F. (1990) Identification of a human insulinoma cDNA encoding a novel mammalian protein structurally related to the yeast dibasic processing protease Kex2. *J. Biol. Chem.*, 265: 2997–3000.

Smyth, D.G., Maruthainar, K., Darby, N.J. et al. (1989) Catalysis of slow C-terminal processing reactions by carboxypeptidase H. *J. Neurochem.*, 53: 489–493.

Song, L. and Fricker, L. (1995a) Processing of procarboxypeptidase E into carboxypeptidase E occurs in secretory vesicles. *J. Neurochem.*, 65: 444–453.

Song, L. and Fricker, L.D. (1995b) Purification and characterization of carboxypeptidase D, a novel carboxypeptidase E-like enzyme, from bovine pituitary. *J. Biol. Chem.*, 270: 25007–25013.

Song, L. and Fricker, L.D. (1996) Tissue distribution and characterization of soluble and membrane-bound forms of metallocarboxypeptidase D. *J. Biol. Chem.*, 271: 28884–28889.

Sower, S.A., Chiang, Y.A., Lova, S. et al. (1993) Primary structure and biological activity of a third gonadotropin-releasing hormone from lamprey brain. *Endocrinology*, 132: 1125–1131.

Steiner, D.F. (1998) The proprotein convertases. *Curr. Opin. Chem. Biol.*, 2:31–39.

Stoffers, D.A., Green, C.B.-R. and Eipper, B.A. (1989) Alternative mRNA splicing generates multiple forms of peptidylglycine α-amidating monooxygenase in rat atrium. *Proc. Natl. Acad. Sci. USA*, 86: 735–739.

Stoffers, D.A., Ouafik, L.H. and Eipper, B.A. (1991) Characterization of novel mRNAs encoding enzymes involved in peptide α-amidation. *J. Biol. Chem.*, 266: 1701–1707.

Suzuki, K., Gamble, R.K. and Sower, S.A. (2000) Multiple transcripts encoding lamprey gonadotropin-releasing hormone-I precursors. *J. Mol. Endocrinol.*, 24: 365–376.

Sykes, P.A., Watson, S.J., Temple, J.S. et al. (1999) Evidence for tissue-specific forms of glutaminyl cyclase. *FEBS Lett.*, 455: 159–161.

Tadros, H., Chretien, M. and Mbikay, M. (2001) The testicular germ-cell protease PC4 is also expressed in macrophage-like cells of the ovary. *Reprod. Immunol.*, 49: 133–152.

Takahashi, K., Okamoto, H., Seino, H. et al. (1990) Peptidylglycine α-amidating reaction: evidence for a two-step mechanism involving a stable intermediate at neutral pH. *Biochem. Biophys. Res. Commun.*, 169: 524–530.

Tan, F., Rehli, M., Krause, S.W. et al. (1997) Sequence of human carboxypeptidase D reveals it to be a member of the regulatory carboxypeptidase family with three tandem active site domains. *Biochem. J.*, 327: 81–87.

Thim, L., Hansen, M.T., Norris, K. et al. (1986) Secretion and processing of insulin precursors in yeast. *Proc. Natl. Acad. Sci. USA*, 83: 6766–6770.

Thomas, G.B., Cummins, J.T., Doughton, B.W. et al. (1988a) Gonadotropin-releasing hormone associated peptide (GAP) and putative processed GAP peptides do not release luteinizing hormone or follicle-stimulating hormone or inhibit prolactin secretion in sheep. *Neuroendocrinology*, 48: 342–350.

Thomas, G., Thorne, B.A., Thomas, L. et al. (1988b) Yeast KEX2 endopeptidase correctly cleaves a neuroendocrine prohormone in mammalian cells. *Science*, 241: 226–230.

Thorner, M.O., Vance, M.L., Laws, E.R. Jr. et al. (1998) The anterior pituitary. In: J.D. Wilson, D.W. Foster, H.M. Kronenberg and P.R. Larsen (Eds.), *Williams Textbook of Endocrinology*, 9th Edition. Saunders, Philadelphia, pp. 249–340.

Vacher, P., Mariot, P., Dufy-Barbe, L. et al. (1991) The gonadotropin-releasing hormone associated peptide reduces calcium entry in prolactin-secreting cells. *Endocrinology*, 128: 285–294.

Valença, M.M., Johnston, C.A., Ching, M. et al. (1987) Evidence for a negative ultrashort feedback mechanism operating on the luteinizing hormone-releasing hormone neuronal system. *Endocrinology*, 121: 2256–2259.

Valença, M.M., Wetsel, W.C., Culler, M.D. et al. (1988) Differential secretion of of proLHRH fragments in response to [K$^+$], prostaglandin E2, and C kinase activation. *Mol. Cell. Endocrinol.*, 55: 95–100.

van Chuoi, M.T., Vacher, P. and Dufy, B. (1993) GnRH-associated peptide decreases cyclic AMP accumulation in the GH3 pituitary cell line. *Neuroendocrinology*, 58: 251–257.

Varlamov, O., Leiter, E.H. and Fricker, L. (1996) Induced and spontaneous mutations at Ser^{202} of carboxypeptidase E. *J. Biol. Chem.*, 271: 13981–13986.

Varlamov, O., Eng, F.J., Novikova, E.G. et al. (1999) Localization of metallocarboxypeptidase D in AtT-20 cells: potential role in prohormone processing. *J. Biol. Chem.*, 274: 14759–14767.

Vishnuvardhan, D., Connolly, K., Cain, B. et al. (2000) PC2 and

7B2 null mice demonstrate that PC2 is essential for normal pro-CCK processing. *Biochem. Biophys. Res. Commun.*, 273: 188–191.

Voigt, P., Ma, Y.J., Gonzalez, D. et al. (1996) Neural and glial-mediated effects of growth factors acting via tyrosine kinase receptors on luteinizing hormone-releasing hormone neurons. *Endocrinology*, 137: 2593–2605.

Utsunomiya, N., Ohagi, S., Sanke, T. et al. (1998) Organization of the human carboxypeptidase E gene and molecular scanning for mutations in Japanese subjects with NIDDM or obesity. *Diabetologia*, 41: 701–705.

Westphal, C.H., Muller, L., Zhou, A. et al. (1999) The neuroendocrine protein 7B2 is required for peptide hormone processing in vivo and provides a novel mechanism for pituitary Cushing's disease. *Cell*, 96: 689–700.

Wetsel, W.C. (1995) Immortalized hypothalamic luteinizing hormone-releasing hormone (LHRH) neurons: a new tool for dissecting the molecular and cellular basis of LHRH physiology. *Cell. Mol. Neurobiol.*, 15: 43–78.

Wetsel, W.C. and Negro-Vilar, A. (1989) Testosterone selectively influences protein kinase C-coupled secretion of prolutcinizing hormone-releasing hormone-derived peptides. *Endocrinology*, 125: 538–547.

Wetsel, W.C., Culler, M.D., Johnston, C.A. et al. (1988) Processing of the luteinizing hormone-releasing hormone precursor in the preoptic area and hypothalamus of the rat. *Mol. Endocrinol.*, 2: 22–31.

Wetsel, W.C., Mellon, P.L., Weiner, R.I. et al. (1991) Metabolism of pro-luteinizing hormone-releasing hormone in immortalized hypothalamic neurons. *Endocrinology*, 129: 1584–1595.

Wetsel, W.C., Eraly, S.A., Whyte, D.B. et al. (1993) Regulation of gonadotropin-releasing hormone by protein kinase-A and -C in immortalized hypothalamic neurons. *Endocrinology*, 132: 2360–2370.

Wetsel, W.C., Liposits, Z., Seidah, N.G. et al. (1995) Expression of candidate pro-GnRH processing enzymes in rat hypothalamus and an immortalized hypothalamic neuronal cell line. *Neuroendocrinology*, 62: 166–177.

Wetsel, W.C., Hill, D.F. and Ojeda, S.R. (1996) Basic fibroblast growth factor regulates the conversion of pro-luteinizing hormone-releasing hormone (Pro-LHRH) to LHRH in immortalized hypothalamic neurons. *Endocrinology*, 137: 2606–2616.

White, R.B., Eisen, J.A., Kasten, T.L. et al. (1998a) Second gene for gonadotropin-releasing hormone in humans. *Proc. Natl. Acad. Sci. USA*, 95: 305–309.

White, R.B., Urbanski, H.F. and Fernald, R.D. (1998b) A second gene for gonadotropin-releasing hormone is expressed in the rhesus macaque. *Soc. Neurosci.*, 24: 1609 (abstract).

Wormald, P.J., Abrahamson, M.J., Seeburg, P.H. et al. (1989) Prolactin-inhibiting activity of GnRH associated peptide in cultured human pituitary cells. *Clin. Endocrinol.*, 30: 149–155.

Xin, X., Varlamov, O., Day, R. et al. (1997) Cloning and sequence analysis of cDNA encoding rat carboxypeptidase D. *DNA Cell Biol.*, 16: 897–909.

Yahalom, D., Chen, A., Ben-Aroya, N. et al. (1999) The gonadotropin-releasing hormone family of neuropeptides in the brain of human, bovine, and rat: identification of a third isoform. *Febs Lett.*, 463: 289–294.

Yoo, M.S., Kang, H.M., Choi, H.S. et al. (2000) Molecular cloning, distribution and pharmacological characterization of a novel gonadotropin-releasing hormone ([Trp^8] GnRH) in frog brain. *Mol. Cell. Endocrinol.*, 164: 197–204.

Yu, W.H., Seeburg, P.H., Nikolics, K. et al. (1988) Gonadotropin-releasing hormone associated peptide exerts a prolactin-inhibiting and weak gonadotropin-releasing activity in vivo. *Endocrinology*, 123: 390–395.

Yu, W.H., Arisawa, M., Millar, R.P. et al. (1989) Effects of gonadotropin-releasing hormone associated peptides (GAP) on the release of luteinizing hormone (LH), follicle-stimulating hormone (FSH) and prolactin (PRL) in vivo. *Peptides*, 10: 1133–1138.

Zheng, M., Streck, R.D., Scott, R.E. et al. (1994) The developmental expression in rat of proteases furin, PC1, PC2, and carboxypeptidase E: implications for early maturation of proteolytic processig capacity. *J. Neurosci.*, 14: 4656–4673.

Zhu, X. and Lindberg, I. (1995) 7B2 facilitates the maturation of proPC2 in neuroendocrine cells and is required for the expression of enzymatic activity. *J. Cell Biol.*, 129: 1641–1650.

I.S. Parhar (Ed.)
Progress in Brain Research, Vol. 141
© 2002 Elsevier Science B.V. All rights reserved

CHAPTER 18

The GnRH neuron: molecular aspects of migration, gene expression and regulation

Andrew Wolfe [1,*], Helen H. Kim [2], Sally Radovick [1]

[1] *Department of Pediatrics, and* [2] *Department of Obstetrics and Gynecology, University of Chicago, 5839 South Maryland Avenue, MC5053, Chicago, IL 60637, USA*

Introduction

The release of gonadotropin-releasing hormone (GnRH) from the GnRH neurons of the basal hypothalamus directs the synthesis and secretion of the pituitary reproductive hormones, luteinizing hormone (LH) and follicle-stimulating hormone (FSH) from the anterior pituitary. As the ultimate neural regulators of the hypothalamic–pituitary–gonadal axis, GnRH neurons integrate a variety of environmental, psychological and homeostatic cues into a signal that effects reproductive function.

GnRH neuronal migration and anatomy

Unlike other hypothalamic neuropeptide producing cells, GnRH neurons do not arise from the developing basal forebrain, they originate in the olfactory placode in mammals and migrate across the nasal cavity and cribriform plate into the forebrain (Schwanzel-Fukuda et al., 1987; Wray et al., 1989). In mice, for example, the GnRH neurons are derived from the neural ectoderm and are born on day 10.5 pc, and begin to express GnRH mRNA and protein at between day 10.75 and 11.5 pc. GnRH neurons

* Correspondence to: A. Wolfe, University of Chicago, Department of Pediatrics, 5839 South Maryland Ave. MC5053, Chicago, IL 60637, USA. Tel.: +1-773-834-8912; Fax: +1-773-702-0443;
E-mail: awolfe@peds.bsd.uchicago.edu

then migrate across the olfactory cavity to the forebrain between day 12.5 pc and day 16.5 pc (Wray et al., 1989; Wu et al., 1997). GnRH neurons migrating across the nasal cavity appear to be guided, in part, by the polysialic acid-rich form of the neural cell adhesion molecule (PSA-NCAM) (Yoshida et al., 1999) in association with the vomeronasal nerves (VNN). Other investigators have posited that factors such as GABA (Fueshko et al., 1998), adhesion related kinase (Fang et al., 1998) or peripherin (Wray et al., 1994) might play a role in the guidance of the GnRH neurons. The initial migration to the forebrain vesicles follows the vomeronasal and terminal nerves, however, after entering the rostral forebrain, GnRH neurons become disassociated from the caudal branch of the VNN. At this time it is thought that GnRH neurons are no longer guided by an, as yet, defined anatomical structure (He et al., 1989) Instead, neuronal migration may trail behind as axonal migration, directed by chemoattractive signals from the medial basal hypothalamus (Rogers et al., 1997) is occurring.

The lack of migration of GnRH neurons into the forebrain can result in reproductive dysfunction. For example, some cases of hypogonadotropic hypogonadism are due to a deletion of an X-linked gene referred to as anosmin-1 or Kalig (Schwanzel-Fukuda et al., 1989; Franco et al., 1991; Wierman et al., 1992). This disorder, Kallmann's Syndrome, results in a lack of proper neuronal migration in humans. The Kalig locus contains four fibronectin type III repeats (Legouis et al., 1991, 1994), a motif associated with adhesion molecules (Yokosaki et al., 1998). Forms of idiopathic hypogonadotropic hypogonadism not associated with Kallmann's syndrome have been described in humans, but the mechanisms of disregulation of the GnRH neurons remain unknown (Waldstreicher et al., 1996; Quinton et al., 1997). Certainly, a lack of developmental progression of GnRH neurons due to the absence of proteins important for migration may result in some of these cases of idiopathic hypogonadotropic hypogonadism.

A mouse model exists for a non-Kallmann's-like associated hypogonadotropic hypogonadism. The hpg mouse was identified as reproductively dysfunctional, and contained a large genetic deletion that included much of the GnRH gene (Mason et al., 1986). Neurons expressing the truncated form of the GnRH gene are still present in the usual locations in the forebrain indicating that migration of the neurons is unaffected by the absence of the GnRH decapeptide in these animals (Livne et al., 1993b). Transplantation of hypothalamic grafts (Rogers et al., 1998), or GnRH expressing cell lines (Kokoris et al., 1988; Silverman et al., 1992) into the hypothalami of the hpg mice rescued the phenotypic hypogonadism. This further underscores the central importance of the GnRH neurons as the primary regulators of pituitary gonadotropin secretion. Suprisingly, despite the efforts of a number of investigators, no mutation of the GnRH gene has been identified to date in humans.

It should be stressed that it is unclear whether GnRH neurons are merely passively responding to signals in their immediate milieu, or whether local signals trigger changes in gene expression in GnRH neurons that are ultimately responsible for migration, and for changes in cellular morphology or GnRH gene expression. There is evidence for the latter model in another neuronal system; the developmental organization of the cerebellum. Normally granule cells migrate along radial glia to their final location in the cerebellum. In the mutant *weaver* mouse the granule cells fail to migrate to their final location in the cerebellum (Rakic and Sidman, 1973a,b). Studies in chimeric mice suggested that the defect in the *weaver* mouse was intrinsic to the migrating granule cell (Goldowitz and Mullen, 1982; Goldowitz, 1989) Evidence for these types

of changes during development in GnRH neurons comes from studies looking at differences in gene expression in GN10 versus GT1 GnRH expressing cells (Fang et al., 1998; Allen et al., 1999). The first is derived from an olfactory tumor (migratory GnRH neurons), and the latter from an hypothalamic tumor (post-migratory GnRH neurons). These studies identified more than 10 factors differentially expressed in the two cell lines including an adhesion related kinase, Ark, produced in GN10 cells but not in GT1 cells that may have an anti-apoptotic function (Fang et al., 1998; Allen et al., 1999). Recent studies from Kramer and Wray have used single cell PCR of identified migratory and post-migratory GnRH neurons from hypothalamic explants to identify a factor called nasal embryonic LHRH factor (NELF) that appears to play a role in olfactory neuronal migration (Kramer and Wray, 2000). Additional evidence comes from a biochemical analysis of GnRH neurons during development showing that the growth and plasticity associated peptide, GAP43, was expressed at much higher levels in neurons within the nasal septum than in neurons that have already migrated into the forebrain (Livne et al., 1993a). Taken together, these studies indicate that there are additional differentiation events that occur following the onset of GnRH gene expression at day 10.75 pc.

In the adult mouse, the vast majority of GnRH containing neurons are located in the hypothalamus at the level of the organum vasculosum of the lamina terminalis (OVLT) and the pre-optic area (POA), but some GnRH neurons have also been localized in the cerebral cortex and the limbic system (King et al., 1982; Barry et al., 1985; Schwanzel-Fukuda et al., 1987). In addition, GnRH neurons are scattered within the olfactory bulbs. Therefore, it appears as if GnRH neurons are strewn along the migratory pathway to the basal hypothalamus. It is not clear what signals are responsible for the organization of the microarchitecture of the mammalian GnRH neurons in the basal forebrain, although there is some evidence that there are important functional groupings within the scattered GnRH neurons. For example there are quiescent subpopulations of GnRH neurons activated during various physiological events, such as during the preovulatory surge in female mammals (Hiatt et al., 1992; Porkka-Heiskanen et al., 1994), that have a specific anatomical arrangement.

Subpopulations of GnRH neurons

In the hypothalamus, during a variety of physiologic stimuli, GnRH neurons have been shown to be 'recruited' into increasing their expression of GnRH (Dutlow et al., 1992; Hiatt et al., 1992; Porkka-Heiskanen et al., 1994; Gore et al., 1999; Gore, 1998). For example, during the preovulatory GnRH surge in rats, it appears that a normally quiescent population of GnRH neurons is activated (Hiatt et al., 1992; Porkka-Heiskanen et al., 1994; Rajendren, 2001). Although these differences in GnRH neuronal activation may be a result of variations in the afferent inputs to each population of GnRH neurons, the pattern of release of GnRH from the median eminence during the GnRH surge is radically different from that seen during non-surge periods (Levine and Ramirez, 1982; Moenter et al., 1991), and may reflect profound differences in the physiology of the surge generating, and the non-surge generating populations of GnRH neurons. The intracellular events required for activation of GnRH neurons during the surge are unknown.

Another instance where GnRH neurons are recruited to produce higher levels of GnRH gene expression is during the period preceding the onset of puberty in the mouse (Gore et al., 1999; Wolfe et al., 1995), and rat (Gore, 1998), perhaps to prepare the neurons for the increased levels of GnRH secretion that occurs during puberty. Increases in GnRH gene expression after the onset of puberty when compared with prepubertal animals have been observed in the hamster (Parfitt et al., 1999). Again, it is not clear what factors (neurotransmitters, growth factors, second messengers or transcription factors) are involved in these puberty-related changes in GnRH

gene expression although there is some evidence that the amino acid family of neurotransmitters [GABA (Mitsushima et al., 1994; Terasawa et al., 1999) or NMDA (Gore et al., 1996; Zamorano et al., 1998; Terasawa et al., 1999)] may play a role in regulating the initiation of puberty.

Skynner et al. have indicated that, indeed, there might be independently derived populations of GnRH neurons located in the brain (Skynner et al., 1999). That is, GnRH neurons derived from regions other than the olfactory placode are present in the mouse brain during early development and persist until the third week of postnatal life. It is not clear whether these ectopically expressed GnRH neurons subserve a functional role in the mouse.

GnRH expressing neuronal cell lines

As a means of developing an in vitro model for the study of GnRH neuronal cell activity, immortalized GnRH-expressing neuronal cell lines were created by targeted tumorigenesis (Mellon et al., 1990; Radovick et al., 1991). Our laboratory targeted the expression of the simian virus 40 T antigen (SV40-Tag) to the GnRH neurons with 1131 bps of the human GnRH gene 5'-upstream regulatory sequence. One of these animals developed an olfactory tumor (Radovick et al., 1991) from which several GnRH immunoreactive cell lines (NLT, GN10, GN11) were subsequently derived. Characterization of the NLT and GN11 cells demonstrated that these cells display neuronal morphology and neuron specific markers [Table 1; (Fischer et al., 1991; Mavilia et al., 1993)]. Solution hybridization-RNase protection assays indicate that NLT cells express proGnRH mRNA at higher levels than GN11 cells. Radioimmunoassays

TABLE 1

Comparison of GN11 and NLT cells

Cell line	Tag expression	Neuronal markers	Relative GnRH mRNA levels	Predominant GnRH mRNA species	GnRH secretion by RIA
Gn11	+	+	0.03	Splice variant	1.5 pg/ml
NLT	+	+	1	Full length	20 pg/ml

Both GN11 and NLT cells express neuronal markers such as Map-2 and Tau, and express the T antigen oncogene. The predominant mRNA species in the GN11 cells is a splice variant in which the second exon is excised. NLT cells primarily express the full length variant. NLT cells release more GnRH peptide than GN11 cells most likely due to the higher relative levels of the full length GnRH mRNA species seen in NLT cells reveal that both cell lines are able to synthesize and secrete GnRH. However, NLT cells secrete about 10 times higher levels of GnRH than GN11 cells [Table 1; (Zhen et al., 1997a)]. We have identified several factors that are differentially expressed in these two cell lines that may contribute to their differences in phenotype (Zhen et al., 1997a). Our laboratory has shown that the POU homeodomain transcription factor Brn-2 is present in NLT cells, but not in GN11 cells [Fig. 1; (Wolfe et al., 2002)] Transfection of GN11 cells with a Brn-2 expression vector stimulates mouse (Fig. 2) and human (Fig. 3) GnRH gene promoter activity. In addition, our laboratory has identified a novel factor, NLT expression factor-1 (NLT-EF1) that is more highly expressed in NLT cells than GN11 cells (Fig. 4). Interestingly, NLT-EF1 shares homology with a mucin-type glycoprotein that stimulates motility of karitinocytes during wound regeneration (Scholl et al., 1999). We are presently exploring whether NLT-EF1 regulates motility of GnRH neurons.

The GT1 cell lines developed by Pamela Mellon's laboratory were harvested from an hypothalamic tumor and may represent a more mature GnRH neuronal cell line (Mellon et al., 1990). Indeed, one group has used a polymerase chain reaction (PCR) based differential display system to identify genes expressed differentially in olfactory bulb derived GN11 cells and hypothalamus derived GT1 cells (Fang et al., 1998; Allen et al., 1999). These genes have been hypothesized to produce proteins important for the regulation of migration and for preventing apoptosis. The three GT1 cell lines that have been primarily used are the GT1-1, GT1-3 and GT1-7 cell lines (Mellon et al., 1990). These cell lines, like the cell lines obtained in our laboratory from the olfactory tumor, possess significant physiological, electrophysiological, morphological, and molecular differences with each other (Mellon et al., 1990; Wetsel et al., 1991; Martinez et al., 1992; Bosma, 1993; Weiner and Martinez, 1993; Poletti et al., 1994).

The GnRH gene

The GnRH neurons are particularly difficult to study due both to their low abundance in the brain [about 800 neurons in the mouse (Wray et al., 1989)], and

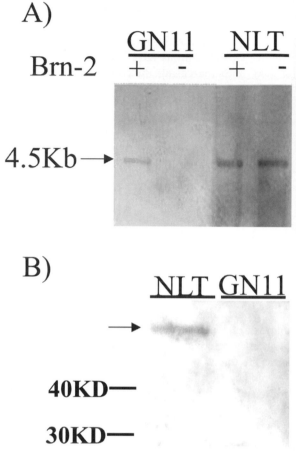

Fig. 1. Brn-2 mRNA and protein are found in GnRH neurons. (A) Northern blot analysis of mRNA obtained from the GN11 or NLT cell lines transfected with either a Brn-2 expression vector (+) or a control vector (−). A digoxigenin labeled riboprobe was found to hybridize at a size of 4.5 kb to mRNA of both GN11 and NLT cells transfected with Brn-2pSG5, and to NLT cells transfected with pSG5, but not to GN11 cells transfected with pSG5. (B) Western blot of nuclear extract proteins from NLT and GN11 cells. Size markers are indicated on the left. A band is observed in the NLT lane (arrow) that runs at about 54 kDa. No band was observed in the GN11 lane. (C) Coronal, whole mount tissue section of a mouse brain at the level of the OVLT. A black precipitate indicates Brn-2 mRNA, and GnRH peptide is labeled with a Cy-3 phosphoramidite (fluorescing). Brn-2 mRNA is observed in the cell body of the GnRH neuron (arrow).

to their diffuse and widely scattered distribution. Despite these obstacles the GnRH gene has been cloned in a number of species (Bond et al., 1989; Hayflick et al., 1989; Radovick et al., 1990; Kepa et al., 1992), and has been found to be about 4 kb

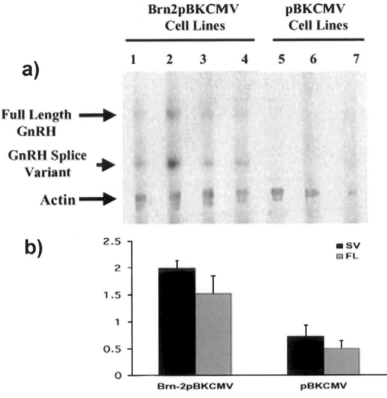

Fig. 2. Brn-2 Stimulates mGnRH Gene Expression. (A) RNase protection assay that was performed on mRNA obtained from stable transfection of GN11 cell lines with either the empty pBKCMV expression vector (lanes 5–7) or the Brn-2pBKCMV expression vector (lanes 1–4). GnRH full length mRNA, the truncated splice variant and actin mRNA are indicated by arrows to the left. Two separate RPAs were performed on the mRNA and the values of both the full length GnRH transcript (FL: Solid bars) and the splice variant (SV: shaded bar) averaged and shown in (B) where values are reported as arbitrary units of GnRH corrected for arbitrary units of actin. In (A) the GnRH bands are scanned from a film after 4 days exposure at −70°C. All quantification of band intensity was performed with the phosphorimager. Background levels for each band were subtracted from band intensity. Images of actin bands are from phosphorimager to reduce intensity for better visualization, and are superimposed on scanned image of film.

in length and to contain four exons. The first exon consists of the 5′ untranslated region, the second exon encodes the signal sequence of GnRH, the GnRH decapeptide, and the first 11 amino acids of GnRH-associated peptide (GAP), and the last two exons encode the remaining amino acids of GAP as well as the 3′ untranslated region. Low levels of GnRH immunoreactivity have also been localized in the placenta, the gonads and the mammary glands (Seeburg et al., 1987). Interestingly, placental GnRH transcription is initiated from a different promoter initiation site than that used for transcription of hypothalamic GnRH mRNA (Dong et al., 1993). In addition, the hypothalamic and placental GnRH genes undergo differential RNA splicing, with the first intron being retained in the placental mRNA and removed in the hypothalamic mRNA (Radovick et al., 1990; Dong et al., 1993). It is unclear what the significance is of the extrahypothalamic expression of GnRH. In the mouse hypothalamus, studies have shown that two differential splicing events can occur. One form of mRNA contains all four exons while the other does not contain the second exon, which encodes the GnRH decapeptide (Zhen et al., 1997a). The factors responsible for this differential splicing are unknown, but clearly a better understanding of the transcription factors that regulate these different splicing events will advance our understanding of both the cell-specific expression and the regulated expression of the hGnRH gene.

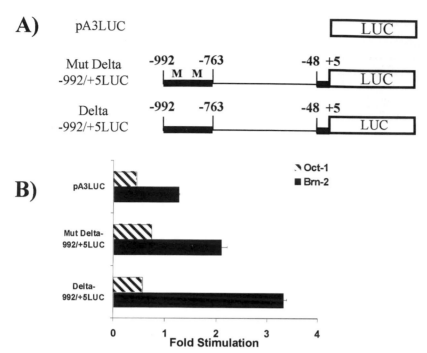

Fig. 3. Brn-2 Stimulates hGnRH Gene Expression. (A) Schematics of constructs used in transient transfection experiments. pA3LUC, Δ−992/+5LUC (labeled Delta−992/+5LUC) and Mut Delta−992/+5LUC. M indicates mutated Brn-2 consensus binding sites in the TSE. (B) Transient transfections of GN11 cells using Lipofectamine Plus reagent (Gibco, BRL). Cells were transfected with reporter vector (Δ−992/+5LUC [labeled Delta−992/+5LUC], Mut Delta−992/+5LUC [the same as Δ−992/+5LUC except with mutations of both Brn-2 consensus sites], and the empty pA3LUC) and 0.5µg of empty pSG5 expression vector, Brn-2pSG5 or Oct-1pSG5. Data are reported as fold stimulation by Brn-2 or Oct-1 as compared to pSG5 + SE.

Cell-specific expression of the GnRH gene

Several in vitro studies have used mouse-derived immortalized GnRH-secreting neuronal cell lines to investigate the neuronal expression of the rat and mouse GnRH gene. Transient transfection studies using the rat GnRH (rGnRH) gene promoter identified a 173 bp proximal promoter region (Eraly and Mellon, 1995) and a 300 bp region located 1.8 kb upstream from the transcription start site (Whyte et al., 1995) that conferred cell-specific expression. Both these sites were reported to be important for the appropriate neuronal expression of the rGnRH gene in vitro (Nelson et al., 2000).

In vivo, GnRH neurons are dispersed and are influenced by growth factors, steroids, and neurotransmitters secreted by various adjacent cell types. Because in vitro studies are unlikely to reflect the elaborate intricacy of in vivo GnRH gene regulation, in vivo models have been developed to study the

regulation of the GnRH gene. A transgenic mouse study, using a *lacZ* reporter, identified the critical elements for expression of the mGnRH gene between −2.1 kb and −1.7 kb of the distal 5' sequence (Pape et al., 1999).

As a model to study the cell-specific expression of the GnRH gene, our laboratory has used a transgenic mouse model containing gene constructs consisting of promoter deletion fragments of the GnRH gene fused to the luciferase reporter gene (LUC). To identify promoter elements of the human GnRH (hGnRH) gene that were essential for cell-specific expression the following promoter deletion fragments were used; −3828/ +5, −1131/+5, −992/+5, −795/+5, and −484/ +5 (Reporter genes named 3828LUC, 1131LUC, 992LUC, 795LUC and 484LUC, respectively). Tissue panels were performed on each of these mouse lines to ascertain whether reporter gene expression (LUC) was confined to GnRH neuron containing re-

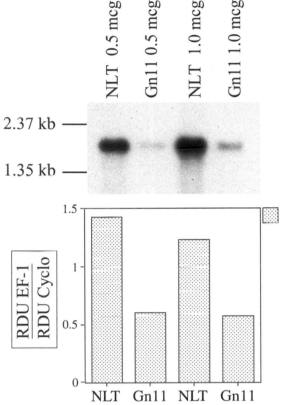

NLT 0.5 mcg

Gn11 0.5 mcg

NLT 1.0 mcg

Gn11 1.0 mcg

2.37 kb —

1.35 kb —

$\dfrac{\text{RDU EF-1}}{\text{RDU Cyclo}}$

1.5

1

0.5

0

NLT Gn11 NLT Gn11

Fig. 4. NLT-EF1 mRNA is More Abundant in NLT Cells Than GN11 cells. Northern blot analysis of mRNA from GN11 and NLT cells using a ^{32}P-labeled cDNA probe for NLT-EF1. 0.5 μg or 1 μg of mRNA was used as indicated. Size marker is indicated on the left. Quantitative representation of band intensity is shown in the graph. Values are given as band intensity of the NLT-EF1 band divided by the band intensity of a cyclophilin standard. NLT-EF1 mRNA levels were found to be between 2–3 fold higher in NLT cells than in GN11 cells.

gions. When promoter fragments larger than 992 bps were used [Fig. 5 for 992LUC mice, data from 3828LUC and 1131LUC mice in (Wolfe et al., 1995)] relatively high levels of luciferase enzyme activity, as measured in a luminometer, were observed in the hypothalamus and olfactory bulbs. Low to undetectable levels of luciferase activity were measured in tissue homogenates taken from other central and peripheral tissues. Double labeling immunohistochemistry was performed to confirm colocalization of the GnRH peptide and the luciferase protein in the hypothalamus (Fig. 6). These studies indicate that promoter elements essential for the cell-specific

expression of the hGnRH gene are located within the first 992 bps of the promoter.

When mice containing either the 795LUC (Fig. 7) or 484LUC transgene were examined, no luciferase activity was observed in any of the tissue homogenates. These studies demonstrate that promoter elements sufficient to target gene expression to GnRH neurons are not located within the first 795 bps of the hGnRH gene promoter. These results indicate that promoter elements of the hGnRH gene between −992 and −795 are essential for cell-specific expression of the gene.

To determine whether this promoter region contained elements that were sufficient for cell-specific expression an additional transgenic animal line was created. These mice contained a reporter gene with a promoter fragment from −992 to −763 fused to a minimal 48 bp hGnRH promoter fragment (Fig. 8a). When tissue panels from these mice were examined, relatively high levels of luciferase expression were observed in the hypothalamus and olfactory bulbs but not in other tissues. In whole, these studies demonstrate that the −992/−795 hGnRH promoter fragment is both essential and sufficient to direct gene expression to GnRH neurons.

Interestingly, the human and mouse cell-specific regions bear little homology to each other and neither share homology with the critical rat promoter regions identified in the in vitro studies. To reconcile these discrepant observations, a similar model system was used to identify the promoter elements of the mGnRH gene that are important for cell-specific expression (Kim et al., 2002). We constructed transgenic mice with various fragments of the mouse GnRH gene promoter fused to the luciferase reporter gene as a marker for in vivo mGnRH gene promoter activity. All three 5′ fragments of the mGnRH promoter that we studied, −3446/+23 bp, −2078/+23 bp, and −1005/+28 bp, targeted expression of the luciferase transgene to the hypothalamus (Fig. 9). Distal to the proximal −173 bp conserved region, there is approximately 50% overall homology between the proximal −1005 bp of the human and mouse GnRH promoters. As discussed above, studies of the hGnRH gene in transgenic mice localized a hypothalamic-specific element between −992 bp and −795 bp of the hGnRH gene. Within the human hypothalamic-specific element, there are several re-

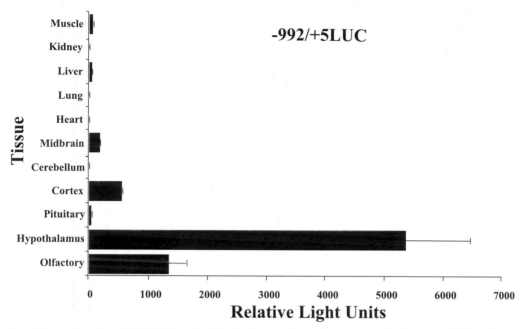

Fig. 5. Tissue Panel from 992LUC Mice. Relative luciferase activity (Shown as relative light units, RLU) of tissues taken from transgenic mice containing the −992/+5LUC transgene. The luciferase reporter is regulated by the −992/+5 hGnRH promoter fragment. All tissue levels are the mean + SE of four adult animals from transgenic line 26 except the hypothalamic values, which are the means + SE of 6 mice. (B) Relative luciferase activity (RLU) of tissues taken from transgenic mice containing the −992/+5LUC transgene. All tissue levels are the mean + SE of three adult animals from transgenic line 64 except the hypothalamic and olfactory values, which are the means + SE of 16 mice. (C) Averaged data from three tissue panels used in Figure 1B where the values are corrected for protein levels by dividing the RLU by mg protein.

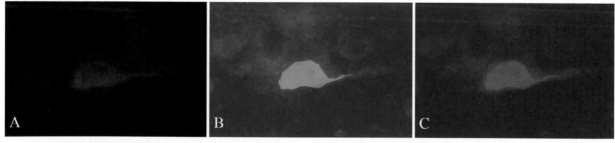

Fig. 6. Luciferase is colocalized with GnRH in 992LUC mice. Double fluorescence histochemistry of mouse brain sections from a 992LUC mouse. Shown is a neuron containing GnRH peptide (shown as red in (A), luciferase protein (shown as green in B). (C) is an overlay of A and B showing colocalization of the two proteins. Cy3 (red) and Alexa Fluoro 488 (green) were visualized using the appropriate filters.

gions that share greater homology with the mGnRH gene, and further study of these homologous regions may elucidate the critical factors that target hypothalamic GnRH expression.

Our findings using the luciferase reporter and the mouse GnRH promoter differ from a recent transgenic mouse study in which the mGnRH promoter fused to a lacZ reporter localized the critical ele-

ments for expression of the mGnRH gene between −2.1 kb and −1.7 kb of the promoter (Pape et al., 1999). In these mice, deletion of mGnRH promoter sequences 5′ to 1.7 kb resulted in a complete absence of detectable β-galactosidase expression within the brain, whereas we detected luciferase expression in transgenic mice bearing the −1005 mGnRH-LUC transgene. In the adult, the proximal −1005 bp of

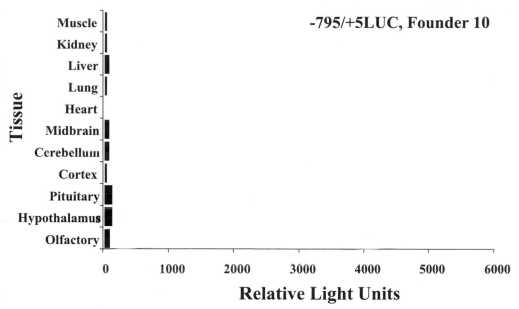

Fig. 7. Tissue Panel from a 795LUC mouse. Relative luciferase activity of tissues taken from a representative transgenic mouse containing the −795/+5LUC transgene (one of 8 founders examined). The luciferase reporter is regulated by the −795/+5 hGnRH promoter fragment.

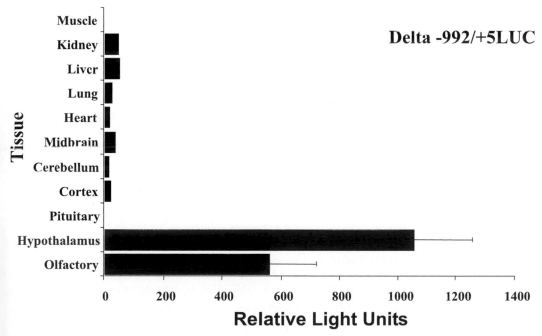

Fig. 8. Tissue Panel from a Δ−992/+5LUC Mouse. Relative luciferase activity of tissues taken from a representative transgenic mouse containing the Δ−992/+5LUC transgene (one of 3 positive founders examined). The luciferase reporter is regulated by the −992/−763/−48/+5 hGnRH promoter fragments. The hypothalamic and olfactory values are the average of animals from all three founders + SE.

252

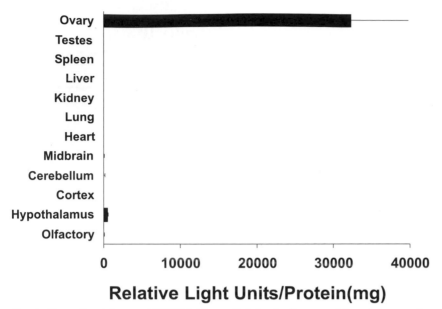

Relative Light Units/Protein(mg)

Fig. 9. Tissue Panel from a 1005LUC Mouse. Relative luciferase activity of tissues taken from a representative transgenic mouse containing the −1005/+5LUC transgene (one of 3 positive founders examined). The luciferase reporter is regulated by the −1005/+23 mGnRH promoter fragment. The hypothalamic and olfactory values are the average of animals from all three founders + SE.

the mGnRH targeted luciferase to the hypothalamus in the −1005 mGnRH-LUC mice, but expression levels were lower (474 ± 86 RLU) compared with the levels seen in the −2078 mGnRH-LUC mice (1,688 ± 444 RLU). Perhaps, the mGnRH promoter region between −2.1 kb and −1.7 kb, although not critical for targeting hypothalamic expression, contains sequences essential for enhancing hypothalamic expression of GnRH.

The mGnRH promoter fragments containing −2078 bp and −1005 bp of 5' sequence targeted luciferase transgene expression to the hypothalamus, but at lower levels than the transgenic mice generated with −3446 bp of mGnRH promoter. In general, comparison of transgene expression levels between different founder lines must be done cautiously since transgene expression levels are affected by the chromosomal integration site and transgene copy number (al Shawi et al., 1990). Nevertheless, transgenic mice derived from 8 different embryos bearing the −2078 mGnRH-LUC transgene were examined. It is striking that even in the transgenic line that expressed luciferase at the highest level, luciferase activity in the hypothalamus was approximately 5 fold lower than was seen in the transgenic mice bearing the −3446 mGnRH-LUC transgene. These data would

suggest that an enhancer for the in vivo expression of hypothalamic mGnRH is contained in the mGnRH promoter region between −3446 bp and −2078 bp.

Our in vivo observations corroborate the in vitro studies performed in the GT1-7 cell line. The rat neuron-specific GnRH enhancer, located between −1863 and −1571 bp, shares 90% homology to the region of the mouse GnRH promoter located between −2384 bp and −2081 bp. Although the region between −2384 bp and −2081 bp was not found to be essential for targeting mGnRH expression to the hypothalamus in our mGnRH-LUC mice, deletion of sequences 5' to −2078 bp resulted in a dramatic decrease in hypothalamic luciferase expression levels. These results would support the hypothesis that critical enhancer sequences for the in vivo expression of hypothalamic mGnRH are located between −2384 bp and −2081 bp of the mGnRH promoter.

In vitro studies with the rGnRH promoter in GT1-7 cells have also identified several transcription factors that interact with these promoter regions to regulate rGnRH expression: C/EBP (Belsham and Mellon, 2000), GATA (Lawson et al., 1996 A.D.; Lawson and Mellon, 1998 A.D.), Oct-1 (Clark and Mellon, 1995; Eraly et al., 1998), Otx 2 (Kelley et al., 2000), SCIP/Oct-6/Tst-1 (Wierman et al., 1997).

The rat and the mouse proximal promoter regions are highly conserved between the rat and mouse, and in vitro studies with the mouse GnRH (mGnRH) gene promoter suggest that Oct-1 may also regulate the neuronal expression of the mGnRH gene (Chandran et al., 1999). Analysis of the corresponding mouse gene sequences, between −2384 bp and −2081 bp, however, reveal differences in the presumed transcription factor recognition sites. There is some unpublished evidence that these changes may eliminate binding to the mGnRH promoter, in vitro (Chandran and DeFranco, 1999). Further study is needed to determine whether Oct-1 also has a role in enhancing the in vivo expression of mGnRH and whether any of the above transcription factors, or an as yet undescribed transcription factor(s), are involved.

Non-neuronal expression of GnRH

More recently, an extra-pituitary role for GnRH has been appreciated. Low levels of GnRH expression have been found in peripheral reproductive tissues, such as placenta (Radovick et al., 1990), breast, ovary and testes (Dong et al., 1993). There is increasing evidence that ovarian GnRH directly modulates ovarian function in a paracrine or autocrine manner. GnRH has been demonstrated to have both stimulatory and inhibitory influences on ovarian cell differentiation, steroidogenesis, and oocyte maturation (Hsueh and Erickson, 1979a,b; Knecht et al., 1985; Richards, 1994).

Interestingly, when the mGnRH promoter region distal to −2078 bp was deleted in our mice bearing the mouse GnRH-luciferase transgene, high levels of ovarian luciferase were detected, and even higher levels of ovarian luciferase were detected in mice bearing the −1005 mGnRH-LUC transgene. Our studies demonstrate that sequences contained within the proximal −1005 bp of the mGnRH promoter are sufficient to target ovarian, as well as hypothalamic, expression of mGnRH. Our data also suggest that an ovarian GnRH repressor element may be located in the distal region of the mGnRH promoter between −3446 bp and −2078 bp since deletion of this region unmasks luciferase expression in the ovaries of transgenic mice bearing the mGnRH-LUC transgene. We speculate that repressor proteins in the ovary normally interact with this ovarian GnRH repressor element and permit GnRH expression only in certain physiologic situations.

Our finding of high levels of ovarian luciferase expression was initially surprising since no ovarian luciferase activity was detected in our previous studies using the hGnRH promoter to direct luciferase expression (Wolfe et al., 1995, 2002). This discrepancy may be due to species-specific differences in the role of GnRH in follicular development or in the ovarian proteins that may interact to regulate ovarian GnRH expression. Alternatively, it is possible that an ovarian GnRH repressor element exists in the human GnRH promoter as well, but has not yet been identified. Further studies may reveal the identity of the specific repressor proteins and the specific physiologic states in which ovarian GnRH expression is increased.

Historically, the study of the molecular elements important for the development and regulation of the GnRH neuron has been challenging. Generally, investigators have relied on anatomical methods or direct or indirect measurement of GnRH release from the hypothalamus either in vivo or in vitro. The development of homogeneous GnRH neuronal cell lines has allowed for the direct examination of the neurons in previously impossible ways. Electrophysiology (Bosma, 1993; LeBeau et al., 2000; Van Goor et al., 2000), signal transduction (Bruder et al., 1992; Zhen et al., 1997b; Poletti et al., 2001), gene expression (Kepa et al., 1996; Yeo et al., 1996; Lawson et al., 1998; Nelson et al., 2000) and protein translation (Gore et al., 1995; Sosnowski et al., 2000) studies have all been performed using cell lines. Transgenic animal models have provided a new and valuable tool for the study of the GnRH neuron. The direct, in vivo, analysis of cell-specific expression (Wolfe et al., 1995; Pape et al., 1999; Suter et al., 2000; Lawson et al., 2002; Wolfe et al., 2002) of the GnRH gene and regulation of the GnRH neuron (Wolfe et al., 1995; Lawson et al., 2002) have been performed. Present technologies exist to produce GnRH neuron-specific deletions of genes and even to control the timing of expression. These models will provide a wealth of information about factors important for GnRH neuronal migration and development. They also will help determine what pathways are used to transmit information about homeostatic, psychological and environmental status to the GnRH neuron.

References

al Shawi, R., Kinnaird, J., Burke, J. and Bishop, J.O. (1990) Expression of a foreign gene in a line of transgenic mice is modulated by a chromosomal position effect. *Mol. Cell. Biol.*, 10: 1192–1198.

Allen, M.P., Zeng, C., Schneider, K., Xiong, X., Meintzer, M.K., Bellosta, P., Basilico, C., Varnum, B., Heidenreich, K.A. and Wierman, M.E. (1999) Growth arrest-specific gene 6 (Gas6)/adhesion related kinase (Ark) signaling promotes gonadotropin-releasing hormone neuronal survival via extracellular signal-regulated kinase (ERK) and Akt. *Mol. Endocrinol.*, 13: 191–201.

Barry, J., Hoffman, G.E. and Wray, S. (1985) LHRH-containing systems. In: A. Björklund and T. Hökfelt (Eds.), *Handbook of Chemical Neuroanatomy. Vol. 4: GABA and Neuropeptides in the CNS, Part 1*. Elsevier, Amsterdam, pp. 166–215.

Belsham, D.D. and Mellon, P.L. (2000) Transcription factors Oct-1 and C/EBPbeta (CCAAT/enhancer-binding protein-beta) are involved in the glutamate/nitric oxide/cyclic-guanosine 5′-monophosphate-mediated repression of mediated repression of gonadotropin-releasing hormone gene expression. *Mol. Endocrinol.*, 14: 212–228.

Bond, C.T., Hayflick, J.S., Seeburg, P.H. and Adelman, J.P. (1989) The rat gonadotropin-releasing hormone: SH locus: structure and hypothalamic expression. *Mol. Endocrinol.*, 3(8): 1257–1262.

Bosma, M.M. (1993) Ion channel properties and episodic activity in isolated immortalized gonadotropin-releasing hormone (GnRH) neurons. *J. Membr. Biol.*, 136: 85–96.

Bruder, J.M., Krebs, W.D., Nett, T.M. and Wierman, M.E. (1992) Phorbol ester activation of the protein kinase C pathway inhibits gonadotropin-releasing hormone gene expression. *Endocrinology*, 131: 2552–2558.

Chandran, U.R. and DeFranco, D.B. (1999) Regulation of gonadotropin-releasing hormone gene transcription. *Behav. Brain Res.*, 105: 29–36.

Chandran, U.R., Warren, B.S., Baumann, C.T., Hager, G.L. and DeFranco, D.B. (1999) The glucocorticoid receptor is tethered to DNA-bound Oct-1 at the mouse gonadotropin-releasing hormone distal negative glucocorticoid response element. *J. Biol. Chem.*, 274: 2372–2378.

Clark, M.E. and Mellon, P.L. (1995) The POU homeodomain transcription factor Oct-1 is essential for activity of the gonadotropin-releasing hormone neuron-specific enhancer. *Mol. Cell. Biol.*, 15(11): 6169–6177.

Dong, K.-W., Yu, K.-L. and Roberts, J.L. (1993) Identification of a major up-stream transcription start site for the human progonadotropin-releasing hormone gene used in reproductive tissues and cell lines. *Mol. Endocrinol.*, 7: 1654–1666.

Dutlow, C.M., Rachman, J., Jacobs, T.W. and Millar, R.P. (1992) Prepubertal increases in gonadotropin-releasing hormone mRNA, gonadotropin-releasing hormone precursor, and subsequent maturation of precurso processing in male rats. *J. Clin. Invest.*, 90: 2496–2501.

Eraly, S.A. and Mellon, P.L. (1995) Regulation of gonadotropin-releasing hormone transcription by protein kinase C is medi-

ated by evolutionarily conserved promoter-proximal elements. *Mol. Endocrinol.*, 9: 848–859.

Eraly, S.A., Nelson, S.B., Huang, K.M. and Mellon, P.L. (1998) Oct-1 binds promoter elements required for transcription of the GnRH gene. *Mol. Endocrinol.*, 12: 469–481.

Fang, Z., Xiong, X., James, A., Gordon, D.F. and Wierman, M.E. (1998) Identification of novel factors that regulate GnRH gene expression and neuronal migration. *Endocrinology*, 139: 3654–3657.

Fischer, I., Richter-Landsberg, C. and Safaei, R. (1991) Regulation of microtubule associated protein 2 (MAP2) expression by nerve growth factor in PC12 cells. *Exp. Cell Res.*, 194: 195–201.

Franco, B., Guioli, S., Pragliola, A., Incerti, B., Bardoni, B., Tonlorenzi, R., Carrozzo, R., Maestrini, E., Pieretti, M. and Taillon-Miller, P. (1991) A gene deleted in Kallmann's syndrome shares homology with neural cell adhesion and axonal path-finding molecules. *Nature*, 353: 529–536.

Fueshko, S.M., Key, S. and Wray, S. (1998) GABA inhibits migration of luteinizing hormone-releasing hormone neurons in embryonic olfactory explants. *J. Neurosci.*, 18: 2560–2569.

Goldowitz, D. (1989) The weaver granuloprival phenotype is due to intrinsic action of the mutant locus in granule cells: evidence from homozygous weaver chimeras. *Neuron*, 2: 1565–1575.

Goldowitz, D. and Mullen, R.J. (1982) Granule cell as a site of gene action in the weaver mouse cerebellum: evidence from heterozygous mutant chimeras. *J. Neurosci.*, 2: 1474–1485.

Gore, A.C. (1998) Diurnal rhythmicity of gonadotropin-releasing hormone gene expression in the rat. *Neuroendocrinology*, 68: 257–263.

Gore, A.C., Ho, A. and Roberts, J.L. (1995) Translational efficiency of gonadotropin-releasing hormone messenger ribonucleic acid is negatively regulated by phorbol ester in GT1-7 cells. *Endocrinology*, 136: 1620–1625.

Gore, A.C., Wu, T.J., Rosenberg, J.J. and Roberts, J.L. (1996) Gonadotropin-releasing hormone and NMDA receptor gene expression and colocalization change during puberty in female rats. *J. Neurosci.*, 16: 5281–5289.

Gore, A.C., Roberts, J.L. and Gibson, M.J. (1999) Mechanism for the regulation of gonadotropin-releasing hormone gene expression in the developing mouse. *Endocrinology*, 140: 2280–2287.

Hayflick, J.S., Adelman, J.P. and Seeburg, P.H. (1989) The complete nucleotide sequence of the human gonadotropin-releasing hormone gene. *Nucleic Acid Res.*, 17(15): 6403–6404.

He, X., Treacy, M.N., Simmons, D.M., Ingraham, H.A., Swanson, L.W. and Rosenfeld, M.G. (1989) Expression of a large family of POU-domain regulatory genes in mammalian brain development. *Nature*, 340: 35–41 [erratum appeared in *Nature* (1989), 340(6235): 662].

Hiatt, E.S., Brunetta, P.G., Seiler, G.R., Barney, S.A., Selles, W.D., Wooledge, K.H. and King, J.C. (1992) Subgroups of luteinizing hormone-releasing hormone perikary defined by computer analysis in the basal forebrain of intact female rats. *Endocrinology*, 130(2): 1030–1043.

Hsueh, A.J. and Erickson, G.F. (1979a) Extra-pituitary inhibition of testicular function by luteinising hormone releasing hormone. *Nature*, 281: 66–67.

Hsueh, A.J. and Erickson, G.F. (1979b) Extrapituitary action of gonadotropin-releasing hormone: direct inhibition ovarian steroidogenesis. *Science*, 204: 854–855.

Kelley, C.G., Lavorgna, G., Clark, M.E., Boncinelli, E. and Mellon, P.L. (2000) The Otx2 homeoprotein regulates expression from the gonadotropin-releasing hormone proximal promoter. *Mol. Endocrinol.*, 14: 1246–1256.

Kepa, J.K., Wang, C., Neeley, C.I., Raynolds, M.V., Gordon, D.F., Wood, W.M. and Wierman, M.E. (1992) Structure of the rat gonadotropin-releasing hormone (rGnRH) gene promoter and functional analysis in hypothalamic cells. *Nucleic Acid Res.*, 20(6): 1393–1399.

Kepa, J.K., Spaulding, A.J., Jacobsen, B.M., Fang, Z., Xiong, X., Radovick, S. and Wierman, M.E. (1996) Structure of the distal human gonadotropin releasing hormone (hGnrh) gene promoter and functional analysis in Gt1-7 neuronal cells. *Nucleic Acids Res.*, 24: 3614–3620.

Kim, H.H., Wolfe, A., Smith, G.R., Tobet, S.A. and Radovick, S. (2002) Promoter sequences targeting tissue-specific gene expression of hypothalamic and ovarian gonadotropin-releasing hormone in vivo. *J. Biol. Chem.*, 277: 5194–5202.

King, J.C., Tobet, S.A., Snavely, F.L. and Arimura, A.A. (1982) LHRH Imminopositive Cells and Their Projections to the Median Eminence and Organum Vasculosum of the Lamina Teminalis. *J. Comp. Neurol.*, 209: 287–300.

Knecht, M., Ranta, T., Feng, P., Shinohara, O. and Catt, K.J. (1985) Gonadotropin-releasing hormone as a modulator of ovarian function. *J. Steroid Biochem.*, 23: 771–778.

Kokoris, G.J., Lam, N.Y., Ferin, M., Silverman, A.J. and Gibson, M.J. (1988) Transplanted gonadotropin-releasing hormone neurons promote pulsatile luteinizing hormone secretion in congenitally hypogonadal (hpg) male mice. *Neuroendocrinology*, 48: 45–52.

Kramer, P.R. and Wray, S. (2000) Novel gene expressed in nasal region influences outgrowth of olfactory axons and migration of luteinizing hormone-releasing hormone (LHRH) neurons. *Genes Dev.*, 14: 1824–1834.

Lawson, M.A. and Mellon, P.L. (1998) Expression of GATA-4 in migrating gonadotropin-releasing neurons of the developing mouse [In Process Citation]. *Mol. Cell. Endocrinol.*, 140: 157–161.

Lawson, M.A., Whyte, D.B. and Mellon, P.L. (1996) GATA factors are essential for activity of the neuron-specific enhancer of the gonadotropin-releasing hormone gene. *Mol. Cell. Biol. 1996 Jul*, 16: 3596–3605.

Lawson, M.A., Buhain, A.R., Jovenal, J.C. and Mellon, P.L. (1998) Multiple factors interacting at the GATA sites of the gonadotropin-releasing hormone neuron-specific enhancer regulate gene expression. *Mol. Endocrinol.*, 12: 364–377.

Lawson, M.A., Macconell, L.A., Kim, J., Powl, B.T., Nelson, S.B. and Mellon, P.L. (2002) Neuron-specific expression in vivo by defined transcription regulatory elements of the GnRH gene. *Endocrinology*, 143: 1404–1412.

LeBeau, A.P., Van Goor, F., Stojilkovic, S.S. and Sherman, A. (2000) Modeling of membrane excitability in gonadotropin-releasing hormone-secreting hypothalamic neurons regulated by Ca^{2+}-mobilizing and adenylyl cyclase-coupled receptors. *J. Neurosci.*, 20: 9290–9297.

Legouis, R., Hardelin, J.-P., Levilliers, J., Claverie, J.-M., Compain, S., Wunderie, V., Millasseau, P., Le Paslier, D., Cohen, D., Caterina, D., Bougueleret, L., Delemarre-Van de Waal, H., Lutfalla, G., Weissenbach, J. and Petit, C. (1991) The candidate gene for the X-linked Kallmann syndrome encodes a protein related to adhesion molecules. *Cell*, 423 435.

Legouis, R., Cohen-Salmon, M., Del Castillo, I. and Petit, C. (1994) Isolation and characterization of the gene responsible for the X chromosome-linked Kallmann syndrome. *Biomed. Pharmacother.*, 48: 241–246.

Levine, J.E. and Ramirez, V.D. (1982) Luteinizing hormone-releasing hormone release during the rat estrous cycle and after ovariectomy, as estimated with push–pull cannulae. *Endocrinology*, 111: 1439–1448.

Livne, I., Gibson, M.J. and Silverman, A.J. (1993a) Biochemical differentiation and intercellular interactions of migratory gonadotropin-releasing hormone (GnRH) cells in the mouse. *Dev. Biol.*, 159: 643–656.

Livne, I., Gibson, M.J. and Silverman, A.J. (1993b) Gonadotropin-releasing hormone (GnRH) neurons in the hypogonadal mouse elaborate normal projections despite their biosynthetic deficiency. *Neurosci. Lett.*, 151: 229–233.

Martinez, d.l.E., Choi, A.L. and Weiner, R.I. (1992) Generation and synchronization of gonadotropin-releasing hormone (GnRH) pulses: intrinsic properties of the GT1-1 GnRH neuronal cell line. *Proc. Natl. Acad. Sci. USA*, 89: 1852–1855.

Mason, A.J., Hayflick, J.S., Zoeller, R.T., Young, W.S., Phillips, H.S., Nikolics, K. and Seeburg, P.H. (1986) A deletion truncating the gonadotropin-releasing hormone gene is responsible for hypogonadism in the hpg mouse. *Science*, 234: 1366–1371.

Mavilia, C., Couchie, D., Mattei, M.G., Nivez, M.P. and Nunez, J. (1993) High and low molecular weight tau proteins are differentially expressed from a single gene. *J. Neurochem.*, 61: 1073–1081.

Mellon, P.L., Windle, J.J., Goldsmith, P.C., Padula, C.A., Roberts, J.L. and Weiner, R.I. (1990) Immortalization of hypothalamic gnrh neurons by genetically targeted tumorigenesis. *Neuron*, 5: 1–10.

Mitsushima, D., Hei, D.L. and Terasawa, E. (1994) gamma-Aminobutyric acid is an inhibitory neurotransmitter restricting the release of luteinizing hormone-releasing hormone before the onset of puberty. *Proc. Natl. Acad. Sci. USA*, 91: 395–399.

Moenter, S.M., Caraty, A., Locatelli, A. and Karsch, F.J. (1991) Pattern of gonadotropin-releasing hormone (gnrh) secretion leading up to ovulation in the ewe: existence of a preovulatory GnRH surge. *Endocrinology*, 129(3): 1175–1182.

Nelson, S.B., Lawson, M.A., Kelley, C.G. and Mellon, P.L. (2000) Neuron-specific expression of the rat gonadotropin-releasing hormone gene is conferred by interactions of a defined promoter element with the enhancer in GT1-7 cells. *Mol. Endocrinol.*, 14: 1509–1522.

Pape, J.R., Skynner, M.J., Allen, N.D. and Herbison, A.E. (1999)

256

Transgenics identify distal 5'- and 3'-sequences specifying gonadotropin-releasing hormone expression in adult mice. *Mol. Endocrinol.*, 13: 2203–2211.

Parfitt, D.B., Thompson, R.C., Richardson, H.N., Romeo, R.D. and Sisk, C.L. (1999) GnRH mRNA increases with puberty in the male syrian hamster brain [In Process Citation]. *J. Neuroendocrinol.*, 11: 621–627.

Poletti, A., Melcangi, R.C., Negri-Cesi, P., Maggi, R. and Martini, L. (1994) Steroid binding and metabolism in the luteinizing hormone-releasing hormone-producing neuronal cell line GT1-1. *Endocrinology*, 135(6): 2623–2628.

Poletti, A., Rampoldi, A., Piccioni, F., Volpi, S., Simeoni, S., Zanisi, M. and Martini, L. (2001) 5Alpha-reductase type 2 and androgen receptor expression in gonadotropin releasing hormone GT1-1 cells. *J. Neuroendocrinol.*, 13: 353–357.

Porkka-Heiskanen, T., Urban, J.H., Turek, F.W. and Levine, J.E. (1994) Gene expression in a subpopulation of luteinizing hormone-releasing hormone (LHRH) neurons prior to the preovuluatory gonadotropin surge. *J. Neurosci.*, 14(9): 5548–5558.

Quinton, R., Hasan, W., Grant, W., Thrasivoulou, C., Quiney, R.E., Besser, G.M. and Bouloux, P.M. (1997) Gonadotropin-releasing hormone immunoreactivity in the nasal epithelia of adults with Kallmann's syndrome and isolated hypogonadotropic hypogonadism and in the early midtrimester human fetus. *J. Clin. Endocrinol. Metab.*, 82: 309–314.

Radovick, S., Wondisford, F.E., Nakayama, Y., Yamada, M., Cutler Jr. G.B. and Weintraub, B.D. (1990) Isolation and characterization of the human gonadotropin-releasing hormone gene in the hypothalamus and placenta. *Mol. Endocrinol.*, 4: 476–480.

Radovick, S., Wray, S., Lee, E., Nicols, D.K., Nakayama, Y., Weintraub, B.D., Westphal, H., Cutler Jr., G.B. and Wondisford, F.E. (1991) Migratory arrest of gonadotropin-releasing hormone neurons in transgenic mice. *Proc. Natl. Acad. Sci. USA*, 88: 3402–3406.

Rajendren, G. (2001) Subsets of gonadotropin-releasing hormone (GnRH) neurons are activated during a steroid-induced luteinizing hormone surge and mating in mice: a combined retrograde tracing double immunohistochemical study. *Brain Res.*, 918: 74–79.

Rakic, P. and Sidman, R.L. (1973a) Sequence of developmental abnormalities leading to granule cell deficit in cerebellar cortex of weaver mutant mice. *J. Comp. Neurol.*, 152: 103–132.

Rakic, P. and Sidman, R.L. (1973b) Weaver mutant mouse cerebellum: defective neuronal migration secondary to abnormality of Bergmann glia. *Proc. Natl. Acad. Sci. USA*, 70: 240–244.

Richards, J.S. (1994) Hormonal control of gene expression in the ovary. *Endocrinol. Rev.*, 15: 725–751.

Rogers, M.C., Silverman, A.J. and Gibson, M.J. (1997) Gonadotropin-releasing hormone axons target the median eminence: in vitro evidence for diffusible chemoattractive signals from the mediobasal hypothalamus. *Endocrinology*, 138: 3956–3966.

Rogers, M.C., Silverman, A.J. and Gibson, M.J. (1998) Preoptic area grafts implanted in mammillary bodies of hypogonadal

mice: patterns of GnRH neuronal projections. *Exp. Neurol.*, 151: 265–272.

Scholl, F.G., Gamallo, C., Vilar?o S. and Quintanilla, M. (1999) Identification of PA2.26 antigen as a novel cell-surface mucin-type glycoprotein that induces plasma membrane extensions and increased motility in keratinocytes. *J. Cell Sci.*, 112: 4601–4613.

Schwanzel-Fukuda, M., Garcia, M.S., Morrell, J.I. and Pfaff, D.W. (1987) Distribution of luteinizing hormone-releasing hormone in the nervus terminalis and brain of the mouse detected by immunocytochemistry. *J. Comp. Neurol.*, 255(2): 231–244.

Schwanzel-Fukuda, M., Bick, D. and Pfaff, D.W. (1989) Luteinizing hormone-releasing hormone (LHRH)-expressing cells do not migrate normally in an inherited hypogonadal (Kallmann) syndrome. *Mol. Brain Res.*, 6: 311–326.

Seeburg, P.H., Mason, A.J., Stewart, T.A. and Nikolics, K. (1987) The mammalian GnRH gene and its pivotal role in reproduction. *Rec. Prog. Horm. Res.*, 41: 69–98.

Silverman, A.J., Roberts, J.L., Dong, K.W., Miller, G.M. and Gibson, M.J. (1992) Intrahypothalamic injection of a cell line secreting gonadotropin-releasing hormone results in cellular differentiation and reversal of hypogonadism in mutant mice. *Proc. Natl. Acad. Sci. USA*, 89: 10668–10672.

Skynner, M.J., Slater, R., Sim, J.A., Allen, N.D. and Herbison, A.E. (1999) Promoter transgenics reveal multiple gonadotropin-releasing hormone-I-expressing cell populations of different embryological origin in mouse brain. *J. Neurosci.*, 19: 5955–5966.

Sosnowski, R., Mellon, P.L. and Lawson, M.A. (2000) Activation of translation in pituitary gonadotrope cells by gonadotropin-releasing hormone. *Mol. Endocrinol.*, 14: 1811–1819.

Suter, K.J., Song, W.J., Sampson, T.L., Wuarin, J.P., Saunders, J.T., Dudek, F.E. and Moenter, S.M. (2000) Genetic targeting of green fluorescent protein to gonadotropin-releasing hormone neurons: characterization of whole-cell electrophysiological properties and morphology. *Endocrinology*, 141: 412–419.

Terasawa, E., Luchansky, L.L., Kasuya, E. and Nyberg, C.L. (1999) An increase in glutamate release follows a decrease in gamma aminobutyric acid and the pubertal increase in luteinizing hormone releasing hormone release in the female rhesus monkeys. *J. Neuroendocrinol.*, 11: 275–282.

Van Goor, F., Krsmanovic, L.Z., Catt, K.J. and Stojilkovic, S.S. (2000) Autocrine regulation of calcium influx and gonadotropin-releasing hormone secretion in hypothalamic neurons. *Biochem. Cell. Biol.*, 78: 359–370.

Waldstreicher, J., Seminara, S.B., Jameson, J.L., Geyer, A., Nachtigall, L.B., Boepple, P.A., Holmes, L.B. and Crowley, W.F.J. (1996) The genetic and clinical heterogeneity of gonadotropin-releasing hormone deficiency in the human. *J. Clin. Endocrinol. Metab.*, 81: 4388–4395.

Weiner, R.I. and Martinez, d.l.E. (1993) Pulsatile release of gonadotrophin releasing hormone (GnRH) is an intrinsic property of GT1 GnRH neuronal cell lines. *Hum. Reprod.*, 8(2): 13–17.

Wetsel, W.C., Mellon, P.L., Weiner, R.I. and Negro-Vilar, A.

(1991) Metabolism of pro-luteinizing hormone-releasing hormone in immortalized hypothalamic neurons. *Endocrinology*, 129: 1584–1595.

Whyte, D.B., Lawson, M.A., Belsham, D.D., Eraly, S.A., Bond, C.T., Adelman, J.P. and Mellon, P.L. (1995) A neuron-specific enhancer targets expression of the gonadotropin-releasing hormone gene to hypothalamic neurosecretory neurons. *Mol. Endocrinol.*, 9: 467–477.

Wierman, M.E., Kepa, J.K., Sun, W., Gordon, D.F. and Wood, W.M. (1992) Estrogen negatively regulates rat gonadotropin releasing hormone (rGnRH) promoter activity in transfected placental cells. *Mol. Cell. Endocrinol.*, 86: 1–10.

Wierman, M.E., Xiong, X., Kepa, J.K., Spaulding, A.J., Jacobsen, B.M., Fang, Z., Nilaver, G. and Ojeda, S.R. (1997) Repression of gonadotropin-releasing hormone promoter activity by the POU homeodomain transcription factor SCIP/Oct-6/Tst-1: a regulatory mechanism of phenotype expression? *Mol. Cell. Biol.*, 17: 1652–1665.

Wolfe, A.M., Wray, S., Westphal, H. and Radovick, S. (1995) Cell-specific expression of the human gonadotropin-releasing hormone gene in transgenic animals. *J. Biol. Chem.*, 271: 20018–20023.

Wolfe, A., Kim, H.S.D., Tobet, S.A. and Radovick, S. (2002) Identification of a discrete promoter region of the human GnRH gene that is sufficient for directing neuron specific expression: A role for POU homeodomain transcription factors. *Mol. Endocrinol.*, 16: 435–449.

Wray, S., Grant, P. and Gainer, H. (1989) Evidence that cells expressing luteinizing hormone-releasing hormone mrna in the mouse are derived from progenitor cells in the olfactory placode. *Proc. Natl. Acad. Sci. USA*, 86: 8132–8136.

Wray, S., Key, S., Qualls, R. and Fueshko, S.M. (1994) A subset of peripherin positive olfactory axons delineates the luteinizing hormone releasing hormone neuronal migratory pathway in developing mouse. *Dev. Biol.*, 166: 349–354.

Wu, T.J., Gibson, M.J., Rogers, M.C. and Silverman, A.J. (1997) New observations on the development of the gonadotropin-releasing hormone system in the mouse. *J. Neurobiol.*, 33: 983–998.

Yeo, T.T., Gore, A.C., Jakubowski, M., Dong, K.W., Blum, M. and Roberts, J.L. (1996) Characterization of gonadotropin-releasing hormone gene transcripts in a mouse hypothalamic neuronal GT1 cell line. *Brain Res. Mol. Brain Res.*, 42: 255–262.

Yokosaki, Y., Matsuura, N., Higashiyama, S., Murakami, I., Obara, M., Yamakido, M., Shigeto, N., Chen, J. and Sheppard, D. (1998) Identification of the ligand binding site for the integrin alpha9 beta1 in the third fibronectin type III repeat of tenascin-C. *J. Biol. Chem.*, 273: 11423–11428.

Yoshida, K., Rutishauser, U., Crandall, J.E. and Schwarting, G.A. (1999) Polysialic acid facilitates migration of luteinizing hormone-releasing hormone neurons on vomeronasal axons. *J. Neurosci.*, 19: 794–801.

Zamorano, P.L., Mahesh, V.B., De Sevilla, L. and Brann, D.W. (1998) Excitatory amino acid receptors and puberty. *Steroids*, 63: 268–270.

Zhen, S., Dunn, I.C., Wray, S., Liu, Y., Chappell, P.E., Levine, J.E. and Radovick, S. (1997a) An alternative gonadotropin-releasing hormone (GnRH) RNA splicing product found in cultured GnRH neurons and mouse hypothalamus. *J. Biol. Chem.*, 272: 12620–12625.

Zhen, S., Zakaria, M., Wolfe, A. and Radovick, S. (1997b) Regulation of gonadotropin-releasing hormone (GnRH) gene expression by insulin-like growth factor I in a cultured GnRH-expressing neuronal cell line. *Mol. Endocrinol.*, 11: 1145–1155.

I.S. Parhar (Ed.)
Progress in Brain Research, Vol. 141
© 2002 Elsevier Science B.V. All rights reserved

CHAPTER 19

Physiology and release activity of GnRH neurons

Yoshitaka Oka [*]

Misaki Marine Biological Station, Graduate School of Science, The University of Tokyo, Kanagawa 238-0225, Japan

Introduction

The gonadotropin-releasing hormone (GnRH), which was first isolated from the mammalian hypothalamus, was originally identified as a hypophysiotrophic decapeptide hormone that facilitates gonadotropin release from the pituitary gonadotropes. Recent immunocytochemical and in situ hybridization techniques have greatly advanced our knowledge of the anatomical features of the GnRH neuronal systems (Oka, 1997 for review). These anatomical studies have shown that the GnRH neurons not only synthesize and transport releasable peptides to the median eminence (hypothalamic system) but also project widely in various brain regions (extrahypothalamic system). On the other hand, combined HPLC chromatography and peptide biochemistry have revealed a wide variety of molecular species of GnRH peptides among different vertebrate species. Furthermore, it has also been recognized quite recently that there are also diversity of GnRH receptors, and the functional and evolutionary significance of such diversity of GnRH ligand and receptors has been attracting more and more attention. Although the function of the hypothalamic GnRH system has been well studied and is exactly what the name implies, i.e., facilitation of the release of gonadotropins, the functional significance of the extrahypothalamic

GnRH systems is elusive and has only poorly been studied yet.

We have been interested in the biological significance of an enigmatic extrahypothalamic GnRH system, the terminal nerve (TN)–GnRH system, of vertebrate brains. Although we still do not have a complete picture of the functional significance of TN-GnRH system, we now have several lines of evidence to suggest that TN-GnRH system serves as a neuromodulatory system that controls the motivational or arousal state of the animal. For the biological analysis of this system using multidisciplinary techniques, we have been using the brain of a tropical fish, the dwarf gourami (*Colisa lalia*), because the GnRH neurons of this fish are large (20 ~ 30 μm in diameter) and make a tight cell cluster without intercalating glial cells just beneath the meningeal membrane so that one can readily record the activities of a single GnRH neuron in a whole brain in vitro preparation using sharp microelectrodes as well as patch pipettes under visual guidance. Thus, it is possible to study, in a semi-intact whole brain, the physiology, morphology, and cell biology of single GnRH neurons by taking advantage of the in vitro experimental conditions. Because the GnRH neuronal cell bodies in most vertebrate species are small (about 10 μm in diameter) and diffusely distributed in various brain regions, it has been extremely difficult to study various features of single GnRH neurons (Kelly et al., 1984). Therefore, the dwarf gourami clearly has experimental advantages over other vertebrates to study the cellular physiology and morphology of GnRH neurons. Although recent technological development has enabled electrical recording of GFP-labeled hypothalamic GnRH

[*] Correspondence to: Y. Oka, Misaki Marine Biological Station, Graduate School of Science, The University of Tokyo, Kanagawa 238-0225, Japan. Tel.: +81-468-81-4105; Fax: +81-468-81-7944;
E-mail: okay@mmbs.s.u-tokyo.ac.jp

neurons visualized in transgenic mice brain slices (Skynner et al., 1999; Spergel et al., 1999; Suter et al., 2000a,b), this necessitates a laborious preparation and does not seem to be suitable for stable recording. In the dwarf gourami, we have also found out that the TN-GnRH neurons exhibit exocytotic peptide release activity even from the cell bodies (Oka and Ichikawa, 1991, 1992; Ishizaki and Oka, unpublished data); this is of great advantage because the peptides are usually released from small nerve terminals or varicosities, which are very difficult to access for direct measurements of their electrical or release activities. Thus, we believe that the whole brain in vitro preparation of the dwarf gourami is ideal for studying the molecular and cellular mechanisms of peptidergic neuromodulation as well as exocytotic peptide release mechanisms in general. Last but not least, the dwarf gouramis are tropical fish and it is very easy to get sexually mature fish and observe the characteristic reproductive behaviors throughout the year; this is advantageous for the neuroethological analysis, which is essential to know the biological significance of the peptidergic neuromodulation. In this chapter, I will summarize our recent research progress in the physiology and release activity of GnRH neurons, especially focusing on the terminal nerve (TN) GnRH system.

Terminal nerve (TN)–GnRH system, and the diversity of GnRH peptides and neuronal systems

Specific antibodies against various neurotransmitter substances and peptide hormones and highly sensitive immunocytochemical techniques have become available in 1970's, and these enabled us to identify not only the precise location of neurons that produce transmitters and hormones but also the trajectories of their neurites in detail. Owing to these technical advancements, the catalogue of GnRH neuronal systems of vertebrates has been expanding to cover various vertebrate phyla from cyclostomes to mammals including humans (see Silverman et al., 1994). These immunocytochemical studies suggested the presence of extrahypothalamic GnRH systems other than the conventional hypothalamic hypophysiotropic (septo-preoptico-infundibular) system. Schwanzel-Fukuda and Silverman (1980) first reported that the rostral part of such extrahypothalamic GnRH cell groups

belongs to the terminal nerve (TN). The TN was first described anatomically as the last (terminal), and the supernumerary, macroscopically identifiable cranial nerve number zero in elasmobranchs by Fritsch (see Demski and Schwanzel-Fukuda, 1987). The name 'nervus terminalis' seems to have derived from the association of a portion of this nerve with the rostral end of the lamina terminalis. Although the TN was subsequently identified in various other vertebrates including human embryos, teleosts, urodele amphibians, and so on, very little attention had been paid to it until the immunocytochemical report of Schwanzel-Fukuda and Silverman (1980). Shortly after this report, Demski and Northcutt (1983) and Springer (1983) reported on very exciting findings about the anatomy and function of the TN of the goldfish (*Carassius auratus*). Demski and Northcutt (1983) reported, by using the HRP (horse-radish peroxidase; they did not use GnRH immunohistochemistry) tract tracing method, that the TN cells which are located at the rostral base of the olfactory bulb project to the retina, supracommissural part of the area ventralis of the telencephalon (Vs) and the preoptic area. They could elicit sperm release by electrical stimulation of the optic nerve, which they supposed should antidromically stimulate the retinopetal TN fibers and lead to the collateral activation of the preoptic area and Vs. It had been suggested in teleosts that the preoptic area and Vs are important for the facilitation of sexual behaviors (Koyama et al., 1984; Satou et al., 1984). Putting these pieces of evidence together, they proposed a very adventurous hypothesis about the function of the terminal nerve; the terminal nerve is a new chemosensory system in vertebrates, and the pheromonal stimulation facilitates sexual behavior via its projection to the preoptic area. However, this possibility was later doubted by Fujita et al. (1991). They stimulated the olfactory epithelium of the goldfish with identified sex pheromones ($17\alpha,20\beta$-dihydroxy-4-pregnen-3-one and prostaglandin $F2\alpha$) and extracellularly recorded the spontaneous activity of the TN cell and the mitral cell (the principal neuron in the olfactory bulb which receives synaptic inputs from the olfactory receptor cells and projects to the secondary olfactory center) of the olfactory bulb. Whereas the mitral cells responded to the pheromonal stimulation, the TN cells did not.

Thus, it is least probable that the TN-GnRH system is chemosensory. We have also shown behaviorally that the total lesion of bilateral TN-GnRH neurons in the male dwarf gourami did not interfere with the overall performance of the sexual behaviors, although it affected the motivational or arousal state of the fish (Yamamoto et al., 1997; also see below). Interestingly, it has been found in some species that the TN-GnRH cell neurites are present in the olfactory nerve and can be traced to the lamina propria of the olfactory epithelium (Eisthen et al., 2000; Wirsig-Wiechmann and Oka, 2002). Furthermore, Eisthen et al. (2000) found that GnRH peptide can modulate the sensitivity of olfactory receptor neurons via modulation of inward Na^+ channels and outward K^+ channels. Demski and Northcutt (1983) may have labeled these neurites retrogradely from the olfactory epithelium. On the other hand, Springer (1983) reported on the anatomy of the goldfish TN including its retinopetal projection by using the cobalt chloride tract-tracing method. Subsequent immunohistochemical studies have clearly demonstrated that these retinopetal projections of TN are GnRH-immunoreactive (Münz and Stumpf, 1981; Stell et al., 1984; Oka and Ichikawa, 1990; Uchiyama, 1990). The presence of GnRH-immunoreactive retinopetal projection proved to be general anatomical characteristics of teleost TN-GnRH systems. Interestingly, the GnRH-immunoreactive retinopetal projection system seems to be unique to teleosts among vertebrate species. With the exception of the retinopetal projections, the presence of GnRH-immunoreactive TN system has been recognized in all vertebrate species reported to date (Demski and Schwanzel-Fukuda, 1987).

In addition to the TN-GnRH system, another extrahypothalamic GnRH system, with their cell bodies in the midbrain and their axons projecting widely in the caudal part of the brain, has been reported in various vertebrate species. For example, in elasmobranchs, teleosts, amphibians, reptiles, birds, and primitive mammals, midbrain GnRH neuronal system has been reported, all of which are immunoreactive to chicken II GnRH (see below). On the other hand, TN and POA–GnRH neurons seem to express various forms of GnRH according to the species. Therefore, the midbrain GnRH neuronal system may be the most evolutionarily conservative one (Muske,

1997; Sherwood et al., 1997). From the anatomical similarities (location of cell bodies and their projection routes) between the chicken II-ir midbrain GnRH neurons and the nucleus of the fasciculus medialis longitudinalis (nFLM), some authors have described them as identical. However, Yamamoto et al. (1998a) have examined this possibility experimentally by double-labeling experiments, in which the nFLM and their descending axons were labeled by retrograde transport of biocytin injected into the spinal cord, and the GnRH-ir profiles were labeled by chicken II specific antibody. As a result, they found that, although the two systems are anatomically juxtaposed with each other, they are separate entities. The cell bodies of nFLM are large and possess thick and straight spinal descending axons, while the midbrain chicken II-GnRH neurons also have large cell bodies but have thin and tortuous axons with many branches.

Although the function of hypothalamic GnRH system is clear, the GnRH neuronal cell bodies in most vertebrate species are small and diffusely distributed in various brain regions, and their axons do not usually form distinct bundles that lead to the median eminence, where GnRH peptides are supposed to be released from the axon terminals into the portal vessels and transported to the gonadotropes to stimulate the release of gonadotropins. Therefore, it has been and still is very difficult to identify the cells of origins of the hypophysiotropic GnRH neurons. Teleosts lack the structure that is equivalent to the median eminence, and the axons of all the hypophysiotropic hormone-producing cells project directly to the pituitary gland. Interestingly, in some teleosts including the dwarf gourami, hypophysiotropic POA-GnRH neurons make distinctive cell clusters in the preoptic area and project their axons to the pituitary, forming a robust axon bundle (Oka and Ichikawa, 1990). Furthermore, we found that, about two weeks after bilateral electrolytic lesions in the TN-GnRH neurons, the POA-GnRH neurons and their axonal projections to the pituitary could be observed in isolation (Yamamoto et al., 1995, 1997). By taking advantage of the whole brain in vitro preparation of the dwarf gourami, injection of axonal tracers such as neurobiotin or DiI into the pituitary became much easier compared to that in vivo, and the neurons of origins that project to the pituitary could be retrogradely labeled, and the double-labeling with GnRH immunocytochemistry clearly demonstrated that the POA-GnRH neurons are the main population of GnRH neurons that project to the pituitary and thus can function actually as a gonadotropin-releasing hormone system (Yamamoto et al., 1998b).

In addition to the studies of anatomically different GnRH neuronal systems, there have been another series of studies aimed at clarifying the diversity and evolution of various molecular forms of GnRH peptides (reviewed by Sherwood et al., 1997). GnRH is composed of only ten amino acid residues, but some of them are variable and more than fourteen (Montaner et al., 2001; the number of variants are still increasing) different molecular species of GnRH peptides have been sequenced thus far (for reviews, see Sherwood et al., 1993, 1997). Moreover, in most classes of vertebrates, several types of GnRH have been found in the brain of single species. For example, salmon GnRH, seabream GnRH, and chicken II GnRH have been found in the many teleost brains. Even in humans, where the mammalian GnRH (originally identified LHRH) has been regarded as one and the only type of GnRH peptide in the brain, expression of a second gene for GnRH that is equivalent to chicken II GnRH was reported (White et al., 1998). They further classified GnRH peptides into GnRH I to III and discussed the evolutionary relationships among them. According to them, GnRH I contains releasing forms produced in the hypothalamus, GnRH II contains evolutionarily conservative mesencephalic forms, and GnRH III contains telencephalic forms that are generally found in the terminal nerve system. They suggest that these three types of GnRH peptides belong to evolutionarily distinct GnRH groups. As discussed later, GnRH have long been suggested to be responsible for the generation of long slow EPSP by inhibiting K^+ M current in the sympathetic ganglion cells of the bullfrog. Interestingly, it has been reported that GnRH II (chicken II GnRH) is expressed in the frog sympathetic ganglions (Troskie et al., 1998). They further found a GnRH receptor selective for chicken II GnRH in the ganglia, and suggested that endogenous chicken II GnRH may play a role in synaptic transmission (or neuromodulation of K^+ M current) in the sympathetic ganglia via a receptor specific for chicken II GnRH.

The anatomical correspondence between the morphologically as well as functionally divergent multiple GnRH systems and a variety of molecular species of GnRH peptides had been rather difficult to demonstrate experimentally until recently. This is mainly because the peptidergic neurons are generally small and scattered in the brain and are hard to label with axonal tracers or make local brain lesions to allow the specific group of GnRH neurons to degenerate. For this kind of analysis, it has proved to be much more advantageous to use the teleost brains than to use the mammalian or other vertebrate brains (see Kim et al., 1997; Oka, 1997; Parhar, 1997). Thus, correspondence between the anatomy of GnRH system and the molecular species has been clearly demonstrated experimentally in the dwarf gourami (Yamamoto et al., 1995, 1997; Ishizaki et al., 2002). As already mentioned, the TN-GnRH neurons of the dwarf gourami make tight cell clusters on both sides of the brain in the transitional area between the olfactory bulb and the telencephalon. Therefore, it is possible to make local electrolytic brain lesions in the TN-GnRH neuron clusters (ganglion of the terminal nerve) on both sides of the brain and make them degenerate. By combining local brain lesioning with immunohistochemistry using antibodies specific to different GnRH molecules, it became possible to differentiate the projection areas of multiple GnRH neuronal systems. Because of the technical difficulty of selectively lesioning specific group of GnRH neurons, comparable anatomical studies have not been carried out in other vertebrate brains. However, recent progress in molecular biological techniques have enabled cloning and sequencing genes coding different molecular variety of GnRH peptides, and it became possible to demonstrate the distribution of neurons expressing mRNAs specific to each type of GnRH molecule by using in situ hybridization techniques (Parhar, 1997). Thus, a growing body of evidence suggests the presence of at least three anatomically and functionally different GnRH systems throughout the vertebrate species (Fig. 1), although the cyclostomes seem to be exceptional (Sower, 1997). For example, in teleosts such as dwarf gourami (Yamamoto et al., 1995; Ishizaki et al., 2002), tilapia (Parhar, 1997), and seabream (Okuzawa et al., 1997), general agreement has been reached, by using specific

Multiple GnRH systems

A Terminal nerve
GnRH system
(sGnRH)

B Midbrain GnRH
system
(CIIGnRH)

C POA GnRH system
(sbGnRH)

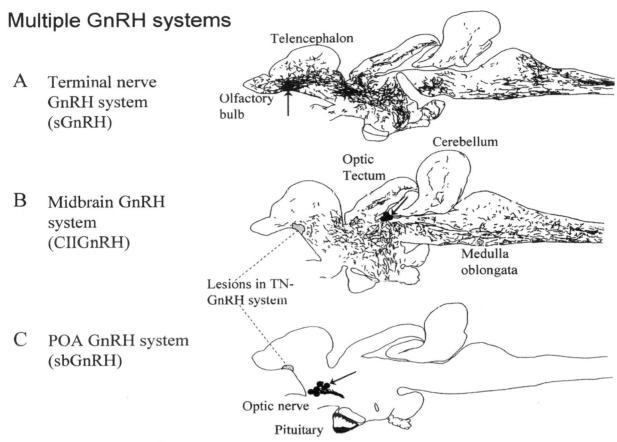

Fig. 1. Schematic illustrations of the distribution of cell bodies and fibers of the TN- (A), midbrain- (B), and POA-GnRH (C) systems, based on the combined local brain lesioning and immunocytochemistry using antisera specific for different molecular species of GnRH. It is now known that each system is immunoreactive to sGnRH, cIIGnRH, and sbGnRH, respectively. The arrows indicate the location of cell bodies, and the stippled areas indicate brain lesions in the TN-GnRH cell bodies. (Modified from Yamamoto et al., 1995.)

immunohistochemistry and in situ hybridization, that the TN-GnRH system expresses salmon GnRH, the POA-GnRH system seabream GnRH, and the midbrain GnRH system chicken II GnRH. From these and other lines of evidence, we suggest that there are basically three different GnRH neuronal systems with different anatomy, molecular species, and functions. (1) The POA (preoptic, hypothalamo–hypophysial)–GnRH system (Fig. 1C) that projects almost exclusively to the median eminence (directly to the pituitary gland in the case of teleosts) but very sparsely to the brain. (2) The TN-GnRH system (Fig. 1A) that projects widely throughout the brain from the olfactory bulb to the spinal cord but never to the pituitary gland (Oka and Ichikawa, 1990; Oka and Matsushima, 1993; Yamamoto et al.,

1995). From this anatomical feature, the TN-GnRH system should not be directly involved in the control of pituitary functions. Instead, it is suggested to function as an important neuromodulatory system (see below, and Oka, 1997; Oka and Abe, 2002). The finding in the goldfish that disruption of TN-GnRH system does not affect the ovarian development or ovulation (Kobayashi et al., 1994; Kim et al., 2001) strongly supports this. (3) The midbrain GnRH system (Fig. 1B) that is considered to be the most evolutionarily conservative GnRH system in the vertebrates. All the midbrain GnRH systems reported thus far are immunoreactive to chicken II type of GnRH peptide, and they also project widely in the brain, especially in the caudal brain structures, but never to the pituitary gland, and thus seems to have some neuromodulatory functions similar to the TN-GnRH system.

Here we would like to stress that it would not be a fruitful discussion to correlate specific molecular species of GnRH peptide or GnRH receptor types with specific functions as has been frequently done, since the relationship between the molecular species and the anatomy of GnRH system varies according to the animal species except for the conservative linkage between the midbrain GnRH system and the chicken II GnRH (see above). Furthermore, the ligand–receptor relationships in GnRH neuronal systems do not seem to be so specific so that it often happens that the GnRH species that should not be considered as physiologically active peptide, as judged from its projection pattern, can often evoke effective responses when administered exogenously in vivo or in vitro (King and Millar, 1997). Therefore, we strongly argue that it is not the type of GnRH molecule or GnRH receptors that determine the function of the particular GnRH neuronal system but the specific projection pattern of the GnRH neurons, i.e., the projection pattern or the distribution areas of neurites of each GnRH neuronal system.

Recent findings on the GnRH neuronal systems in the ancestral chordates, ascidians, are noteworthy from the evolutionary viewpoint of the multiple GnRH neuronal systems of vertebrates. The presence of GnRH-like substance was first reported by immunocytochemical studies using vertebrate GnRH antisera (Georges and Dubois, 1980; Mackie, 1995). We made a precise mapping of GnRH-immunoreactive neurons in the central as well as in peripheral nervous systems of *Ciona* (Tsutsui et al., 1998). We carried out immunohistochemistry using anti-salmon GnRH and anti-chicken II GnRH antisera and found GnRH-immunoreactive structures in the neural ganglion, which is considered to be the central nervous system of ascidians, and along the inner wall of the dorsal blood sinus, which is a thick blood cavity running from the cerebral ganglion towards the gonads along the gonoducts (spermiduct and oviduct; the ascidians are hermaphrodites), as well as on the surface of the ovary. Although the existence of pituitary homologs and the hypothalamo–hypophysial–gonadal axis in the ascidians is still controversial, such findings of ascidian GnRH neuronal system may shed light on the evolution of

vertebrate GnRH nervous system. There are some recent interesting findings on the ascidian GnRH molecules. Powell et al. (1996) reported two new molecular species of GnRH that are specifically produced in the nervous system of ascidians (Tunicate I and Tunicate II). Di Fiore et al. (2000) found mammalian GnRH and chicken I GnRH in the gonad of the tunicate, *Ciona*, and showed that these GnRH peptides facilitate synthesis and release of sex steroids in the gonad of *Ciona* as well as release of LH from the rat pituitary. Both of these GnRH peptides are expressed in the hypophysiotropic POA GnRH neurons, and it is interesting that this type of GnRH peptides already existed in the ancestral chordate, ascidians. These results led them to conclude that the amino acid sequence and function of these two molecular forms of GnRH peptides have been well conserved during evolution of chordates.

Morphology and physiology of single GnRH neurons

Because of the technical difficulties of recording from single GnRH neurons, there has been very limited information on the morphological and physiological characteristics of individual GnRH neurons. By taking advantage of the accessibility of single GnRH neurons in the whole brain in vitro preparation of the dwarf gourami, we succeeded in recording and labeling single GnRH neurons intracellularly (Oka, 1992; Oka and Matsushima, 1993). The morphology and physiology of single GnRH neurons have already been well described in detail (see Oka, 1997 for review), and these will not be repeated here in detail. In summary, most of the TN-GnRH cells showed endogenous slow (1–7 Hz) regular pacemaker potentials (Fig. 2). The intrinsic nature of this activity was demonstrated by the voltage dependence of the pacemaker frequency, rhythm resetting, and persistence of rhythmicity after synaptic isolation. Only a small number of TN cells showed either irregular or bursting discharge patterns. Anatomical observation of intracellularly labeled cells (Fig. 3) showed that, regardless of discharge patterns, all the TN cells had multiple axonal branches which project to those areas where we had previously demonstrated dense GnRH-immunoreactive fibers (Oka and Ichikawa, 1990; Yamamoto et al., 1995). In-

Intracellular recording of TN-GnRH neurons

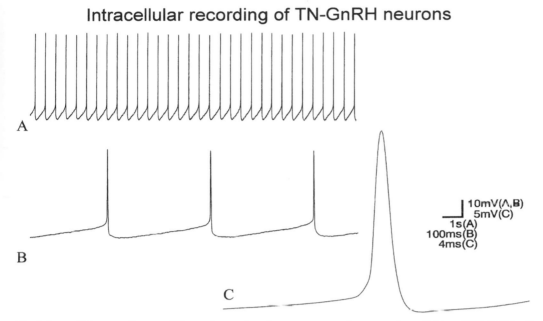

Fig. 2. Intracellular recordings on different time scales of spontaneous regular pacemaker activities of TN-GnRH cells in a whole brain in vitro preparation. It is characterized by regular and slow-frequency spikes with rather long durations. During interspike intervals, smooth and gradual depolarizations, which are similar to the cardiac pacemaker potentials, are evident. (Modified from Oka and Matsushima, 1993.)

terestingly, in the recordings where several different cells were recorded from the same preparations, they showed similar pacemaker frequencies. Therefore, it was suggested that the frequencies of pacemaker activities are related to the physiological conditions of the fish, which are controlled by the hormonal or environmental factors. Furthermore, the morphological features of TN GnRH neurons are relevant for the regulation of excitability and/or transmitter release (see below) of target neurons in a wide variety of brain regions simultaneously via multiple axonal branches (Fig. 3). Thus, TN-GnRH cells possess the morphological and physiological characteristics relevant to function as a neuromodulator.

Mechanisms of generation and modulation of pacemaker potentials

From the results thus far obtained, the pacemaker potential and its frequency or pattern seems to be important for TN-GnRH neurons as a neuromodulator. Therefore, we focused on the generation and modulation mechanisms of pacemaker frequencies (Oka, 1995, 1996; Abe and Oka, 1999, 2000, 2002).

We first found a novel TTX-resistant persistent Na^+ current, $I_{Na(slow)}$, which supplies the persistent depolarizing drive and plays an important role in the generation of pacemaker potentials in TN-GnRH cells (Oka, 1995). This current was discovered when the pacemaker potentials were resistant to tetrodotoxin (TTX) but were readily blocked by substituting Na^+-impermeant ions (tetramethylammonium or choline) for Na^+ in the perfusing solution, and the resting membrane potential became more hyperpolarized than the control level. $I_{Na(slow)}$ current was further characterized by using the patch voltage clamp recording (Oka, 1996). $I_{Na(slow)}$ currents could be isolated pharmacologically by blocking K^+ currents, Ca^{2+} currents, and conventional fast Na^+ current. The current was characterized by resistance to TTX blockade, dependence on external Na^+, slow activation, very slow and little inactivation, and wide overlap of activation and inactivation curves (window currents) near the resting potential. These characteristics are distinct from those of conventional fast Na^+ current, and are relevant for the generation of persistent inward currents necessary for the pacemaker activity of TN-GnRH cells.

Intracellular staining of TN-GnRH neurons; reconstruction from serial sections

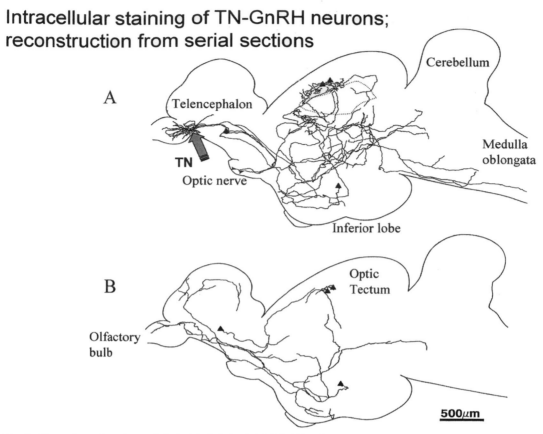

Fig. 3. Illustrations of an intracellularly labeled single TN-GnRH cell reconstructed from serial sagittal sections and are drawn on a representative parasagittal plane. Biocytin was injected into a single neuron unilaterally (the arrow in A). Labeled neurites are seen widely in the brain from the olfactory bulb to the medulla and the rostral spinal cord, and from the medial to the lateral ends. In addition, some axons cross to the contralateral side (B) at various levels, e.g., in the anterior commissure, posterior commissure, tectal commissure, and ventral tegmentum (filled triangles), and project widely on the contralateral side. (Modified from Oka and Matsushima, 1993.)

We next searched for the candidate outward currents that should counteract the persistent depolarizing drive supplied by $I_{Na(slow)}$. We demonstrated, by using the whole-cell voltage clamp recording, that the TN-GnRH cells have at least four types of voltage-dependent K^+ currents; (1) 4-aminopyridine (4AP)-sensitive K^+ current, (2) Tetraethylammonium (TEA)-sensitive K^+ current, and (3), (4) two types of TEA- and 4AP-resistant K^+ currents. Among these, the second, TEA-sensitive K^+ current evoked from a holding potential of -100 mV was slowly activating, long-lasting and showed comparatively low-threshold of activation. This current was only partially inactivated at steady state of -60 to -40 mV, which is equivalent to the resting membrane potential of the TN-GnRH neurons. Furthermore, in current-

clamp recordings bath application of TEA together with TTX reversibly blocked the pacemaker potentials. Therefore, we concluded that the TEA-sensitive K^+ current, $I_{K(V)}$, is the most likely candidate that contributes to the repolarizing phase of the pacemaker potentials of TN-GnRH cells.

From these results, we concluded that the basic pattern of pacemaker activities of the TN-GnRH neurons, especially their subthreshold component, is generated in the following manner. When the TN-GnRH cells are at a resting potential, considerable amount of $I_{Na(slow)}$ is non-inactivated and supplies the persistent depolarizing drive, and the membrane potentials gradually depolarizes. When the membrane potential reaches the activation threshold for the $I_{K(V)}$, outward current develops, and the net flux of current reverses to outward. Then, the membrane potential becomes hyperpolarized and deactivates the K^+ current, and the next cycle begins. In addition, we have recently found out that the voltage-dependent N-type Ca^{2+} current sensitive to ω-Conotoxin GVIA and the voltage-independent Ca^{2+} influx sensitive to Zn^{2+} or SK&F96365 (store-operated Ca^{2+} current, SOC) may be also involved in the generation of pacemaker potentials (Abe and Oka, 2002; Oya et al., 2001). Considering the activation kinetics, the N-type Ca^{2+} current are suggested to be involved in the action potential phase (suprathreshold phase) of the pacemaker activities, together with the TTX-sensitive conventional $I_{Na(fast)}$. The SOC probably contributes to the generation of background persistent inward current together with $I_{Na(slow)}$.

We next searched for possible candidates that modulate the pacemaker activity of GnRH neurons (Abe and Oka, 2000). We found that the salmon GnRH (sGnRH), which is the same molecular species of GnRH produced by TN-GnRH neurons themselves, affects the pacemaker activities as follows (Fig. 4).

In Ringer solution, TN-GnRH neuron showed slow regular beating discharge. During the bath application of sGnRH, the firing frequency of pacemaker activity was transiently decreased (early phase), and subsequently increased (late phase). It should be noted that this modulation occurred without detectable membrane potential changes in most cases, which precludes the possibility that the modulation of pacemaker activity, decrease and then increase of frequency, may be simply caused by hyperpolarization and depolarization of the membrane, respectively. This biphasic modulation of pacemaker activity was also evoked by bath application of another kind of GnRH peptide (mammalian GnRH) but was not evoked by inactive GnRH analog (GnRH antagonist alone) and was inhibited or attenuated by GnRH antagonist. These results strongly suggested that modulation by GnRH peptide of pacemaker activity of TN-GnRH neurons is caused by GnRH receptor activation, although there does not seem to be a selectivity of different molecular species of GnRH for the receptor activation. This is in agreement with the report that GnRH receptors of non-mammalian species respond to any types of GnRH peptides (King and Millar, 1997). This does not, however, mean that GnRH released from the GnRH neurons that belong to the other GnRH systems (other molecular species of GnRH) activates the TN-GnRH neurons under physiological conditions, because immunoreactive fibers of the other GnRH systems are not distributed near the TN-GnRH neurons (Yamamoto et al., 1995).

In order to elucidate possible mechanisms underlying this modulation of pacemaker activity, we next examined whether a G-protein-coupled process was involved in this modulation. There are alternative possible mechanisms to explain the biphasic modulations of the frequency of pacemaker activity of TN-GnRH neurons. First, GnRH receptors may exist on the cell surface of TN-GnRH neurons, and the pacemaker activity of TN-GnRH neurons may be directly modulated by the downstream cell signaling pathways. Second, GnRH receptors may exist on the cell surface of non-GnRH neurons, and the pacemaker activity of TN-GnRH neurons may be indirectly modulated by these neurons. To test these alternative possibilities, we dialyzed the cell with GDP-β-S, a GDP derivative that is a competitive inhibitor of many G-protein-mediated processes, by including it in the patch pipette solution, since it has been already established that the GnRH receptors are members of the G-protein-coupled receptors (Stojilkovic et al., 1994a). After the diffusion of GDP-β-S into the TN-GnRH neuron, bath application of sGnRH failed to evoke any modulation of the firing frequency of pacemaker activity. From these results, we suggested that G-protein-coupled process mediates this biphasic modulation of the pacemaker activity by sGnRH in TN-GnRH neurons. It has been reported that GT1-7 cells express GnRH receptors (Krsmanovic et al., 1993; Stojilkovic et al., 1994a,b). Moreover, GnRH neurons of hypothalamic culture have been shown by double immunostaining to co-express GnRH and GnRH receptors (Krsmanovic et al., 1999). Taken together, it is highly possible that the GnRH receptor exists on the cell surface of TN-GnRH neurons and plays a triggering role in modulating the ion channel(s) underlying the pacemaker activity via G-protein-mediated signaling pathways. To confirm this possibility, the molecular nature of the GnRH receptors expressed on the cell surface of TN-GnRH neurons should be identified by using in

Biphasic modulation of pacemaker potentials by sGnRH

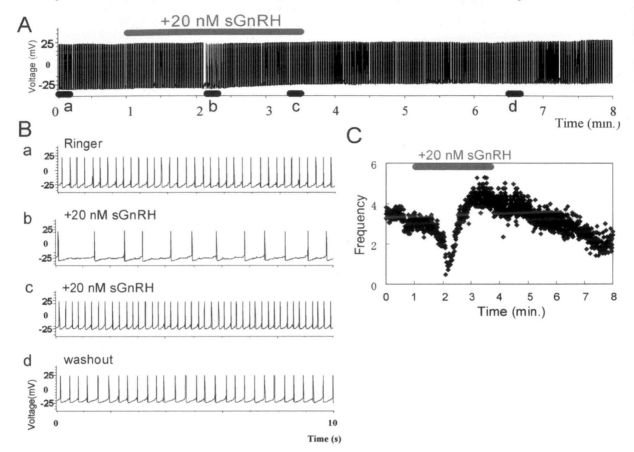

Fig. 4. Modulation of pacemaker frequency by sGnRH. (A) In a current-clamp whole cell recording from a TN-GnRH neuron, bath application of sGnRH, the same molecular species of GnRH peptide produced by TN-GnRH neurons themselves, biphasically modulated their pacemaker activity. (B) Bath application of sGnRH transiently decreased (b) and subsequently increased the frequency of pacemaker activity (c). Following washout, the firing frequency of pacemaker activity recovered (d). (C) Frequency of pacemaker activity plotted against the time course. (Modified from Abe and Oka, 2000.)

situ hybridization and patch-RT-PCR methods in the future study.

Since the basic rhythm of the pacemaker potential was found to be generated by the interplay of $I_{Na(slow)}$ and $I_{K(V)}$ currents (see above), we expected that sGnRH may modulate $I_{Na(slow)}$ or $I_{K(V)}$ or both. However, our preliminary experiments showed that the bath application of sGnRH failed to evoke any noticeable modulation of the $I_{Na(slow)}$ or $I_{K(V)}$. Therefore, we suspected that Ca^{2+} currents that are suggested to be also involved in the generation of pacemaker activities (see above) may be modulated for the facilitation of pacemaker potentials, and so

we examined such possibilities (Abe and Oka, 2002; Oya et al., 2001). We used several specific blockers for each type of voltage-dependent Ca^{2+} currents (Nifedipine for L-type, ω-Conotoxin GVIA for N-type, and ω-Agatoxin TK for P/Q type) as well as those for the voltage-independent SOC (Zn^{2+} and SK&F96365). We found that the N-type Ca^{2+} currents and the SOC are involved in the facilitation of pacemaker potentials by sGnRH (Oya et al., 2001).

Then we isolated Ca^{2+} currents by the whole cell voltage clamp recording of TN-GnRH neurons (Abe and Oka, 2002). We could identify at least two kinds of Ca^{2+} currents; transient low voltage-

Effects of sGnRH on Ca²⁺ currents in TN-GnRH neurons

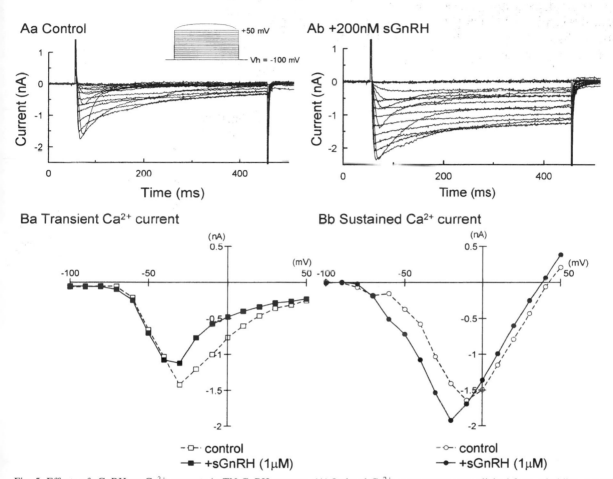

Fig. 5. Effects of sGnRH on Ca²⁺ currents in TN-GnRH neurons. (Λ) Isolated Ca²⁺ current responses elicited from a holding potential of −100 mV before (a) and after the addition of 200 nM sGnRH to the bath solutions (b). (B) I/V curves were constructed by plotting the averaged current amplitudes ($n = 10$) evoked from holding potentials of −100 mV before and after the addition of 200 nM sGnRH to the bath solution. The current–voltage relations of transient current (a) and sustained current (b) are shown. The error bars are not indicated for clarity. (Modified from Abe and Oka, 2002.)

activated (LVA) currents and persistent high voltage-activated (HVA) currents. When we applied sGnRH to the perfusing solution, we found that the persistent HVA current is affected by sGnRH. The activation of HVA current shifted to more hyperpolarized potentials, and the current amplitude was also increased by sGnRH (Fig. 5). From these results, we suggested that a kind of HVA Ca²⁺ current is modulated (facilitated) by sGnRH and contributes to the late-phase increase of pacemaker potentials. Although we have not yet examined the specific Ca²⁺ chan-

nel blockers in the voltage clamp recording, it is highly possible that this HVA Ca²⁺ current is the ω-Conotoxin-sensitive N-type Ca²⁺ current studied in the current clamp recording. We also examined the mechanisms responsible for the early phase decrease of pacemaker potentials by GnRH. Since the ryanodine receptor antagonist Ruthenium Red and IP₃ receptor antagonist heparin introduced into the cell by including them in the pipette solution disrupted the early-phase decrease of pacemaker potentials but left the late-phase increase intact, it was suggested that the early-phase decrease was brought about through the activation of Ca^{2+}-dependent K^+ currents by Ca^{2+} released from the intracellular stores. This was confirmed by current clamp recording in which apamin, a specific blocker of small conductance Ca^{2+}-dependent K^+ currents, inhibited the early phase decrease of pacemaker potentials (Abe and Oka, 2002). Furthermore, in voltage clamp recordings, we found transient potassium current that is dependent on the presence of external Ca^{2+} ions, which is kinetically different from the 4AP-sensitive potassium A-current that we previously found in this neuron (Abe and Oka, 1999). We assumed that this current corresponds to the apamin-sensitive $I_{K(Ca)}$ and checked whether this tentative $I_{K(Ca)}$ could be modulated by sGnRH or not. Repetitive 200 ms test pulses were continuously applied to the cell during the recording to evoke the tentative $I_{K(Ca)}$, and sGnRH was perfused for certain periods. Although the current showed some degree of rundown during the recording, the current amplitude was clearly increased during the application of sGnRH. Thus, we suggest that this tentative $I_{K(Ca)}$ that is facilitated by sGnRH contributes to the early phase transient decrease of frequency of pacemaker potentials.

As we have described so far, the pacemaker potential of TN-GnRH neurons are frequency-modulated by sGnRH biphasically, and the modulation consists of the transient decrease and subsequent increase in the frequency of pacemaker potentials. Such biphasic modulations of the electrical activities have been reported for the changes in membrane potentials of clonal GH3 cell lines induced by TRH (Ozawa and Sand, 1986), and those of gonadotropes and immortalized GnRH cell line (GT1-7 cell) induced by GnRH (Zheng et al., 1997; van Goor et al., 1999). In such cases, it has been suggested that a transient hyperpolarization arises from the activation of Ca^{2+}-activated K^+ currents induced by Ca^{2+} released from the intracellular stores by receptor activation. It is also suggested that Ca^{2+} influx through voltage-independent channels that are activated by depletion of Ca^{2+} stores (SOC) may be activated after this and are involved in the increased pacemaker frequency in the later phase (van Goor et al., 1999). As reported in the adrenal chromaffin cells, this kind of calcium influx may further be involved in the release of GnRH peptides (see Fomina and Nowycky,

1999). Similar mechanisms may exist in TN-GnRH neurons. Future studies should analyze the changes of $[Ca^{2+}]_i$ induced by GnRH application and analyze the properties of target voltage-dependent and/or voltage-independent channel(s) modulated by the GnRH-induced signaling pathways.

What, then, is the physiological significance of such modulations of pacemaker activity by sGnRH upon TN-GnRH neurons? TN-GnRH neurons of the dwarf gourami make tight cell clusters without intervening glial cells (Oka and Ichikawa, 1990; Oka and Matsushima, 1993; Oka, 1997), and the possibility of active exocytotic release from the cell body and its vicinity (somatodendritic release of GnRH) has been suggested (Oka and Ichikawa, 1991, 1992; also see below). Somatodendritic release of neurohypophysial peptides by exocytosis from the hypothalamic magnocellular neurons has also been reported (Pow and Morris, 1989). On the other hand, there are studies showing that GnRH receptors are widely distributed throughout the brain (Stojilkovic et al., 1994b; Jennes et al., 1997). In addition, considerable overlap of the brain areas that contain GnRH-producing cells and those that exhibit expression of GnRH receptor mRNA, has been reported (Jennes et al., 1996). From these observations and ours, it is suggested that GnRH released from GnRH neurons facilitates the activities of their own (autocrine) and/or neighboring GnRH neurons (paracrine) and may cause synchronized positive feedback facilitation of multiple GnRH neurons. It has been well established that the release of GnRH from the hypothalamus is pulsatile, and this pulsatile release of GnRH is essential for the control of hypothalamo–pituitary–gonadal axis and hence the regulation of normal reproductive function in mammals (Terasawa, 1998; Terasawa et al., 1999). Although the TN-GnRH and POA-GnRH systems may be anatomically and functionally different, it is tempting to assume that the autocrine or paracrine mechanisms may also underlie the pulsatile release of GnRH. It is also known that in oxytocin neurons the release of oxytocin from single neuron into the brain environment stimulates its own activity and thus further release (Freund-Mercier and Richard, 1984; Moos et al., 1984). Therefore, this mechanism is probably common to all neurosecretory neurons or secretory cells, whose synchronized facilitation of firing leads to facilitated release.

Neuromodulatory action of GnRH peptides

Modulation of ion channels by GnRH

We originally proposed the idea that the TN-GnRH system functions as a neuromodulator system in the central nervous system (Oka, 1992; Oka and Matsushima, 1993), from a series of studies using the dwarf gourami TN-GnRH system. However, experimental evidence for a neuromodulatory action of GnRH in the sympathetic ganglia (causing late slow EPSP) dates back to the early 70s (see Jan and Jan, 1983 for review). Jan and Jan reported that GnRH is co-released with Ach from the sympathetic ganglion cells, diffuse for some distance, and evokes late slow EPSP in the postganglionic cells (Jan and Jan, 1983). Later, it was found out that this extremely slow and long-lasting potential change results from the closing of non-inactivating K^+ M current that is active at the resting membrane potential range (Brown, 1988). Although the exact cellular mechanisms of this M current inhibition has yet to be completely understood (Marrion, 1997), its physiological significance is suggested to be the increase in the excitability of the postganglionic neuron membranes. It has been shown that, owing to the membrane potential-clamping effect of the M current, the postganglionic neurons discharge on-to-one action potentials in response to the preganglionic fiber stimulation. During the late slow EPSP, when the M currents are inhibited, the postganglionic neurons now respond with persistent action potential responses.

It has been reported recently that GnRH peptide is also involved in the neuromodulation of Ca^{2+} currents. The Ca^{2+} currents have been classified into L, T, N, P/Q, R types based on the kinetic properties of channel opening-closing and on the specificity of blockage by natural neurotoxins. Among others, the N and P/Q types of Ca^{2+} currents have been reported to be modulated by GnRH (Elmslie et al., 1990), and the signal transduction mechanisms of the modulation have been well studied (Marrion, 1997). Since the N and P/Q type Ca^{2+} channels are localized in presynaptic active zones and are involved in the Ca^{2+} influx necessary to the transmitter release, it may be suggested that GnRH peptide modulates the transmitter release via modulation of these Ca^{2+} currents. However, this has not yet been demonstrated experimentally and will be an exciting future problem to be examined.

On the other hand, much less information on the possible neuromodulator functions of GnRH is available in the central nervous system. In the rat hippocampus, GnRH induces a long-lasting depolarization associated with increased input resistance (decreased membrane conductance), a decrease in the after-hyperpolarization following a train of action potentials, and a reduction in accommodation of repetitive cell discharge (Wong et al., 1990). However, ion channel modulation and its signal transduction mechanisms in the central nervous system have not been well studied probably because of its technical difficulties compared with the peripheral ganglion neurons. As already mentioned above, we have recently found out that the GnRH peptide produced by the TN-GnRH neurons themselves acts on the TN-GnRH neurons as a neuromodulator to change their ionic channel properties (Abe and Oka, 2000). Since the TN-GnRH neurons project their neurites in wide areas of the brain (Oka and Matsushima, 1993), and the GnRH receptors are also widely distributed in the brain (Stojilkovic et al., 1994b; Jennes et al., 1997), it is suggested that the GnRH peptides released from the neurites in wide areas of the brain may act as neuromodulators via similar mechanisms.

Modulation of neuronal functions by GnRH

We have thus far discussed the neuromodulatory function of GnRH peptides on the ionic channel properties, and now let us discuss the modulation by GnRH peptides of neural functions as a whole. In the teleost retina, which receives dense projection of TN-GnRH fibers arising from TN cells, TN-GnRH fibers are known to synapse on dopaminergic interplexiform cells (Zucker and Dowling, 1987), and GnRH and other neuropeptides affect the retinal ganglion cell activity (Stell et al., 1984; Walker and Stell, 1986). Umino and Dowling (1991) reported that when the retina was superfused with Ringer's solution containing GnRH, horizontal cells depolarized, and their response to small spots increased, whereas their responses to full-field lights decreased. Their results suggested that GnRH acts by stimulating the release of dopamine from interplexiform cells. Furthermore, Behrens et al. (1993) reported that GnRH elicits light adaptive formation of horizontal cell spinules in vitro by stimulating the dopaminergic interplexiform cells. Thus, the TN-GnRH fibers projecting to the retina seem to have a definitive physiological function. It will be an interesting future project to study neuromodulatory action of GnRH on GnRH target cells in the brain (such as olfactory bulb, ventral telencephalon, etc.) and its relation to various spontaneous discharge modes by taking advantage of the whole brain in vitro preparation of the dwarf gourami.

Recently, Eisthen et al. (Eisthen et al., 2000) reported on very interesting results showing that GnRH peptide modulates the sensitivity of olfactory receptor neurons by modulating their Na^+ and K^+ channel properties. They used the olfactory receptor neurons from mudpuppies (*Necturus maculosus*) and did voltage-clamped whole-cell recordings to examine the effects of GnRH on voltage-activated currents in olfactory receptor neurons from epithelial slices. They found that GnRH slowly but reversibly increases the magnitude, but does not alter the kinetics, of a TTX-sensitive inward current (most probably conventional Na^+ current involved in the generation of action potentials) and a certain outward currents. This effect appeared to be seasonal, with more animals responding to GnRH during the courtship and mating season. Taken together, they suggest that GnRH increases the excitability of olfactory receptor neurons and that the terminal nerve functions to modulate the odorant sensitivity of olfactory receptor neurons. Although this hypothesis needs to be further tested by examining the olfactory responses to the odorants and the currents need to be isolated for more rigorous electrophysiological analysis, it will surely provide very attracting further working hypotheses to be tested in the future studies.

Behavioral functions of TN-GnRH system

Unfortunately, we still do not know much about the behavioral consequence of neuromodulation by TN-GnRH system. Wirsig and Leonard (1987) were the first to report effects of terminal nerve lesions on the sexual behaviors of vertebrates. They reported in the male hamster that surgical ablation of the terminal nerve affects male sexual behavior. However, since they did not use the GnRH immunohistochemistry, the extent of the lesion of the TN-GnRH system as a whole was not clear. We (Yamamoto et al., 1997) took advantage of the morphological features of the dwarf gourami TN-GnRH neurons (the cell bodies are large and make tight cell clusters near the ventral surface of the brain) and examined the behavioral effects of specific and complete lesion of the TN-GnRH system. The sexual behavior of the dwarf gourami consists of several readily quantifiable behavioral repertoire or patterns. We quantified the male sexual behavior patterns during one-hour mating trials. We then placed electrolytic lesions in the bilateral clusters of TN-GnRH neurons of the male fish and allowed them to survive for two weeks. All of the GnRH fibers originating from the TN-GnRH neurons had disappeared when the brains were examined by immunohistochemistry after behavioral tests. When some parameters of sexual behavior patterns before and one/two weeks after the operations were compared, we observed a delicate but characteristic behavioral impairment. The occurrence of one-hour mating trials during which the male failed to perform nest building at all was significantly higher in totally or partially TN-lesioned fish compared to intact, sham-operated, or olfactory nerve-cut controls. This means that they became less motivated to start the whole sequence of behavior. On the other hand, these lesions did not affect the overall incidence of the sexual behavior patterns, once the behavioral sequence was triggered. These results seem to suggest that TN GnRH system is involved in the control of the threshold for the initiation of the nest building behavior, although it is not indispensable for the general performance of reproductive behavior, once the behavioral sequence is triggered. Thus, the TN-GnRH system has a delicate control on the motivational or arousal state of the animal in general. We have recently found out in our preliminary behavioral study that the frequency of the male nest-building behavior remains at an elevated state for three or four hours after the end of an hour of pairing with a sexually mature female partner; the males become highly motivated for the nest-building behavior, and the frequency of this behavior can be used as a highly reliable and reproducible index of motivated state of the fish. By using this behavioral parameter, we are now examining if the activation

273

of TN-GnRH system is involved in this motivated state.

Fine structural evidence for exocytotic release activity of TN-GnRH neurons

As already described above, the axons of the TN-GnRH neurons are distributed throughout the brain from the olfactory bulb to the spinal cord. Therefore, GnRH peptides should be able to modulate the excitabilities and/or the transmitter release of the target neurons simultaneously via the widely projecting axonal branches, if GnRH peptides are released from these axons. Macromolecules such as peptides are released from the neurons via exocytosis, and we now turn our topic to the exocytotic release of GnRH from the TN-GnRH neurons. The ultrastructure of the TN-GnRH cells has been described by Matsutani and Uchiyama (1986) and Oka and Ichikawa (1991, 1992), and the TN-GnRH neuronal cell bodies have been shown to exhibit characteristics similar to those of other peptide-synthesizing cells, that is, highly indented nucleus, stacks of well-developed rough endoplasmic reticulum and Golgi apparatus, numerous membrane-bound dense-cored vesicles (DCVs), and large electron dense droplets (Matsutani and Uchiyama, 1986; Demski and Fields, 1988; Oka and Ichikawa, 1991). GnRH immunoreactivity has been demonstrated in these dense-cored vesicles in the cell bodies by means of immunoelectron microscopy (Oka and Ichikawa, 1992). Thus, it is suggested that the cell bodies contain releasable GnRH peptides. Interestingly, coated pits or vesicles, which are associated with endocytotic membrane retrieval after an active exocytosis, were frequently distributed beneath the plasma membrane of the cell body and small somatic processes. The fine structural evidence thus suggested a somatodendritic release of GnRH. Furthermore, the cell bodies were closely apposed with each other, and they were closely apposed without intervening glial cells, and there was no structural diffusional barrier. The morphological evidence thus suggested the presence of some kind of paracrine or autocrine activity of GnRH neurons by GnRH peptides. The exocytotic profiles of the DCVs are usually elusive under a conventional EM because of their short time-course. It is to be expected that GnRH peptides released by exocytosis diffuse into the intercellular space and act on the GnRH neuron itself or the neighboring GnRH neurons.

In the GnRH fiber varicosities, GnRH immunoreactivities have also been demonstrated in DCVs (Oka and Ichikawa, 1992), but none of these structures showed any evidence of the presence of synaptic active zones, that is, postsynaptic densities, widened synaptic clefts, or accumulation of synaptic vesicles. Thus, it may be hypothesized that TN-GnRH cells secrete GnRH non-synaptically from DCV-containing fiber varicosities as well as soma and that it exerts its modulatory action on GnRH receptors located on nearby as well as distant target neurons. Recently, evidence favoring the idea of non-synaptic release of neuropeptides is accumulating (Thureson-Klein and Klein, 1990). Similar mechanisms have also been suggested for GnRH release in the midbrain central gray of the rat (Buma, 1989). Such mode of peptidergic action may be relevant for long-lasting and wide-spread neuromodulation.

RIA measurement of GnRH release from the brain slices

In order to know the release activity of GnRH peptides from the GnRH systems, we measured the release of GnRH peptides into the medium from the brain–pituitary slices by the radio-immunoassay (RIA) (Ishizaki et al., 2002). To measure the GnRH release activities from different GnRH systems, and to examine whether there are differences between them, we conducted a static incubation of brain–pituitary slices under various conditions, and GnRH released into the incubation medium was measured by RIA. The slices were divided into two parts, the one containing GnRH neurons in the preoptic area and axon terminals in the pituitary (POA-GnRH slices), and the other containing the cell bodies and fibers of TN-GnRH neurons and midbrain tegmentum–GnRH neurons (TN-GnRH slices). We demonstrated that the GnRH release was evoked by high $[K^+]_o$ depolarizing stimuli (in both slices) via Ca^{2+} influx through voltage-gated Ca^{2+} channels. From the results of experiments using specific Ca^{2+} channel blockers, it was suggested that the GnRH release from POA-GnRH slices induced by depolarization is mainly dependent on the Ca^{2+} influx through ω-conotoxin-sensitive N-type Ca^{2+} channels, and that from TN-GnRH slices induced by depolarization is mainly dependent on both nifedipine-sensitive L- and ω-conotoxin-sensitive N-type Ca^{2+} channels. In either slice, ω-agatoxin-sensitive P/Q type Ca^{2+} channels were not involved. The most prominent difference between the GnRH release from the POA-GnRH and TN-GnRH slices, however, was the presence of sexual difference in the GnRH release only in the POA-GnRH slices. In TN-GnRH slices, our evidence suggested that the store-operated Ca^{2+} influx (SOC) may be involved in the basal (spontaneous and unstimulated) GnRH release. This is interesting, since our recent patch current-clamp studies in TN-GNRH neurons suggest the involvement of SOC in the generation as well as modulation of pacemaker activities of TN-GnRH neurons (Oya et al., 2001; see above). Finally, we found that glutamate application significantly increased the GnRH release in the TN-GnRH but not in the POA-GnRH slices in a dose dependent manner. In accordance with these results, glutamate applications increased the frequency of pacemaker potentials of TN-GnRH cells in a dose dependent manner. Furthermore, in our recent preliminary study, we are accumulating evidence for the existence of ionotropic as well as metabotropic glutamate receptors and their depolarizing effects in TN-GnRH neurons.

New approaches to study GnRH release activities of single GnRH neurons; electrochemistry using carbon fiber electrodes (CFEs)

Although GnRH content in brain and GnRH release from brain slices have been measured with radioimmunoassay (Okuzawa et al., 1990; Yu et al., 1991), it has been difficult to measure GnRH release in real-time. On the other hand, electrochemical techniques using carbon fiber electrodes (CFE) have been recently developed to detect the exocytotic release of neurotransmitters such as catecholamines (Leszczyszyn et al., 1990) and serotonin (Alvarez de Toledo et al., 1993). These techniques can detect electroactive transmitters or hormones released from neurons or secretory cells in real-time. It is also reported that some amino acids such as Trp and Tyr are electroactive (readily oxidizable), and the small peptides that contain these amino acids can be electroactive (Bennett et al., 1981). Because sGnRH

contains three electroactive amino acid residues (two Trp and one Tyr residues), we expected that sGnRH would be electroactive and could be detected by CFE from the GnRH axon terminals in the pituitary (Ishizaki and Oka, 2001).

We devised a CFE by inserting a single carbon fiber of 7 μm diameter into a glass capillary, and the glass capillary was pulled to form a microelectrode by a vertical pipette puller. The carbon fiber was insulated, and the electrode was filled with 2 M KCl and was connected to the patch-clamp amplifier. Having decided the voltage dependence of redox current of synthetic sGnRH solution by a cyclic voltammetry (the oxidation current appeared at the potential higher than 600–700 mV and was maximum at about 900–1000 mV), the holding potential (Vh) of the CFE was set at 900 mV for amperometric recording. sGnRH solution was pressure-ejected to the electrode tip using a puffer pipette, and the oxidation current of sGnRH was recorded. Amperometric currents were recorded in response to the puffer application of sGnRH solution. The amperometric currents could be measured at Vh higher than 600 mV, and they increased up to the voltage limit of the recording system, 1000 mV. The oxidation current peak of sGnRH was in good agreement with those of Trp and Tyr, which are electroactive residues in sGnRH (Bennett et al., 1981; Paras and Kennedy, 1995). Since the oxidation peaks of other electroactive substances (e.g. serotonin at 620 mV and dopamine at 600 mV; Kruk and O'Connor, 1995) are lower than that of sGnRH, we were fairly sure that the current we were measuring was that of sGnRH and not contaminating currents, if the amperometric current recorded at 900 mV disappear at the Vh lower than 600–700 mV.

The dose-response was examined by recording amperometric currents to various concentrations of sGnRH solutions. The Vh was held at 900 mV, and sGnRH was applied by puffer pipette. As a control, Ringer solution was pressure-ejected to the tip of the CFE before each experimental recording, and the control current values were subtracted from each experimental data. The detection limit was between 10^{-5} M and 10^{-6} M in four CFEs tested. Although this concentration may be higher than the physiological serum concentration of GnRH, the local GnRH concentration close to the nerve terminals or release sites should exceed these concentrations. Therefore, the CFE method described here should be able to detect, in real-time, the release of GnRH in a region very close to the GnRH release sites such as the GnRH axon terminals in the teleost pituitary.

We then used the brain–pituitary slices similar to the one used for the RIA for recording sGnRH release activity by amperometry. It has been reported that the dorsal region of the pituitary contains many GnRH-immunoreactive fibers and terminals originating from POA-GnRH neurons in the dwarf gourami

(Maejima et al., 1994; Yamamoto et al., 1998b). Because the POA-pituitary slices of the dwarf gourami release high amount of GnRH in response to depolarizing stimuli in radioimmunoassays (Ishizaki et al., 2002; see above), it is to be expected that the local concentration of GnRH released from the pituitary will be high enough to be recorded by amperometry. The tip of the CFE was positioned so as to lightly touch the surface of the dorsal region of the pituitary (illustrated in Fig. 6A).

Fig. 6 shows an amperometric current recording

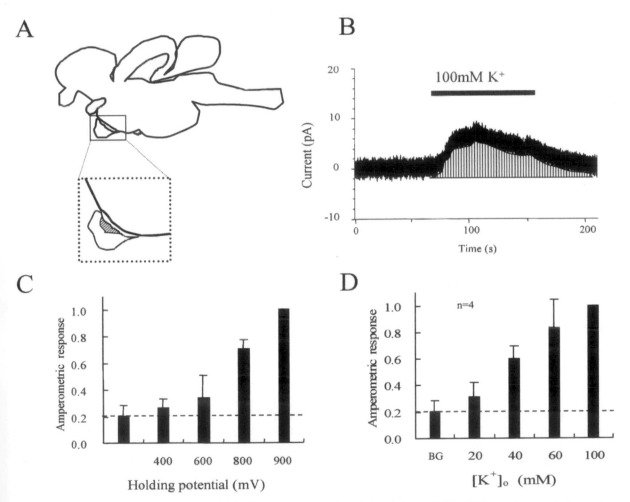

Fig. 6. Amperometric recording of release activities from the axon terminals in the pituitary. (A) The CFE was placed in the hatched area in the pituitary slice where the axon terminals from GnRH neurons are dense. (B) Amperometric recording (Vh = 900 mV) from the pituitary slice stimulated by high K+ solution (100 mM). (C) Relationship between the Vh of the electrode and the amperometric current that was recorded from the pituitary slices stimulated by 100 mM K+. (D) $[K^+]_o$-dependence of release activity of the pituitary slice (Vh = 900 mV). The data are normalized against the value for the maximal concentration of $[K^+]_o$ (100 mM). (Modified from Ishizaki and Oka, 2001.)

from the brain–pituitary slice preparation. Vh of CFE was held at 900 mV. A bulk current response was observed after bath application of a Ringer solution containing high K^+ concentration (100 mM) to the slice for 90 seconds. The time integral of this amperometric current, the charge transferred by oxidation, was calculated (Fig. 6B), and the release activity as measured by these values was dependent on the Vh of CFE (Fig. 6C). The amperometric response was also dependent on $[K^+]_o$, i.e., the degree of membrane depolarization (Fig. 6D). Because the amperometric response at the Vh lower than 600 mV was comparable to the background level, it was concluded that the contamination of the oxidation currents of catecholamines or serotonin was almost negligible. Since the voltage dependence of the amperometric current was comparable to that of the amperometric current in response to synthetic sGnRH, it is strongly suggested that the amperometric current in Fig. 6 is mainly attributed to the oxidation current of sGnRH peptide released from the axon terminals of POA-GnRH neurons in the pituitary. Although there are other axon terminals that secrete small peptide hormones such as isotocin and vasotocin in the teleost pituitary, it has been shown that the dorsal region of the pituitary where CFE was placed has the heaviest projection of GnRH-immunoreactive fibers but not isotocin-immunoreactive fibers in the dwarf gourami (Maejima et al., 1994). Furthermore, isotocin and vasotocin are considered to be very weakly electroactive, since they only contain one Tyr residue.

Thus, we succeeded in developing a new method that enables a real-time measurement of GnRH peptide release activity from the axon terminals in the pituitary in real-time using brain-pituitary slice preparations (Ishizaki and Oka, 2001). Although the currents recorded from the pituitary were not amperometric spikes but bulk currents, which may be attributed to the release from a large number of axon terminals, this technique will prove to be a powerful new tool for the study of GnRH release and may be applied to the real-time recording of release activity in other parts of the brain or in cell cultures that release GnRH. If the present method can be applied to the study of release activity of GnRH neurons in the brain, it may help us understand the functional significance of the multifunctional GnRH systems.

Working hypotheses

Fig. 7 illustrates our working hypothesis concerning the generation and modulation of pacemaker potentials of TN-GnRH cells. The TN-GnRH cells show regular pacemaker activities whose basic subthreshold rhythm is dependent on the interplay between the TTX-resistant persistent sodium current, $I_{Na(slow)}$, and a TEA-sensitive voltage-dependent potassium current, $I_{K(V)}$, persistent depolarizing drive and counteracting hyperpolarization, respectively. The store-operated Ca^{2+} current (SOC) may also partly contribute to the persistent inward current drive for the generation of pacemaker potentials. In addition, ω-conotoxin-sensitive N-type Ca^{2+} current(s) also seem to be involved in the suprathreshold (action potential) phase of the pacemaker potentials. The TN-GnRH neurons release GnRH not only from the varicosities and axon terminals but also from the somatodendritic areas. GnRH peptides released from the somatodendritic areas of the GnRH neurons facilitate the activities of their own (autocrine) and/or neighboring GnRH neurons (paracrine) and may cause synchronized positive feedback facilitation of multiple GnRH neurons. During this kind of modulation, the released GnRH peptide binds to the G-protein-coupled GnRH receptors in the cell membrane of TN-GnRH neurons and may function in the following manner.

(1) GnRH receptor activation facilitates Ca^{2+} release from intracellular Ca^{2+} store. The increased intracellular Ca^{2+} activates apamin-sensitive Ca^{2+}-dependent K^+ current(s) and decreases the frequency of pacemaker potentials transiently.

(2) The downstream signaling pathway somehow increases ω-conotoxin-sensitive N-type Ca^{2+} current(s) and hence the frequency of pacemaker potentials. Alternatively, it is also possible that Ca^{2+} influx through voltage-independent channels that are activated by depletion of Ca^{2+} stores (SOC) may be activated after (1) and are involved in the late phase increase of pacemaker frequencies (see van Goor et al., 1999). This kind of calcium influx may further be involved in the release of GnRH peptides (Fomina and Nowycky, 1999).

Similarly, changes in the physiological conditions of the fish that are triggered by environmental, pheromonal, hormonal factors, etc., probably act on

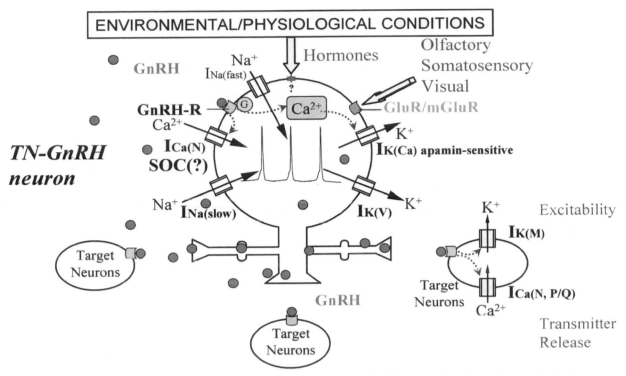

Fig. 7. Diagram illustrating the neuromodulator functions of TN-GnRH neurons, including some hypotheses. See text for details.

the GnRH neurons via hormones and/or neurotransmitters, and the properties of the ionic channels that underlie pacemaker activities may be modified by some kind of signal transduction mechanisms similar to the one described above. In this respect, it is interesting to note that some of the candidate GnRH neurons in ascidian neural ganglion (Tsutsui et al., 1998) may respond directly or indirectly to light stimuli (Tsutsui and Oka, 2000). In the dwarf gourami, Yamamoto and Ito (2000) reported on anatomical evidence for the somatosensory and visual inputs to the TN-GnRH neurons from the nucleus tegmentoterminalis in the midbrain tegmentum, and olfactory inputs from the olfactory bulb and the primary olfactory projection areas in the telencephalon. On the other hand, the TN-GnRH neurons are suggested to have various kinds of hormone receptors besides GnRH receptors; it has been reported in Tilapia (*Oreochromis niloticus*) that TN-GnRH neurons express receptors for thyroid hormone, testosterone, cortisol, etc., and these hormones are involved in the regulation of GnRH genes in TN-GnRH neurons (Soga et al., 1998; Parhar et al., 2000).

There are plenty of reports to show that GnRH peptide modulates several kinds of ion channels, e.g., potassium M current or N and P type I_{Ca} (see above). Since these currents are involved in the control of neuronal excitability and transmitter release, respectively, TN-GnRH cells may modulate the excitability of target neurons or presynaptic release of transmitters in wide brain regions simultaneously via extensive, multiple axonal branches. We have actually measured the exocytotic release of GnRH peptides from the brain slices containing the cell bodies and axons of TN-GnRH neurons by RIA and found that N-type Ca^{2+} currents and SOC, which we find are involved in the generation and modulation of pacemaker activities of GnRH neurons, are also involved in the control of GnRH release from the TN-GnRH slices. We hope that we can apply the real-time electrochemical measurement of GnRH release using a carbon fiber electrode (CFE) to measure the changes in GnRH release according to the changes in pacemaker frequencies of TN-GnRH neurons in near future. Finally, these cellular events may lead to a delicate control of the motivational or arousal state in various aspects of the animal behavior.

As we have seen in this paper, there are multiple GnRH neuronal systems with different morphology, function, and molecular component, i.e., TN- and midbrain GnRH systems, which are considered to be neuromodulatory, and the hypophysiotropic POA GnRH system, which is the originally defined 'gonadotropin-releasing' system. In spite of these differences, the GnRH neuronal systems may have some basic characteristics in common as peptidergic neurons. Through a proper and deliberate synthesis of results obtained by using these diverse GnRH neuronal systems and multidisciplinary techniques, general principles concerning not only the GnRH neuronal system but also the peptidergic neuronal system as a whole will sure to emerge. I believe that the teleost brain will greatly contribute to this exciting field of research.

Abbreviations

4AP	4-aminopyridine
ACh	acetyl choline
CFE	carbon fiber electrode
DCV	dense-cored vesicle
EM	electron microscopy
EPSP	excitatory postsynaptic potential
GFP	green fluorescent protein
GnRH	gonadotropin-releasing hormone
HRP	horse-radish peroxidase
nFLM	nucleus of the fasciculus medialis longitudinalis
POA	preoptic area
SOC	store-operated Ca^{2+} current
TN	terminal nerve
TTX	tetrodotoxin
Vh	holding potential

Acknowledgements

I would like to thank Dr. Yamamoto who contributed to the immunohistochemical and behavioral studies, Dr. Abe who contributed to the electrophysiological studies, and Ms. Ishizaki who contributed to the real-time CFE measurement of GnRH release. We would like to extend our gratitude to all of our colleagues and friends for help and discussion. This research was supported by grant-in-aid for Fundamental Scientific Research from Ministry of Education, Culture, Sports, Science and Technology (MEXT) of Japan to Y. Oka (#10554050 and #12440237).

References

Abe, H. and Oka, Y. (1999) Characterization of K^+ currents underlying pacemaker potentials of fish gonadotropin-releasing hormone cells. *J. Neurophysiol.*, 81: 643–653.

Abe, H. and Oka, Y. (2000) Modulation of pacemaker activity by salmon gonadotropin-releasing hormone (sGnRH) in terminal nerve (TN)-GnRH neurons. *J. Neurophysiol.*, 83: 3196–3200.

Abe, H. and Oka, Y. (2002) Mechanisms of the modulation of pacemaker activity by GnRH peptides in the terminal nerve-GnRH neurons. *Zool. Sci.*, 19: 111–128.

Alvarez de Toledo, G., Fernandez-Chacon, R. and Fernandez, J.M. (1993) Release of secretory products during transient vesicle fusion. *Nature*, 363: 554–558.

Amano, M., Oka, Y., Aida, K., Okumoto, N., Kawashima, S. and Hasegawa, Y. (1991) Immunocytochemical demonstration of salmon GnRH and chicken GnRH-II in the brain of masu salmon, *Oncorhynchus masou. J. Comp. Neurol.*, 314: 587–597.

Behrens, U.D., Douglas, R.H. and Wagner, H.J. (1993) Gonadotropin-releasing hormone, a neuropeptide of efferent projections to the teleost retina induces light-adaptive spinule formation on horizontal cell dendrites in dark-adapted preparations kept in vitro. *Neurosci. Lett.*, 164: 59–62.

Bennett, G.W., Brazell, M.P. and Marsden, C.A. (1981) Electrochemistry of neuropeptides: A possible method for assay and in vivo detection. *Life Sci.*, 29: 1001–1007.

Brown, D.A. (1988) M-currents: an update. *Trends Neurosci.*, 11: 294–299.

Buma, P. (1989) Characterization of luteinizing hormone-releasing hormone fibres in the mesencephalic central grey substance of the rat. *Neuroendocrinol.*, 49: 623–630.

Demski, L.S. and Fields, R.D. (1988) Dense-cored vesicle-containing components of the terminal nerve of sharks and rays. *J. Comp. Neurol.*, 278: 604–614.

Demski, L. and Northcutt, R.G. (1983) The terminal nerve: A new chemosensory system in vertebrates? *Science*, 220: 435–437.

Demski, L.S. and Schwanzel-Fukuda, M. (1987) The terminal nerve (nervus terminalis): structure, function, and evolution. *Ann. N.Y. Acad. Sci.*, 519: 1–468.

Di Fiore, M.M., Rastogi, R.K., Cecilliani, F., Messi, E., Botte, V., Botte, L., Pinelli, C., D'Aniello, B. and D'Aniello, A. (2000) Mammalian and chicken I forms of gonadotropin-relrasing hormone in the gonads of a protochordate, *Ciona intestinalis. Proc. Natl. Acad. Sci. USA*, 97: 2343–2348.

Eisthen, H.L., Delay, R.J., Wirsig-Wiechmann, C.R. and Dionne, V.E. (2000) Neuromodulatory effects of gonadotropin releasing hormone on olfactory receptor neurons. *J. Neurosci.*, 20: 3947–3955.

Elmslie, K.S., Zhou, W. and Jones, S.W. (1990) LHRH and GTP-γ-S modify calcium current activation in bullfrog sympathetic neurons. *Neuron*, 5: 75–80.

Fomina, A.F. and Nowycky, M.C. (1999) A current activated on depletion of intracellular Ca^{2+} stores can regulate exocytosis in adrenal chromaffin cells. *J. Neurosci.*, 19: 3711–3722.

Freund-Mercier, M.J. and Richard, P. (1984) Electrophysiological evidence for facilitatory control of oxytocin neurons by oxytocin during suckling in the rat. *J. Physiol.* 352: 447–466.

Fujita, I., Sorensen, P.W., Stacey, N.E. and Hara, T.J. (1991) The olfactory system, not the terminal nerve, functions as the primary chemosensory pathway mediating responses to sex pheromones in male goldfish. *Brain Behav. Evol.*, 38: 313–321.

Georges, D. and Dubois, M.P. (1980) Mise en évidence par des techniques d'immunofluorescence d'un antigène de type LII-RH dans le système nerveux de *Ciona* intestinalis (tunicier ascidiacé). *C.R. Acad. Sc. Paris*, 290: 29–31.

Ishizaki, M. and Oka, Y. (2001) Amperometric recording of gonadotropin-releasing hormone release activity in the pituitary of the dwarf gourami (teleost) brain–pituitary slices. *Neurosci. Lett.*, 299: 121–124.

Ishizaki, M., Iigo, M., Amano, M., Yamamoto, N. and Oka, Y. (2002) Different modes of gonadotropin-releasing hormone (GnRH) release from multiple GnRH systems as revealed by radioimmunoassay using brain slices of a teleost, the dwarf gourami (*Colisa lalia*). Submitted.

Jan, L.Y. and Jan, Y.N. (1983) A LHRH-like peptidergic neurotransmitter capable of 'action at a distance' in autonomic ganglia. *Trends Neurosci.*, 6: 320–325.

Jennes, L., McShane, T., Brame, B. and Centers, A. (1996) Dynamic changes in gonadotropin releasing hormone receptor mRNA content in the mediobasal hypothalamus during the rat estrous cycle. *J. Neuroendocrinol.*, 8: 275–281.

Jennes, L., Centers, A. and Eyigor, O. (1997) GnRH receptors in the rat central nervous system. In: I.S. Parhar and Y. Sakuma (Eds.), *GnRH Neurons: Genes to Behavior*. Brain Shuppan Publishers, Tokyo, pp. 79–95.

Kelly, M.J., Ronnekleiv, O.K. and Eskay, R.L. (1984) Identification of estrogen-responsive LHRH neurons in the guinea pig hypothalamus. *Brain Res. Bull.*, 12: 399–407.

Kim, M.-H., Amano, M., Suetake, H., Kobayashi, M. and Aida, K. (1997) GnRH neurons and gonadal maturation in masu salmon and goldfish. In: I.S. Parhar and Y. Sakuma (Eds.), *GnRH Neurons: Genes to Behavior*. Brain Shuppan Publishers, Tokyo, pp. 313–324.

Kim, M., Kobayashi, M., Oka, Y., Amano, M., Kawashima, S. and Aida, K. (2001) Effects of olfactory tract section on the immunohistochemical distribution of brain GnRH in the female goldfish, *Carassius auratus. Zool. Sci.*, 18: 241–248.

King, J.A. and Millar, R.P. (1997) Coordinated evolution of GnRHs and their receptors. In: I.S. Parhar and Y. Sakuma (Eds.), *GnRH Neurons: Genes to Behavior*. Brain Shuppan Publishers, Tokyo, pp. 51–77.

Kobayashi, M., Amano, M., Kim, M., Furukawa, K., Hasegawa, Y. and Aida, K. (1994) Gonadotropin-releasing hormones of terminal nerve origin are not essential to ovarian development and ovulation in goldfish. *Gen. Comp. Endocrinol.*, 95: 192–200.

Koyama, Y., Satou, M., Oka, Y. and Ueda, K. (1984) Involvement of the telencephalic hemispheres and the preoptic area in sexual behavior of the male goldfish, *Carassius auratus*: A brain-lesion study. *Behav. Neural Biol.*, 40: 70–86.

Krsmanovic, L.Z., Stojilkovic, S.S., Mertz, L.M., Tomic, M. and Catt, K.J. (1993) Expression of gonadotropin-releasing hormone receptors and autocrine regulation of neuropeptide release in immortalized hypothalamic neurons. *Proc. Natl. Acad. Sci. USA*, 90: 3908–3912.

Krsmanovic, L.Z., Martinez-Fuentes, A.J., Arora, K.K., Mores, N., Navaro, C.E., Chen, H-.C., Stojilkovic, S.S. and Catt, K.J. (1999) Autocrine regulation of gonadotropin-releasing hormone secretion in cultured hypothalamic neurons. *Endocrinology*, 140(3): 1423–1431.

Kruk, Z.L. and O'Connor, J.J. (1995) Fast electrochemical studies in isolated tissues. *Tr. Pharmacol. Sci.*, 16: 145–149.

Leszczyszyn, D.J., Jankowski, J.A., Viveros, O.H., Diliberto Jr., E.J., Near, J.A. and Wightman, R.M. (1990) Nicotinic receptor-mediated catecholamine secretion from individual chromaffin cells. *J. Biol. Chem.*, 265: 14736–14737.

Mackie, G.O. (1995) On the visceral nervous system of *Ciona. J. Mar. Biol. Assoc. UK*, 75:141–151.

Maejima, K., Oka, Y., Park, M.K. and Kawashima, S. (1994) Immunohistochemical double-labeling study of gonadotropin-releasing hormone (GnRH)-immunoreactive cells and oxytocin-immunoreactive cells in the preoptic area of the dwarf gourami, *Colisa lalia. Neurosci. Res.*, 20: 189–193.

Marrion, N.V. (1997) Control of M-current. *Ann. Rev. Physiol.*, 59: 483–504.

Matsutani, S. and Uchiyama, H.I. (1986) Cytoarchitecture, synaptic organization and fiber connections of the nucleus olfactoretinalis in a teleost (*Navodon modestus*). *Brain Res.*, 373: 126–138.

Montaner, A.D., Park, M.K., Fischer, W.H., Craig, A.G., Chang, J.P., Somoza, G.M., Rivier, J.E. and Sherwood, N.M. (2001) Primary structure of a novel gonadotropin-releasing hormone in the brain of a teleost, pejerrey. *Endocrinology*, 142: 1453–1460.

Moos, F., Freund-Mercier, M.J., Guerne, J.M., Stoeckel, M.E. and Richard, P. (1984) Release of oxytocin and vasopressin by magnocellular nuclei in vitro: Specific facilitatory effect of oxytocin on its own release. *J. Endocrinol.*, 102: 63–72.

Muske, L.E. (1997) Ontogeny, phylogeny and neuroanatomical organization of multiple molecular forms of GnRH. In: I.S. Parhar and Y. Sakuma (Eds.), *GnRH Neurons: Genes to Behavior*. Brain Shuppan Publishers, Tokyo, pp. 145–180.

Münz, H. and Stumpf, W.E.J. (1981) LHRH systems in the brain of platyfish. *Brain Res.*, 221: 1–13.

Oka, Y. (1992) Gonadotropin-releasing hormone (GnRH) cells of the terminal nerve as a model neuromodulator system. *Neurosci. Lett.*, 142: 119–122.

Oka, Y. (1995) Tetrodotoxin-resistant persistent Na$^+$ current underlying pacemaker potentials of fish gonadotrophin-releasing hormone neurones. *J. Physiol.*, 482: 1–6.

Oka, Y. (1996) Characterization of TTX-resistant persistent Na$^+$

280

current underlying pacemaker potentials of fish gonadotropin-releasing hormone (GnRH) neurons. *J. Neurophysiol.*, 75: 2397–2404.

Oka, Y. (1997) The gonadotropin-releasing hormone (GnRH) neuronal system of fish brain as a model system for the study of peptidergic neuromodulation. In: I.S. Parhar and Y. Sakuma (Eds.), *GnRH Neurons: Genes to Behavior*. Brain Shuppan Publishers, Tokyo, pp. 245–276.

Oka, Y. and Abe, H. (2002) Physiology of GnRH neurons and modulation of their activities by GnRH. In: R.J. Handa, S. Hayashi, E. Terasawa, and M. Kawata (Eds.), *Neuroplasticity, Development, and Steroid Hormone Action*. CRC Press, Tokyo, pp. 191–203.

Oka, Y. and Ichikawa, M. (1990) Gonadotropin-releasing hormone (GnRH) immunoreactive system in the brain of the dwarf gourami (*Colisa lalia*) as revealed by light microscopic immunocytochemistry using a monoclonal antibody to common amino acid sequence of GnRH. *J. Comp. Neurol.*, 300: 511–522.

Oka, Y. and Ichikawa, M. (1991) Ultrastructure of the ganglion cells of the terminal nerve in the dwarf gourami (*Colisa lalia*). *J. Comp. Neurol.*, 304: 161–171.

Oka, Y. and Ichikawa, M. (1992) Ultrastructural characterization of gonadotropin-releasing hormone (GnRH)-immunoreactive terminal nerve cells in the dwarf gourami. *Neurosci. Lett.*, 140: 200–202.

Oka, Y. and Matsushima, T. (1993) Gonadotropin-releasing hormone (GnRH)-immunoreactive terminal nerve cells have intrinsic rhythmicity and project widely in the brain. *J. Neurosci.*, 13: 2161–2176.

Okuzawa, K., Amano, M., Kobayashi, M., Aida, K., Hanyu, I., Hasegawa, Y. and Miyamoto, K. (1990) Differences in salmon GnRH and chicken GnRH-II contents in discrete brain areas of male and female rainbow trout according to age and stage of maturity. *Gen. Comp. Endocrinol.*, 80: 116–126.

Okuzawa, K., Granneman, J., Bogerd, J., Goos, H.J.T., Zohar, Y. and Kagawa, H. (1997) Distinct expression of GnRH genes in the red seabream brain. *Fish Physiol. Biochem.*, 17: 71–79.

Oya, T., Abe, H. and Oka, Y. (2001) Involvement of Ca^{2+} entry mechanisms in the generation and modulation of pacemaker activities in the TN-GnRH neurons. *Zool. Sci.* 18: 107.

Ozawa, S. and Sand, O. (1986) Electrophysiology of excitable endocrine cells. *Physiol. Rev.*, 66(4): 887–952.

Paras, C.D. and Kennedy, R.T. (1995) Electrochemical detection of exocytosis at single rat melanotrophs. *Anal. Chem.*, 67: 3633–3637.

Parhar, I.S. (1997) GnRH in tilapia: three genes, three origins and their roles. In: I.S. Parhar and Y. Sakuma (Eds.), *GnRH Neurons: Genes to Behavior*. Brain Shuppan Publishers, Tokyo, pp. 99–122.

Parhar, I.S., Soga, T. and Sakuma, Y. (2000) Thyroid hormone and estrogen regulate brain region-specific messenger ribonucleic acids encoding three gonadotropin-releasing hormone genes in sexually immature male fish, *Oreochromis niloticus*. *Endocrinology*, 141: 1618–1626.

Pow, D.V. and Morris, J.F. (1989) Dendrites of hypothalamic magnocellular neurons release neurohypophysial peptides by exocytosis. *Neuroscience*, 32: 435–439.

Powell, J.F.F., Reska-Skinner, S.M., Prakash, M.O., Fischer, W.H., Park, M., Rivier, J.E., Craig, A.G., Mackie, G.O. and Sherwood, N.M. (1996) Two new forms of gonadotropin-releasing hormone in a protochordate and the evolutionary implications. *Proc. Natl. Acad. Sci. USA*, 93: 10461–10464.

Satou, M., Oka, Y., Kusunoki, M., Matsushima, T., Kato, M., Fujita, I. and Ueda, K. (1984) Telencephalic and preoptic areas integrate sexual behavior in hime salmon (landlocked red salmon, *Oncorhynchus nerka*): Results of electrical brain stimulation experiments. *Physiol. Behav.*, 33: 441–447.

Schwanzel-Fukuda, M.S. and Silverman, A.-J. (1980) The nervus terminalis of the guinea pig: a new luteinizing hormone-releasing hormone (LHRH) neuronal system. *J. Comp. Neurol.*, 191: 213–225.

Sherwood, N.M., Lovejoy, D.A. and Coe, I.R. (1993) Origin of mammalian gonadotropin-releasing hormones. *Endocr. Rev.*, 14: 241–254.

Sherwood, N.M., Von Schalburg, K. and Lescheid, D.W. (1997) Origin and evolution of GnRH in vertebrates and invertebrates. In: Y. Sakuma and I. Parhar (Eds.), *GnRH Neurons: Genes to Behavior*. Brain Shuppan, Tokyo, pp. 3–25.

Silverman, A.-J., Livne, I. and Witkin, J.W. (1994) The gonadotropin-releasing hormone (GnRH) neuronal systems: Immunocytochemistry and in situ hybridization. In: E. Knobil and J.D. Neil (Eds.), *The Physiology of Reproduction, 2nd Edition*. Raven Press, New York, NY, pp. 1683–1709.

Skynner, M.J., Slater, R., Sim, J.A., Allen, N.D. and Herbison, A.E. (1999) Promoter transgenics reveal multiple gonadotropin-releasing hormone-I-expressing cell populations of different embryonic origin in mouse brain. *J. Neurosci.*, 19: 5955–5966.

Soga, T., Sakuma, Y. and Parhar, I.S. (1998) Testosterone differentially regulates expression of GnRH messenger RNAs in the terminal nerve, preoptic and midbrain of male tilapia. *Mol. Brain Res.*, 60: 13–20.

Sower, S.A. (1997) Evolution of GnRH in fish of ancient origins. In: I.S. Parhar and Y. Sakuma (Eds.), *GnRH Neurons: Genes to Behavior*. Brain Shuppan Publishers, Tokyo, pp. 27–49.

Spergel, D.J., Krüth, U., Haneley, D.F., Sprengel, R. and Seeburg, P.H. (1999) GABA and glutamate-activated channels in green fluorescent protein-tagged gonadotropin-releasing hormone neurons in transgenic mice. *J. Neurosci.*, 19: 2037–2050.

Springer, A.D. (1983) Centrifugal innervation of goldfish retina from ganglion cells of the nervus terminalis. *J. Comp. Neurol.*, 214: 404–415.

Stell, W.K., Walker, S.E. and Chohan, K.S.B. (1984) The goldfish nervus terminalis: A luteinizing hormone-releasing hormone and molluscan cardioexcitatory peptide immunoreactive olfactoretinal pathway. *Proc. Natl. Acad. Sci.*, 81: 940–944.

Stojilkovic, S.S., Krsmanovic, L.Z., Spergel, D.J. and Catt, K.J. (1994a) Gonadotropin-releasing hormone neurons: Intrinsic pulsatility and receptor-mediated regulation. *Trends Endocrinol. Metab.*, 5: 201–209.

Stojilkovic, S.S., Reinhart, J. and Catt, K.J. (1994b) Gonadotropin-releasing hormone receptors: Structure and signal transduction pathways. *Endocrinol. Rev.*, 15(4): 462–499.

Suter, K.J., Song, W.J., Sampson, T.L., Wuarin, J.-P., Saunders, J.T., Dudek, F.E. and Moenter, S.M. (2000a) Genetic targeting of green fluorescent protein to GnRH neurons: characterization of whole-cell electrophysiological properties and morphology. *Endocrinology*, 141: 412–419.

Suter, K.J., Wuarin, J.-P., Smith, B.N., Dudek, F.E. and Moenter, S.M. (2000b) Whole-cell recordings from preoptic/hypothalamic slices reveal burst firing in gonadotropin-releasing hormone neurons identified with green fluorescent protein in transgenic mice. *Endocrinology*, 141: 3731–3736.

Terasawa, E. (1998) Cellular mechanism of pulsatile LHRH release. *Gen. Comp. Endocrinol.*, 112: 283–295.

Terasawa, E., Schanhofer, W.K., Keen, K.L. and Luchansky, L. (1999) Intracellular Ca^{2+} oscillations in luteinizing hormone-releasing hormone neurons derived from the embryonic olfactory placode of the rhesus monkey. *J. Neurosci.*, 19: 5898–5909.

Thureson-Klein, A.K. and Klein, R.L. (1990) Exocytosis from neuronal large dense-cored vesicles. *Int. Rev. Cytol.*, 121: 67–126.

Troskie, B., Illing, N., Rumbak, E., Sun, Y.-M., Hapgood, J., Sealfon, S., Conklin, D. and Millar, R. (1998) Identification of three putative GnRH receptor subtypes in vertebrates. *Gen. Comp. Endocrinol.*, 112: 296–302.

Tsutsui, H. and Oka, Y. (2000) Photosensitive neurons in the GnRH neuron-rich area of the cerebral ganglion in an ascidian, *Ciona savignyi*. *Biol. Bull.*, 198: 26–28.

Tsutsui, H., Yamamoto, N., Ito, H. and Oka, Y. (1998) GnRH-immunoreactive neuronal system in the presumptive ancestral chordate, *Ciona intestinalis* (Ascidian). *Gen. Comp. Endocrinol.*, 112: 426–432.

Uchiyama, H. (1990) Immunohistochemical subpopulations of retinopetal neurons in the nucleus olfactoretinalis in a teleost, the whitespotted greenling (*Hexagrammos stelleri*). *J. Comp. Neurol.*, 293: 54–62.

Umino, O. and Dowling, J.E. (1991) Dopamine release from interplexiform cells in the retina: effects of GnRH, FMRF-amide, bicuculine, and enkephalin on horizontal cell activity. *J. Neurosci.*, 11: 3034–3046.

van Goor, F., Krsmanovic, L.Z., Catt, K.J. and Stojilckvic, S.S. (1999) Coordinate regulation of gonadotropin-releasing hormone neuronal firing patterns by cytosolic calcium and store depletion. *Proc. Natl. Acad. Sci. USA*, 96: 4101–4106.

Walker, S.E. and Stell, W.K. (1986) Gonadotropin-releasing hormone (GnRH), molluscan cardioexcitatory peptide (FMRF-amide), enkephalin and related neuropeptides affect goldfish retinal ganglion cell activity. *Brain Res.*, 384: 262–273.

White, R.B., Eisen, J.A., Kasten, T.L. and Fernald, R.D. (1998) Second gene for gonadotropin-releasing hormone in humans. *Proc. Natl. Acad. Sci. USA*, 95: 305–309.

Wirsig, C.R. and Leonard, C.M. (1987) Terminal nerve damage impairs the mating behavior of the male hamster. *Brain Res.*, 417: 293–303.

Wirsig-Wiechmann, C.R. and Oka, Y. (2002) The terminal nerve ganglion cells project to the olfactory mucosa in the dwarf gourami brain. *Neurosci. Res.*, 44: 337–341.

Wong, M., Eaton, M.J. and Moss, R.L. (1990) Electrophysiological actions of luteinizing hormone-releasing hormone: intracellular studies in the rat hippocampal slice preparation. *Synapse*, 5: 65–70.

Yamamoto, N. and Ito, H. (2000) Afferent sources to the ganglion of the terminal nerve in teleosts. *J. Comp. Neurol.*, 428: 355–375.

Yamamoto, N., Oka, Y., Amano, M., Aida, K., Hasegawa, Y. and Kawashima, S. (1995) Multiple gonadotropin-releasing hormone (GnRH) immunoreactive systems in the brain of the dwarf gourami, *Colisa lalia*: immunohistochemistry and radioimmunoassay. *J. Comp. Neurol.*, 355: 354–368.

Yamamoto, N., Oka, Y. and Kawashima, S. (1997) Lesions of gonadotropin-releasing hormone (GnRH)-immunoreactive terminal nerve cells: effects on the reproductive behavior of male dwarf gouramis. *Neuroendocrinology*, 65: 403–412.

Yamamoto, N., Oka, Y., Yoshimoto, M., Sawai, N., Albert, J.S. and Ito, H. (1998a) Gonadotropin-releasing hormone neurons in the gourami midbrain: a double labeling study by immunocytochemistry and tracer injection. *Neurosci. Lett.*, 240: 50–52.

Yamamoto, N., Parhar, I.S., Sawai, N., Oka, Y. and Ito, H. (1998b) Preoptic gonadotropin-releasing hormone (GnRH) neurons innervate the pituitary in teleosts. *Neurosci. Res.*, 31: 31–38.

Yu, K.L., Rosenblum, P.M. and Peter, R.E. (1991) In vitro release of gonadotropin-releasing hormone from the brain preoptic-anterior hypothalamic region and pituitary of female goldfish. *Gen. Comp. Endocrinol.*, 81: 256–267.

Zheng, L., Krsmanovic, L.Z., Vergara, L.A., Catt, K.J. and Stojilkovic, S.S. (1997) Dependence of intracellular signaling and neurosecretion on phospholipase D activation in immortalized gonadotropin-releasing hormone neurons. *Proc. Natl. Acad. Sci. USA*, 94: 1573–1578.

Zucker, C.L. and Dowling, J.E. (1987) Centrifugal fibres synapse on dopaminergic interplexiform cells in the teleost retina. *Nature*, 300: 166–168.

I.S. Parhar (Ed.)
Progress in Brain Research, Vol. 141
© 2002 Elsevier Science B.V. All rights reserved

CHAPTER 20

A role for non-neuronal cells in synchronization of intracellular calcium oscillations in primate LHRH neurons

Ei Terasawa [1,2,*], Trevor A. Richter [1], Kim L. Keen [1]

[1] *Wisconsin National Primate Research Center, and* [2] *Department of Pediatrics, University of Wisconsin, Madison, WI 53715-1261, USA*

Introduction

Until the discovery of the LHRH molecule in 1971 by Schally and his collaborators (Schally et al., 1971), the hypothalamus remained a black box in the hypothalamo–pituitary–gonadal axis. In the last three decades, a quantum leap in our understanding on the hypothalamic control of reproductive function has been made. In primates, approximately 2000 LHRH neurons originate from the olfactory placode during the early embryonic stages and settle down in the septum–preoptic regions and basal hypothalamus (Anthony et al., 1984; Ronnekleiv and Resko, 1990; Quanbeck et al., 1997). Of these 2000 LHRH neurons, only a subset distributed in the medial basal hypothalamus is necessary and sufficient for sustaining reproductive function, since only lesion, but not complete deafferentation, of the medial basal hypothalamus interferes with ovulatory cycles (Krey et al., 1975; Plant et al., 1978).

LHRH is released in a pulsatile manner into the portal circulation (Carmel et al., 1976; Clarke and Cummins, 1982). An increase in the frequency and amplitude of LHRH pulses are essential for the onset of puberty (Terasawa and Fernandez, 2001) and the maintenance of regular LHRH pulsatility is important for normal reproductive function (Knobil, 1980,

1988). A study using multiunit activity recording from the medial basal hypothalamus indicates that pulsatile LHRH release is a consequence of synchronized activity of individual LHRH neurons. An increase in single unit spike components extracted from the multiunit 'LHRH pulse-generator' activity (Wilson et al., 1984) by cluster analysis occurs as a simultaneous increase in the firing rate of many individual hypothalamic neurons rather than the recruitment of new bursting neurons (Cardenas et al., 1993).

The mechanism underlying LHRH synchronization is currently unknown. This is because, unlike oxytocin and vasopressin neurons, LHRH neurons do not form a nucleus in the hypothalamus (Silverman, 1988) and there is little evidence for physical contact between LHRH neurons, such as synapses or syncitium formation, that would allow the synchronization of LHRH neurons (Witkin, 1999). It is, therefore, hypothesized that non-neuronal elements, such as glia, may coordinate activity among LHRH neurons (Terasawa, 1995, 2001). Here we report evidence that non-neuronal cells in cultures containing LHRH neurons derived from the embryonic olfactory placode participate in the synchronization of LHRH neurons.

LHRH release from cultured neurons

Previously, we have shown that cultures from the olfactory placode and terminal nerve region obtained from monkey embryos at embryonic day (E) 35–37 contain LHRH neurons, which are easily identifi-

---

* Correspondence to: E. Terasawa, Wisconsin National Primate Research Center, 1223 Capitol Court, Madison, WI 53715-1299, USA. Tel.: +1-608-263-3579; Fax: +1-608-263-3524; E-mail: terasawa@primate.wisc.edu

able because of their unique appearance (Terasawa et al., 1993). These cultures also contain non-neuronal cells, such as epithelial cells and fibroblasts, but there are few non-LHRH neurons and no glia (Terasawa et al., 1993). Cultured LHRH neurons release the decapeptide in a pulsatile manner at intervals of approximately an hour (Terasawa et al., 1999a), similar to that observed in vivo (Dierschke et al., 1970; Knobil, 1980; Gearing and Terasawa, 1988) and LHRH release is dependent on the presence of extracellular calcium, $[Ca^{2+}]_e$ (Terasawa et al., 1999a). LHRH release is also induced by depolarization stimuli, such as challenge with high K^+ and the Na^+ channel opener, veratridine (Terasawa et al., 1999a). Ca^{2+} enters LHRH neurons through voltage-sensitive L-type Ca^{2+} channels and stimulates LHRH release (Terasawa et al., 1999a). Pulsatile release of LHRH was also reported in placode cultures from sheep and rats (Duittoz and Batailler, 2000; Funabashi et al., 2001) as well as in GT1 cells (Krsmanovic et al., 1992; Martinez de la Escalera et al., 1992; Wetsel et al., 1992). Interestingly, the interval of pulsatile LHRH release in vitro is species specific, as reported in vivo, i.e., approximately 60 min in sheep (Duittoz and Batailler, 2000) and 30 min in rats (Funabashi et al., 2001) and GT1 cells, which are of mouse origin (Krsmanovic et al., 1992; Martinez de la Escalera et al., 1992; Wetsel et al., 1992).

Synchronization of intracellular Ca^{2+} oscillations in LHRH neurons and non-neuronal cells

Individual LHRH neurons exhibit periodic increases in intracellular Ca^{2+}, $[Ca^{2+}]_i$, concentrations (Charles and Hales, 1995; Charles et al., 1996; Terasawa et al., 1999b; Moore and Wray, 2000; Nunez et al., 2000) and an $[Ca^{2+}]_i$ increase is preceded by increases in electrical firing activity in GT1 cells (Costantin and Charles, 1999). Moreover, $[Ca^{2+}]_i$ oscillations in cultured primate LHRH neurons synchronize at intervals of approximately 60 min (Terasawa et al., 1999b). Similar synchronization of $[Ca^{2+}]_i$ oscillations in GT1 cells also has been reported with much shorter intervals (Charles et al., 1996). Electrical activities recorded by a low impedance multiple plate electrode recording system reveal that many GT1 neurons also exhibit synchronous neural activities (Funabashi et al., 2001;

Nunemaker et al., 2001). However, since in the hypothalamus the LHRH neurosecretory system that releases LHRH in a pulsatile manner is intermingled with other neurons and glia, unlike GT1 cells cultures, the non-LHRH cells in addition to LHRH neurons in our cultures allow us to address the question of whether the activity of only LHRH neurons is synchronized or whether non-LHRH cells are involved in this phenomenon.

A previous study (Terasawa et al., 1999b) indicated that olfactory placode cultures containing LHRH neurons exhibit periodic synchronization of $[Ca^{2+}]_i$ oscillations (Fig. 1). A recent study (Richter et al., 2002) further indicated that both LHRH neurons and non-neuronal cells in the same cultures exhibited spontaneous $[Ca^{2+}]_i$ oscillations with similar interpeak intervals (IPI). Moreover, $[Ca^{2+}]_i$ peaks in individual LHRH neurons and non-neuronal cells were periodically synchronized across the cell population (Fig. 1; Richter et al., 2002). In fact, $[Ca^{2+}]_i$ peaks in many cells often occurred within a narrow window of time (~20 sec, Terasawa et al., 1999b; Richter et al., 2002). When synchronization occurred in the majority of the cell population (80–100% of cells), the interval of synchronization was ~60 min (Fig. 1; Richter et al., 2002), which is similar to the interval of LHRH release in vitro (Terasawa et al., 1999a) and in vivo (Knobil, 1988). The amplitude of highly synchronized $[Ca^{2+}]_i$ peaks (80–100% of cells in a population) was significantly ($p < 0.05$) larger than that of unsynchronized peaks (Richter et al., 2002), and the IPI immediately after the highly synchronized $[Ca^{2+}]_i$ peaks among LHRH neurons was greater than for other $[Ca^{2+}]_i$ peaks (Fig. 2). The IPI immediately following highly synchronized $[Ca^{2+}]_i$ pulses among non-neuronal cells did not differ from non-synchronized peaks (Richter et al., 2002). These observations suggest that synchronization of $[Ca^{2+}]_i$ oscillations is an organized cell activity through which the cells in the network communicate with each other for a certain function, such as LHRH neurosecretion.

Synchronization of oscillations in $[Ca^{2+}]_i$ appeared as intercellular Ca^{2+} waves that spread across fields containing both LHRH neurons and non-neuronal cells. However, propagation of $[Ca^{2+}]_i$ oscillations in LHRH neurons differs from that in non-LHRH cells: the average speed at which Ca^{2+} waves

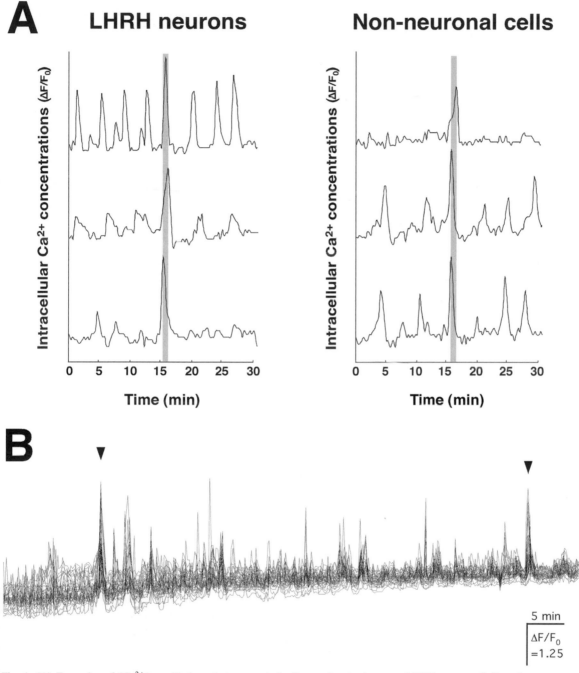

Fig. 1. (A) Examples of $[Ca^{2+}]_i$ oscillations that are periodically synchronized among LHRH neurons (left) and non-neuronal cells (right), from the same olfactory placode culture. The traces represent changes in $[Ca^{2+}]_i$ concentrations in 3 LHRH neurons and 3 non-neuronal cells obtained simultaneously by ratiometric measurements using fura2-AM. The shaded regions indicate narrow windows of time within which each of the cells exhibited a $[Ca^{2+}]_i$ peak. (B) An example of synchronization of $[Ca^{2+}]_i$ peaks occurring at an interval of ~60 min. Arrowheads indicate synchronization of $[Ca^{2+}]_i$ peaks among >80% of 72 individual cells ($n = 34$ LHRH neurons, 38 non-neuronal cells) in the culture. (Based on Richter et al., 2002.)

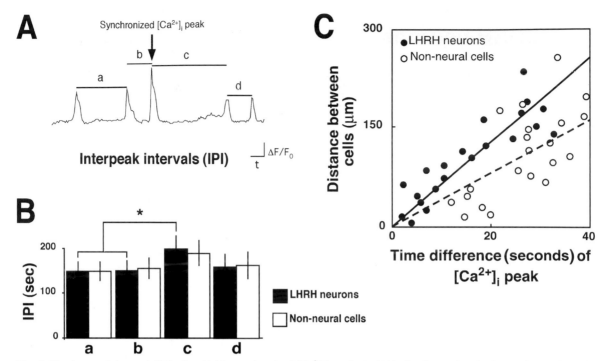

Fig. 2. The interpeak interval (IPI) after highly synchronized $[Ca^{2+}]_i$ peaks (>80% of cells synchronized) was longer than other IPIs. (A) The IPI was calculated for the two peaks in $[Ca^{2+}]_i$ that immediately preceded (a, b) or immediately followed (c, d) a peak that was highly synchronized among cells (arrow). The tracing is from a representative individual cell. (B) Mean (±SE) IPI for LHRH neurons (solid bars) and non-neuronal cells (open bars) for the IPIs (a–d) illustrated in A. Data represent synchronizations from 7 different cultures ($n = 628$ cells). *, $p < 0.05$, ANOVA with Fisher's PLSD post hoc test. (C) Ca^{2+} waves that were associated with synchronized oscillations in $[Ca^{2+}]_i$ spread more rapidly in LHRH neurons than in non-neuronal cells. Data represent the distance between cells vs. the time at which a peak in $[Ca^{2+}]_i$ occurred during synchronization. To generate this plot, we grouped data in bins according to the mean distance between cells (0–10 μm, 10–20 μm, . . . , >200 μm; $n = 21$ bins). For LHRH neurons (closed symbols, solid line), $Y = 7.05X$, $r^2 = 0.67$, $p < 0.001$. For non-neuronal cells (open symbols, broken line), $Y = 4.48X$, $r^2 = 0.41$, $p = 0.001$. The slope of the regression for LHRH neurons was significantly ($p < 0.01$, Student's t-test, $n = 21$) greater than for non-neuronal cells, indicating that Ca^{2+} waves were propagated more rapidly in LHRH neurons than in non-neuronal cells (17.6 μm/s in LHRH neurons vs. 11.1 μm/s in non-neuronal cells). (Based on Richter et al., 2002.)

propagated among LHRH neurons (17.6 μm/s) was significantly ($P < 0.01$) faster than that among non-neuronal cells (11.1 μm/s, Fig. 2; Richter et al., 2002). Similar propagation of intercellular signals in GT1 cells has been also reported: An elevation of $[Ca^{2+}]_i$ originating in a cell is propagated to adjacent cells (Charles et al., 1996) and neuronal excitation monitored by a voltage-sensitive dye periodically spreads across adjacent cells (Hiruma et al., 1997). Moreover, highly synchronized $[Ca^{2+}]_i$ oscillations and intercellular Ca^{2+} waves occur only in cultures that contain LHRH neurons and are absent from cultures that do not contain LHRH neurons (Richter et al., 2002). Although the source initiating the synchronized $[Ca^{2+}]_i$ oscillations in our cultures is un-

clear at this time, this observation clearly suggests that a LHRH neuron or group of LHRH neurons is responsible for initiating synchronization.

Ca^{2+} waves occur in many different types of cell, including glia, neurons, epithelial cells, endothelial cells, hepatocytes, and pancreatic β-cells (Cornell-Bell et al., 1990; Sanderson et al., 1990; Boitano et al., 1992; Nathanson et al., 1995; Charles et al., 1996; Cao et al., 1997; Newman and Zahs, 1997; Harris-White et al., 1998; Guthrie et al., 1999; Cotrina et al., 2000). Because neuronal activity can induce Ca^{2+} waves in glia (Dani et al., 1992), and glial Ca^{2+} waves induce Ca^{2+} transients and electrical activity in neurons (Nedergaard, 1994; Hassinger et al., 1996; Newman and Zahs, 1997), the concept has been developed that the calcium wave is a mode of intercellular communication not only among glial cells or neurons, but also between neurons and glia. It is possible, therefore, that in our cultures non-neuronal cells play a role similar to that of glial cells in the brain and participate in synchronizing LHRH neural activity.

Cellular elements participating in the synchronization of LHRH neural activity

It has been reported that perikarya of LHRH neurons are covered by glial sheaths (Witkin et al., 1991) and LHRH neuroterminals are intimately associated with glial end feet (King and Rubin, 1994). Previously, we hypothesized that glia play an important role in LHRH pulsatility (Terasawa, 1995). This hypothesis was based upon the facts that (1) pulsatile LHRH release occurs within the rat stalk–median eminence (Maeda et al., 1995; Purnelle et al., 1997), where an abundance of neuroterminals and glial cells are present, but not perikarya of LHRH neurons (Silverman, 1988), (2) neuropeptide Y (NPY) and norepinephrine (NE) are also released in a pulsatile manner that is coupled with LHRH pulses (Terasawa et al., 1988; Woller et al., 1992), but LHRH neurons receive little direct synaptic input from NPY and NE neurons (Thind et al., 1993), suggesting a role for glial communication among those neuronal systems, (3) glia play a critical role in release of oxytocin and vasopressin (Hatton, 1985), and (4) glia appear to play a role in a part of the signaling pathway of neurons (Parpura et al., 1994; Araque et al., 1999; Haydon, 2000, 2001; Ullian et al., 2001).

In our cultures, synchronization of $[Ca^{2+}]_i$ oscillations among cells is propagated as an intercellular Ca^{2+} wave across fields of cells containing both LHRH neurons and non-neuronal cells. A similar synchronization pattern of electrical activity in simulated networks of coupled cells, in which sparse, random activation of individual cells occurs spontaneously to produce intercellular activity waves, has been reported (Lewis and Rinzel, 2000). Since these authors show that in theoretical model systems the development of intercellular waves requires some form of coupling among neighboring cells (Lewis and Rinzel, 2000), the propagation of Ca^{2+} waves through fields including LHRH neurons and non-neuronal cells in our cultures is interpreted to mean

that LHRH neurons are functionally coupled to non-neuronal cells. Similar integrated signaling between neurons and non-neuronal cells has also been shown in cultures of rat forebrain tissue, where astrocytes are able to transmit Ca^{2+} signals to neurons (Nedergaard, 1994). Thus, non-neuronal cells, such as glia, might be a crucial component of the LHRH neurosecretory system in vivo, providing an indirect coupling mechanism to facilitate the synchronization of isolated LHRH neurons.

Possible chemical mechanisms mediating interneuronal signals

As candidate intercellular messengers, nitric oxide (NO, Snyder, 1992; Schuman and Madison, 1994), ATP (Osipchuk and Cahalan, 1992; Newman and Zahs, 1997; Guthrie et al., 1999; Cotrina et al., 2000), inositol 1,4,5-trisphosphate (IP$_3$, Charles et al., 1996), glutamate (Parpura et al., 1994), as well as cations, such as K^+ and Ca^{2+} (Nedergaard, 1994), have been described.

NO, a free radical gas produced during the conversion of arginine to citrulline in the presence of NO synthase, is involved in control of LHRH pulsatility (Brann and Mahesh, 1997; Kawakami et al., 1998). NO induces LHRH release in GT-1 cells (Sortino et al., 1994) and NO synthase mRNA is present in GT1 cells (Mahachoklertwuttana et al., 1994). An in vivo study in our laboratory has also shown that NO can induce both NPY and LHRH release from the stalk–median eminence: Using a push–pull perfusion method in the rhesus monkey, infusion of L-arginine, the precursor for NO production, stimulated NPY and LHRH release, while D-arginine (as a control) failed to cause any increase in release (Terasawa and Nyberg, 1997). Although we found the presence of nicotinamide adenine dinucleotide phosphate (NADPH)-diaphorase, which is a subtype of NO synthase (Wang et al., 1997), in fibers located in the infundibular nucleus and median eminence region, and in perikarya located in the perimammillary region of the hypothalamus (phenotypes of fibers and perikarya yet to be clarified) in rhesus monkeys (unpublished observation), the physiological role of these cells is yet to be determined.

Intercellular diffusion of various small molecules such as ions, second messengers and metabolites between cells can occur through transmembrane channels, namely gap junctions, without direct physical contact of the cells. Gap junctions are particularly important for intercellular communication, because they are widespread and permit rapid intercellular transit. The presence of dye-coupling between GT1 cells has been reported (Liposits et al., 1991; Wetsel et al., 1992), and connexin 26, a protein associated with gap junctions, was found in GT1 cells (Krsmanovic et al., 1992; Matesic et al., 1993; Hu et al., 1999). LHRH neurons in the hypothalamus (Hosny and Jennes, 1998) and in our cultures (Terasawa, unpublished observation) are immunopositive to connexin 32, another protein associated with gap junctions. Moreover, in our cultures Lucifer yellow injected into single non-neuronal cells diffuses into neighboring non-neuronal cells as well as to LHRH neurons (Richter, Wolfgang, and Terasawa, unpublished data). Finally, blocking gap junctions with octanol or carbenoxolone abolishes intercellular Ca^{2+} waves and synchronous electrical activity in GT1 cells (Charles et al., 1996; Funabashi et al., 2001).

Recent preliminary studies suggest that the synchronization of $[Ca^{2+}]_i$ oscillations in our cultures may be mediated by extracellular adenosine triphosphate (ATP) diffusing through gap junctions and/or purinergic receptors, P2X subtypes: (1) ATP, but neither ADP nor AMP, induced $[Ca^{2+}]_i$ increases with a dose-responsive manner, and (2) ATP resulted in a significant increase in LHRH release. Nonetheless, the observation that the speed of Ca^{2+} waves is faster among neurons than non-neuronal cells (Fig. 2; Richter et al., 2002) suggests that neurons are equipped with additional mechanisms for intercellular communication, such as synaptic neurotransmission. Although we have yet to confirm the presence of synapses between LHRH neurons in our cultures, relatively immature synapses between GT1 neurons have been reported (Liposits et al., 1991).

It has been shown in GT1 cells that depolarization always precedes $[Ca^{2+}]_i$ increases (Costantin and Charles, 1999). This important observation suggests that Ca^{2+} waves (i.e., synchronization) are propagated by initial depolarization and subsequent chemical diffusion of substances, such as NO, ATP, IP_3, K^+, and Ca^{2+}, as well as synaptic transmission. Further studies are needed in this regard.

Synchronization of $[Ca^{2+}]_i$ oscillations and LHRH neurosecretion

We have reported that highly synchronized $[Ca^{2+}]_i$ oscillations occur at a frequency similar to LHRH release in vitro (Terasawa et al., 1999b) and in vivo (Knobil, 1988). A recent study suggests that synchronization of $[Ca^{2+}]_i$ oscillations in GT1 neurons is associated with increases in membrane area, which is indicative of neurosecretion (Vazquez-Martinez et al., 2001). Therefore, we hypothesize that highly synchronized oscillations in $[Ca^{2+}]_i$ could provide a stimulus for LHRH neurosecretion. Since neurosecretion requires an increase in $[Ca^{2+}]_i$, the increase in $[Ca^{2+}]_i$ that occurs during a highly synchronous event could produce a suprathreshold elevation of $[Ca^{2+}]_i$ that triggers neurosecretion. The finding in a previous study that the amplitude of $[Ca^{2+}]_i$ peaks of synchronized oscillations was significantly larger than non-synchronized peaks supports this hypothesis. In fact, the larger peak amplitude might reflect an increase in $[Ca^{2+}]_i$ that is not only sufficient (i.e. a suprathreshold stimulus) to stimulate Ca^{2+}-dependent neurosecretion, but would also cause concomitant increases in $[Ca^{2+}]_i$ in neighboring cells, thereby transmitting a neurosecretion-inducing stimulus to other neurons. In addition, the observation that the IPI following a highly synchronous $[Ca^{2+}]_i$ peak is greater than IPIs at other times might be due to the additional time required for ion pumps on external and internal membranes to redistribute intracellular Ca^{2+} that has been released into the cytoplasm from endoplasmic reticulum and mitochondrial stores following a large amplitude, highly synchronized $[Ca^{2+}]_i$ increase. The simultaneous measurement of LHRH neurosecretion and $[Ca^{2+}]_i$ oscillations in single cells is needed to test the aforementioned hypothesis.

Conclusion

The study of the mechanism of LHRH pulse generation is important, since abnormalities in pulsatile LHRH release are associated with reproductive disorders in humans (Crowley et al., 1985; Spratt et al., 1987). Primate olfactory placode cultures which contain LHRH neurons and other cells exhibit hourly rhythms of LHRH release and synchronization of

$[Ca^{2+}]_i$ oscillations. Synchronization of $[Ca^{2+}]_i$ oscillations occurs among LHRH neurons and non-LHRH cells, suggesting that LHRH neurons form a functional network with non-neuronal cells for pulsatile LHRH release. Moreover, synchronization of $[Ca^{2+}]_i$ oscillations in non-LHRH cells occurs only when LHRH neurons are present in cultures, indicating that LHRH neurons are responsible for the initiation of synchronization. Translating these observations into events in the hypothalamus, we speculate that a similar mechanism may exist between LHRH neurons and glia in the medial basal hypothalamus for LHRH release in vivo. Much work, however, remains including elucidating: (1) the mechanism generating periodic $[Ca^{2+}]_i$ oscillations, (2) the mechanism determining the timing of $[Ca^{2+}]_i$ synchronization in LHRH neurons, (3) the molecule(s) responsible for intercellular communication among LHRH neurons and non-LHRH cells including glia, and (4) the intracellular mechanism that translates LHRH neuronal activity into neurosecretion. It is our hope that the answers to these questions will be forthcoming in the near future.

Acknowledgements

A portion of this review was based on studies supported by NIH grants HD15433, HD11533 and RR00167 (publication number 42-004 from the Wisconsin Regional Primate Research Center).

References

Anthony, E.L., King, J.C. and Stopa, E.G. (1984) Immunocytochemical localization of LHRH in the median eminence, infundibular stalk, and neurohypophysis. Evidence for multiple sites of releasing hormone secretion in humans and other mammals. *Cell Tissue Res.*, 236: 5–14.

Araque, A., Parpura, V., Sanzgiri, R.P. and Haydon, P.G. (1999) Tripartite synapses: glia, the unacknowledged partner. *Trends Neurosci.*, 22: 208–215.

Boitano, S., Dirksen, E.R. and Sanderson, M.J. (1992) Intercellular propagation of calcium waves mediated by inositol trisphosphate. *Science*, 258: 292–295.

Brann, D.W. and Mahesh, V.B. (1997) Excitatory amino acids: evidence for a role in the control of reproduction and anterior pituitary hormone secretion. *Endocrinol. Rev.*, 18: 678–700.

Cao, D., Lin, G., Westphale, E.M., Beyer, E.C. and Steinberg, T.H. (1997) Mechanisms for the coordination of intercellular calcium signaling in insulin-secreting cells. *J. Cell Sci.*, 110: 497–504.

Cardenas, H., Ordog, T., O'Byrne, K.T. and Knobil, E. (1993) Single unit components of the hypothalamic multiunit electrical activity associated with the central signal generator that directs the pulsatile secretion of gonadotropic hormones. *Proc. Natl. Acad. Sci. USA*, 90: 9630–9634.

Carmel, P.W., Araki, S. and Ferin, M. (1976) Pituitary stalk portal blood collection in rhesus monkeys: Evidence of pulsatile release of gonadotropin-releasing hormone (GnRH). *Endocrinology*, 99: 243–248.

Charles, A.C. and Hales, T.G. (1995) Mechanisms of spontaneous calcium oscillations and action potentials in immortalized hypothalamic (GT1-7) neurons. *J. Neurophysiol.*, 73: 56–64.

Charles, A.C., Kodali, S.K. and Tyndale, R.F. (1996) Intracellular calcium wave in neurons. *Mol. Cell. Neurosci.*, 7: 337–353.

Clarke, I.J. and Cummins, J.T. (1982) The temporal relationship between gonadotropin releasing hormone (GnRH) and luteinizing hormone (LH) secretion in the ovariectomized ewe. *Endocrinology*, 111: 1737–1739.

Cornell-Bell, A.H., Finkbeiner, S.M., Cooper, M.S. and Smith, S.J. (1990) Glutamate induces calcium waves in cultured astrocytes: long-range glial signaling. *Science*, 247: 470–473.

Cotrina, M.L., Lin, H.-C., López-García, J.C., Naus, C.C.G. and Nedergaad, M. (2000). ATP-mediated glial signaling. *J. Neurosci.*, 20: 2835–2844.

Costantin, J.L. and Charles, A.C. (1999) Spontaneous action potentials initiate rhythmic intercellular calcium waves in immortalized hypothalamic (GT1-1) neurons. *J. Neurophysiol.*, 82: 429–435.

Crowley Jr., W.F., Filicori, M., Spratt, D.T. and Santoro, N.F. (1985). The physiology of gonadotropin-releasing hormone (GnRH) secretion in men and women. *Rec. Prog. Horm. Res.*, 41: 473–531.

Dani, J.W., Chernjavsky, A. and Smith, S.J. (1992) Neuronal activity triggers calcium waves in hippocampal astrocyte networks. *Neuron*, 8: 429–440.

Dierschke, D.J., Bhattacharya, A.N., Atkinson, L.E. and Knobil, E. (1970) Circhoral oscillations of plasma LH levels in the ovariectomized rhesus monkey. *Endocrinology*, 87: 850–853.

Duittoz, A.H. and Batailler, M. (2000) Pulsatile LHRH secretion from primary culture of sheep olfactory placode explants. *J. Reprod. Fert.*, 120: 391–396.

Gearing, M. and Terasawa, E. (1988) Luteinizing hormone-releasing hormone (LHRH) neuroterminals mapped using the push–pull perfusion method in the rhesus monkey. *Brain Res. Bull.*, 21: 117–121.

Guthrie, P.B., Knappenberger, J., Segal, M., Bennett, M.V., Charles, A.C. and Kater, S.B. (1999) ATP released from astrocytes mediates glial calcium waves. *J. Neurosci.*, 19: 520–528.

Funabashi, T., Suyama, K., Uemura, T., Hirose, M., Hirahara, F. and Kimura, F. (2001) Immortalized gonadotropin-releasing hormone neurons (GT1-7 cells) exhibit synchronous bursts of action potentials. *Neuroendocrinology*, 73: 157–165.

Harris-White, M.E., Zanotti, S.A., Frautschy, S.A. and Charles, A.C. (1998) Spiral intercellular calcium waves in hippocampal slice cultures. *J. Neurophysiol.*, 79: 1045–1052.

Hassinger, T.D., Guthrie, P.B., Atkinson, P.B., Bennett, M.V. and Kater, S.B. (1996) An extracellular signaling component in propagation of astrocytic calcium waves. *Proc. Natl. Acad. Sci. USA*, 93: 13268–13273.

Hatton, G.I. (1985) Reversible synapse formation and modulation of cellular relationships in the adult hypothalamus under physiological conditions. In: C.W. Cotman (Ed.), *Synaptic Plasticity*. Guilford Press, New York, NY, pp. 373–404.

Haydon, P.G. (2000) Neuroglial networks: neurons and glia talk to each other. *Curr. Biol.*, 10: R712–714.

Haydon, P.G. (2001) GLIA: listening and talking to the synapse. *Nat. Rev. Neurosci.*, 2: 185–193.

Hiruma, H., Uemura, T. and Kimura, F. (1997) Neuronal synchronization and ionic mechanisms for propagation of excitation in the functional network of immortalized GT1-7 neurons: Optical imaging with a voltage-sensitive dye. *J. Neuroendocrinol.*, 9: 835–840.

Hosny, S. and Jennes, L. (1998) Identification of gap junctional connexin-32 mRNA and protein in gonadotropin-releasing hormone neurons of the female rats. *Neuroendocrinology*, 67: 101–108.

Hu, L., Olson, A.J., Weiner, R.I. and Goldsmith, P.C. (1999) Connexin 26 expression and extensive gap junctional coupling in cultures of GT1-7 cells secreting gonadotropin-releasing hormone. *Neuroendocrinology*, 70: 221–227.

Kawakami, S., Hirunagi, K., Ichikawa, M., Tsukamura, H. and Maeda, K. (1998) Evidence for terminal regulation of GnRH release by excitatory amino acids in the median eminence in female rats: a dual immunoelectron microscopic study. *Endocrinology*, 139: 1458–1461.

King, J.C. and Rubin, B.S. (1994) Dynamic changes in LHRH neurovascular terminals with various endocrine conditions in adults. *Horm. Behav.*, 28: 349–356.

Knobil, E. (1980) The neuroendocrine control of the menstrual cycle. *Recent Prog. Horm. Res.*, 36: 53–88.

Knobil, E. (1988) The neuroendocrine control of ovulation. *Hum. Reprod.*, 3: 469–472.

Krey, L.C., Butler, W.R. and Knobil, E. (1975) Surgical disconnection of the medial basal hypothalamus and pituitary function in the rhesus monkey. I. Gonadotropin secretion. *Endocrinology*, 96: 1073–1087.

Krsmanovic, L.Z., Stojilkovic, S.S., Merelli, F., Dufour, S.M., Virmani, M.A. and Catt, K.J. (1992) Calcium signaling and episodic secretion of gonadotropin-releasing hormone in hypothalamic neurons. *Proc. Natl. Acad. Sci. USA*, 89: 8462–8466.

Lewis, T.J. and Rinzel, J. (2000) Self-organized synchronous oscillations in a network of excitable cells coupled by gap junctions. *Network*, 11: 299–320.

Liposits, Z., Merchenthaler, I., Wetsel, W.C., Reid, J.J., Mellon, P.L., Weiner, R.I. and Negro-Vilar, A. (1991) Morphological characterization of immortalized hypothalamic neurons synthesizing luteinizing hormone-releasing hormone. *Endocrinology*, 129: 1575–1583.

Maeda, K.I., Tsukamura, H., Ohkura, S., Kawakami, S., Nagabukuro, H. and Yodoyama, A. (1995) The LHRH pulse generator: a mediobasal hypothalamic location. *Neurosci. Biobehav. Rev.*, 19: 427–437.

Mahachoklertwuttana, P., Sanchez, J., Kaplan, S.L. and Grumbach, M.M. (1994) N-methyl-D-aspartate (NMDA) receptors mediate the release of gonadotropin-releasing hormone (GnRH) by NMDA in a hypothalamic GnRH neuronal cell line (GT 1-1). *Endocrinology*, 134: 1023–1030.

Martinez de la Escalera, G., Choi, A.L. and Weiner, R.I. (1992) Generation and synchronization of gonadotropin-releasing hormone (GnRH) pulses: Intrinsic properties of the GT1-1 GnRH neuronal cell line. *Proc. Natl. Acad. Sci. USA*, 89: 1852–1855.

Matesic, M.F., Germak, J.A., Dupont, E. and Vadhuker, B.V. (1993) Immortalized hypothalamic luteinizing hormone-releasing hormone neurons express a connexin 26-like protein and display functional gap junction coupling assayed by fluorescence recovery after photobleaching. *Neuroendocrinology*, 58: 485–492.

Moore Jr., J.P. and Wray, S. (2000) Luteinizing hormone-releasing hormone (LHRH) biosynthesis and secretion in embryonic LHRH. *Endocrinology*, 141: 4486–4495.

Nathanson, M.H., Burgstahler, A.D., Mennone, A., Fallon, M.B., Gonzalez, C.B. and Saez, J.C. (1995) Ca^{2+} waves are organized among hepatocytes in the intact organ. *Am. J. Physiol.*, 269: G167–171.

Nedergaard, M. (1994) Direct signaling from astrocytes to neurons in cultures of mammalian brain cells. *Science*, 263: 1768–1771.

Newman, E.A. and Zahs, K.R. (1997) Calcium waves in retinal glial cells. *Science*, 275: 844–847.

Nunemaker, C.S., DeFazio, R.A., Geusz, M.E., Herzog, E.D., Pitts, G.R. and Moenter, S.M. (2001) Long-term recordings of networks of immortalized GnRH neurons reveal episodic patterns of electrical activity. *J. Neurophysiol.*, 86: 86–93.

Nunez, L., Villalobos, C., Boockfor, F.R. and Frawley, L.S. (2000) The relationship between pulsatile secretion and calcium dynamics in single, living GnRH neurons. *Endocrinology*, 141: 2012–2017.

Osipchuk, Y. and Cahalan, M. (1992) Cell-to-cell spread of calcium signals mediated by ATP receptors in mast cells. *Nature*, 359: 241–244.

Parpura, V., Basarsky, T.A., Liu, F., Jeftinija, K., Jeftinija, S. and Haydon, P.G. (1994) Glutamate-mediated astrocyte-neuron signaling. *Nature*, 369: 744–747.

Plant, T.M., Nakai, Y., Belchetz, P., Keogh, E. and Knobil, E. (1978) The sites of action of estradiol and phentolamine in the inhibition of the pulsatile, circhoral discharges of LH in the rhesus monkey (*Macaca mulatta*). *Endocrinology*, 102: 1015–1018.

Purnelle, G., Gerard, A., Czajkowski, V. and Bourguignon, J.P. (1997) Pulsatile secretion of gonadotropin-releasing hormone by rat hypothalamic explants of GnRH neurons without cell bodies. *Neuroendocrinology*, 66: 305–312.

Quanbeck, C., Sherwood, N.M., Millar, R.P. and Terasawa, E. (1997) Two populations of luteinizing hormone-releasing hormone neurons in the forebrain of the rhesus macaque during embryonic development. *J. Comp. Neurol.*, 380: 293–309.

Richter, T.A., Keen, K.L. and Terasawa, E. (2002) Synchronization of Ca^{2+} oscillations among primate LHRH neurons and non-neuronal cells in vitro. *J. Neurophysiol.*, 88: 1559–1567.

Ronnekleiv, O.K. and Resko, J.A. (1990) Ontogeny of gonadotropin-releasing hormone-containing neurons in early fetal development of rhesus macaques. *Endocrinology*, 126: 498–511.

Sanderson, M.J., Charles, A.C. and Dirksen, E.R. (1990) Mechanical stimulation and intercellular communication increases intracellular Ca^{2+} in epithelial cells. *Cell Regul.*, 1: 585–596.

Schally, A.V., Arimura, A., Kastin, A.J., Matsuo, H., Baba, Y., Redding, T.W., Nair, R.M., Debeljuk, L. and White, W.F. (1971) Gonadotropin-releasing hormone: One polypeptide regulates secretion of luteinizing and follicle-stimulating hormones. *Science*, 173: 1036–1038.

Schuman, E.M. and Madison, D.V. (1994) Nitric oxide and synaptic function. *Annu. Rev. Neurosci.*, 17: 153–183.

Silverman, A.J. (1988) The gonadotropin-releasing hormone (GnRH) neuronal systems: Immunocytochemistry. In: E. Knobil and J.D. Neill (Eds.), *The Physiology of Reproduction*. Raven Press, New York, NY, pp. 1283–1304.

Snyder, S.H. (1992) Nitric oxide: first in a new class of neurotransmitters. *Science*, 257: 494–496.

Sortino, M.A., Aleppo, G., Scapagnini, U. and Canonico, P.L. (1994) Involvement of nitric oxide in the regulation of gonadotropin-releasing hormone release from the GT1-1 neuronal cell line. *Endocrinology*, 134: 1782–1787.

Spratt, D.I., Finkelstein, J.S., Butler, J.P., Badger, T.M. and Crowley Jr., W.F. (1987) Effects of increasing the frequency of low doses of gonadotropin-releasing hormone (GnRH) on gonadotropin secretion in GnRH-deficient men. *J. Clin. Endocrinol. Metab.*, 64: 1179–1186.

Terasawa, E. (1995) Control of luteinizing hormone releasing hormone pulse generation in nonhuman primates. *Cell. Mol. Neurobiol.*, 15: 141–164.

Terasawa, E. (2001) Luteinizing hormone-releasing hormone (LHRH) neurons: mechanism of pulsatile LHRH release. *Vit. Horm.*, 63: 91–130.

Terasawa, E. and Fernandez, D.L. (2001) Neurobiological mechanisms of the onset of puberty in primates. *Endocr. Rev.*, 22: 111–151.

Terasawa, E. and Nyberg, C.L. (1997) LHRH pulse generation in the monkey. In vivo and in vitro studies. In: K. Maeda, H. Tsukamura and A. Yokoyama (Eds.), *Neural Control of Reproduction: Physiology and Behavior*. Karger, Basel, pp. 57–70.

Terasawa, E., Krook, C., Hei, D.L., Gearing, M., Schultz, N.J. and Davis, G.A. (1988) Norepinephrine is a possible neurotransmitter stimulating pulsatile release of luteinizing hormone releasing hormone in the rhesus monkey. *Endocrinology*, 123: 1808–1816.

Terasawa, E., Quanbeck, C.D., Schulz, C.A., Burich, A.J., Luchansky, L.L. and Claude, P. (1993) A primary cell culture system of luteinizing hormone releasing hormone (LHRH) neurons derived from fetal olfactory placode in the rhesus monkey. *Endocrinology*, 133: 2379–2390.

Terasawa, E., Keen, K.L., Mogi, K. and Claude, P. (1999a) Pulsatile release of luteinizing hormone-releasing hormone in cultured LHRH neurons derived from the embryonic olfactory placode of the rhesus monkey. *Endocrinology*, 140: 1432–1441.

Terasawa, E., Schanhofer, W.K., Keen, K.L. and Luchansky, L.L. (1999b) Intracellular Ca^{2+} oscillations in luteinizing hormone-releasing hormone (LHRH) cells derived from the embryonic olfactory placode of the rhesus monkey. *J. Neurosci.*, 19: 5898–5909.

Thind, K.K., Boggan, J.E. and Goldsmith, P.C. (1993) Neuropeptide Y system of the female monkey hypothalamus: retrograde tracing and immunostaining. *Neuroendocrinology*, 57: 289–298.

Ullian, E.M., Sapperstein, S.K., Christopherson, K.S. and Barres, B.A. (2001) Control of synapse number by glia. *Science*, 291: 657–661.

Vazquez-Martinez, R., Shorte, S.L., Boockfor, F.R. and Frawley, L.S. (2001) Synchronized exocytotic bursts from gonadotropin-releasing hormone-expressing cells: dual control by intrinsic cellular pulsatility and gap junctional communication. *Endocrinology*, 142: 2095–2101.

Wang, H., Christian, H.C. and Morris, J.F. (1997). Dissociation of nitric oxide synthase immunoreactivity and NADPH-diaphorase enzyme activity in rat pituitary. *J. Endocrinol.*, 154: R7–R11.

Wetsel, W.C., Valenca, M.M., Merchenthaler, I., Liposits, Z., Lopez, F.J., Weiner, R.I., Mellon, P.L. and Negro-Vilar, A. (1992) Intrinsic pulsatile secretory activity of immortalized luteinizing hormone-releasing hormone secreting neurons. *Proc. Natl. Acad. Sci. USA*, 89: 4149–4153.

Wilson, R.C., Kesner, J.S., Kaufman, J.M., Uemura, T., Akema, T. and Knobil, E. (1984) Central electrophysiologic correlates of pulsatile luteinizing hormone secretion in the rhesus monkey. *Neuroendocrinology*, 39: 256–260.

Witkin, J.W. (1999) Synchronized neuronal networks: The GnRH system. *Microsc. Res. Tech.*, 44: 11–18.

Witkin, J.W., Ferin, M., Popilskis, S.J. and Silverman, A.J. (1991) Effects of gonadal steroids on the ultrastructure of GnRH neurons in the rhesus monkey: synaptic input and glial apposition. *Endocrinology*, 129: 1083–1092.

Woller, M.J., McDonald, J.K., Reboussin, D.M. and Terasawa, E. (1992) Neuropeptide Y is a neuromodulator of pulsatile LHRH release in the gonadectomized rhesus monkey. *Endocrinology*, 430: 2333–2342.

I.S. Parhar (Ed.)
Progress in Brain Research, Vol. 141

CHAPTER 21

GnRH in the regulation of female rat sexual behavior

Y. Sakuma[*]

Department of Physiology, Nippon Medical School, Tokyo 113-8602, Japan

Introduction

The potency of gonadotropin-releasing hormone (GnRH) to enhance the lordosis reflex, principal estrogen-sensitive component of female rat sexual behavior, was demonstrated soon after the isolation of the mammalian decapeptide and the determination of its amino-acid sequence (Moss and McCann, 1973; Pfaff, 1973). It has been hypothesized that GnRH may play a role in synchronizing the timing of ovulation and behavioral receptivity. GnRH prolongs estrogen-induced receptivity in the ovariectomized ewe (Caraty et al., 2002). At the same time, the existence of different subsets of mammalian GnRH neurons have been suggested to be involved in the neuroendocrine and behavioral regulation in the mouse (Skynner et al., 1999b; Rajendren, 2001). Besides, more than a dozen isoforms of GnRH, which are conserved by 50–90% in their amino-acid sequence, are known across species (Sherwood et al., 1993). In the brain of many vertebrates, two or three isoforms of GnRH are expressed along with unique receptors (King and Millar, 1995). In the brain of the mouse, rat and primate, two isoforms of GnRH (GnRH-II and III) have been identified in addition to the original mammalian decapeptide, which is now sometimes referred to as GnRH-I (Lescheid et al., 1997; Chen et al., 1998; White et al., 1998). GnRH-II or [His^5,Trp^7,Tyr^8] GnRH, isolated from

chicken brain (King and Millar, 1984; Miyamoto et al., 1984), is expressed in almost all vertebrate classes. Chromatographic elution pattern and antigenicity suggest that GnRH-III is similar to salmon GnRH (sGnRH, [Trp^7,Leu^8]GnRH), which is ubiquitous in the terminal nerve ganglia of many teleosts (Parhar et al., 2000). In addition to these natural isoforms of GnRH, many GnRH analogs have been synthesized in attempts to exploit them as potent drugs to control fertility.

Expression of GnRH and its message is enhanced in the preoptic and limbic structures in the female rat in estrus (Malik et al., 1991). Estrogen enhances the expression of both GnRH receptor message and molecule (Jennes and Conn, 1994). Although the lack of nuclear estrogen receptor α (ERα), which is critically involved in the induction of the lordosis reflex (Ogawa et al., 1998), rules out possible direct regulation of the expression of GnRH by estrogen (Watson et al., 1992; Leng, 1999), changes in brain content of GnRH during the estrous cycle (Malik et al., 1991), persistent estrous females (Ma et al., 1990) or those changes that follow estrogen treatment of the ovariectomized females (Malik et al., 1991; Petersen et al., 1995) suggest transsynaptic or other non-genomic mechanism for estrogen to regulate GnRH turnover. Estrogen-dependent regulation of GnRH may also depend on ERβ, which is co-localized in certain GnRH neurons in the rat preoptic area (Kallo et al., 2001). Thus, the theory that GnRH mediates and synchronizes neuroendocrine and behavioral events may be valid in principle, albeit its mechanism of action appears to be far more complicated than it was thought initially.

---

[*] Correspondence to: Y. Sakuma, Department of Physiology, Nippon Medical School, Sendagi 1, Bunkyo, Tokyo 113, Japan. Tel.: +81-(3)-3824-6640; Fax: +81-(3)-5685-3055; E-Mail: ysakuma@nms.ac.jp

Neural actions of GnRH

Neural actions of GnRH underlie changes in behavioral expression induced by this peptide. GnRH is present in dense core vesicles of presynaptic terminals in the diagonal band and preoptic area, suggesting that the peptide probably acts as a transmitter in these structures (Chen et al., 1990). GnRH modulates the cholinergic transmission in frog neuromuscular junction (Akasu et al., 1983). Behavioral and neural effects of GnRH have been demonstrated by a variety of methods in the preoptic, medial basal hypothalamic and limbic structures. GnRH bindings were found in high concentrations in the dorsal hippocampus, amygdala, septum, and subiculum and in low amounts in the hypothalamus. Generally, GnRH molecule and receptor distribution is in good correlation in these structures (Jennes and Conn, 1994).

We have shown that microiontophoresis of GnRH causes excitation in many neurons in the female rat hypothalamus and preoptic area (Kawakami and Sakuma, 1974). Neurons in brain slices taken from the female rat ventromedial nucleus and preoptic area responded to GnRH in perfusate *in vitro* experiments (Kow and Pfaff, 1988). A direct evidence for a modulatory role of GnRH in the synaptic transmission has been provided by intracellular recordings from postganglionic neurons in the bullfrog sympathetic ganglia, in which a GnRH-like peptide functions as the transmitter for the exceptionally slow excitatory postsynaptic potential that lasts several minutes (Kuffler, 1980).

A brief overview on the effects of GnRH on neuronal activities in discrete brain structures, that have been associated with the regulation of female rat sexual behavior, follows.

Preoptic area

In the rat and in many vertebrates, a major population of GnRH neurons has been identified in the diagonal band–medial septum–medial preoptic continuum. GnRH neurons in these rostral structures have been emphasized not only in the neuroendocrine regulation of the anterior pituitary but also experiments in the ferrets or hypogonadal mice associate these neurons with behavioral regulation. This view is supported by observations that approximately 50% of the GnRH neurons project to the median eminence and are capable of neuroendocrine control, while others project elsewhere and are apparently involved in peptidergic neurotransmission (Jennes and Stumpf, 1986; Silverman et al., 1987). The fact that at the time of the preovulatory gonadotropin surge, Fos expression (Lee et al., 1990) or GnRH gene activation (Porkka-Heiskanen et al., 1994) occurs in a limited number of GnRH neurons also collaborates this interpretation.

Hypogonadal female mice, that are genetically deficient in GnRH neurons, show no sexual activity and are infertile. Implants of fetal brain grafts that contain GnRH neurons in the third ventricle induce vaginal opening and persistent vaginal estrus, ovarian, and uterine development and increased gonadotropin secretion. When these females are mated with normal males, they show comparable levels of lordosis as seen in normal female mice in estrus (Gibson et al., 1987). Immunocytochemistry revealed innervation of the median eminence, but not the midbrain central gray by axons originating from preoptic implants (Saitoh et al., 1992). Further elaboration is needed on the role played by GnRH and analogs in the midbrain, in light of the detection of GnRH-II in this region (Rissman et al., 1995).

Midbrain central gray

In the female rat, GnRH-immunoreactive nerve fibers descend in the midbrain central gray and continue as far caudal as the ponto-mesencephalic junction (Liposits and Setalo, 1980; Buma, 1989). Autoradiography and radioligand assays revealed substantial binding of GnRH in the central gray (Badr and Pelletier, 1987; Jennes et al., 1988). The numbers of central gray neurons that are excited by the microiontophoresis of GnRH increase following estrogen treatment in the ovariectomized female rat (Chan et al., 1985; Schiess et al., 1987). Taken together with morphological demonstration of estrogen-receptor positive neurons in the central gray (Murphy and Hoffman, 2001), the results of the microiontophoresis study suggest an estrogen-dependent regulation of GnRH receptors in the central gray. The cellular mechanisms for this enhanced sensitivity are unclear, but a distinct possibility is that GnRH receptor function may be enhanced by either increase in number or coupling to GnRH, as has been suggested for pituitary gonadotrophs (Bauer-Dantoin et al., 1995; Ortmann and Diedrich, 1999). Indirect modulation of the neuronal activity cannot be ruled out as a route for GnRH action. The excitatory GnRH effect on central gray neurons in perfused midbrain slices *in vitro* has been attributed to modulation of GABAergic or adrenergic transmission (Ogawa et al., 1992). In any case, evidence abounds that GnRH promotes the lordosis reflex via its excitatory action on neurons in the midbrain central gray (Sakuma and Pfaff, 1980; Riskind and Moss, 1983a,b; Sakuma and Pfaff, 1983; Sirinathsinghji, 1984), similar to electrical stimulation of this structure (Sakuma and Pfaff, 1979).

The central gray is not a single entity and contains neural components that inhibit the lordosis reflex. Contrary to the dorsal subdivision of the central gray (Sakuma and Pfaff, 1979), electrical stimulation of the ventral central gray disrupts lordosis (Arendash and Gorski, 1983). The effect is probably due to the activation of axons of passage that originate in the ventral tegmental area (Sakamoto et al., 1993) and it is unlikely that GnRH interacts with these axons. Functional dichotomy between the ventral and dorsal subdivisions is also known for pain control (Cannon et al., 1982) or autonomic regulation (Lovick, 1991).

The limbic structures

GnRH receptors are densely expressed in the medial amygdala, hippocampus and septum (Badr and Pelletier, 1987; Jennes et al., 1988). Microiontophoresis of GnRH causes excitatory response in a subset of individual neurons in the medial amygdala and the effect is augmented by estrogen (Dudley et al., 1990). In some cells, electrical stimulation of the medial septum similarly excites their activity. Although estrogen enhances the sensitivity of individual neurons to GnRH, it decreases the percentage of the neurons that are excited by septal stimulation (Dudley et al., 1990). On the other hand, bilateral lesion of the medial amygdala diminished the lordosis reflex, simultaneously with a reduction of septal preoptic GnRH neurons that exhibit Fos immunoreactivity (Pfaus et al., 1996). Taken together, these findings suggest that GnRHergic neurons, that employ GnRH as an excitatory neurotransmitter or neuromodulator,

reside in the septal–preoptic continuum and project to the medial amygdala. They may constitute a neural substrate for olfactory enhancement of a facilitatory system of the lordosis reflex (Rajendren and Moss, 1993).

Hippocampal pyramidal cells, particularly those in the CA1 and CA3 regions, express GnRH receptors (Jennes et al., 1988). Intracellular recordings from CA1 neurons revealed that GnRH decapeptide and its behaviorally active fragment Ac-LHRH[5-10] induce long-lasting depolarization (Chen et al., 1993). However, septo-preoptic GnRH neurons do not project to the hippocampus (Dudley et al., 1992). A more stable C-fragment catabolites derived from the hydroxylated GnRH, which also possesses behavioral actions, has been proposed as a possible natural ligand for hippocampal GnRH receptors (Gautron et al., 1993).

Ventral tegmental area

The presence of fibers with GnRH immunoreactivity (Siverman and Krey, 1978; Merchenthaler et al., 1984) and GnRH agonist binding sites (Jennes et al., 1988) in the ventral tegmental area suggest a role for GnRH in the neural transmission in this structure. Indeed, microinfusion of GnRH into the ventral tegmental area enhances the lordosis reflex (Sirinathsinghji et al., 1986). Although tentative assumption at that time was that GnRH accomplished the behavioral effect through neuronal activation of the ventral tegmental area, results of our recent behavioral (Hasegawa et al., 1991) and electrophysiological (Hasegawa and Sakuma, 1993) studies suggest otherwise. Inhibition of neuronal activity in the ventral tegmental area caused by microiontophoresis of GnRH, that confirms results of the behavioral study (Suga et al., 1997), will be discussed later in this chapter.

GnRH receptors in the pituitary and brain

Species differences and receptor subtypes

GnRH receptors with characteristics similar to those in the anterior pituitary have been identified in the rat brain. However, as much as 9 subtypes of GnRH, which differ in amino-acid contents and hypophysiotropic potencies are known between different species (King and Millar, 1995). Mammalian and chicken-II GnRH molecules are co-localized in the musk shrew brain (Rissman et al., 1995). In addition to mammalian GnRH, GnRH-II has been detected in the rat midbrain (Parhar, unpublished observation). Thus, co-localization of different GnRH species in the rat brain appears to be a distinct possibility.

It was noted that the behavioral effect does not require the entire amino-acid sequence of the decapeptide, but a fragment Ac-GnRH$^{5-10}$ possesses a comparable potency as GnRH decapeptide to augment lordosis reflex in estrogen-treated ovariectomized female rats (Dudley and Moss, 1988). In addition to GnRH decapeptide and its C-terminal fragments, several GnRH analogs modulate the lordosis reflex. It has been shown that the behavioral effects of analogs do not always correlate straightforwardly with their hypophysiotrpic actions in the female (Dudley et al., 1981; Zadina et al., 1981; Sakuma and Pfaff, 1983; Moss and Dudley, 1990) as well as male (Bhasin et al., 1988) rats. In an *in vitro* system, certain analog stimulates inositol phospholipid turnover and increases intracellular Ca^{2+} as does the decapeptide, but does not induce gonadotropin secretion in pituitary gonadotrophs (Levy et al., 1990).

Some behaviorally active analogs, that inhibit pituitary gonadotropin secretion, often produce agonistic responses when applied to neurons by microiontophoresis. Microiontophoresis of Ac-GnRH$^{5-10}$, which lacks hypophysiotropic activity, induces changes similar to those induced by GnRH decapeptide in the medial amygdala (Dudley et al., 1990), midbrain central gray (Chan et al., 1985; Schiess et al., 1987) and hippocampus (Chen et al., 1993). Microiontophoresis of a fragment without behavioral action to neurons in these structures showed no effect at all.

These observations imply multitudes of GnRH molecules and corresponding receptors in the brain that employs different GnRH molecules as their ligands.

GnRH action on neuronal circuitry for female rat sexual behavior

GnRH and its fragments and analogs have been infused into the medial preoptic area, the ventro-

medial hypothalamus (Dudley and Moss, 1988) or the midbrain central gray (Sakuma and Pfaff, 1980; Riskind and Moss, 1983b; Sakuma and Pfaff, 1983; Sirinathsinghji, 1984; Dudley and Moss, 1988) to study their effects on the lordosis reflex. It is of interest that D-Phe^2, D-Ala^6-GnRH, an antagonist analog for pituitary gonadotropin secretion (Beattie et al., 1975), was more potent to promote the lordosis reflex than GnRH decapeptide when applied to the medial preoptic area (Suga and Sakuma, unpublished observation) or the midbrain central gray (Sakuma and Pfaff, 1983) (Fig. 1).

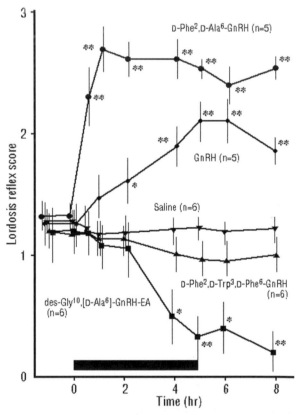

Fig. 1. Time course of changes in the lordosis reflex score induced by GnRH or its analogs in agar-saline gel. Cannulae containing GnRH or analogs were placed in the midbrain central gray for a 5-hour period, as indicated by the solid bar at the bottom. des-Gly10,[D-Ala6]-GnRH-EA is an agonist, and D-Phe2,D-Ala6-GnRH and D-Phe2,D-Trp3,D-Phe6-GnRH are antagonists interms of their effects on pituitary LH release. Vertical bars denote SEM; n = number of rats. $*p < 0.05$; $**p < 0.01$ when compared to blank (agar-saline) cannula at each time point (two-way analysis of variance and Student–Newman–Keuls test). Figure reproduced, with permission, from Suga et al., 1997).

The restoration of reproductive capability in the hypogonadal mice by implants of preoptic GnRH cells has been attributed to the innervation of the median eminence by axons of the implanted GnRH cells (Gibson et al., 1987; Saitoh et al., 1992). Neural basis for the emergence of the lordosis reflex in these animals remains unknown.

A lesion study implicates the dorsal hippocampus in the facilitation of the lordosis reflex (Cameron et al., 1979). The effect has been attributed to interactions with progesterone-sensitive or serotonergic neural system (Franck and Ward, 1981). However, further scrutiny is needed to establish causal relationship between the GnRH-induced changes in the neuronal activity in the hippocampus (Chen et al., 1993), that presumably depend on GnRH receptors in this structure (Jennes et al., 1988), and the regulation of female rat sexual behavior.

The ventral tegmental area as a relay for the GnRH effect

Electrical stimulation of the midbrain ventral tegmental area disrupted the lordosis reflex without afflicting proceptive components of female rat sexual behavior (Hasegawa et al., 1991) as did the medial preoptic stimulation (Takeo et al., 1993). Although the ascending, primarily dopaminergic projection toward the forebrain is widely acknowledged to be the major route for the ventral tegmental area to accomplish its behavioral and emotional regulation (Kato and Sakuma, 2000), results from both behavioral (Hasegawa et al., 1991) and electrophysiological (Sakamoto et al., 1993) experiments implicate nondopaminergic (Fallon et al., 1984; Kalen et al., 1988), descending (Beckstead et al., 1979; Swanson, 1982) efferents in the disruption of the lordosis reflex.

We have shown that estrogen decreases the excitability of efferents of the ventral tegmental area that descend in the ventral central gray toward the lower brain stem (Hasegawa and Sakuma, 1993). Furthermore, these ventral tegmental neurons are innervated by preoptic neurons that are similarly inhibited by estrogen (Sakamoto et al., 1993). Taken together, these observations lead to a conclusion that estrogen promotes lordosis reflex by disinhibition of the preoptic and ventral tegmental neuronal links.

The midbrain ventral tegmental area has been positively associated with GnRH-dependent regulation of the female rat sexual behavior. GnRH infusion into the ventral tegmental area potentiates the lordosis reflex. The effect persists in animals with neurotoxic deletions of dopaminergic neurons (Sirinathsinghji et al., 1986). The GnRH-induced facilitation of the lordosis reflex, therefore, does not depend on dopaminergic neurons as in the stimulus-bound inhibition of the reflex in the pimozide-treated animals (Hasegawa et al., 1991). The opposite effects of GnRH infusion and local electrical stimulation suggest that GnRH suppresses the activity of the descending, estrogen-sensitive neurons in the ventral tegmental area.

Microiontophoresis of GnRH and its behaviorally active analog D-Phe^2, D-Ala^6-GnRH in the ventral tegmental area support this theory (Fig. 2). These substances predominantly suppress the activity of neurons that are antidromically activated from the central gray, in which axons of non-dopaminergic neurons in the ventral tegmental area have been shown to descend toward the lower brain stem (Beckstead et al., 1979).

Neuronal sensitivity to GnRH depends critically on estrogen. In the estrogen-primed ovariectomized rats, 8 identified neurons were inhibited by GnRH without exception, whereas GnRH induced no inhibition and excited 3 among 10 identified neurons in the ovariectomized rats without estrogen treatment. The behaviorally active analog D-Phe^2, D-Ala^6-GnRH, that enhances lordosis reflex, inhibited 3 and excited other 3 neurons in the estrogen-treated ovariectomized rats, but lacked both behavioral and neuronal effects in the absence of estrogen. Other neural effects of estrogen in the ventral tegmental area include a decreased spontaneous discharge rate and an increase in the antidromic activation threshold, both of which bolster the theory that inhibition of neuronal transmission via the ventral tegmental area may be causally related to the enhancement of the lordosis reflex; GnRH and estrogen-regulated expression of GnRH receptors may mediate the inhibitory effects of estrogen.

Conclusion

Since GnRH was first described in 1970s as a neuronal modulator in the regulation of the lordosis

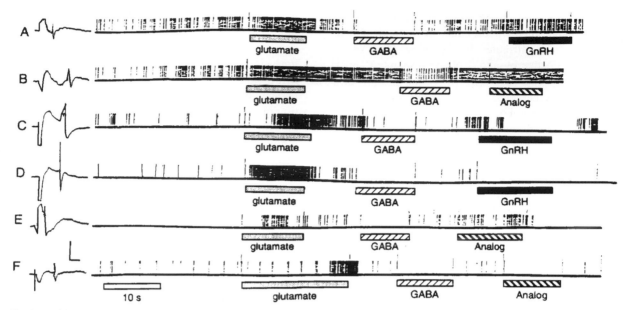

Fig. 2. Antidromic action potentials and response profiles of microiontophoresis of 6 representative neurons. Neurons in ovariectomized non-treated rats had higher firing rates and showed no response to GnRH (A) and to D-Phe^2,D-Ala^6-GnRH (Analog) (B). (C–F), recordings from ovariectomized, estrogen-primed rats. Typical GnRH-induced inhibition of the spontaneous firing in two identified neurons (C, D), with different time courses. (E), a 'silent' neuron which was activated by the Analog; (F), an excitation caused by the analog in a different neuron. Calibrations for the antidromic potentials, 5 ms and 5 mV; scale for the chart recordings, 10 s. Figure reproduced, with permission, from Sakuma and Pfaff (1983).

reflex in the female rat in estrus, many other peptides, amines and other molecules have been added to the list of substances that affects this behavior. However, in our understanding of the control of the lordosis reflex, GnRH ranks high among these substances. This applies to our knowledge on both site and molecular form of the neurotropic action. However, some question persists as more answers are provided by behavioral and neurophysiological studies on the roles of GnRH in the regulation of female rat sexual behavior.

Among the issues discussed in this review, it may well be questioned whether the estrogen-sensitive preoptic projections to the ventral tegmental area employ GnRH as neurotransmitter. Although results from behavioral and neurophysiological experiments favor this possibility, it contradicts to an accepted wisdom that GnRH neurons do not express ERα. GnRH neurons with midbrain projections may make an exception for this consensus, or GnRHergic interneurons in the midbrain, that evade neurophysiological analysis, may intervene. Another possibility is that the estrogen effects may be accomplished via

ERβ or through non-genomic routes. Co-localization of ERβ and GnRH has been shown in certain preoptic neurons (Skynner et al., 1999a; Kallo et al., 2001). An emerging hypothesis implicates the binding of estrogen to presumed membrane receptors, or interactions with the binding of ligands to their receptors (Raab et al., 1999; Clarke et al., 2000). Experiments are being undertaken to determine possible interactions between GnRH and estrogen via these routes.

References

Akasu, T., Kojima, M. and Koketsu, K. (1983) Luteinizing hormone-releasing hormone modulates nicotinic ACh-receptor sensitivity in amphibian cholinergic transmission. *Brain Res.*, 279: 347–351.

Arendash, G.W. and Gorski, R.A. (1983) Suppression of lordotic responsiveness in the female rat during mesencephalic electrical stimulation. *Pharmacol. Biochem. Behav.*, 19: 351–357.

Badr, M. and Pelletier, G. (1987) Characterization and autoradiographic localization of LHRH receptors in the rat brain. *Synapse*, 1: 567–571.

Bauer-Dantoin, A.C., Weiss, J. and Jameson, J.L. (1995) Roles of estrogen, progesterone, and gonadotropin-releasing hormone (GnRH) in the control of pituitary GnRH receptor gene expression at the time of the preovulatory gonadotropin surges. *Endocrinology*, 136: 1014–1019.

Beattie, C.W., Corbin, A., Foell, T.J., Garsky, V., McKinley, W.A., Rees, R.W., Sarantakis, D. and Yardley, J.P. (1975) Luteinizing hormone-releasing hormone: Antiovulatory activity of analogs substituted in positions 2 and 6. *J. Med. Chem.*, 18: 1247–1250.

Beckstead, R.M., Domesick, V.B. and Nauta, W.J. (1979) Efferent connections of the substantia nigra and ventral tegmental area in the rat. *Brain Res.*, 175: 191–217.

Bhasin, S., Fielder, T., Peacock, N., Sod-Moriah, U.A. and Swerdloff, R.S. (1988) Dissociating antifertility effects of GnRH-antagonist from its adverse effects on mating behavior in male rats. *Am. J. Physiol.*, 254: E84–E91.

Buma, P. (1989) Characterization of luteinizing hormone-releasing hormone fibres in the mesencephalic central grey substance of the rat. *Neuroendocrinology*, 49: 623–630.

Cameron, W.R., Gage III, F.H., Hitt, J.C. and Popolow, H.B. (1979) The influence of the anterodorsal hippocampus on feminine sexual behavior in rats. *Behav. Neural. Biol.*, 27: 72–86.

Cannon, J.T., Prieto, G.J., Lee, A. and Liebeskind, J.C. (1982) Evidence for opioid and non-opioid forms of stimulation-produced analgesia in the rat. *Brain Res.*, 243: 315–321.

Caraty, A., Delaleu, B., Chesneau, D. and Fabre-Nys, C. (2002) Sequential role of E_2 and GnRH for the expression of estrous behavior in ewes. *Endocrinology*, 143: 139–145.

Chan, A., Dudley, C.A. and Moss, R.L. (1985) Hormonal modulation of the responsiveness of midbrain central gray neurons to LH-RH. *Neuroendocrinology*, 41: 163–168.

Chen, A., Yahalom, D., Ben-Aroya, N., Kaganovsky E., Okon, E. and Koch, Y. (1998) A second isoform of gonadotropin-releasing hormone is present in the brain of human and rodents. *FEBS Lett.*, 435: 199–203.

Chen, W.P., Witkin, J.W. and Silverman, A.J. (1990) Sexual Dimorphism in the Synaptic Input to Gonadotropin Releasing Hormone Neurons. *Endocrinology*, 126: 695–702.

Chen, Y., Wong, M. and Moss, R.L. (1993) Effects of a biologically active LHRH fragment on CA1 hippocampal neurons. *Peptides*, 14: 1079–1081.

Clarke, C.H., Norfleet, A.M., Clarke, M.S., Watson, C.S., Cunningham, K.A. and Thomas, M.L. (2000) Perimembrane localization of the estrogen receptor alpha protein in neuronal processes of cultured hippocampal neurons. *Neuroendocrinology*, 71: 34–42.

Dudley, C.A. and Moss, R.L. (1988) Facilitation of lordosis in female rats by CNS-site specific infusions of an LH-RH fragment-Ac-LH-RH-(5–10). *Brain Res.*, 441: 161–167.

Dudley, C.A., Vale, W., Rivier, J. and Moss, R.L. (1981) The effect of LHRH antagonist analogs and an antibody to LHRH on mating behavior in female rats. *Peptides*, 2: 393–396.

Dudley, C.A., Lee, Y. and Moss, R.L. (1990) Electrophysiological identification of a pathway from the septal area to the medial amygdala: Sensitivity to estrogen and luteinizing hormone-releasing hormone. *Synapse*, 6: 161–168.

Dudley, C.A., Sudderth, S.B. and Moss, R.L. (1992) LHRH neurons in the medial septal-diagonal band-preoptic area do not project directly to the hippocampus: A double-labeling immunohistochemical study. *Synapse*, 12: 139–146.

Fallon, J.H., Schmued, L.C., Wang, C., Miller, R. and Banales, G. (1984) Neurons in the ventral tegmentum have separate populations projecting to telencephalon and inferior olive are histochemically different, and may receive direct visual input. *Brain Res.*, 321: 332–336.

Franck, J.A.E. and Ward, I.L. (1981) Intralimbic progesterone and methysergide facilitate lordotic behavior in estrogen-primed female rats. *Neuroendocrinology*, 32: 50–56.

Gautron, J.P., Pattou, E., Leblanc, P., L'Heritier, A. and Kordon, C. (1993) Preferential distribution of C-terminal fragments of [hydroxyproline⁹]LHRH in the rat hippocampus and olfactory bulb. *Neuroendocrinology*, 58: 240–250.

Gibson, M.J., Moscovitz, H.C., Kokoris, G.J. and Silverman, A.-J. (1987) Female sexual behavior in hypogonadal mice with GnRH-containing brain grafts. *Horm. Behav.*, 21: 211–222.

Hasegawa, T. and Sakuma, Y. (1993) Developmental effect of testosterone on estrogen sensitivity of the rat preoptic neurons with axons to the ventral tegmental area. *Brain Res.*, 611: 1–6.

Hasegawa, T., Takeo, T., Akitsu, H., Hoshina, Y. and Sakuma, Y. (1991) Interruption of the lordosis reflex of female rats by ventral midbrain stimulation. *Physiol. Behav.*, 50: 1033–1038.

Jennes, L. and Conn, P.M. (1994) Gonadotropin-releasing hormone and its receptors in rat brain. *Front. Neuroendocrinol.*, 15: 51–77.

Jennes, L. and Stumpf, W.E. (1986) Gonadotropin-releasing hormone immunoreactive neurons with access to fenestrated capillaries in mouse brain. *Neuroscience*, 18: 403–416.

Jennes, L., Dalati, B. and Conn, P.M. (1988) Distribution of gonadotropin releasing hormone agonist binding sites in the rat central nervous system. *Brain Res.*, 452: 156–164.

Kalen, P., Skagerberg, G. and Lindvall, O. (1988) Projections from the ventral tegmental area and mesencephalic raphe to the dorsal raphe nucleus in the rat: Evidence for a minor dopaminergic component. *Exp. Brain Res.*, 73: 69–77.

Kallo, I., Butler, J.A., Barkovics-Kallo, M., Goubillon, M.L. and Coen, C.W. (2001) Oestrogen receptor beta-immunoreactivity in gonadotropin releasing hormone-expressing neurones: Regulation by oestrogen. *J. Neuroendocrinol.*, 13: 741–748.

Kato, A. and Sakuma, Y. (2000) Neuronal activity in female rat preoptic area associated with sexually motivated behavior. *Brain Res.*, 862: 90–102.

Kawakami, M. and Sakuma, Y. (1974) Responses of hypothalamic neurons to the microiontophoresis of LH-RH, LH and FSH under various levels of circulating ovarian hormones. *Neuroendocrinology*, 15: 290–307.

King, J.A. and Millar, R.P. (1984) Isolation and structural characterization of chicken hypothalamic luteinizing hormone releasing hormone. *J. Exp. Zool.*, 232: 419–423.

King, J.A. and Millar, R.P. (1995) Evolutionary aspects of gonadotropin-releasing hormone and its receptor. *Cell. Mol. Neurobiol.*, 15: 5–23.

Kow, L.-M. and Pfaff, D.W. (1988) Transmitter and peptide

actions on hypothalamic neurons in vitro: Implications for lordosis. *Brain Res. Bull.*, 20: 857–861.

Kuffler, S.W. (1980) Slow synaptic responses in autonomic ganglia and the pursuit of a peptidergic transmitter. *J. Exp. Biol.*, 89: 257–286.

Lee, W.S., Smith, M.S. and Hoffman, G.E. (1990) Luteinizing hormone-releasing hormone neurons express Fos protein during the proestrous surge of luteinizing hormone. *Proc. Natl. Acad. Sci. USA*, 87: 5163–5167.

Leng, G. (1999) Might oestrogen act directly on GnRH neurones? *J. Neuroendocrinol.*, 11: 323–324.

Lescheid, D.W., Terasawa, E., Abler, L.A., Urbanski, H.F., Warby, C.M., Millar, R.P. and Sherwood, N.M. (1997) A second form of gonadotropin-releasing hormone (GnRH) with characteristics of chicken GnRH-II is present in the primate brain. *Endocrinology*, 138: 5618–5629.

Levy, A., Lightman, S.L., Hoyland, J., Rawlings, S. and Mason, W.T. (1990) A gonadotropin-releasing hormone (GnRH) antagonist distinguishes 3 populations of GnRH analog-responsive cells in human and rat pituitary in vitro and produces an acute increase in intracellular Ca^{2+}-concentration without inducing gonadotropin secretion. *Mol. Endocrinol.*, 4: 678–684.

Liposits, Z. and Setalo, G. (1980) Descending luteinizing hormone-releasing hormone (LH-RH) nerve fibers to the midbrain of the rat. *Neurosci. Lett.*, 20: 1–4.

Lovick, T.A. (1991) Central nervous system integration of pain control and autonomic function. *News Physiol. Sci.*, 6: 82–86.

Ma, Y.J., Kelly, M.J. and Ronnekleiv, O.K. (1990) Pro-gonadotropin-releasing hormone (ProGnRH) and GnRH content in the preoptic area and the basal hypothalamus of anterior medial preoptic nucleus/suprachiasmatic nucleus-lesioned persistent estrous rats. *Endocrinology*, 127: 2654–2664.

Malik, K.F., Silverman, A.-J. and Morrell, J.I. (1991) Gonadotropin-releasing hormone messenger RNA in the rat: Distribution and neuronal content over the estrous cycle and after castration of males. *Anat. Rec.*, 231: 457–466.

Merchenthaler, I., Gorcs, T., Setalo, G., Petrusz, P. and Flerko, B. (1984) Gonadotropin-releasing hormone (GnRH) neurons and pathways in the rat brain. *Cell Tissue Res.*, 237: 15–29.

Miyamoto, K., Hasegawa, Y., Nomura, M., Igarashi, M., Kangawa, K. and Matsuo, H. (1984) Identification of the second gonadotropin-releasing hormone in chicken hypothalamus: Evidence that gonadotropin secretion is probably controlled by two distinct gonadotropin-releasing hormones in avian species. *Proc. Natl. Acad. Sci. USA*, 81: 3874–3878.

Moss, R.L. and Dudley, C.A. (1990) Differential effects of a luteinizing-hormone-releasing hormone (LHRH) antagonist analogue on lordosis behavior induced by LHRH and the LHRH fragment Ac-LHRH[5–10]. *Neuroendocrinology*, 52: 138–142.

Moss, R.L. and McCann, S.M. (1973) Induction of mating behavior in rats by luteinizing hormone-releasing factor. *Science (Wash.)*, 181: 177–179.

Murphy, A.Z. and Hoffman, G.E. (2001) Distribution of gonadal steroid receptor-containing neurons in the preoptic–periaqueductal gray–brainstem pathway: A potential circuit for the initiation of male sexual behavior. *J. Comp. Neurol.*, 438: 191–212.

Ogawa, S., Kow, L.-M. and Pfaff, D.W. (1992) Effects of lordosis-relevant neuropeptides on midbrain periaqueductal gray neuronal activity in vitro. *Peptides*, 13: 965–975.

Ogawa, S., Eng, V., Taylor, J., Lubahn, D.B., Korach, K.S. and Pfaff, D.W. (1998) Roles of estrogen receptor alpha gene expression in reproduction-related behaviors in female mice. *Endocrinology*, 139: 5070–5081.

Ortmann, O. and Diedrich, K. (1999) Pituitary and extrapituitary actions of gonadotrophin-releasing hormone and its analogues. *Hum. Reprod.*, 14: 194–206.

Parhar, I.S., Soga, T. and Sakuma, Y. (2000) Thyroid hormone and estrogen regulate brain region-specific messenger ribonucleic acids encoding three gonadotropin-releasing hormone genes in sexually immature male fish, *Oreochromis niloticus*. *Endocrinology*, 141: 1618–1626.

Petersen, S.L., McCrone, S., Keller, M. and Shores, S. (1995) Effects of estrogen and progesterone on luteinizing hormone-releasing hormone messenger ribonucleic acid levels: Consideration of temporal and neuroanatomical variables. *Endocrinology*, 136: 3604–3610.

Pfaff, D.W. (1973) Luteinizing hormone-releasing factor potentiates lordosis behavior in hypophysectomized ovariectomized female rats. *Science (Wash.)*, 182: 1148–1149.

Pfaus, J.G., Marcangione, C., Smith, W.J., Manitt, C. and Abillamaa, H. (1996) Differential induction of Fos in the female rat brain following different amounts of vaginocervical stimulation: Modulation by steroid hormones. *Brain Res.*, 741: 314–330.

Porkka-Heiskanen, T., Urban, J.H., Turek, F.W. and Levine, J.E. (1994) Gene expression in a subpopulation of luteinizing hormone-releasing hormone (LHRH) neurons prior to the preovulatory gonadotropin surge. *J. Neurosci.*, 14: 5548–5558.

Raab, H., Karolczak, M., Reisert, I. and Beyer, C. (1999) Ontogenetic expression and splicing of estrogen receptor-alpha and beta mRNA in the rat midbrain. *Neurosci. Lett.*, 275: 21–24.

Rajendren, G. (2001) Subsets of gonadotropin-releasing hormone (GnRH) neurons are activated during a steroid-induced luteinizing hormone surge and mating in mice: A combined retrograde tracing double immunohistochemical study. *Brain Res.*, 918: 74–79.

Rajendren, G. and Moss, R.L. (1993) The role of the medial nucleus of amygdala in the mating-induced enhancement of lordosis in female rats: the interaction with luteinizing hormone-releasing hormone neuronal system. *Brain Res.*, 617: 81–86.

Riskind, P. and Moss, R.L. (1983a) Effects of lesions of putative LHRH-containing pathways and midbrain nuclei on lordotic behavior and luteinizing hormone release in ovariectomized rats. *Brain Res. Bull.*, 11: 493–500.

Riskind, P. and Moss, R.L. (1983b) Midbrain LHRH infusions enhance lordotic behavior in ovariectomized estrogen-primed rats independently of a hypothalamic responsiveness to LHRH. *Brain Res. Bull.*, 11: 481–485.

Rissman, E.F., Alones, V.E., Craig-Veit, C.B. and Millam, J.R.

(1995) Distribution of chicken-II gonadotropin-releasing hormone in mammalian brain. *J. Comp. Neurol.*, 357: 524–531.

Saitoh, Y., Gibson, M.J. and Silverman, A.J. (1992) Targeting of gonadotropin-releasing hormone axons from preoptic area grafts to the median eminence. *J. Neurosci. Res.*, 33: 379–391.

Sakamoto, Y., Suga, S. and Sakuma, Y. (1993) Estrogen-sensitive neurons in the female rat ventral tegmental area: A dual route for the hormone action. *J. Neurophysiol.*, 70: 1469–1475.

Sakuma, Y. and Pfaff, D.W. (1979) Facilitation of female reproductive behavior from mesencephalic central gray in the rat. *Am. J. Physiol.*, 237: R278–R284.

Sakuma, Y. and Pfaff, D.W. (1980) LH-RH in the mesencephalic central grey can potentiate lordosis reflex of female rats. *Nature (London)*, 283: 566–567.

Sakuma, Y. and Pfaff, D.W. (1983) Modulation of the lordosis reflex of female rats by LHRH, its antiserum and analogs in the mesencephalic central gray. *Neuroendocrinology*, 36: 218–224.

Schiess, M.C., Dudley, C.A. and Moss, R.L. (1987) Estrogen priming affects the sensitivity of midbrain central gray neurons to microiontophoretically applied LHRH but not beta-endorphin. *Neuroendocrinology*, 46: 24–31.

Sherwood, N.M., Lovejoy, D.A. and Coe, I.R. (1993) Origin of mammalian gonadotropin-releasing hormones. *Endocrinol. Rev.*, 14: 241–254.

Siverman, A.-J. and Krey, L.C. (1978) The luteinizing hormone-releasing hormone (LH-RH) neuronal networks of the guinea pig brain. I. Intra- and extra-hypothalamic projections. *Brain Res.*, 157: 233–246.

Silverman, A.J., Jhamandas, J. and Renaud, L.P. (1987) Localization of luteinizing hormone-releasing hormone (LHRH) neurons that project to the median eminence. *J. Neurosci.*, 7: 2312–2319.

Sirinathsinghji, D.J. (1984) Modulation of lordosis behavior of female rats by naloxone, β-endorphin and its antiserum in the mesencephalic central gray: possible mediation via GnRH. *Neuroendocrinology*, 39: 222–230.

Sirinathsinghji, D.J.S., Whittington, P.E. and Audsley, A.R. (1986) Regulation of mating behaviour in the female rat by genadotropin-releasing hormone in the ventral tegmental area: Effects of selective destruction of the A10 dopamine neurones. *Brain Res.*, 374: 167–173.

Skynner, M.J., Sim, J.A. and Herbison, A.E. (1999a) Detection of estrogen receptor α and β messenger ribonucleic acids in adult gonadotropin-releasing hormone neurons. *Endocrinology*, 140: 5195–5201.

Skynner, M.J., Slater, R., Sim, J.A., Allen, N.D. and Herbison, A.E. (1999b) Promoter transgenics reveal multiple gonadotropin-releasing hormone-I-expressing cell populations of different embryological origin in mouse brain. *J. Neurosci.*, 19: 5955–5966.

Suga, S., Akaishi, T. and Sakuma, Y. (1997) GnRH inhibits neuronal activity in the ventral tegmental area of the estrogen-primed ovariectomized rat. *Neurosci. Lett.*, 228: 13–16.

Swanson, L.W. (1982) The projections of the ventral tegmental area and adjacent regions: A combined fluorescent retrograde tracer and immunofluorescence study in the rat. *Brain Res. Bull.*, 9: 321–353.

Takeo, T., Chiba, Y. and Sakuma, Y. (1993) Suppression of the lordosis reflex of female rats by efferents of the medial preoptic area. *Physiol. Behav.*, 53: 831–838.

Watson, R.E., Langub, M.C. and Landis, J.W. (1992) Further evidence that most luteinizing hormone-releasing hormone neurons are not directly estrogen-responsive: Simultaneous localization of luteinizing hormone-releasing hormone and estrogen receptor immunoreactivity in the guinea-pig brain. *J. Neuroendocrinol.*, 4: 311–317.

White, R.B., Eisen, J.A., Kasten, T.L. and Fernald, R.D. (1998) Second gene for gonadotropin-releasing hormone in humans. *Proc. Natl. Acad. Sci. USA*, 95: 305–309.

Zadina, J.E., Kastin, A.J., Fabre, L.A. and Coy, D.H. (1981) Facilitation of sexual receptivity in the rat by an ovulation-inhibiting analog of LHRH. *Pharmacol. Biochem. Behav.*, 15: 961–964.

I.S. Parhar (Ed.)
Progress in Brain Research, Vol. 141
© 2002 Elsevier Science B.V. All rights reserved

CHAPTER 22

Nutrition, reproduction, and behavior

Jennifer L. Temple, Emilie F. Rissman[*]

Program in Neuroscience and Department of Biology, University of Virginia, Charlottesville, VA 22903, USA

Introduction

The relationship between nutrition and reproduction has been demonstrated in species from every mammalian order, including primates (Wade et al., 1996). Successful pregnancy and lactation require tremendous amounts of energy. Ovulatory cycles, mating behavior, steroidogenesis, and lactation are inhibited by caloric restriction, especially in small mammals (reviewed in Schneider and Wade, 2000). This reproductive inhibition is caused by negative energy balance. For example, females who exercise excessively without a compensatory increase in food intake often become amenorrheic (Sanborn et al., 1982; Williams et al., 2001). Negative energy balance inhibits reproduction and available energy is allocated to immediate survival needs, such as maintaining body temperature, metabolism, and foraging (Bronson, 2000; Fig. 1). This insures that reproduction is timed to coincide with resource abundance, which is favorable to survival of offspring.

Interactions between nutrition and reproduction

Gonadotropin-releasing hormone

Reproductive inhibition in response to undernutrition correlates with alterations in GnRH pulsatility

---
[*] Correspondence to: E.F. Rissman, Department of Biochemistry and Molecular Genetics, University of Virginia Medical School, P.O. Box 800733, Charlottesville, VA 22908, USA. Tel.: +1-434-982-5611; Fax: +1-434-243-8433;
E-mail: rissman@virginia.edu.

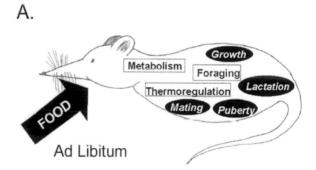

A.

Ad Libitum

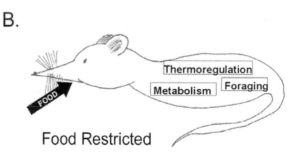

B.

Food Restricted

Fig. 1. A schematic drawing of a well-fed (A) and a food restricted (B) musk shrew. Words inside the squares are survival needs and italicized words inside the black ovals are reproductive activities. Notice that when food intake is limited, reproductive activities are suppressed so that available energy can be shifted to meet survival needs.

(reviewed in Schneider and Wade, 2000). Food restriction inhibits the frequency of pulsatile release of GnRH (I'Anson et al., 2000). This decrease in GnRH pulse frequency is correlated with an overall decrease in luteinizing hormone (LH) pulse frequency and a decrease in plasma LH levels (Bronson, 1988;

McGuire et al., 1996; Adam et al., 1997). When glycolysis is blocked, in place of food restriction, LH pulse frequency is also inhibited in sheep (Bucholtz et al., 1996). Conversely, the suppression of gonadotropin release and ovulatory cycles caused by food restriction can be reversed by pulsatile delivery of GnRH (Foster and Olster, 1985; Bronson, 1986; Armstrong and Britt, 1987; Cameron and Nosbisch, 1991; Cameron, 1996). While GnRH pulsatility appears to be one level of regulation, it is probably not the only aspect of GnRH function on which food restriction acts. For example, caloric restriction may affect GnRH production, pituitary sensitivity to GnRH, and/or steroid negative-feedback on GnRH.

In food restricted ewes maintained at a prepubertal weight (growth-restricted), a higher percentage of GnRH-immunoreactive (GnRH-ir) neurons are present in the medial basal hypothalamus (MBH) as compared to *ad libitum* fed lambs (I'Anson et al., 1997). This neuronal distribution is more similar to that of the fetal lamb than the adult. Because food restriction was imposed after birth, it is unlikely that it affected migration of GnRH neurons. Instead, it is likely that the MBH population of GnRH neurons is particularly susceptible to increased activation of inhibitory neurotransmitter pathways, and thus GnRH release is inhibited from cells in this region. A decline in GnRH release could have resulted in peptide accumulation in the GnRH cell bodies. Yet, there were no differences in GnRH content in the MBH between restricted and *ad libitum* fed ewes. Higher levels of GnRH might be expected if there was an accumulation of GnRH in MBH cells and terminals (Ebling et al., 1990). However, one problem with peptide content studies is the loss of cellular resolution. Thus, measurements of peptide content in regions containing both cells and fibers may not reveal more subtle changes in GnRH neurons. A recent study in male prairie voles showed that prolonged (1–3 weeks) food restriction increased GnRH-ir cell number in the POA, GnRH-ir soma size, and GnRH fiber density (Kriegsfeld et al., 2001). These data support the idea that GnRH release is inhibited by food restriction.

In addition to alterations in GnRH peptide content, nutritional manipulations may affect transcription of GnRH mRNA. Gruenewald and Matsumoto

(1993) used in situ hybridization to examine levels of preproGnRH mRNA expression in food deprived (60 hours) and *ad libitum* fed male rats. Food deprived rats had significantly fewer cells expressing GnRH mRNA, but the amount of GnRH mRNA per cell did not differ between the groups. While this study suggests that there are alterations in GnRH mRNA after food restriction, other studies have produced conflicting results. Nappi and Rivest (1997) found no differences in the level of GnRH gene expression per cell between fed and fasted cycling female rats. In fasted female rats GnRH mRNA levels correlated with the presence or absence of estrous cycling. Non-cycling females had lower levels of GnRH expression per cell. This result reveals a relationship between GnRH gene expression and estrous cycling that is not necessarily directly related to nutritional status. A review by Polkowska (1996) also suggests that, although food restriction delays puberty in ewes, when the GnRH-ir neurons are examined, their numbers are normal as is their morphology. Although these data are equivocal, the studies highlighted here suggest subtle effects of food restriction on GnRH production. In addition, differences in GnRH production and regulation after food restriction are species and sex-specific.

Luteinizing hormone

Downstream of GnRH, the HPG axis may be susceptible to caloric restriction via altered LH secretion. Release of LH can be influenced by food intake indirectly due to deficient signaling from the brain, or directly, due to effects on pituitary GnRH receptors, LH or FSH production, and/or gonadotropin release. Food restricted female animals tend to have lower plasma levels of LH and FSH than *ad libitum* fed animals (Bergendahl et al., 1989; Foster et al., 1989). In both men and women, GnRH treatment produced elevated levels of plasma LH and FSH in fasted, but not control, individuals (Rojdmark, 1987). Similarly, ovariectomized (OVX), steroid primed food restricted (60% of *ad lib* intake for 9–11 days) rats had higher plasma LH and FSH in response to low-dose administration of GnRH than *ad libitum* fed controls (Lively and Piacsek, 1988). This increase in pituitary sensitivity to GnRH is interesting and further supports the idea that the GnRH neuron pulse generator is the main level at which nutrition regulates fertility.

Another measure of food restriction effects on the HPG axis is pituitary LH and FSH content. Female rats maintained on a restricted diet (10 grams of food daily) for 60 days had lower pituitary LH and FSH content, lower serum LH, FSH and E_2 levels, and smaller and fewer gonadotropin positive pituitary cells compared to *ad libitum* fed controls (Kotsuji et al., 1992). These effects were reversed by daily administration of GnRH. Because GnRH corrected the effects of food restriction, it is likely that the primary deficit is in endogenous GnRH production and/or release. Food restricted male rats (60% of *ad libitum* for 6 weeks) had lower anterior pituitary content of LHβ and FSHβ mRNAs than *ad libitum* fed controls (Hans et al., 1998). Shorter duration of underfeeding (four days instead of six weeks) produced a 42% decrease in pituitary content of the gonadotropin alpha-subunit as compared to *ad libitum* fed controls, but no difference in either of the β subunits were noted (Bergendahl et al., 1989). These contradictory findings could be related to differences in the severity and duration of the food restriction. However, despite the discrepancies in the data, both studies showed that food restriction led to a decrease in pituitary gonadotropin mRNAs. In sum, nutritional manipulations affect multiple levels of the HPG axis, from GnRH production and release to gonadotropin subunit expression and production.

Negative feedback

Negative feedback of steroid hormones on the HPG axis is crucial for its proper functioning. Estradiol and testosterone (T) act at both the pituitary and the hypothalamus to decrease LH release, presumably by decreasing GnRH secretion (reviewed in Brown, 1994). Food restriction can increase sensitivity to steroid negative feedback. Food deprivation for 48 hours in cycling rats decreased plasma LH and lowered the frequency and amplitude of the LH pulse (Cagampang et al., 1990). This same manipulation had no effect in ovariectomized rats, even when the food deprivation was extended to 72 hours (Cagampang et al., 1990). A similar effect was reported in gonad-intact male rats, where food restriction to 50% of *ad libitum* for 7 days decreased LH levels compared to *ad libitum* fed controls. The same restriction paradigm had no effect in castrated rats (Dong et al., 1994). These data suggest that food restriction only inhibits the GnRH pulse generator when gonads are intact. It is possible that the initial release from negative feedback following gonadectomy produces such high levels of LH that acute food restriction is not sufficient to produce notable differences. Bergendahl and Hutaneiemi (1994) found that a 5-day fast suppressed LH secretion in sham-operated and acutely castrated rats, but castration 2 weeks prior to food deprivation stimulated LH release. Thus, when gonads are intact, the negative effects of food deprivation dominate, but without steroids present, food deprivation may have stimulatory effects.

Another important component of steroid negative feedback is estrogen action via neural estrogen receptors. Li et al. (1994) demonstrated that ovariectomized hamsters fasted for 48 hours had fewer estrogen receptor immunoreactive (ER-ir) cells in the ventromedial nucleus of the hypothalamus (VMN) and more ER-ir cells in the POA relative to *ad libitum* fed controls. In rats the same effect of acute food deprivation was noted in the VMN (Marin-Bivens et al., 2000). This decrease in ER-ir cells in the VMN probably causes the decrease in lordosis duration after food deprivation, which persists even in the presence of exogenous E_2 (Dickerman et al., 1993). In the POA, an increase in ER-ir cell number may, in part, mediate the increased sensitivity of GnRH to steroids, but at this point, this remains speculation (Li et al., 1994).

Mating behavior

Mating behavior may be the most important level of regulation for reproduction. An animal must mate in order to reproduce, thus, inhibiting mating behavior is the ultimate way to suppress reproduction during unfavorable environmental conditions. In species with estrous cycles, behavioral inhibition can be indirect, via decreases in circulating steroid hormone levels. However, the neural mechanisms, such as regulation of steroid receptor containing cells, that control mating behavior can also be influenced by undernutrition. In food deprived female Syrian hamsters, the sexual posture, lordosis, can be elicited by treatment with estradiol and progesterone, however

306

the duration of lordosis is significantly shortened by food deprivation relative to *ad libitum* fed females (Dickerman et al., 1993). This same behavioral inhibition can be elicited by pharmacological blockade of metabolic pathways in place of food deprivation. In ovariectomized female hamsters treated with estradiol and progesterone, female mating behavior is disrupted when glycolysis and fatty acid oxidation are blocked simultaneously (Dickerman et al., 1993; Jones and Lubbers, 2001). Perhaps exogenous steroid hormone treatment cannot completely reverse the effects of metabolic fuel limitation on mating behavior because of the simultaneous decrease in estrogen receptor containing cells in the VMN. Thus, in addition to affecting various aspects of the HPG axis, undernutrition also inhibits mating behavior.

Gaps in our knowledge

One important gap in the literature reviewed is the identity of the primary signals from food that trigger the cascade of downstream events that eventually culminate in adjustments in reproductive status to match available energy (Schneider and Wade, 2000). There are many putative factors ranging from metabolic fuels and gut peptides to hormones and neurotransmitters. Some appear to be more relevant than others and there is variation among species in their relative importance (reviewed in Kalra and Kalra, 1996). In addition to not knowing what the signal is, there is also some question about the brain area(s) where these peripheral signals act. The hypothalamus is obviously one area that is affected by undernutrition and in addition to regulating sex behavior, this region plays a role in food intake, motor activity, and ingestive behavior (Williams et al., 2000). Perhaps the hypothalamus is directly sensitive to nutritional status, however, there may be other brain regions, such as the hindbrain, that detect signals reflecting metabolic fuel availability first and then relay information to the hypothalamus (Schneider and Zhu, 1994; Schneider et al., 1995; Grill et al., 1998).

Another aspect of nutrition and reproduction that has not been explored is the GnRH pulse generator. Many of the studies mentioned here show that GnRH pulsatility is affected by undernutrition. This could be due to changes in GnRH production, in GnRH re-

lease, and/or GnRH pulse generation. One aspect yet to be explored is the effect that undernutrition may have on electrical activity of GnRH neurons. With the engineering of transgenic mice that have green fluorescent protein-labeled GnRH neurons; visualization and recording from these cells is simplified (Suter et al., 2000).

Many alterations in the GnRH system after nutritional manipulation have been well-documented. To date, however, no one has linked these changes to behavioral deficits. There is a large literature on extrapituitary effects of GnRH. Intracerebroventricular GnRH administration to ovariectomized/hypophysectomized female rats facilitates sexual receptivity (Pfaff, 1973; Moss and Foreman, 1976). In addition, only 6 amino acids from the GnRH decapeptide are necessary for these effects (Dudley and Moss, 1988; Fernandez-Fewell and Meredith, 1995). What remains to be determined is if the deficits in mating behavior are functionally related to changes in GnRH following food restriction.

The musk shrew

Musk shrews display mating-induced ovulation. Our laboratory has been working with this animal model for over 15 years in order to study environmental regulation of reproduction. Two of the primary stimuli that affect reproduction in this species are mating (Rissman, 1992) and nutritional status (Gill and Rissman, 1996; Temple and Rissman, 2000a). Because both puberty and ovulation are induced by mating, we can precisely control these two neuroendocrine events and study how they are influenced by food manipulations.

Musk shrews have several characteristics that allow us to answer questions that would be difficult using a rodent model. First, they are induced ovulators and do not display any hormonal or behavioral estrous cycles. Instead females usually mate and ovulate whenever they come in contact with a male (Dryden, 1969). Thus, we can use gonad-intact females for our studies because ovaries are in a quiescent state until multiple matings occur. Second, mating behavior is not dependent on high circulating levels of steroid hormones in females. Unlike rodents, plasma E_2 is nearly undetectable prior to mating in this species (Fortune et al., 1992). Instead, testosterone (T) is the primary steroid hormone secreted from musk shrew ovaries and adrenal glands (Rissman and Bronson, 1987). Although ovaries and adrenals must be intact for mating to occur, the levels of T do not change over a cycle as steroid hormones do in other species (Rissman and Bronson, 1987; Fortman et al., 1992). This offers two distinct advantages. First, experimental manipulations can be applied without concern for disruption of cyclic secretion of hormones. This also allows us to look at direct effects of environmental factors on behavior, instead of indirect effects on behavior due to changes in plasma concentrations of steroid hormones. Second, musk shrews are very sensitive to small reductions in food intake and respond rapidly to reinstatement of food after a mild food restriction (Gill and Rissman, 1996; Temple and Rissman, 2000a). This allows us to examine primary metabolic signals that influence mating behavior and HPG axis function.

Behavioral regulation of GnRH

Mating-induced ovulation is a reproductive strategy that is used by females of many species. In most well-studied species, the female comes into estrus, which is characterized by high plasma levels of estradiol. Estrus is stimulated by environmental cues, such as photoperiod in ferrets and rabbits, or contact with male conspecifics in voles (reviewed in Spies et al., 1997). Once plasma estradiol levels are high, females will mate and mating induces ovulation. Mating induced ovulation in the musk shrew has several features that differ from the 'classic' case described above. First, mating is not dependent upon high levels of estradiol, thus female shrews do not come into estrus, per se (Rissman et al., 1987; Rissman and Crews, 1988). Instead, female musk shrews mate whenever they come in contact with a male, despite having undetectable levels of plasma estradiol (Rissman et al., 1987; Fortune et al., 1992). Second, musk shrews rarely become pregnant after their first mating bout (Clendenon and Rissman, 1990; Rissman, 1992). Instead, two or three matings, separated by at least 24 hours, are required to induce ovulation (Rissman, 1992).

Mating has a significant effect on GnRH production in the female musk shrew. GnRH content increases in forebrain nuclei 24–48 hours after the initial mating bout (Dellovade et al., 1995b; Fig. 2). In addition, if this mating bout does not result in ovulation, GnRH content remains high until the female copulates again. Once the female ovulates, GnRH content drops significantly, suggesting that GnRH has been released. In addition, the number of forebrain GnRH immunoreactive cells is increased after mating as well as after brief interactions with males that do not include mating (Dellovade and Rissman,

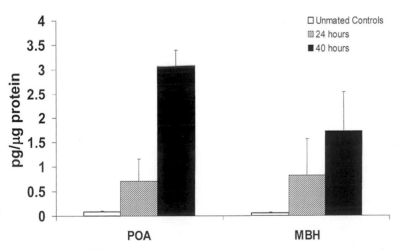

Fig. 2. Mean ± SEM amount of GnRH (pg/μg protein) in neural tissue taken from the preoptic area (POA) and the medial basal hypothalamus (MBH) from females who either received no mating or were sacrificed 24 or 40 hours after a single ejaculation was received. Redrawn from Dellovade et al., 1995b.

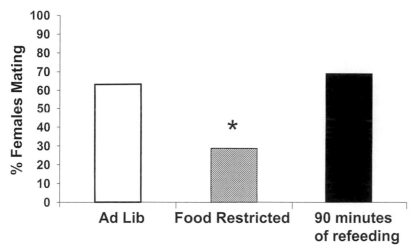

Fig. 3. Percentage of females receiving 5 placed intromissions during a mating test. Subjects were either fed *ad libitum* ($n = 16$), food restricted to 60% of *ad libitum* ($n = 15$), or food restricted to 60% of *ad libitum* and re-fed 90 minutes prior to the mating bout ($n = 15$). * = significantly different from other groups ($p < 0.05$). Redrawn from Temple and Rissman, 2000a.

1994; Dellovade et al., 1995a,b). These data suggest that mating directly regulates several aspects of GnRH system functioning, including GnRH production and/or release.

Nutritional regulation of mating behavior

Because mating behavior is integral for regulation of ovulation and GnRH in this species, and because it has been well-established in other species that successful reproduction depends on sufficient food availability, we investigated the consequences of nutritional manipulation on mating behavior in female musk shrews. In one study 48 hours of food-restriction significantly decreased the percentage of females displaying mating behavior (Gill and Rissman, 1996). Next, we focused on recovery of function after food restriction. Because female musk shrews do not rely on peak steroid hormone levels for mating, we hypothesized that restoration of mating behavior after food restriction would occur rapidly. We examined mating behavior in female shrews that were either fed *ad libitum*, food restricted to 60% of their average *ad libitum* food intake for 48 hours, or food restricted as just described, but given *ad libitum* access to food 90 minutes prior to mating. We found that food restriction significantly reduced the percentage of females that displayed mating behavior (Temple and Rissman, 2000a). More importantly,

this behavioral deficit was completely reversed by only 90 minutes of re-feeding (Fig. 3). These effects of nutritional manipulation on mating could have been due to either changes in circulating steroid hormone levels and/or changes in androgen receptor or estrogen receptor immunoreactive cell numbers. We have shown that neither of these is the case. Food restriction and re-feeding had no effect on plasma testosterone concentrations. Moreover, testosterone administration was ineffective in reversing the behavioral effects of food restriction (Temple and Rissman, 2000a).

Although plasma steroid hormone concentrations were not affected by nutritional manipulation, it is possible that neural steroid hormone receptors could have been. Food deprivation decreases the number of ER-ir cell in the VMN of female hamsters (Li et al., 1994). In order to explore this hypothesis, we counted the number of ER-ir and androgen receptor immunoreactive (AR-ir) cells in the POA and VMN of females that were either fed *ad libitum*, food restricted for 48 hours, or food restricted and re-fed for 90 minutes prior to sacrifice. There were no differences in any of these conditions in numbers of ER-ir or AR-ir cells in either the POA or VMN (Temple and Rissman, 1999; Figs. 4 and 5). These findings lead us to conclude that there is an alternative, non-steroidal mechanism by which nutritional manipulations affect mating behavior.

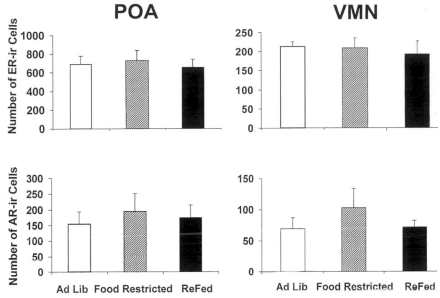

Fig. 4. Mean ± SEM number of estrogen receptor immunoreactive (ER-ir; top panel) and androgen receptor immunoreactive (AR-ir; bottom panel) cells in the preoptic area (POA; left panel) and the ventromedial nucleus of the hypothalamus (VMN; right panel). Females were either fed *ad libitum*, food restricted to 60% of *ad libitum*, or food restricted to 60% of *ad libitum* and re-fed 90 minutes prior to the mating bout (*n* = 8 per feeding condition). There were no significant differences among any of the feeding conditions in either area of the brain.

These data demonstrate that mating behavior can be influenced by acute metabolic fuel availability. The rapidity of behavioral recovery in this species is faster than anything reported in the literature. In addition, because these effects are occurring in the absence of changes in steroid hormone concentrations or neural steroid hormone receptors, the musk shrew is a good model species for the study of non-steroidal mediators of nutritional infertility. Because mating behavior is a salient, non-invasive measure of reproduction, this model is also excellent to assess primary signals from food that affect reproduction.

Metabolic cues and mating behavior

One unanswered question in this field is which primary signals, associated with food, initiate and support reproductive processes? We hypothesized that signals generated by the oxidation of metabolic fuels were being utilized to reinstate mating behavior in musk shrews. To test this hypothesis, we blocked individual metabolic fuel pathways during the 90 minute refeeding period and examined the effects on behavioral recovery of mating behavior. When

either glycolysis or fatty acid oxidation were inhibited individually, the refeeding-induced restoration of mating behavior was suppressed (Temple et al., 2002). Therefore both glycolysis and fatty acid oxidation are necessary for behavioral recovery. Next, either glucose or fat were given to food restricted females to determine whether either metabolic fuel substrate could support mating after food restriction. We found that neither fuel alone was sufficient (Temple et al., 2002). These data suggest that glucose and fatty acids are likely primary signals for mating behavior. In addition, because both fuels appear to be necessary simultaneously, it is likely that they are both acting to increase the levels of a common signal (such as ATP) that lies downstream of metabolic fuel oxidation.

Nutritional regulation of GnRH

To determine if nutritional status has direct effects on GnRH, we examined three aspects of GnRH system function in females from each of our feeding conditions (*ad libitum* fed, food restricted, and food restricted and re-fed for 90 minutes). First, we quan-

310

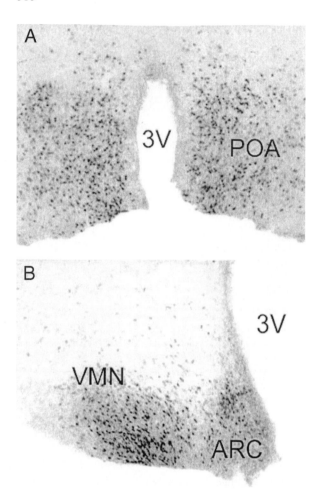

Fig. 5. Photomicrographs of estrogen receptor immunoreactive staining in the preoptic area (A) and the ventromedial nucleus of the hypothalamus (B). Photographs were taken at 20× magnification. 3V = third ventricle, POA = preoptic area, ARC = arcuate nucleus, VMN = ventromedial nucleus of the hypothalamus.

tified proGnRH immunoreactive neurons in several areas of the forebrain. Food restricted females had significantly more proGnRH-ir cells in the POA than either the *ad libitum* fed or re-fed females (Temple and Rissman, 2000b; Fig. 6). One interpretation of these data is that food restriction decreases the rate at which proGnRH is processed to the GnRH decapeptide, thus increasing the amount of peptide available for detection by immunocytochemistry. In addition, because the re-fed group of females had numbers of immunoreactive cells that were comparable to *ad libitum* fed females, we hypothesized that re-feeding

increases the rate of GnRH processing. There are at least four enzymatic processing steps involved in the conversion of proGnRH to the mature and functional decapeptide form (Wetsel et al., 1995). Each of these enzymes has the potential to modify GnRH abundance in response to nutritional signals. We have also examined the effects of nutritional status on GnRH content in the POA and in the medial basal hypothalamus. Levels of GnRH peptide are uniform in the POA regardless of feeding condition. Yet, food restricted females had a 30-fold increase in the amount of GnRH present in the MBH as compared to *ad lib* and re-fed females (Temple and Rissman, 2000b; Fig. 7). This suggests that food restriction inhibits GnRH release from its terminals, and that re-feeding rapidly reverses this effect. We also examined the effects of nutritional status on pituitary responsiveness to exogenous GnRH. Nearly 100% of the females in all feeding conditions ovulated 20–24 hours after GnRH treatment. However, food restricted females had significantly fewer corpora lutea, suggesting significantly fewer ova ovulated, as compared to *ad libitum* and re-fed females (Temple and Rissman, 2000b). In the musk shrew food restriction affects every level of the HPG axis that we examined. This is not surprising and, in fact, this corresponds very well with what we know from other species. What was surprising is that in this species, the HPG axis deficits brought on by food restriction were all reversed after 90 minutes of re-feeding.

Future directions

Currently, we are examining the link between the undernutrition-induced behavioral deficits and the GnRH system. We have shown that undernutrition decreases the availability and/or release of GnRH, and also inhibits mating behavior. Because GnRH has effects on mating behavior in both male and female mammals (Zadina et al., 1981; Fernandez-Fewell and Meredith, 1995), we have asked if these two consequences of food limitation are related. In the musk shrew, intracerebroventricular (i.c.v.) administration of mammalian GnRH facilitates sexual receptivity in females. Administration of a GnRH antagonist into the third ventricle of ovariectomized, steroid-primed rats inhibits lordosis (Dudley et al., 1981; Moss and Dudley, 1990). In addition, GnRH

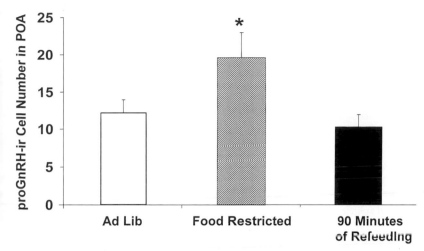

Fig. 6. The mean ± SEM number of proGnRH immunoreactive (proGnRH-ir) cells in the preoptic area (POA) of females who were either fed *ad libitum* ($n = 6$), food restricted to 60% of *ad libitum* ($n = 7$), or food restricted to 60% of *ad libitum* and re-fed 90 minutes prior to the mating bout ($n = 7$). * = significantly different from other groups ($p < 0.05$). Redrawn from Temple and Rissman, 2000b.

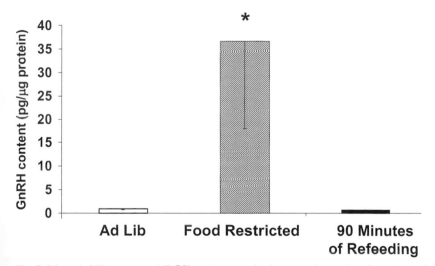

Fig. 7. Mean ± SEM amount of GnRH (pg/µg protein) in neural tissue taken from the medial basal hypothalamus (MBH) from females who were either fed *ad libitum* ($n = 11$), food restricted to 60% of *ad libitum* ($n = 11$), or food restricted to 60% of *ad libitum* and re-fed 90 minutes prior to the mating bout ($n = 13$). * = significantly different from other groups ($p < 0.05$). Redrawn from Temple and Rissman, 2000b.

administration to male hamsters can elicit mating behavior after the removal of the vomeronasal organ (Fernandez-Fewell and Meredith, 1995). We hypothesize that i.c.v. administration of GnRH to food restricted musk shrews will restore mating behavior. While several studies have shown that GnRH release and mating behavior are both affected by nutritional manipulation, ours would be the first to show a functional relationship between these two deficits.

Lessons to be learned from the musk shrew

The strength of the musk shrew model lies in our ability to precisely control the onset of puberty. Because of this, we can examine the effects of environmental stimuli on pubertal maturation with a high degree of precision. In terms of nutritional infertility, the advantages are two-fold. First, we can examine steroid-independent effects of nutrition on both mating behavior and on properties of GnRH neurons in gonad-intact females. In other species, food restriction decreases steroidogenesis, thus the kinds of studies we do would have to be done in ovariectomized, steroid-treated females. This type of manipulation is useful, but there are limitations on how results are interpreted and their ecological relevance. Second, because restoration of reproductive function occurs rapidly in musk shrews, we can examine the primary signals from food intake that affect mating behavior and the HPG axis during a finite time period. This is a problem that is unsolved in part because behavioral restoration after food restriction takes a minimum of 6 hours after re-feeding, even in steroid treated females (Jones and Lubbers, 2001). After this much time has passed, primary signals have already acted on multiple downstream targets.

Our data show that the musk shrew is a good model for examining how nutrition affects reproduction. Musk shrews are extremely sensitive to small and rapid fluctuations in food intake. In addition, because we can work with gonadally intact animals, we can examine the consequences of nutritional manipulation in a physiologically relevant manner. The studies highlighted here provide a comparative perspective on mechanisms underlying nutrition and reproduction interactions in mammals.

References

Adam, C.L., Findlay, P.A., Kyle, C.E., Young, P. and Mercer (1997) Effect of chronic food restriction on pulsatile luteinizing hormone secretion and hypothalamic neuropeptide Y gene expression in castrate male sheep. *J. Endocrinol.*, 152(2): 329–337.

Armstrong and Britt (1987) Nutritionally-induced anestrus in gilts: metabolic and endocrine changes associated with cessation and resumption of estrous cycles. *J. Anim. Sci.*, 65(2): 508–523.

Bergendahl, M. and Hutaneiemi, I. (1994) The time since castration influences the effects of short-term starvation on gonadotropin secretion in male rats. *J. Endocrinol.*, 143: 209–219.

Bergendahl, M., Perheentupa and Hutaniemi, I. (1989) Effect of short-term starvation on reproductive hormone gene expression, secretion and receptor levels in male rats. *J. Endocrinol.*, 121(3): 409–417.

Bronson, F.H. (1986) Food-restricted, prepubertal, female rats: Rapid recovery of luteinizing hormone pulsing with excess food, full recovery of pubertal development with

gonadotropin-releasing hormone. *Endocrinology*, 118: 2483–2487.

Bronson, F.H. (1988) Effect of food manipulation on the GnRH–LH–estradiol axis of young female rats. *Am. J. Physiol.*, 252: R140–R144.

Bronson, F.H. (2000) Puberty and energy reserves: A walk on the wild side. In: K. Wallen and J.E. Schneider (Eds.), *Reproduction in Context*. MIT Press, pp. 15–34.

Brown, R.E. (1994) An Introduction to Neuroendocrinology. Cambridge University Press, UK.

Bucholtz, D.C., Vidwans, N.M., Herbosa, C.G., Schillo, K.K. and Foster, D.L. (1996) Metabolic interfaces between growth and reproduction V. Pulsatile luteinizing hormone secretion is dependent on glucose availability. *Endocrinology*, 137: 601–607.

Cagampang, F.R.A., Maeda, K.I., Yokoyama, A., Ota, K. (1990) Effect of food deprivation on pulsatile LH release in the cycling and ovariectomized female rat. *Horm. Met. Res.*, 22: 269–272.

Cameron, J.L. (1996) Nutritional determinants of puberty. *Nutrit. Rev.*, 54: S17–S22.

Cameron, J.L. and Nosbisch, C. (1991) Suppression of pulsatile luteinizing hormone and testosterone secretion during short term food restriction in the adult male rhesus monkey (*Macaca mulatta*). *Endocrinology*, 128(3): 1532–1540.

Clendenon, A.L. and Rissman, E.F. (1990) Prolonged copulatory behavior facilitates pregnancy success in the musk shrew. *Physiol. Behav.*, 47: 831–835.

Dellovade, T.L. and Rissman, E.F. (1994) Gonadotropin-releasing hormone-immunoreactive cell numbers change in response to social interactions. *Endocrinology*, 134(5): 2189–2197.

Dellovade, T.L., Hunter, E. and Rissman, E.F. (1995a) Interactions with males promote rapid changes in gonadotropin-releasing hormone immunoreactive cells. *Neuroendocrinology*, 62: 385–395.

Dellovade, T.L., Ottinger, M.A. and Rissman, E.F. (1995b) Mating alters gonadotropin-releasing hormone cell number and content. *Endocrinology*, 136: 1648–1657.

Dickerman, R.W., Li, H.Y. and Wade, G.N. (1993) Decreased availability of metabolic fuels suppresses estrous behavior in Syrian hamsters. *Am. J. Physiol.*, 264: R568–R572.

Dong, O., Bergendahl, M., Huhtaniemi, I. and Handelsman, D.J. (1994) Effect of undernutrition on pulsatile luteinizing hormone (LH) secretion in castrate and intact male rats using an ultrasensitive immunofluorometric LH assay. *Endocrinology*, 135(2): 745–750.

Dryden, G.L. (1969) Reproduction in the *Suncus murinus*. *J. Reprod. Fertil. Supp.*, 6: 377–396.

Dudley, C.A. and Moss, R.L. (1988) Facilitation of lordosis in female rats by CNS-site specific infusions of an LH-RH fragment, Ac-LH-RH-(5–10). *Brain Res.*, 16; 441(1–2): 161–167.

Dudley, C.A., Vale, W., Rivier, J. and Moss, R.L. (1981) The effect of LHRH antagonist analogs and an antibody to LHRH on mating behavior in female rats. *Peptides*, 2(4): 393–396.

Ebling, F.J., Wood, R.I., Karsch, F.J., Vannerson, L.A., Suttie,

J.M., Bucholtz, D.C., Schall, R.E. and Foster, D.L. (1990) Metabolic interfaces between growth and reproduction. III. Central mechanisms controlling pulsatile luteinizing hormone secretion in the nutritionally growth-limited female lamb. *Endocrinology*, 126(5): 2719–2727.

Fernandez-Fewell, G.D. and Meredith, M. (1995) Facilitation of mating behavior in male hamsters by LHRH and AcLHRH5-10: interaction with the vomeronasal system. *Physiol. Behav.*, 57(2): 213–221.

Fortman, M., Dellovade, T.L. and Rissman, E.F. (1992) Adrenal contribution to the induction of sexual behavior in the female musk shrew. *Horm. Behav.*, 26: 76–86.

Fortune, J.E., Eppig, J.J. and Rissman, E.F. (1992) Mating stimulates estradiol production by ovaries of the musk shrew (*Suncus murinus*). *Biol. Reprod.*, 46: 885–891.

Foster, D.L. and Olster, D.H. (1985) Effect of restricted nutrition on puberty in the lamb: patterns of tonic luteinizing hormone (LH) secretion and competency of the LH surge system. *Endocrinology*, 116(1): 375–381.

Foster, D.L., Ebling, F.J.P., Micka, A.F., Bucholtz, D.C., Wood, R.I., Suttie, J.M. and Fenner, D.E. (1989) Metabolic interfaces between growth and reproduction I. Nutritional modulation of gonadotropin, prolactin, and growth hormone secretion on the growth-limited female lamb. *Endocrinology*, 138: 855–858.

Gill, C.J. and Rissman, E.F. (1996) Female sexual behavior is inhibited by short- and long-term food restriction. *Physiol. Behav.*, 61(3): 387–394.

Grill, H.J., Ginsberg, A.B., Seeley, R.J. and Kaplan, J.M. (1998) Brainstem application of melanocortin receptor ligands produces long-lasting effects on feeding and body weight. *J. Neurosci.*, 18(23): 10128–10135.

Gruenewald, D.A. and Matsumoto, A.M. (1993) Reduced gonadotropin-releasing hormone gene expression with fasting in the male rat brain. *Endocrinology*, 132: 480–482.

Hans, E.S., Lu, D.H. and Nelson, J.F. (1998) Food restriction differentially affects mRNAs encoding the major anterior pituitary tropic hormones. *J. Gerontol. A Biol. Sci. Med. Sci.*, 53(5): B322–B329.

I'Anson, H., Terry, S.K., Lehman, M.H. and Foster, D.L. (1997) Regional differences in the distribution of gonadotropin-releasing hormone cells between rapidly growing and growth-restricted prepubertal female sheep. *Endocrinology*, 138: 230–236.

I'Anson, H., Manning, J.M., Herbosa, C.G., Pelt, J., Friedman, J.R., Wood, R.I., Bucholtz, D.C. and Foster, D.L. (2000) Central inhibition of gonadotropin-releasing hormone secretion in the growth-restricted hypogonadotropic female sheep. *Endocrinology*, 141(2): 520–527.

Jones, J.E. and Lubbers, L.S. (2001) Suppression and recovery of estrous behavior in Syrian hamsters after changes in metabolic fuel availability. *Am. J. Physiol. Regulatory Integrative Comp. Physiol.*, 280: R1393–R1398.

Kalra, S.P. and Kalra, P.S. (1996) Nutritional infertility: the role of the interconnected hypothalamic neuropeptide Y-galanin-opioid network. *Front. Neuroendocrinol.*, 17: 371–401.

Kotsuji, F., Hosokawa, K. and Tominaga, T. (1992) Daily administration of gonadotrophin-releasing hormone increases pituitary gonadotroph number and pituitary gonadotrophin content, but not serum gonadotrophin levels, in female rats on day 1 of dioestrus. *J. Endocrinol.*, 132(3): 395–400.

Kriegsfeld, L.J., Ranalli, N.J., Trasy, A.G. and Nelson, R.J. (2001) Food restriction affects the gonadotropin releasing hormone neuronal system of male prairie voles (*Microtus ochrogaster*). *J. Neuroendocrinol.*, 13(9): 791–798.

Li, H.Y., Wade, G.N. and Blaustein, J.D. (1994) Manipulations of metabolic fuel availability alter estrous behavior and neural estrogen receptor immunoreactivity in Syrian hamsters. *Endocrinology*, 135(1): 240–247.

Lively, K.M. and Piacsek, B.E. (1988) Gonadotropin-releasing hormone sensitivity in underfed prepubertal female rats. *Am. J. Physiol.*, 255: E482–E487.

Marin-Bivens, C.L., Jones, J.E. and Wade, G.N. (2000) Acute fasting decreases sexual receptivity and estrogen receptor-α immunoreactivity in adult female rats. *Soc. Neurosci. Abstracts*, 472.7.

McGuire, M.K., Myers, T.R., Butler, W.R. and Rasmussen, K.M. (1996) Naloxone administration does not relieve the inhibition of gonadotropin release in food-restricted lactating rats. *J. Nutr.*, 126: 2113–2119.

Moss, R.L. and Dudley, C.A. (1990) Differential effects of a luteinizing-hormone-releasing hormone (LHRH) antagonist analogue on lordosis behavior induced by LHRH and the LHRH fragment Ac-LHRH5-10. *Neuroendocrinology*, 52(2): 138–142.

Moss, R.L. and Foreman, M.M. (1976) Potentiation of lordosis behavior by intrahypothalamic infusion of synthetic luteinizing hormone-releasing hormone. *Neuroendocrinology*, 20(2): 176–181.

Nappi, R.E. and Rivest, S. (1997) Effect if immune and metabolic challenges on the luteinizing hormone-releasing hormone neuronal system in cycling female rats: an evaluation at the transcriptional level. *Endocrinology*, 138: 1374–1384.

Pfaff, D.W. (1973) Luteinizing hormone-releasing factor potentiates lordosis behavior in hypophysectomized ovariectomized female rats. *Science*, 14; 182(117): 1148–1149.

Polkowska, J. (1996) Stress and nutritional influences on GnRH and somatostatin neuronal systems in the ewe. *Acta Neurobiol. Exp.*, 56: 797–806.

Rissman, E.F. (1992) Mating induces puberty in the female musk shrew. *Biol. Reprod.*, 47: 473–477.

Rissman, E.F. and Bronson, F.H. (1987) Role of the ovary and adrenal gland in the sexual behavior of the musk shrew, *Suncus murinus. Biol. Reprod.*, 36: 664–668.

Rissman, E.F. and Crews, D. (1988) Hormonal correlates of sexual behavior in the female musk shrew: The role of estradiol. *Physiol. Behav.*, 44: 1–7.

Rissman, E.F., Nelson, R.J., Blank, J.L. and Bronson, F.H. (1987) Reproductive response of a tropical mammal the musk shrew (*Suncus murinus*) to photoperiod. *J. Reprod. Fertil.*, 81: 563–566.

Rojdmark, S. (1987) Influence of short-term fasting on the pituitary–testicular axis in normal men. *Horm. Res.*, 25: 140–146.

Sanborn, C.F., Martin, B.J. and Wagner, W.W. (1982) Is ath-

314

letic amenorrhea specific to runners? *Am. J. Obstet. Gynecol.*, 143(8): 859–861.

Schneider, J.E. and Wade, G.N. (2000) Inhibition of reproduction in service of energy balance. In: K. Wallen and J.E. Schneider (Eds.), *Reproduction in Context*. MIT Press, pp. 35–82.

Schneider, J.E. and Zhu, Y. (1994) Caudal brain stem plays a role in metabolic control of estrous cycles in Syrian hamsters. *Brain Res.*, 661(1–2): 70–74.

Schneider, J.E. et al. (1995) Glucoprivic treatments that induce anestrus, but do not affect food intake, increase FOS-like immunoreactivity in the area postrema and nucleus of the solitary tract in Syrian hamsters. *Brain Res.*, 698(1–2): 107–113.

Spies, H.G., Pau, K.Y. and Yang, S.C. (1997) Coital and estrogen signals: a contrast in the preovulatory neuroendocrine networks of rabbits and rhesus monkeys. *Biol. Reprod.*, 56(2): 310–319.

Suter, K.J., Wuarin, J.P., Smith, B.N., Dudele, F.E. and Moenter, S.M. (2000) Whole-cell recordings from preoptic/hypothalamic slices reveal burst firing in gonadotropin-releasing hormone neurons identified with green fluorescent protein in transgenic mice. *Endocrinology*, 141(10): 3731–3736.

Temple, J.L. and Rissman, E.F. (1999) Effects of food restriction and brief refeeding on proGnRH immunoreactivity and on estrogen receptor immunoreactivity in the female musk shrew (*Suncus murinus*). *Soc. Neurosci. Abstracts*, 479.19.

Temple, J.L. and Rissman, E.F. (2000a) Brief refeeding restores

reproductive redness in food restricted female musk shrews (*Suncus murinus*). *Horm. Behav.*, 38: 21–28.

Temple, J.L. and Rissman, E.F. (2000b) Acute refeeding reverses food-restriction-induced hypothalamic–pituitary–gonadal axis deficits. *Biol. Reprod.*

Temple, J.L., Schneider, J.E., Scott, D.K., Korutz, A. and Rissman, E.F. (2002) Mating behavior is regulated by acute changes in metabolic fuels. *Am. J. Physiol. Regul. Integr. Comp. Physiol.*, 282(3): R782–R790.

Wade, G.N., Schneider, J.E. and Li, H.Y. (1996) Control of fertility by metabolic cues. *Am. J. Physiol.*, 270: E1–E19.

Wetsel, W.C., Liposits, Z., Seidah, N.G. and Collins, S. (1995) Expression of candidate pro-GnRH processing enzymes in rat hypothalamus and an immortalized hypothalamic neuronal cell line. *Neuroendocrinology*, 62(2): 166–177.

Williams, G., Harrold, J.A. and Cutler, D.J. (2000) The hypothalamus and the regulation of energy homeostasis: lifting the lid on a black box. *Proc. Nutr. Soc.*, 59(3): 385–396.

Williams, N.I., Caston-Balderrama, A.L., Helmreich, D.L., Parfitt, D.B., Nosbisch, C. and Cameron, J.L. (2001) Longitudinal changes in reproductive hormones and menstrual cyclicity in cynomolgus monkeys during strenuous exercise training: abrupt transition to exercise-induced amenorrhea. *Endocrinology*, 142(6): 2381–2389.

Zadina, J.E., Kastin, A.J., Fabre, L.A. and Coy, D.H. (1981) Facilitation of sexual receptivity in the rat by an ovulation-inhibiting analog of LHRH. *Pharmacol. Biochem. Behav.*, 15(6): 961–964.

I.S. Parhar (Ed.)
Progress in Brain Research, Vol. 141
© 2002 Published by Elsevier Science B.V.

CHAPTER 23

GnRH, brain mast cells and behavior

Ann-Judith Silverman [1,*], Lori Asarian [3], Mona Khalil [2], Rae Silver [1,3,4]

[1] *Department of Anatomy and Cell Biology, and* [2] *Department of Biochemistry and Molecular Biophysics, Columbia University College of Physicians and Surgeons, New York, NY 10032, USA*
[3] *Department of Psychology, Columbia University, and* [4] *Department of Psychology, Barnard College, New York, NY 10027, USA*

Introduction

Mast cells, potent unicellular migratory glands, are found in all tissues and organs of the body including the brain and reproductive tract. In this chapter we will show that GnRH is found in mast cell granules. We will also present evidence that the brain mast cell population increases under specific behavioral and endocrine states, including those associated with reproduction. Where pertinent we will also discuss other mast cell–neuroendocrine interactions.

Mast cells are derived from the haematopoietic stem cell of the bone marrow. They circulate as committed precursors and undergo their final phenotypic differentiation in a tissue specific manner. Progenitors of mast cells contain cytoplasmic granules, and express RNAs encoding mast cell-associated proteases, but lack expression of the high-affinity immunoglobulin E receptor (Rodewald et al., 1996). In the ring dove, immature mast cells were identified at the ultrastructural level as early as embryonic day 15. At this age, the cells are present in the pia adjacent to the medial habenula. They have fewer granules which are more electron lucent compared to granules in mature cells (Fig. 1A and B). The staining properties of mast cells change with maturation, reflecting changes in the content of the granules (Zhuang et al.,

1999). Mast cells are highly plastic in that they can change phenotype when transplanted from one environment to another (e.g., from gut mucosa to connective tissue; Kitamura et al., 1987a,b). Depending on their place of residence, alterations in environmental signals and activational state, mast cells store or produce upon activation, a wide range of mediators. Stored classes of mediators include: biogenic amines, bioactive peptides (including gonadotropin releasing hormone (GnRH)), mast cell specific serine proteases, and heparin sulfated proteoglycans. Interestingly, mast cells are the only source of heparin in the body (Humphries et al., 1999). Upon stimulation by immune or non-immune signals they can make and release prostaglandins, leukotrienes, cytokines and growth factors to list but a few of the potential categories (reviewed in Wedemeyer et al., 2000).

Mast cells are long-lived cells and can undergo many rounds of activation, involving degranulation (Fig. 2, see page 319) and replenishment (Dvorak, 1989; Wilhelm et al., 2000; Xiang et al., 2001). Mast cells can release mediators by piecemeal degranulation (Kops et al., 1990; Dvorak, 1992; Wilhelm et al., 2000) resulting in differential release of granular contents. Mast cells also degranulate by compound exocytosis. This involves the fusion of the secretory granule with the plasma membrane, followed by the subsequent fusion of other granules to the first, resulting in the rapid release of soluble granular contents. When this occurs the soluble mediators (e.g., 5-HT) and about 25% of the sulfated proteoglycans diffuse out into the extracellular space. The remain-

---
*Correspondence to: A.-J. Silverman, Department of Anatomy and Cell Biology, Columbia University College of Physicians and Surgeons, New York, NY 10032, USA.

ing material is an insoluble core of heparin sulfated proteoglycans to which the mast cell specific neutral proteases such as chymase (a chymotryptic endopep-

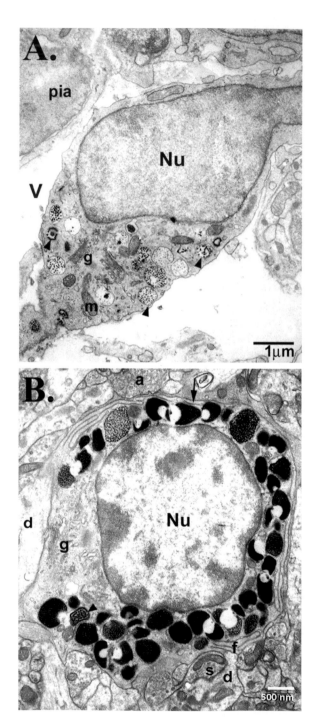

tidase) and tryptase (a trypsin-like exopeptidase) are attached (Kaartinen et al., 1995). Molecules in the mast cell's environment that can bind to heparin and/or to the proteoglycan can be captured by the mast cell, as is the case for such diverse substances as low density lipoprotein (LDL) (Kovanen, 1991) and the chemotherapeutic drug, adriamycin (Crivellato et al., 1997).

Historically there was considerable controversy over the existence of brain mast cells, particularly in the normal human brain (Olsson, 1968). However, mast cells are morphologically and histochemically distinct from other cell types. The occurrence of mast cells in the brain is now indisputable (Dropp, 1976; Ibrahim et al., 1979; Theoharides, 1990; Silver et al., 1996), and a detailed analysis of optimal conditions for visualization of mast cells is available (Florenzano and Bentivoglio, 2000). We have examined the staining properties of brain mast cells in rats, and found that mast cells can be identified with the following 3 markers we have used: serotonin, heparin, and toluidine blue (Fig. 3, see page 319).

GnRH and brain mast cells

Multiple forms of the GnRH decapeptide have been found in the brains of vertebrates. These include the mammalian form, mGnRH (Schally et al., 1971) produced in neurons originating from the mammalian olfactory placode (Schwanzel-Fukuda and Pfaff, 1989; Wray et al., 1989), chicken GnRH-I (cGnRH-I) which differs at position 8 in which glutamine is substituted for an arginine (Gln[8]) (avian placodal form), and chicken GnRH-II (cGnRH-II)

Fig. 1. Electron micrographs of dove brain mast cells. (A) Mast cell in the pia near the choroid plexus of the third ventricle (V) from an E15 dove embryo. Note the elongate shape indicative of cellular migration. Immature granules, with particulate interiors of varying architecture, are indicated with arrowheads. These granules do not contain GnRH nor do they stain with acidic toluidine blue. Nu = nucleus; g = Golgi apparatus. Scale bar = 1 μm. (B) By one month of age, the mast cell granules (arrow) are more mature. They are mostly electron dense with an electron lucent cap typical of avian mast cells. Most cells however do not make GnRH at this time. Note that the mast cell is fully integrated into the brain's Nu = nucleus; g = Golgi apparatus. Scale bar = 500 nm.

which contains amino acids $His^5–Trp^7–Tyr^8$. Antibody absorption studies in dove brain mast cells indicated that mast cell GnRH-like immunoreactivity was abolished when an antibody for mammalian GnRH (LR-1) is preabsorbed with either cGnRH-I or cGnRH-II, while GnRH neurons staining is abolished by cGnRH-I but not cGnRH-II. This suggested that in the dove, the GnRH in mast cells might be a different peptide than that found in GnRH producing neurons.

To test if mast cells could accumulate GnRH via endocytosis, in vivo radiolabeled GnRH was infused into the CSF of the dove (Silverman et al., 1994). Mast cells in the pia mater contacting the CSF as well as those in the medial habenula which borders the CSF were not labeled. The negative data are not likely due to a deficiency in endocytosis as mast cells are capable of endocytosis of many substances including synthetic beads, bacteria, and LDL both in vivo and in vitro (Bacteria — Shin et al., 2000; LDL — Wang et al., 1995). In fact, GnRH immunoreactivity in dove brain mast cells is localized to the Golgi apparatus as well as the secretory granules suggesting *de novo* synthesis (Fig. 4). As noted below, rat mast cells contain GnRH (Fig. 5, see page 319) and preliminary studies in the laboratory suggest that they express GnRH mRNA. We are exploring whether the gene transcript produced is the same as the one found in the hypothalamus.

The presence of GnRH in mast cells is of interest in that the decapeptide has multiple functions besides regulating the pituitary–gonadal axis. It is known to potentiate sexual behavior (Moss and McCann, 1973; Pfaff, 1973) and may function as a neurotransmitter within the brain (Chen et al., 1993). Depending on the species GnRH axons can be found in many brain regions (e.g., amygdala) outside of the septo-hypothalamic pathway to the median eminence (Silverman, 1988). GnRH can also act as a paracrine/autocrine factor and as an immunomodulator (reviewed in Marchetti et al., 1996). Given that mast cells degranulate in the normal brain it is possible that GnRH from this cellular source is active in the CNS.

Other immune cells also synthesize bioactive GnRH peptide and mRNA (Emanuele et al., 1990; Azad et al., 1991; Maier et al., 1992) and GnRH-receptor mRNA (Weesner et al., 1997; Chen et al., 1999). GnRH has a stimulatory effect on lymphocyte proliferation (Marchetti et al., 1989; Morale et al., 1991) and GnRH signaling at the lymphocyte stimulates expression of the interleukin-2 receptor (Batticane et al., 1991; Chen et al., 1999). In addition, GnRH antagonist administration to mice prone to lupus erythematosus decreases antibody production and hematuria, and increases the animal's life span, suggesting that this autoimmune disease is, in part, attributable to excessive expression of immune cell GnRH (Jacobson et al., 1994). Interestingly GnRH and GnRH agonists and antagonists can induce histamine secretion from mast cells, leading to cutaneous anaphylaxis (Rivier et al., 1986; Phillips et al., 1988; Sundaram et al., 1988). This was a clinical problem in the treatment of precocious puberty, which was solved by the redesign of GnRH analogues (Xiang et al., 2001).

Brain mast cells change with the animal's state

The central nervous system with its formidable blood-tissue barrier is an immunologically privileged site. However, this does not mean that the adult brain escapes immune surveillance. For example, activated T cells enter the normal CNS without precipitating an inflammatory response (Hickey, 2001). Similarly, mast cells enter the normal CNS during development (Lambracht-Hall et al., 1990; Zhuang et al., 1999) and new recruitment can be elicited under normal physiological conditions (Zhuang et al., 1993; Yang et al., 1999). Unlike T cells that apparently enter at random sites in the CNS (Hickey, 2001), mast cells are located in specific brain regions in a species specific manner and, as described below, alterations in their number also occur in specific brain regions and nuclei within those regions.

Brain mast cells and reproduction

A correlation between brain mast cells and reproduction was noted by Flood and Kruger (1970) when they found an increase in the number of mast cells (identified by their metachromatic properties to acidic aniline dyes such as toluidine blue) in the brains of hedgehogs during hibernation, a time of gonadal regression, compared to the awake state. Observations from our laboratory made the first direct connection between mast cell number, their activation state, and the animal's reproductive status. In the ring dove, mast cells are found in several brain regions but are concentrated in the epithalamus (medial habenula) (Zhuang et al., 1999). These were first identified by the presence of GnRH-like immunoreactivity (Silver et al., 1992); mast cell identity was later confirmed by the presence of the biogenic amine, histamine, and by ultrastructural characteristics, especially granular morphology and nuclear substructure (Silverman et al., 1994).

In subsequent experiments in the ring dove we concentrated our observations on the medial habenula as mast cell number and activation are reproducibly associated with sex behavior or sex steroids. Courtship in the ring dove activates ovarian follicular development and estrogen–progesterone secretion in females and stimulates androgen production in males (Feder et al., 1977; Silver et al., 1980). Within two hours of pairing with an adult female, GnRH positive mast cells increase dramatically in number in the medial habenula from ~300 to about ~1200 (Zhuang et al., 1993). Similar observations were made in females though the number of brain mast cells was more variable among females than among males (unpublished data from the laboratory). In males the lowest number of mast cells in the habenula was found in castrated birds (Zhuang et al., 1993). Treatment of birds with silastic capsules containing gonadal steroids (testosterone or dihydrotestosterone

in males, estradiol 17-β in females) also resulted in an elevation in mast cell numbers in the medial habenula (Wilhelm et al., 2000). These findings indicate that the mast cell population in the habenula is elevated by gonadal hormones even in the absence of sexual behavior. There was no sexually dimorphism in the response. There was no difference between testosterone and dihydrotestosterone-treated males indicating that the elevation of mast cell numbers did not require aromatization of androgen in the male. Although the medial habenula has been associated with arousal, its role in reproduction per se is unknown. Nonetheless, specific behaviors and hormonal conditions create states favorable for 'entry' of mast cells into this nucleus (see below).

An increase in the adult brain mast cell number is not confined to avian species. Yang et al. (1999) found that mating followed by cohabitation with females resulted in an increase in mast cells positive for GnRH in the brain of the male mice, with the largest increase observed when males were housed with females for 15 to 19 days post-coitum. In the mouse, mast cells are primarily localized at perivascular sites within selected thalamic nuclei including the ventral posterior, lateral dorsal, central medial and lateral and medial geniculate.

As in mice, prolonged pairing of male rats with estrogen–progesterone primed ovariectomized females resulted in elevation of brain mast cell number in specific thalamic nuclei. Mast cells increased in

Fig. 2. These three rat thalamic mast cells were all stained with acidic toluidine blue which reacts with the sulfated proteoglycans of the granules. In (A) this resting (non-stimulated) cell is uniformly stained and the color is a deep blue/purple. This indicates that the cell is filled with granules. In (B) this activated cell has released granules into the neuropil (arrowheads). In (C) the cell is devoid of mature granules (blue/purple) and contains much paler staining (pale purple/pink) granules. This change in color is due to a loss of the proteoglycans upon exocytosis. The large accumulation of stained material (arrow) is most likely the perinuclear Golgi where packaging of new granules is taking place as the cell recovers. Scale bar = 8 μm.

Fig. 3. (A) In this section of the thalamus, the mast cells were double-labeled for 5-HT (red) and heparin (revealed by avidin-Cy3 binding; green). The overlay indicates that both products are in granules in each cell. (B) Next the coverslip was removed and the section stained for TB (B). All the mast cells are visualized with the 3 markers used. Scale bar = 30 μm.

Fig. 5. Double-label immunocytochemistry of brain mast cells in the rat thalamus using antisera directed against mammalian GnRH and 5-HT. Each micrograph is a projection of 7 serially collected 1 μm optical sections. (A) mast cells with GnRH immunoreactivity (red); (B) mast cells with 5-HT immunoreactivity (green); (C) overlay. Scale bar = 10 μm.

Fig. 6. In this micrograph the red fluorescence reveals glial astrocytic fibrillary protein (red) in processes associated with a large thalamic blood vessel. These processes are the last cellular element of the blood brain barrier. The cells (arrow) were isolated from rat peritoneal flush and labeled with CellTracker Green and injected intravenously into a host animal. One hour later they are found in the thalamus on the brain side the blood brain barrier. Scale bar = 25 μm.

319

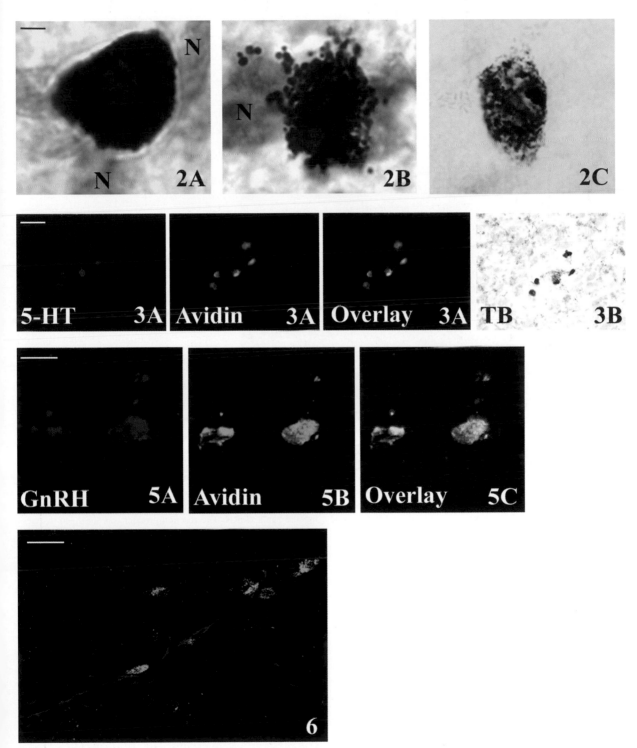

Figs. 2, 3, 5 and 6.

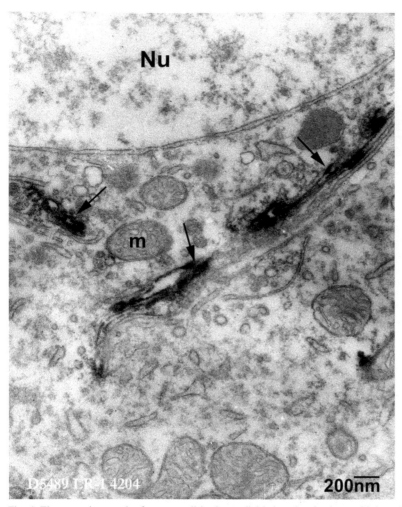

Fig. 4. Electron micrograph of a mast cell in the medial habenula of a 1 mo old dove. The tissue had been processed for ultrastructural demonstration of GnRH immunoreactivity using the LR-1 antibody directed again GnRH. The reaction product was localized to stacks of the Golgi apparatus (arrows). Nu = nucleus; m = mitochondrion. Scale bar = 200 nm.

nuclei of *each* functional thalamic region, i.e., sensory (e.g. VP, Po, dLGN, MGN), motor (e.g. VM), and limbic (e.g. MD, PVT, PT, IAM) areas (Price, 1995). We have found that $28.1 \pm 7.6\%$ of mast cells in unpaired male rats are GnRH-immunoreactive. While the number of mast cells increases with pairing, the proportion of those containing GnRH remains approximately the same (Khalil et al., 2001, and in progress). The increase in the mast cell population in these specific thalamic nuclei may reflect the fact that reproductive pairing involves numerous somatosensory, visual and auditory stimuli, complicated motor responses, and motivational and emo-

tional (i.e. 'limbic') processes. Investigations into the neural control of female copulatory behavior have focused on forebrain pathways (Meisel and Sachs, 1994) and many of the thalamic nuclei are intimately connected with these pathways.

In addition to the mating paradigms noted above, post-partum female rats have more mast cells in their thalami than do age-matched virgin controls. This latter phenomenon is modulated by the complex hormonal changes of pregnancy and parturition or by the stress (see below) of having pups removed for a 4 day period (Silverman et al., 2000).

Brain mast cells and stress

Behavioral manipulations that are independent of re-productive function are also associated with changes in the number of brain mast cells. Immobilization stress (30 minutes) induced intracranial rat mast cell degranulation and elevated mast cell protease levels in the cerebrospinal fluid (Theoharides et al., 1995). This paradigm also increased the permeability of the blood-brain barrier as measured by tracer (^{99}Tc gluceptate) distribution in the brain parenchyma (Esposito et al., 2001). Breakdown in the integrity of the blood-brain barrier induced by brain mast cell activation precedes the onset of clinical symptoms in multiple sclerosis in humans (Goodin et al., 1999) as well as in a rodent model (Kermode et al., 1990). Five day exposure to a novel male rat increased the number of mast cells in the thalamus, perhaps because this pairing was stressful given the potential for aggressive attacks and territorial disputes (Asarian et al., 2002) as has been suggested by experiments in mice (Cirulli et al., 1998).

Mast cells migrate from blood to brain

Cammermeyer (1973) suggested that mast cells may migrate through the area postrema in the dorsolateral part of the medulla to enter the brain. In the dove, we cannot state whether mast cells migrate from the pia mater or across the endothelium of blood vessels to intraparenchymal locations. We have found that in the dove mast cells are located near the tip of the habenula (where the pia mater attaches to the choroid plexus) after 30–60 minutes of courtship and become more widely distributed within the habenula after 2 hours. The apparent rate of movement of activated mast cells into the medial habenula is consistent with the known rate of movement in vitro (up to 180 μm/h) leading us and others to suggest that increases in mast cell numbers reported under this and other physiological and behavioral conditions may be due to migration of new cells into the brain (*vide supra*). Yang et al. (1999) showed that after pairing for short periods (1–7 days post-coitum), mast cells were found in the velum interpositum (choroid plexus of the third ventricle) while after 15–19 days post-coitum they were abundant in the thalamic parenchyma. This led to the hypothesis that mast cells are translocated from the velum interpositum on dorsal surface of the brain and become associated with the blood vessels and thalamic parenchyma from days 1 to 19 post-coitum.

We have demonstrated that in rat mature mast cells can enter the normal adult brain via a vascular route. Mast cells labeled with a vital dye and injected intravenously were found in the parenchyma of the thalamus within one hour (Fig. 6, see page 319). The labeled cells represented 10% of the resident mast cells and were found on the brain side of the vascular basal lamina. Note that the rapid rate at which donor mast cells crossed the blood-brain barrier is consistent with the rate of increase in numbers of mast cell in doves, mice and rats following the various behavioral and endocrine stimulations described above (Silverman et al., 2000).

The signaling mechanisms recruiting mast cells to the brain are still obscure (as they are for other immune system cells). The chemoattraction of mast cells has been attributed to chemokines. Numerous mast cell chemoattractants have been described, including MCP-1, RANTES, TGF-β, IL-3, etc. (Bochner and Schleimer, 2001). None, however, have been shown to be specific to the thalamic or pial regions where we and others have reported increases in mast cell number. A chemokine mechanism has been recently described which may account for the selective aggregation of mast cells in the dorsal thalamus. Endothelial cells in blood vessels that supply the dorsal thalamus and choroid plexus express CX-CR4 receptors, which bind avidly to the chemokine, stromal cell-derived factor 1α (SDF-1) (Banisadr et al., 2000). SDF-1 is a CXC chemokine originally isolated from a bone marrow stromal cell line (Tashiro et al., 1993; Nagasawa et al., 1994). In the brain, SDF-1 induces the migration of microglial cells and astrocytes (Tanabe et al., 1997). This ligand is also a potent mast cell chemoattractant. This chemokine facilitates mast cell migration across human umbilical venous endothelial cells in vitro without inducing degranulation (Lin et al., 2000). Only two arteries, the dorsal thalamic artery and the anterior choroidal artery, supply the dorsal thalamus (Scremin, 1995). Thus, a chemokine specialization of the vasculature of the dorsal thalamus might attract mast cells. This subject is under investigation. What other specialized mechanisms regulate their subsequent entry into the

322

brain parenchyma or otherwise affect their function in these areas are currently unknown.

Another potential mechanism by which mast cells might increase in number is by in situ cell division of resident brain mast cells. Studies using BrDU (a thymedine analogue) in doves, however, found no evidence for mitosis within the mast cell population in the MH, although incorporation was present in neuronal stem cells (unpublished observations from the laboratory).

Mast cells are activated and modulated by reproductive hormones

As mentioned previously, in the ring dove treatment with gonadal steroids results in an increase in the mast cell population in the medial habenula. Pharmacological activation of these cells opens the blood-brain barrier (Zhuang et al., 1996). We therefore tested whether administration of gonadal steroids was accompanied by a change in the activational state of brain mast cells. Electron microscopic analysis revealed that in steroid treated animals a larger percent of their mast cells showed signs of ongoing compound exocytosis or piecemeal degranulation than was seen in control birds treated with cholesterol. Isolated birds and those that had courted for 2 hours had more habenular mast cells in the resting state (Wilhelm et al., 2000). Steroid hormones have been shown to modulate secretion from peritoneal mast cells in vitro. Progesterone triggers selective secretion of 5-hydroxytryptamine and not other mediators (Vliagoftis et al., 1990). Estradiol also augments the release of mast cell mediators in the presence of myelin basic protein (Theoharides et al., 1993), substance P or Compound 48/80 (a polyamine known to stimulate mast cell secretion) (Vliagoftis et al., 1992).

Given these findings it is of interest to determine if mast cells have steroid hormone receptors. Estrogen receptor immunoreactivity is expressed in mast cells located in the human bladder (Pang et al., 1995), human upper airways (Zhao et al., 2001) and rat dura mater (Rozniecki et al., 1999). We have documented using immunocytochemistry that rat thalamic mast cells express estrogen receptor α (unpublished observations from the laboratory). In the upper airways and the bladder of humans, mast cells, but not lymphocytes, macrophages, or other immune cells, express estrogen and progesterone receptors (Zhao et al., 2001).

Conclusion

Our laboratory has demonstrated for the first time that normal behavior is associated with an increase in the brain parenchyma of mast cells and that they contain immunoreactive GnRH. Furthermore, preliminary results suggest that brain mast cells have the ability to synthesize as well as store GnRH. Mast cells have the ability to migrate rapidly into the brain, to degranulate and to release a plethora of mediators, including GnRH, in the neuropil. Their presence in the brain following/during normal behavioral manipulations could represent a significant neural-immune control mechanism modulating normal brain function in specific CNS regions.

Abbreviations: Thalamic Nuclei

dLGN	dorsal lateral geniculate
IAM	interanteromedial
MD	mediodorsal
MGN	medial geniculate
Po	posterior
PT	paratenial
PVT	paraventricular thalamus
VM	ventromedial
VP	ventroposterior

Acknowledgements

This work was supported by NIH grants MH 54088 (AJS), MH 29380 (RS) and T32 DK 07328-23 (MK).

References

Asarian, L., Yousefzadeh, E., Silverman, A.J. and Silver, R. (2002) Stimuli from conspecifics influence brain mast cell population in male rats. Horm. Behav., 42: 1–12.

Azad, N., Emanuele, N.V., Halloran, M.M., Tentler, J. and Kelley, M.R. (1991) Presence of luteinizing hormone-releasing hormone (LHRH) mRNA in rat spleen lymphocytes. Endocrinology, 128: 1679–1681.

Banisadr, G., Dicou, E., Berbar, T., Rostene, W., Lombet, A. and Haour, F. (2000) Characterization and visualization of [125I] stromal cell-derived factor-1 alpha binding to CXCR4 receptors in rat brain and human neuroblastoma cells. *J. Neuroimmunol.*, 110: 151–160.

Batticane, N., Morale, M.C., Gallo, F., Farinella, Z. and Marchetti, B. (1991) Luteinizing hormone-releasing hormone signaling at the lymphocyte involves stimulation of interleukin-2 receptor expression. *Endocrinology*, 129: 277–286.

Bochner, B.S. and Schleimer, R.P. (2001) Mast cells, basophils, and eosinophils: distinct but overlapping pathways for recruitment. *Immunol. Rev.*, 179: 5–15.

Cammermeyer, J. (1973) Migration of mast cells through the area postrema. *J. Hirnforsch.*, 14: 519–526.

Chen, H.F., Jeung, E.B., Stephenson, M. and Leung, P.C. (1999) Human peripheral blood mononuclear cells express gonadotropin-releasing hormone (GnRH), GnRH receptor, and interleukin-2 receptor gamma-chain messenger ribonucleic acids that are regulated by GnRH in vitro. *J. Clin. Endocrinol. Metab.*, 84: 743–750.

Chen, Y., Wong, M. and Moss, R.L. (1993) Effects of a biologically active LHRH fragment on CA1 hippocampal neurons. *Peptides*, 14: 1079–1081.

Cirulli, F., Pistillo, L., De Acetis, L., Alleva, E. and Aloe, L. (1998) Increased number of mast cells in the central nervous system of adult male mice following chronic subordination stress. *Brain Behav. Immun.*, 12: 123–133.

Crivellato, F., Candussioi, L., Decoiti, G., Klugmann, F.B. and Baldini, L. (1997) Adriamycin binds to the matrix of secretory granules during mast cell exocytosis. *Biotech. Histochem.*, 72: 111–116.

Dropp, J.J. (1976) Mast cells in mammalian brain. *Acta Anat. (Basel)*, 94: 1–21.

Dvorak, A.M. (1989) Human mast cells. *Adv. Anat. Embryol. Cell Biol.*, 114: 1–107.

Dvorak, A.M. (1992) Basophils and mast cells: piecemeal degranulation in situ and ex vivo: a possible mechanism for cytokine-induced function in disease. *Immunol. Ser.*, 57: 169–271.

Emanuele, N.V., Emanuele, M.A., Tentler, J., Kirsteins, L., Azad, N. and Lawrence, A.M. (1990) Rat spleen lymphocytes contain an immunoactive and bioactive luteinizing hormone-releasing hormone. *Endocrinology*, 126: 2482–2486.

Esposito, P., Gheorghe, D., Kandere, K., Pang, X., Connolly, R., Jacobson, S. and Theoharides, T.C. (2001) Acute stress increases permeability of the blood-brain barrier through activation of brain mast cells. *Brain Res.*, 888: 117–127.

Feder, H.H., Storey, A., Goodwin, D., Reboulleau, C. and Silver, R. (1977) Testosterone and '5alpha-dihydrotestosterone' levels in peripheral plasma of male and female ring doves (Streptopelia risoria) during and reproductive cycle. *Biol. Reprod.*, 16: 666–677.

Flood, P.R. and Kruger, P.G. (1970) Fine structure of mast cells in the central nervous system of the hedgehog. *Acta Anat. (Basel)*, 75: 443–52.

Florenzano, F. and Bentivoglio, M. (2000) Degranulation, density and distribution of mast cells in the rat thalamus: a light and electron microscopic study in basal conditions and after

intracerebroventricular administration of nerve growth factor. *J. Comp. Neurol.*, 424: 651–669.

Goodin, D.S., Ebers, G.C., Johnson, K.P., Rodriguez, M., Sibley, W.A. and Wolinsky, J.S. (1999) The relationship of multiple sclerosis to physical trauma and psychological stress. *Neurology*, 52: 1737–1745.

Hickey, W.F. (2001) Basic principles of immunological surveillance of the normal central nervous system. *Glia*, 36: 118–124.

Humphries, D.E., Wong, G.W., Friend, D.S., Gurish, M.F., Qiu, W.T., Huang, C., Sharpe, A.H. and Stevens, R.L. (1999) Heparin is essential for the storage of specific granule proteases in mast cells. *Nature*, 400: 769–772.

Ibrahim, M.Z., Al Wirr, M.E. and Bahuth, N. (1979) The mast cells of the mammalian central nervous system. III. Ultrastructural characteristics in the adult rat brain. *Acta Anat. (Basel)*, 104: 134–154.

Jacobson, J.D., Nisula, B.C. and Steinberg, A.D. (1994) Modulation of the expression of murine lupus by gonadotropin-releasing hormone analogs. *Endocrinology*, 134: 2516–2523.

Kaartinen, M., Penttila, A. and Kovanen, P. (1995) Extracellular mast cell granules carry apolipoprotein B-100-containing lipoproteins into phagocytes in human arterial intima: functional coupling of exocytosis and phagocytosis in neighboring cells. *Arterioscler. Thromb. Vasc. Biol.*, 15: 2047–2054.

Kermode, A.G., Thompson, A.J., Tofts, P., MacManus, D.G., Kendall, B.E., Kingsley, D.P., Moseley, I.F., Rudge, P. and McDonald, W.I. (1990) Breakdown of the blood-brain barrier precedes symptoms and other MRI signs of new lesions in multiple sclerosis. Pathogenetic and clinical implications. *Brain*, 113: 1477–1489.

Khalil, M., Asarian, L., Silverman, A.J. and Silver, R. (2001) Gonadotropin-releasing hormone in rat brain mast cells. Society for Behavioral Neuroendocrinology Annual Meeting Abstracts. *Horm. Behav.*, 39: 335.

Kitamura, Y., Kanakura, Y., Fujita, J. and Nakano, T. (1987a) Differentiation and transdifferentiation of mast cells; a unique member of the hematopoietic cell family. *Int. J. Cell Cloning*, 5: 108–121.

Kitamura, Y., Kanakura, Y., Sonoda, S., Asai, H. and Nakano, T. (1987b) Mutual phenotypic changes between connective tissue type and mucosal mast cells. *Int. Arch. Allergy Appl. Immunol.*, 82: 244–248.

Kops, S.K., Theoharides, T.C. and Kashgarian, M.G. (1990) Ultrastructural characteristics of rat peritoneal mast cells undergoing differential release of serotonin without histamine and without degranulation. *Cell Tissue Res.*, 262: 415–424.

Kovanen, P.T. (1991) Mast cell granule-mediated uptake of low density lipoproteins by macrophages: a novel carrier mechanism leading to the formation of foam cells. *Ann. Med.*, 23: 551–559.

Lambracht-Hall, M., Dimitriadou, V. and Theoharides, T.C. (1990) Migration of mast cells in the developing rat brain. *Brain Res. Dev. Brain Res.*, 56: 151–159.

Lin, T.-J., Issekutz, T.B. and Marshall, J.S. (2000) Human mast cells transmigrate through human umbilical vein endothelial

324

monolayers and selectively produce IL-8 in response to stromal cell-derived factor-1α. *J. Immunol.*, 165: 211–220.

Maier, C.C., Marchetti, B., LeBoeuf, R.D. and Blalock, J.E. (1992) Thymocytes express a mRNA that is identical to hypothalamic luteinizing hormone-releasing hormone mRNA. *Cell. Mol. Neurobiol.*, 12: 447–454.

Marchetti, B., Guarcello, V., Moralem, M.C., Bartoloni, G., Raiti, F., Palumbo Jr., G., Farinella, Z., Cordaro, S. and Scapagnini, U. (1989) Luteinizing hormone-releasing hormone (LHRH) agonist restoration of age-associated decline of thymus weight, thymic LHRH receptors, and thymocyte proliferative capacity. *Endocrinology*, 125: 1037–1045.

Marchetti, B., Gallo, F., Farinella, Z., Romeo, C. and Morale, M.C. (1996) Luteinizing hormone-releasing hormone (LHRH) receptors in the neuroendocrine-immune network. Biochemical bases and implications for reproductive physiopathology. *Ann. N.Y. Acad. Sci.*, 784: 209–236.

Meisel, R.L. and Sachs, B.D. (1994) The Physiology of Male Sexual Behavior. In: E. Knobil and J.D. Neill (Eds.), *Physiology of Reproduction, Vol. 2*. Raven Press, New York, NY, pp. 3–105.

Morale, M.C., Batticane, N., Bartoloni, G., Guarcello, V., Farinella, Z., Galasso, M.G. and Marchetti, B. (1991) Blockade of central and peripheral luteinizing hormone-releasing hormone (LHRH) receptors in neonatal rats with a potent LHRH-antagonist inhibits the morphofunctional development of the thymus and maturation of the cell-mediated and humoral immune responses. *Endocrinology*, 128: 1073–1085.

Moss, R.L. and McCann, S.M. (1973) Induction of mating behavior in rats by luteinizing hormone-releasing factor. *Science*, 181: 177–179.

Nagasawa, T., Hirota, S. and Kishimoto, T. (1994) Molecular cloning and structure of a pre-B-cell growth-stimulating factor. *Proc. Natl. Acad. Sci. (USA)*, 91: 2305–2309.

Olsson, Y. (1968) Mast cells in the nervous system. *Int. Rev. Cytol.*, 24: 27–70.

Pang, X., Marchand, J., Sant, G.R., Kream, R.M. and Theoharides, T.C. (1995) Increased number of substance P positive nerve fibres in interstitial cystitis. *Br. J. Urol.*, 75: 744–750.

Pfaff, D.W. (1973) Luteinizing hormone-releasing factor potentiates lordosis behavior in hypophysectomized ovariectomized female rats. *Science*, 182: 1148–1149.

Phillips, A., Hahn, D.W., McGuire, J.L., Ritchie, D., Capetola, R.J., Bowers, C. and Folkers, K. (1988) Evaluation of the anaphylactoid activity of a new LHRH antagonist. *Life Sci.*, 4: 883–888.

Price, J.L. (1995) Thalamus. In: G. Paxinos (Ed.), *The Rat Nervous System*. Academic Press, Sydney, pp. 629–648.

Rivier, J.E., Porter, J., Rivier, C.L., Perrin, M., Corrigan, A., Hook, W.A., Siraganian, R.P. and Vale, W.W. (1986) New effective gonadotropin releasing hormone antagonists with minimal potency for histamine release in vitro. *J. Med. Chem.*, 29: 1846–1851.

Rodewald, H.R., Dessing, M., Dvorak, A.M., Galli, S.J. (1996) Identification of a committed precursor for the mast cell lineage. *Science*, 271: 818–822.

Rozniecki, J.J., Dimitriadou, V., Lambracht-Hall, M., Pang, X.

and Theoharides, T.C. (1999) Morphological and functional demonstration of rat dura mater mast cell–neuron interactions in vitro and in vivo. *Brain Res.*, 849: 1–15.

Schally, A.V., Arimura, A., Baker, Y., Nair, R.M.G., Matsuo, H., Redding, T.W., Debeljuk, L. and White, W.F. (1971) Isolation and properties of the FSH and LH-releasing hormone. *Biochem. Biophys. Res. Commun.*, 43: 393–399.

Schwanzel-Fukuda, M. and Pfaff, D.W. (1989) Origin of luteinizing hormone-releasing hormone neurons. *Nature*, 338: 161–164.

Scremin, O.U. (1995). Cerebral Vascular System. In: G. Paxinos (Ed.), *The Rat Nervous System*. Academic Press, Sydney, pp. 3–35.

Shin, J.S., Gao, Z. and Abraham, S.N. (2000) Involvement of cellular caveolae in bacterial entry into mast cells. *Science*, 285: 785–788.

Silver, R., Goldsmith, A.R. and Follett, B.K. (1980) Plasma luteinizing hormone in male ring doves during the breeding cycle. *Gen. Comp. Endocrinol.*, 42: 19–24.

Silver, R., Ramos, C.L. and Silverman, A.J. (1992) Sexual behavior triggers the appearance of non-neuronal cells containing gonadotropin-releasing hormone-like immunoreactivity. *J. Neuroendocrinol.*, 4: 207–210.

Silver, R., Silverman, A.J., Vitkovic, L. and Lederhendler, I.I. (1996) Mast cells in the brain: evidence and functional significance. *Trends Neurosci.*, 19: 25–31.

Silverman, A.J. (1988). The gonadotropin-releasing hormone (GnRH) neuronal systems: Immunocytochemistry. In: E. Knobil and J. Neill (Eds.), *Physiology of Reproduction*. Raven Press, New York, NY, pp. 1283–1304.

Silverman, A.J., Millar, R.P., King, J.A., Zhuang, X. and Silver, R. (1994) Mast cells with gonadotropin-releasing hormone-like immunoreactivity in the brain of doves. *Proc. Natl. Acad. Sci. (USA)*, 91: 3695–3699.

Silverman, A.J., Sutherland, A.K., Wilhelm, M. and Silver, R. (2000) Mast cells migrate from blood to brain. *J. Neurosci.*, 20: 401–408.

Sundaram, K., Didolkar, A., Thau, R., Chaudhuri, M. and Schmidt, F. (1988) Antagonists of luteinizing hormone releasing hormone bind to rat mast cells and induce histamine release. *Agents Actions*, 25: 307–313.

Tanabe, S., Heesen, M., Yoshizawa, I., Berman, M.A., Luo, Y., Bleul, C.C., Springer, T.A., Okuda, K., Gerard, N. and Dorf, M.E. (1997) Functional expression of the CXC-chemokine receptor-4/fusin on mouse microglial cells and astrocytes. *J. Immunol.*, 159: 905–911.

Tashiro, K., Tada, H., Heilker, R., Shirozu, M., Nakano, T. and Honjo, T. (1993) Signal sequence trap: A cloning strategy for secreted protein and type I membrane proteins. *Science*, 261: 600–603.

Theoharides, T.C. (1990) Mast cells: the immune gate to the brain. *Life Sci.*, 46: 607–617.

Theoharides, T.C., Dimitriadou, V., Letourneau, R., Rozniecki, J.J., Vliagoftis, H. and Boucher, W. (1993) Synergistic action of estradiol and myelin basic protein on mast cell secretion and brain myelin changes resembling early stages of demyelination. *Neuroscience*, 57: 861–871.

Theoharides, T.C., Spanos, C., Pang, X., Alferes, L., Ligris, K., Letourneau, R., Rozniecki, J.J., Webster, E. and Chrousos, G.P. (1995). Stress-induced intracranial mast cell degranulation: a corticotropin-releasing hormone-mediated effect. *Endocrinology*, 136: 5745–5750.

Vliagoftis, H., Dimitriadou, V. and Theoharides, T.C. (1990) Progesterone triggers selective mast cell secretion of 5-hydroxytryptamine. *Int. Arch. Allergy Appl. Immunol.*, 93: 113–119.

Vliagoftis, H., Dimitriadou, V., Boucher, W., Rozniecki, J.J., Correia, I., Raam, S. and Theoharides, T.C. (1992) Estradiol augments while tamoxifen inhibits rat mast cell secretion. *Int. Arch. Allergy Immunol.*, 98: 398–409.

Wang, Y., Lindstedt, K.A. and Kovanen, P.T. (1995) Mast cell granule remnants carry LDL into smooth muscle cells of the synthetic phenotype and induce their conversion into foam cells. *Arterioscler. Thromb. Vasc. Biol.*, 15: 801–810.

Wedemeyer, J., Tsai, M. and Galli, S.J. (2000) Roles of mast cells and basophils in innate and acquired immunity. *Curr. Opin. Immunol.*, 12: 624–631.

Weesner, G.D., Becker, B.A. and Matteri, R.L. (1997) Expression of luteinizing hormone-releasing hormone and its receptor in porcine immune tissues. *Life Sci.*, 61: 1643–1649.

Wilhelm, M., King, B., Silverman, A.J. and Silver, R. (2000) Gonadal steroids regulate the number and activational state of mast cells in the medial habenula *Endocrinology*, 141: 1178–1186.

Wray, S., Nieburgs, A. and Elkabes, S. (1989) Spatiotemporal cell expression of luteinizing hormone-releasing hormone in the prenatal mouse: evidence for an embryonic origin in the olfactory placode. *Brain Res. Dev. Brain Res.*, 46: 309–318.

Xiang, Z., Block, M., Lofman, C. and Nilsson, G. (2001) IgE-mediated mast cell degranulation and recovery monitored by time-lapse photography. *J. Allergy Clin. Immunol.*, 108: 116–121.

Yang, M., Chien, C. and Lu, K. (1999) Morphological, immuno-histochemical and quantitative studies of murine brain mast cells after mating. *Brain Res.*, 846: 30–39.

Zhao, X.J., McKerr, G., Dong, Z., Higgins, C.A., Carson, J., Yang, Z.Q. and Hannigan, B.M. (2001) Expression of oestrogen and progesterone receptors by mast cells alone, but not lymphocytes, macrophages or other immune cells in human upper airways. *Thorax*, 56: 205–211.

Zhuang, X., Silverman, A.J. and Silver, R. (1993) Reproductive behavior, endocrine state, and the distribution of GnRH-like immunoreactive mast cells in dove brain. *Horm. Behav.*, 27: 283–295.

Zhuang, X., Silverman, A.J. and Silver, R. (1996) Brain mast cell degranulation regulates blood-brain barrier. *J. Neurobiol.*, 31: 393–403.

Zhuang, X., Silverman, A.J. and Silver, R. (1999) Distribution and local differentiation of mast cells in the parenchyma of the forebrain. *J. Comp. Neurol.*, 408: 477–488.